DEVELOPMENTS IN MOLECULAR VIROLOGY

Yechiel Becker, Series Editor
Julia Hadar, Managing Editor

Other books in the series:
Becker, Y. (ed.) *Herpesvirus DNA* (1981)
Becker, Y. (ed.), *Replication of Viral and Cellular Genomes* (1984)
Kohn, A. and Fuchs, P. (eds.) *Mechanisms of Viral Pathogenesis: from Gene to Pathogen* (1984)

ANTIVIRAL DRUGS AND INTERFERON

ANTIVIRAL DRUGS AND INTERFERON: THE MOLECULAR BASIS OF THEIR ACTIVITY

Edited by

Yechiel Becker
The Hebrew University of Jerusalem, Israel
Julia Hadar, Managing Editor

Martinus Nijhoff Publishing
a *member of the Kluwer Academic Publishers Group*
Boston/The Hague/Dordrecht/Lancaster

Distributors for North America:
Kluwer Academic Publishers
190 Old Derby Street
Hingham, MA 02043

Distributors Outside North America:
Kluwer Academic Publishers Group
Distribution Centre
P.O. Box 322
3300 AH Dordrecht
The Netherlands

Library of Congress Cataloging in Publication Data
Main entry under title:

Antiviral drugs and interferon.

(Developments in molecular virology)
Includes index.
1. Antiviral agents. 2. Interferon. 3. Herpes simplex virus—Effect of drugs on. 4. Viruses—Reproduction I. Becker, Yechiel. II. Series, [DNLM: 1. Antiviral agents. 2. Interferons. 3. Virus diseases—Drug therapy. 4. Virus replication—Drug effects. W1 DE998DG/QV 268.5 A6334]
RM411.A575 1984 616.9'25061 84-4045
ISBN 0-89838-643-8

Printed in the United States of America

CONTENTS

Drugs Affecting RNA Viruses

Interferon

LIST OF CONTRIBUTORS

Ken Bastow
Department of Pharmacology
University of North Carolina at Chapel Hill
Chapel Hill, North Carolina 27514, USA

Yechiel Becker
Department of Molecular Virology
The Hebrew University
Hadassah Medical School
Post Office Box 1172
Jerusalem, ISRAEL

Meir Bialer
Pharmacy Department
School of Pharmacy
Hebrew University
Jerusalem 91120, ISRAEL

L. A. Caliguiri
Department of Microbiology and Immunology
Albany Medical College of Union University
Albany, New York 12208, USA

Peter Canonico
Department of Antiviral Studies
U. S. Army Medical Research Institute
of Infectious Diseases
Fort Detrick
Frederick, Maryland 21701, USA

P. Chandra
Center of Biological Chemistry
Laboratory of Molecular Biology
University Medical School
Frankfurt 70, WEST GERMANY

Yung-Chi Cheng
Department of Pharmacology and Medicine
The University of North Carolina
Chapel Hill, North Carolina 27514, USA

Ting-Chao Chou
Sloan-Kettering Institute for Cancer Research
Memorial Sloan-Kettering Cancer Center
1275 York Avenue
New York, New York 10021, USA

Erik De Clercq
Rega Institute for Medical Research
Katholieke Universiteit Leuven
B-3000 Leuven, BELGIUM

I. Demirhan
Center of Biological Chemistry
Laboratory of Molecular Biology
University Medical School
Frankfurt 70, WEST GERMANY

David Derse
Department of Pharmacology
University of North Carolina at Chapel Hill
Chapel Hill, North Carolina 27514, USA

U. Ebener
Center of Biological Chemistry
Laboratory of Molecular Biology
University Medical School
Frankfurt 70, WEST GERMANY

Hans J. Eggers
Institut fur Virologie der Universitat zu Koln
Furst-Puckler-Str. 56
D-5000 Koln 41, WEST GERMANY

Gertrude B. Elion
Wellcome Research Laboratories
Burroughs Wellcome Co.
Research Triangle Park, North Carolina 27709, USA

B. Eriksson
Department of Antiviral Chemotherapy
Research and Development Laboratories
Astra Lakemedel AB
Sodertalje, SWEDEN

Jack J. Fox
Sloan-Kettering Institute for Cancer Research
Memorial Sloan-Kettering Cancer Center
1275 York Avenue
New York, New York 10021, USA

Karl Frank
Department of Pharmacology
University of North Carolina at Chapel Hill
Chapel Hill, North Carolina 27514, USA

D. Gericke
Center of Biological Chemistry
University School of Medicine
Theodore- Stern-Kai 7
Frankfurt 70, WEST GERMANY

Allan R. Goldberg
The Rockefeller University
1230 York Avenue
New York, New York 10021, USA

Barry Goz
Department of Pharmacology
University of North Carolina
School of Medicine
Chapel Hill, North Carolina 27514, USA

Alan J. Hay
Division of Virology
National Institute for Medical Research
Mill Hill
London, NW7 1AA, UNITED KINGDOM

Ehud Katz
The Chanock Center for Virology
The Hebrew University
Hadassah Medical School
Jerusalem, ISRAEL

B. Kornhuber
Department of Pediatric Oncology
University Medical School
Frankfurt 70, WEST GERMANY

Robert A. Lamb
The Rockefeller University
1230 York Avenue
New York, New York 10021, USA

A. Martin Lerner
Department of Medicine
Wayne State University
School of Medicine
5980 Cass Avenue
Detroit, Michigan 48202, USA

Carlos Lopez
Sloan-Kettering Institute for Cancer Research
Memorial Sloan-Kettering Cancer Center
1275 York Avenue
New York, New York 10021, USA

Thomas W. North
Department of Biochemistry and Pharmacology
Tufts University Schools of Veterinary Medicine
and Medicine
Boston, Massachusetts 02111, USA

B. Oberg
Department of Antiviral Chemotherapy
Research and Development Laboratories
Astra Lakemedel AB
Sodertalje, SWEDEN

Michael J. Otto
Department of Pharmacology
Yale University School of Medicine
333 Cedar Street
New Haven, Connecticut 06510, USA

William H. Prusoff
Department of Pharmacology
Yale University School of Medicine
333 Cedar Street
New Haven, Connecticut 06510, USA

Michael Revel
Department of Virology
The Weizmann Institute of Science
Rehovot, 76100, ISRAEL

M. Schrom
Department of Microbiology and Immunology
Albany Medical College of Union University
Albany, New York 12208, USA

Pravinkumar B. Sehgal
The Rockefeller University
1230 York Avenue
New York, New York 10021, USA

Ilan Sela
Virus Laboratory
Faculty of Agriculture
Post Office Box 12
Rehovot 76100, ISRAEL

Roberts A. Smith
Department of Chemistry and Biochemistry
University of California at Los Angeles
Los Angeles, California 90024, USA

Igor Tamm
The Rockefeller University
1230 York Avenue
New York, New York 10021, USA

Paul F. Torrence
National Institutes of Health
Building 4, Room 126
Bethesda, Maryland 20205, USA

Kyoichi A. Watanabe
Sloan-Kettering Institute for Cancer Research
Memorial Sloan-Kettering Cancer Center
1275 York Avenue
New York, New York 10021, USA

Mary C. Zambon
Division of Virology
National Institute for Medical Research
Mill Hill
London, NW7 1AA, UNITED KINGDOM

PREFACE

Research on antiviral drugs and their mode of action in infected cells, in animals and in man, has led to a better understanding of the molecular processes involved in virus replication. Screening of large numbers of natural and semisynthetic compounds resulted in the characterization of certain substances that had a limited efficiency as antiviral drugs. A few chemically synthesized compounds were also found to be effective as antiviral agents in the chemotherapy of human virus diseases. A major difficulty in the development of effective antiviral agents has been the lack of selectivity, and toxicity for uninfected cells, of drugs that effectively inhibited virus replication in vitro. Further understanding of the molecular processes of virus replication in infected cells has resulted in the development of new antivirals directed at virus-coded enzymes or proteins. Recent studies on antivirals that are activated by the herpes simplex virus type 1-coded thymidine kinase from a prodrug to an antiviral drug have opened new directions in the development of effective antiviral drugs.

The present book deals with a number of antiviral drugs effective against herpes simplex viruses and provides some insight into the molecular aspects of virus replication. It also throws light on the new approaches to the development of antiviral drugs. The molecular basis of the antiviral activity of new and known drugs and their possible use in chemotherapy of viral disease are presented in this book. All these contributions were written by scientists who have been pioneers in the development of the various antiviral agents.

I wish to thank the authors of the various chapters for their collaboration. Regretfully, a few chapters did not arrive in time for inclusion in this volume. The assistance of Dr. Julia Hadar as managing editor, and Mrs. Esther Herskovics with the secretarial work, is greatly appreciated.

Yechiel Becker

PREFACE

1

INTRODUCTION: CURRENT TRENDS IN THE RESEARCH ON ANTIVIRAL DRUGS

Y. BECKER, Department of Molecular Virology, The Hebrew University, Hadassah Medical School, Post Office Box 1172, Jerusalem, Israel

The need for antiviral agents when viral vaccines are not available

The discovery of antibacterial agents, such as penicillin and streptomycin, led to the expectation that among the fermentation products of molds, substances with antiviral properties would be found. Many naturally occurring antiviral agents are able to inhibit virus replication in cultured cells in vitro (1), but unfortunately, almost all of them were found to be unsuitable for treatment of virus diseases. The alternative approach was to chemically synthesize antiviral compounds. Some of these were effective in inhibiting virus replication in mammalian cells, and a few are in use today for curing virus infections in man (2).

The conventional approach to the prevention of epidemics like poliomyelitis, influenza, and Rift Valley fever, to mention a few, is the development of a potent virus vaccine. This may be done either by inactivation of the infectious virus (e.g. Salk's inactivated poliovirus vaccine) or by development of live attenuated virus strains (e.g. Sabin's live attenuated poliovirus vaccine). Immunization of people prior to or during an epidemic or a pandemic usually succeeds in halting the spread of the virus. The Jennerian type vaccination procedure has been used in an extensive campaign by the World Health Organization and local authorities in recent years to eradicate smallpox. This was preferable to using a prophylactic drug like Marboran (see Chapter 13). The need for antiviral agents arises only when the development of a vaccine is not possible. Viruses belonging to the family Herpesviridae cannot be used for the preparation of inactivated or live virus vaccines, since many of these viruses contain genes that cause cell transformation. Apart from causing diseases and latent infections in man, these viruses have also been implicated in certain human cancers, e.g. Epstein-Barr virus in Burkitt's lymphoma and nasopharyngeal carcinoma, herpes simplex virus type 2 (HSV-2) in cervical carcinoma, and cytomegalovirus in Kaposi's sarcoma.

Thus, antiviral agents against these viruses are of great importance, more so because genital herpes (caused by HSV-2) is spreading rapidly. A substantial part of the present volume is devoted to antiherpes virus drugs, their mode of action and chemotherapeutic value.

Difficulties in the development of influenza virus vaccines have led to the use of amantadine as a prophylactic drug (see Chapter 18). Since amantadine is also used in the treatment of Parkinson's disease, its ability to interact with nerve cells is of concern. Recent developments in genetic engineering have led to the cloning of the hemagglutinin (HA) and neuraminidase (NA) genes of influenza virus in bacterial plasmids, after synthesis of a DNA copy of each gene (3, 4). Such procedures could lead to the preparation of new effective vaccines for immunization of large populations to stop influenza epidemics. This would seem to be a more practical approach than the development of anti-influenza virus drugs.

Broad-spectrum antivirals

The discovery of interferon by Isaacs and Lindenman in 1957 (5) and of its ability to inhibit most viruses in cultured cells in-vitro gave rise to the expectation that a natural broad-spectrum antiviral agent would become available in the near future. Thus, interferon inducers (see Chapter 19), which are chemically synthesized polynucleotide chains, were developed, with the aim of stimulating interferon production in-vivo in a virus-infected individual. However, it was soon discovered that synthetic double-stranded RNA-like interferon inducers are highly toxic. The search for non-toxic interferon inducers still continues (6). The use of interferon for treatment of patients with virus infections was hampered at first by the lack of sufficient quantities of interferon. This was overcome by Cantell and his collaborators (7, 8) who produced limited quantities of human interferon by inducing white blood cells to produce interferon. Human interferon has been used to treat cancer patients, and patients with hepatitis B, herpesviruses, influenza virus and other virus infections (9, 10). These preparations were not very pure and did not always produce the desired effect in patients.

Interferon genes have now been isolated and cloned in bacterial plasmids (11-15). Interferon can be produced in bacteria as well as in human cells, and the way is now open for large-scale production of human interferon. The use of monoclonal antibodies prepared against human interferon as a means of purification will make it possible to obtain highly purified interferon preparations for human trials.

Interferon-like polypeptides are produced by plants (see Chapter 20), and their mode of antiviral activity in plant cells is similar to that of human interferon. Recently, it was reported by I. Sela and his colleagues (16, 17) that human interferon is active in plant cells. Unfortunately, the reverse, namely that plant interferon is effective in human cells, is not true.

The main problem with interferon is its inability to be selectively transported to the cells in which virus replication takes place when administered systemically to a patient. The local induction of interferon in virus-infected tissue may be an effective means of localizing the virus infection and preventing its spread, but targeting of interferon to the specific virus-infected organ is not yet possible.

Antivirals in cancer chemotherapy

Shannon and Schabel (18) have reviewed the effectiveness of antiviral drugs in cancer patients under chemotherapy. In such patients, the cytotoxic anti-cancer drugs cause immunosuppression, rendering the patients more sensitive to virus infections or to reactivation of latent herpesviruses. Some of the antiviral drugs (polynucleotide analogs) known today are effective in inhibiting the retrovirus reverse transcriptase, and mercaptopolycytidylic acid (MPC) has been tested in leukemia patients (see Chapter 12) with encouraging results. A retrovirus, human T-cell leukemia-lymphoma virus (HTLV) isolated from T-cells of leukemia and lymphoma patients, has been implicated in leukemogenesis (19). It is too early to assess the role of antiviral drugs in human leukemia, but future work on the virus etiology of human cancer should provide a lead to the use of antivirals in cancer chemotherapy.

Drugs interfering with cellular functions

Drugs which have the ability to interfere with the function of the host cell RNA polymerase II are capable of inhibiting the replication of viruses that require the cellular enzyme for the synthesis of their mRNA species. Halogenated ribofuranosylbenzimidazoles (see Chapter 14) inhibit the host cell RNA polymerase II. Although the effect of the drugs is not selectively directed against virus-specified molecular events, the study of this group of compounds might add further knowledge about the cell-virus relationship.

Virazole (ribavirin)(1-β-D-ribofuranosyl-1,2,4-triazole-3-carboxamide) has been reported to have broad-spectrum antiviral activity against RNA and DNA viruses (20), but was found to exert a toxic effect on cellular metabolism (Chapter 15). Its ability to alter nucleotide pools and the packaging of mRNA, processes that are not virus-specific, leads to the inhibition of virus replication in the virus-treated cells.

Virus-coded enzymes as targets for antiviral drugs

The ability of most antiviral compounds to interfere with the molecular processes of the host cell has been a major obstacle in the selection of drugs suitable for treating virus diseases. The ideal antiviral drug would interfere only with a virus-coded process and not with the molecular processes in uninfected mammalian cells. Research in molecular virology during the last twenty years has made it possible to define which antiviral drugs would meet these requirements. Thus, antiviral agents can be divided into several categories: (a) antivirals which interfere with cellular processes required by the virus for its replication; (b) antivirals which selectively bind to, or interfere with, virus-coded enzymes and thus inhibit their function; (c) antivirals that bind to the virus nucleic acid and therefore inhibit its expression and function; (d) antivirals which prevent the processing of the viral precursor polypeptides; (e) antivirals which interfere with virus assembly and inhibit the formation of virus progeny; (f) antivirals which modify the viral proteins on the surface of the viral envelope, and thus prevent the virus from infecting new cells.

Among the viral enzymes studied for the development of synthetic and natural antiviral compounds, the viral DNA polymerase has received a great deal of attention (see Chapter 8). Inhibitors which bind to, and inhibit, the viral nucleic acid-synthesizing enzymes would be highly effective as antiviral agents, provided they are selective and inhibit only virus-coded enzymes. However, when inhibitors of the DNA polymerase were studied, it became evident that drug-resistant virus mutants are easily selected in the presence of the inhibitor. From this observation, it can be deduced that in the future it may be necessary to consider the use of two antiviral drugs in treating virus infections, so as to overcome the mutation or the selection of resistant virus strains.

Some of the enzymes coded for by poxviruses are listed in Table 1. These include the DNA-dependent RNA polymerase responsible for the synthesis of the viral messenger RNA, the DNA polymerase and gyrase (enzymes associated with the replication of the viral DNA), and an enzyme involved in the processing of the viral structural proteins inserted into the membrane. Natural substances with antiviral activity, which are fermentation products of molds (Table 1) include distamycin and netropsin which bind to the viral DNA and inhibit the transcription of viral messenger RNA, thus inhibiting the replication of the virus. Novobiocin, which is thought capable of binding to gyrase, might inhibit viral DNA synthesis. Rifampin prevents the synthesis of the viral envelope by inhibiting cleavage of a polypeptide inserted into the envelope. As a result, the viral envelope cannot become functional and the synthesis of infectious virions is completely inhibited.

Distamycin acts by binding to A-T rich sequences in the viral DNA and interferes with the viral and cellular RNA polymerases in the transcription of the DNA templates. However, the drug can be readily dissociated from the DNA,and virus replication can be resumed after its removal.

Unfortunately, these antiviral substances are not specific inhibitors of viral processes and have toxic effects on uninfected cells. Therefore, although they are very useful for studies on the molecular processes of poxvirus replication, they cannot be used as effective antiviral drugs for the treatment of poxvirus infections. The dangers of smallpox have by now markedly declined because of the vaccination campaign of the World Health Organization.

Certain adenovirus strains and hepatitis B virus remain important pathogens in medicine, but no specific antiviral drugs for use against them are yet available (Table 1).

All over the world, herpesviruses are a very important group of human pathogens, and the development of effective antiviral agents against these virus infections are of paramount importance. Table 1 shows a number of antiviral substances developed during the last twenty years to cure herpesvirus infections not only in the eye, but also herpes keratitis in the skin, herpes genitalis, and herpes encephalitis in the brain. Most of these antivirals were designed to inhibit the viral DNA polymerase. Table 1 further shows that distamycin and netropsin inhibit herpesvirus replication by binding to the DNA. The effect of distamycin A (See Chapter 10) and related compounds on the ability of the HSV-coded DNA polymerase to synthesize viral DNA is due to the binding to the viral DNA. The ability of these compounds to bind to cellular DNA makes them unsuitable for antiviral chemotherapy.

Halogenated pyrimidines such as iododeoxyuridine (IUdR) (see Chapter 2),

Table 1. DNA virus-coded enzymes as possible targets for antiviral drugs

Virus	Viral enzyme inhibited	Antiviral agent	Mode of action
Poxvirus	DNA-dependent RNA polymerase	Distamycin Netropsin	Bind to DNA template and inhibit the viral enzyme
	DNA polymerase	Distamycin Netropsin	Bind to DNA template and inhibit viral DNA polymerase
	Gyrase (Topoisomerase)	Novobiocin	Binds to the enzyme— inhibition of DNA synthesis
		Rifampin	Inhibition of cleavage of precursor peptide, abnormal envelope
Iridovirus	DNA polymerase	Phosphono-acetate	Inhibition of DNA polymerase
Adenovirus		None	Viral protein 32K possibly involved in viral DNA synthesis
Hepatitis B	DNA polymerase in Dane particles	None	
Herpes simplex virus	DNA polymerase	Distamycin Netropsin	Binding to viral DNA, prevent DNA synthesis
		F_3T	Modification of template
		Zn^{++}	Binding to enzyme
		Phosphono-acetate Phosphono-formate	Binding to the enzyme site for γ-phosphate of the nucleoside -P-P-P
		Acycloguanosine (Acyclovir) BVDU FIAC	A prodrug activated by phosphorylation by the viral thymidine kinase. Acyclog-PPP binds irreversibly to viral DNA-pol. TK^- virus mutants are resistant to the drugs
	DNAses	None	
	Thymidine kinase	None	
	dUTPase	None	
	Gyrase (Topoisomerase)	Novobiocin	Binds to the enzyme and inhibits DNA synthesis
	Helicase	None	

aranucleosides (see Chapter 3), and trifluorothymidine (F_3T) modify the DNA template and thus prevent DNA replication by the herpesvirus DNA polymerase (Table 1). Zinc ions are also able to bind to the viral DNA polymerase, which is a zinc metallo-enzyme, and thus prevent its activity (21).

Phosphonoacetate (PAA) and phosphonoformate (PFA) (see Chapter 9) both bind to the same site on the DNA polymerase. As a result, the enzyme cannot cleave the pyrophosphate from the nucleoside triphosphate, and its activity is inhibited. The difficulty with these substances is the rapid selection of resistant virus mutants from the population of wild type virus. The mutation is in the DNA polymerase, and the resistant virus is no longer sensitive to the drug. It is possible to isolate HSV mutants which produce a PAA-resistant DNA polymerase, and in cells infected with the mutants, the enzyme is resistant to PAA over a large range of concentrations (22).

Antiviral substances that are activated by the HSV-1 thymidine kinase (see Chapter 4), including acyclovir (Chapter 5), BVDU (Chapter 6) and FIAC (Chapter 7), have a distorted sugar moiety; they thus affect the viral DNA polymerase and inhibit its activity. It is evident from Table 1 that antiviral substances that inhibit the herpesvirus DNA polymerase can interfere with three important sites on the DNA polymerase: the metallo-enzyme site, the phosphate binding site, and the sugar-base binding site.

Acycloguanosine is phosphorylated by the HSV-1-coded thymidine kinase into acycloguanosine triphosphate in infected cells only and is then able to inhibit the viral DNA polymerase. This compound can be regarded as a proviral drug, as can BVDU and FIAC, since even though they may enter all the cells of the body, their activation can occur only in cells infected with HSV-1. The difficulty with this drug is the emergence of TK^--resistant mutants. HSV-2 is less sensitive to acyclovir than HSV-1.

The chemical synthesis of effective antiherpesvirus drugs can be approached as follows: a) modification of the purine and pyrimidine bases or the sugar moiety, b) production of analogs that can be phosphorylated only by the herpesvirus thymidine kinase, and c) production of phosphorylated analogs capable of binding to other enzymes required for viral replication, such as dUTPase, helicase and topoisomerase (Table 1).

A number of the antivirals to herpesviruses discussed in Chapters 2 to 10 have shown promising results in clinical trials in man.

Table 2 describes some antiviral substances effective against RNA viruses. These include guanidine (Chapter 16) and HBB (Chapter 17) that inhibit the

Table 2. RNA virus-coded enzymes as possible targets for antiviral drugs

Virus	Viral enzyme inhibited	Antiviral agent	Mode of action
Influenza	RNA-dependent RNA polymerase	Actinomycin D Amantadine	Inhibition of initiation of RNA synthesis by RNA polymerase
	Neuraminidase	FANA	Inhibition of enzyme activity
Poliovirus	RNA-dependent RNA polymerase	HBB Guanidine	Inhibition of replication complex
	Inhibition of (cellular) peptidases	Zn^{++}	Processing of precursor peptides
Retroviruses	RNA-dependent DNA polymerase	Distamycin A Netropsin	Binding to single-stranded viral DNA
		Rifamycins Streptovaricin Alkaloids	Binding to the viral enzyme
		Oligothymidylate derivatives Vinyl analogs Partially thiolated poly C	Template-primer analogs

poliovirus replication complex, and zinc, which inhibits the peptidases that cleave the viral precursor peptides. In the case of influenza virus, actinomycin D and amantadine (see Chapter 18) inhibit the function of the virus RNA-dependent RNA polymerase. These anticellular substances also prevent the synthesis of the cellular mRNA required for the production of dinucleotides necessary for the initiation of influenza virus RNA polymerase activity. The substance 2-deoxy-2,3-dehydro-N-trifluoroacetylneuraminic acid (FANA) was found to inhibit the viral neuraminidase, and thus prevent the virus from infecting cells (Table 2).

Most synthetic and natural antivirals have toxic side effects, making it impossible to use them in chemotherapy because of potential damage to uninfected tissues. However, studies on the group of antivirals activated by the HSV-1 thymidine kinase have great potential for future developments in antiviral chemotherapy. The synthesis of new antiviral prodrugs that selectively inhibit a specific viral function expressed only in infected cells would allow for wide use of antivirals in the curing of virus diseases. This new approach might add another dimension to the development of antiviral agents, and it holds promise for the safe and effective treatment of virus diseases in man.

REFERENCES

1. Becker, Y. Pharmac. Ther. 10:119-159, 1980.
2. Becker, Y. and Hadar, J. Progr. Med. Virol. 26:1-44, 1980.
3. Porter, A.G., Barber, C., Carey, N.J., Hallewell, R.A., Threlfall, G. and Emtage, J.S. Nature 282:471-477, 1979.
4. Emtage, J.S., Tacon, W.C.A., Catlin, G.H., Jenkins, B., Porter, A.G. and Carey, N.H. Nature 283:171-174.
5. Isaacs, A. and Lindenmann, J. Proc. Roy. Soc. Lond. (Biol.) 147:258-267, 1957.
6. Gordon, J. and Minks, M.A. Microbiol. Rev. 45:244-266, 1981.
7. Strander, H., Mogensen, K.E. and Cantell, K. J. Clin. Microb. 1:116-117, 1975.
8. Mogensen, K.E. and Cantell, K. Pharmac. Therap. 1:369-381, 1977.
9. Dunnick, J.K. and Galasso, G.J. J. Infect. Dis. 139:109-123, 1979.
10. Krim, M. Blood 55:875-884, 1980.
11. Derynck, R., Content, J., DeClercq, E., Volckaert, G., Tavernier, J., Devos, R. and Fiers, E. Nature 285:536-542, 1980.
12. Nagata, S., Taira, H., Hall, A., Johnsrud, L., Streuli, M., Ecsödi, J., Boll, W., Cantell, K. and Weissman, C. Nature 284:316-320, 1980.
13. Taniguchi, T., Mantei, N., Schwarzstein, M., Nagata, S., Muramatsu, M., and Weissman, C. Nature 285:547-549, 1980.
14. Taniguchi, T., Guapente, L., Roberts, T.M., Kimelman, D., Douhan, D. III, and Ptashne, M. Proc. Natl. Acad. Sci. USA 77:5230-5233, 1980.
15. Taniguchi, T., Pang, R.H.L., Yip, Y.K., Henriksen, D. and Vilcek, J. Proc. Natl. Acad. Sci. USA 78:3469-3472, 1981.
16. Orchansky, P., Rubinstein, M. and Sela, I. Proc. Natl. Acad. Sci. USA 79:2278-2280, 1982.
17. Devash, Y., Biggs, S. and Sela, I. Science 216:1415-1416, 1982.
18. Shannon, W.M. and Schabel, F.M. Jr. Pharmac. Therap. 11:263-390, 1980.
19. Manzari, F., Gallo, R.C., Franchini, G., Westin, E., Ceccherini-Nelli, L., Popovic, M. and Wong-Staal, F. Proc. Natl. Acad. Sci. USA 80: 11-15, 1983.
20. Sidwell, R.S., Robins, R.K. and Hillyard, I.W. Pharmac. Therap. 6: 123-146, 1979.
21. Fridlender, B., Chejanovsky, N., and Becker, Y. Virology 84:551-554, 1978.
22. Fridlender, B., Chejanovsky, N., and Becker, Y. Antimicrob. Ag. Chemother. 13:124-127, 1978.

2

ANTIVIRAL ACTIVITY OF IODINATED PYRIMIDINE DEOXYRIBONUCLEOSIDES

MICHAEL J. OTTO[1], BARRY GOZ[2] AND WILLIAM H. PRUSOFF[1] Department of Pharmacology, Yale University School of Medicine, 333 Cedar Street, New Haven, CT 06510, [2]Department of Pharmacology, University of North Carolina, School of Medicine, Chapel Hill, NC 27514.

SUMMARY

The nucleoside analogs AIdUrd, IdUrd and IdCyd exert their antiviral effects after phosphorylation and subsequent incorporation into viral DNA. AIdUrd and IdCyd require the herpesvirus pyrimidine deoxyriboside kinase for the initial phosphorylation, but IdUrd is substrate for both the viral and cellular thymidine kinase. The incorporation into the viral DNA is a critical event in the antiviral action of these analogs. Their incorporation results in altered expression of viral specific RNA and proteins and the production of an abnormal population of virions with lowered specific infectivity and the ability to interfere with the replication of standard herpesvirus.

INTRODUCTION

5-Iodo-2'-Deoxyuridine (IdUrd)

Soon after the synthesis of IdUrd in 1959 (1), Herrmann (2) reported that the compound inhibited the replication in cultured cells of several DNA-containing animal viruses. Since then numerous studies have described inhibition of the replication of almost all DNA viruses (reviewed in Prusoff and Goz, 3). In 1962,

Kaufman and colleagues (4) observed that IdUrd was effective in the treatment of human herpes keratitis and the compound subsequently was approved by the Federal Drug Administration as the first antiviral drug for clinical use. As a deoxythymidine analog, IdUrd or its anabolites can replace or compete with deoxythymidine and its anabolites in all of the relevant reactions examined thus far. Our research, therefore, has focused on the enzymatic steps leading to the ultimate incorporation of IdUrd into DNA in place of deoxythymidine and on the consequences of IdUrd-substituted DNA.

The viricidal action of IdUrd results from its incorporation into viral DNA. In our laboratory, early experiments with bacteriophage were used to study this aspect of the biological action of IdUrd. Bacteriophage can be grown easily in large yield and, at that time, a great deal more was known about the molecular biology of bacterial viruses than animal viruses. Phage (T2 and T4) with IdUrd-substituted DNA were prepared, and the replicative and biochemical properties of these phage were measured (5). The yield of phage particles with IdUrd-substituted DNA was about the same as that of normal or unsubstituted phage. The phage with substituted DNA (greater than 60 percent IdUrd substitution for deoxythymidine) were able to adsorb to the bacteria and inject their DNA, but were unable to induce normal amounts of the phage-specific enzymes 2'-deoxycytidylate hydroxymethylase, dihydrofolate reductase and lysozyme. The reduced enzyme activity of 2'-deoxycytidylate hydroxymethylase was probably due to reduced synthesis of the enzyme rather than synthesis of an altered, less active enzyme (6).

To more readily measure the effect of IdUrd substitution in phage T4 DNA, the genetic technique of marker rescue was used in which cells are dually infected with IdUrd-substituted phage and an amber mutant phage (7). Thirty genes were studied in this manner. These included genes for enzymes involved in DNA synthesis, genes for structural proteins and genes for maturational proteins. By this method, a wide range of gene inactivation by IdUrd was found. Some genes were only reduced in activity by about 20%, others by more than 90%. The viability of the phage and the ability of individual genes to function decreased as a function of increasing IdUrd substitution.

In experiments with adenovirus type 2 (Ad2), the viral structural proteins were labelled with tritiated amino acids. Infection of KB cells in the presence of IdUrd caused a dose-dependent decrease in both total and infectious viral particles (8). Incomplete Ad2 particles were separated from complete particles by equilibrium CsCl buoyant density analysis. At a medium concentration of 28 μM IdUrd, the yield of complete particles had been reduced to about one tenth of control but the yield of infectious particles had been reduced by 4.5 logs. Although a greater proportion of the total particles were incomplete the amount was small compared to the loss of infectivity. Since the specific infectivity (the ratio of infectious virus to the number of particles) was a small fraction, 0.037 percent, of that of the control, the major cause for the reduction of infectious virus by IdUrd was production of uninfectious virus rather than the reduction in total virus particle production. Analysis of the DNA of the completed particles showed that 7-10 percent of deoxythymidine residues had been replaced by IdUrd, and no DNA fragmentation was observed (8).

The polypeptides were analysed by SDS polyacrylamide gels, and the viral coat polypeptides of IdUrd-containing particles were present in the correct proportions relative to control particles with no apparent differences in the polypeptide molecular weights.

IdUrd in the medium did not affect the synthesis of early (2-6 hr) viral proteins, but did markedly decrease the rate of synthesis of late (9-24 hr) viral proteins. Even at 56 μM IdUrd, however, the synthesis of all the viral proteins could be detected, and there was no preferential inhibition of any one viral protein. Pennington (9) has reported a similar differential effect on early and late protein synthesis of vaccinia virus by 5-bromo-2'-deoxyuridine (BrdUrd). Pennington found, however, that the synthesis of only some of the late viral proteins were inhibited.

The data from the Ad2 experiments seem to be in agreement with the results from the phage experiments. The principal lethal site of action of IdUrd is its incorporation into the viral DNA. Additional experiments reported here attempt to characterize the results of this incorporation in order to determine the mechanism by which IdUrd exerts its lethal effect.

5-Iodo-2'-deoxycytidine (ICdR, IdCyd)

Chang and Welch (10) synthesized ICdR in an attempt to obtain a compound with similar properties to IdUrd, but one with a lower rate of metabolism, a greater stability to heat and a greater aqueous solubility. This was achieved. Cooper (11) found that 5-bromo-2'-deoxycytidine is phosphorylated to BrdCMP in extracts of HSV infected cells, but not in extracts of uninfected cells, and he suggested the substrate specificity of BrdCyd for the virus-induced enzyme may provide a basis for the selective chemotherapy of herpetic infections with BrdCyd or IdCyd. The Km of mouse and human cytosol dCyd-kinase for BrdCyd is about 200 times larger than the Km for the normal substrate dCyd. Camiener and Smith (12) found the phosphorylated product, BrdCMP, to be a good substrate for dCMP deaminase. Thus antiviral selectivity can be demonstrated in those infected cell lines which do not have a cytidine deaminase, but can deaminate the halogenated analog at the nucleotide level. Those uninfected cells that have a cytidine deaminase will convert IdCyd to IdUrd and marked cytotoxicity will be observed. Unfortunately human tissues contain very potent cytidine deaminase activity (12), which prevents ICdR from being a selective antiviral agent for humans. Tetrahydrouridine, a transition state analog of dCyd, inhibits the enzyme cytidine deaminase and thereby decreased the toxicity of BrdCyd both _in vitro_ and _in vivo_, and enhanced the efficacy of BrdCyd in the therapy of experimental herpes encephalitis in mice (13).

5-Iodo-5'-amino-2',5'-dideoxyuridine (AIU, AIdUrd)

In an attempt to reduce the toxicity of IdUrd but still retain antiviral activity, the 5'-amino analog of 5-iodo-2'-deoxyuridine was first synthesized by Lin et al. (14,15). The assumption was made that a compound with an amino moiety in the 5'-position would not be incorporated into DNA and therefore would have decreased toxicity. The 5'-amino analog of thymidine, previously synthesized by Horwitz et al. (16) as well as AIdUrd, was found to have good antiviral activity with a highly restricted spectrum of activity but without significant cytotoxicity (14).

AIdUrd was tested in cell culture against a number of DNA and RNA viruses, and only-herpes type viruses are sensitive to AIdUrd: HSV-1 (17), VZV (18), EBV (19), guinea pig herpesvirus (20) and to a lesser extent HSV-2 (20). As with many other nucleoside analogues, AIU has little or no effect against cytomegalovirus, and none against vaccinia, adenovirus and minute virus of mice (MVM) (20). Among the RNA viruses tested, only murine leukemia virus was inhibited by AIdUrd while RSV, polio, Sendai, influenza and measles viruses were not affected

(20). AIU is effective in therapy of herpes simplex keratitis in rabbits (21,22) as well as in oral HSV-2 infection in mice (23), and i.p. HSV-1 infection in mice (24). However it is considerably less potent than IdUrd both *in vitro* and *in vivo*.

An important finding was the total lack of detectable toxicity of AIU in cell culture (20). Seventeen different strains of cells were tested, including cells from murine, avian, simian and human origin. Whereas IdUrd is mutagenic in L5178Y cells, AIdUrd was found to be nonmutagenic (20). Studies *in vivo* with mice have yielded similar results. No toxicity was observed in 1 day, 8 day, 10 week old, 15 week old or adult mice when AIdUrd was given as suspension even at 450 mg/Kg/day for 5 days. It does, however, display a unique lethality when given i.p. in solution, (30 mg/2 ml), to 10 week-old mice but not 15 week-old mice (Sim, personal communication). At very high doses (2 g/Kg/day for 5 days) AIdUrd appears to be mildly immunosuppressive whereas IdUrd at 100 mg/Kg/day is strongly immunosuppressive (25).

Our effort has mainly been to understand the biochemical basis for the antiviral activity of IdUrd and AIdUrd. The approach has been to investigate the effects of the drugs on macromolecular synthesis i.e. protein, RNA and DNA synthesis. Earlier experiments with IdUrd were done with both phage and adenovirus-2, but more recently our attention has focused on IdUrd and AIdUrd and their effects on herpesvirus replication.

EXPERIMENTAL RESULTS

Biochemical Effects of Metabolism of IdUrd and AIdUrd in Herpesvirus-Infected Cells

The enzymology of herpes replication accounts for the therapeutic selectivity of IdUrd and AIdUrd. Many investigators have demonstrated that herpes viruses induce increased deoxythymidine kinase activity (actually a pyrimidine deoxyribonucleoside kinase activity) (26). The Km of thymidine for the herpes-specified deoxythymidine kinase is 0.4-0.8 uM (25,27). Thus, in herpes-infected cells there is higher deoxythymidine kinase activity (up to about 10 fold) and the substrate, deoxythymidine or IdUrd, has a higher affinity for the viral enzyme compared to the cellular enzyme. There is no difference between deoxythymidine and IdUrd in Km and Ki, respectively, relative to either the cellular or viral enzyme (28,29). Thus, infected cells may be expected to phosphorylate and ultimately incorporate more IdUrd into DNA than uninfected cells. In addition, selectivity is favored by the fact that, by definition, the virus is obliged to synthesize DNA to replicate while most cells *in vivo* do not synthesize DNA. A comparison by Ostrander and Cheng (30) of the Km values for nucleoside triphosphates for HSV and cellular DNA polymerase suggested a more efficient binding by the viral enzyme. Also, the Km and Ki for deoxythymidine triphosphate and IdUrd triphosphate, respectively, are the same for the herpes DNA polymerase (30). In our studies of deoxyribonucleotide metabolism in herpes-infected HeLa cells (31), we found also about a 10-fold increase in the deoxythymidine triphosphate pool 8 hr after herpes infection. Since the culture medium did not contain deoxythymidine, the likely source of the increased deoxythymidine triphosphate pool is deoxythymidine monophosphate formed either by the de novo pathway via dUMP, or when the DNase induced by the virus degrades the cellular DNA. If anything, such an increase would be expected to lessen the selectivity of IdUrd through competition for the various deoxythymidine anabolizing enzymes.

Table 1. Metabolism of AIdUrd in infected and uninfected cells

Cell Type	Virus Added	AIU *cpm/10^6 cells
HeLa S_3	-	142
BALB 3T3	-	110
A-9 L-cell	-	189
A-9-L-cell	MVM	168
A-9-L-cell	HSV-1	69,800$^+$
Vero	-	275
Vero	HSV-1	75,000$^+$
Vero	Vaccinia	630
Vero	SV40	375

*cell-associated cpm
$^+$98-100% DNAse I sensitive, 0.14% RNAse sensitive

Subconfluent cell cultures (1 X 10^4 cells/90 mm dish) were infected with the indicated virus at a multiplicity of infection of 10 or mock-infected with an equivalent volume of PBS. After a 60-min adsorption period the inoculum was removed and 5 ml of medium containing 21.3 μM [^{125}I]AIdUrd (4.5 X 10^5 cpm/ml) were added. Twenty-three hours after drug addition the cells were detached by gently scraping, harvested by low speed centrifugation, and washed twice with 5 ml of PBS. The cell pellets were then counted directly in an Intertechnique counter (31).

The molecular basis for the selective anti-herpesvirus activity of AIdUrd has been extensively investigated. It was found that AIdUrd can be phosphorylated by the herpesvirus pyrimidine deoxyribonucleoside kinase but not by the cellular thymidine kinase (32). Chen et al. (32) found that [^{125}I]AIdUrd was metabolized in HSV-1-infected Vero cells to [^{125}I]AIdUTP but not in uninfected Vero cells. A number of cell lines were screened with and without infection by sensitive (HSV-1) or insensitive (MVM, vaccinia, SV40) viruses (Table 1). Only those cells which were infected with HSV-1 accumulated significant acid-insoluble counts in the presence of [^{125}I]AIdUrd. This radioactivity was found to be 98-100% DNAse I-sensitive indicating incorporation of the nucleoside into DNA.

In addition, extracts from infected Vero cells were compared to extracts from uninfected cells for their relative abilities to phosphorylate AIdUrd *in vitro*. Only lysates from infected cells could phosphorylate AIdUrd with AIdUTP being the product which could be recovered from the reaction. The absence of AIdUMP was thought to be due to the extreme lability of the monophosphoramidate, and the absence of the diphosphoramidate, which is very stable (33), is a consequence of rapid enzymic phosphorylation to the triphosphoramidate, a relatively stable compound at physiological pH.

Chen, Summers and Prusoff (unpublished studies) investigated the effect of AIdUrd on herpes simplex-transformed murine $LMTK^-$ cells and on the parental non-transformed cells. They found that the virus-transformed cells, in contrast to the parental strain, were markedly sensitive to inhibition by this analog (Table 2). Whereas the transformed cell lines were strongly inhibited by 0.02 to 0.05 mM AIdUrd, concentrations of 3 mM were required to inhibit the growth of the parental strain.

Table 2. Effect of AIdUrd on the Growth of Parental and HSV-1 Transformed Mouse L Cells

Cells	ED_{50} (mM)
$LMTK^-$	3
A9 (L cells, TK^+, H6PRT)	2
Tf-1	0.100
Tf-5	0.050
Tf-6	0.060
Tf-C3	0.075

Four independently-derived lines of $LMTK^-$ cells transformed to the TK^+ phenotype with UV-irradiated HSV-1 CL 101 by the procedure of Munyon et al. (34) are designated Tf 1, Tf 5, Tf 6, Tf C3. Cells were plated at low density (less than 1% of confluent density) in 96-well microtiter dishes in the presence of various concentrations of AIdUrd. The number of cells per well were determined after four days of growth.

In vitro studies by Chen and Prusoff (27) with AIdUrd and the purified HSV-1 pyrimidine deoxyribonucleoside kinase indicated that AIdUrd was a good competitive inhibitor of the enzyme with respect to both thymidine and deoxycytidine (Table 3). *E. coli* thymidine kinase, however, was unable to phosphorylate AIdUrd under similar conditions.

Table 3. Kinetic constants of the HSV-1-induced thymidine kinase (26)

Substrate	Km (μM)	Ki (μM)	Relative ratio of product formed[c]
Thymidine	0.8		24
Deoxycytidine	700	700[a] (against thymidine)	31
AIdUrd	5.1[b]	4[a] (against deoxycytidine)	1
		5[a] (against thymidine)	

[a]Competitive inhibition. The assay conditions were 75 mM Tris-HCl (pH 7.8), 1.5 mM ATP, 1.5 mM $MgCl_2$, 0.2 mg bovine albumin, 1 mM β-mercaptoethanol, various concentrations of [2-^{14}C] thymidine (56 mCi/mmol), [5-^{3}H] deoxycytidine (6 Ci/mmol) or A[^{125}I]dUrd (5 mCi/mmol in total volume of 0.1 ml.

[b]Based on the formation of A[^{125}I]dUrd diphosphate from A[^{125}I]dUrd. AIdUrd diphosphate was separated by spotting a portion of the enzyme reaction mixture on Whatman number 1 paper followed by developing in propanol/NH_4OH/H_2O, 55/10/55.

[c]Based on the rate of formation of product: thymidine to thymidylate, deoxycytidine to deoxycytidylate; AIdUrd to AIdUDP.

The phosphorylation of AIdUrd by the herpes thymidine kinase and subsequent incorporation into DNA implies the formation of phosphoramidate linkages in the DNA.

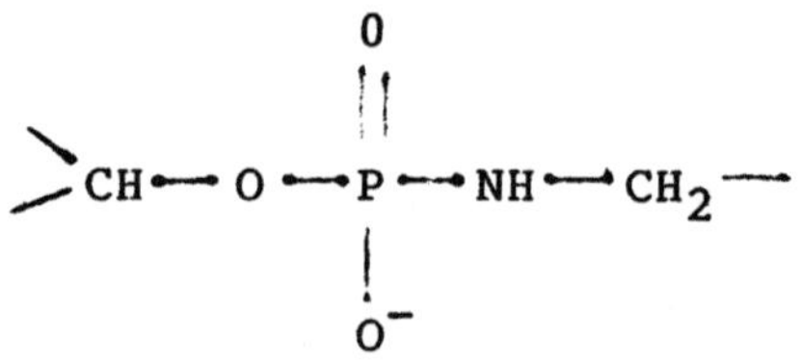

Analysis of this DNA by bouyant density centrifugation,

DNAse I, spleen DNAse II, micrococcal nuclease, and spleen and venom phosphodiesterases clearly indicated that AIdUrd is incorporated internally in the DNA structure (32). AIdUMP containing DNA therefore contains phosphoramidate linkages. The selective HSV-1-induced phosphorylation of AIdUrd and its incorporation into DNA are most likely to be critical events for the expression of the antiviral activity.

An analysis by Fischer et al. (35) of HSV-1 DNA synthesized in AIdUrd or IdUrd-treated cells indicated that the level of inhibition of HSV-1 corresponded most closely with the level of incorporation of the nucleoside and did not correlate with any inhibition of DNA synthesis. In fact subsequent studies (Otto and Prusoff, unpublished) on the level of viral DNA synthesis in AIdUrd – treated Vero cells, using incorporation of [^{32}P]orthophosphate, indicated that there was only a slight reduction in the level of ^{32}P incorporated into viral DNA. However, when [^{3}H]-thymidine or [^{3}H]deoxyadenosine were used as labelled probes significant reduction in incorporation was observed. These differences relate to the marked increase in dCTP, dGTP and dATP pool sizes produced by AIdUrd treatment of the infected cells (Figure 1). A greater pool size would lead to a lower specific activity of added label and result in less radioactivity in the DNA. The increased pool size of dTTP in the HSV-infected cell was actually decreased in half when AIdUrd was present, and the apparent inhibition of DNA synthesis was most likely due to the substitution of AIdUTP for [^{3}H]dTTP in the DNA.

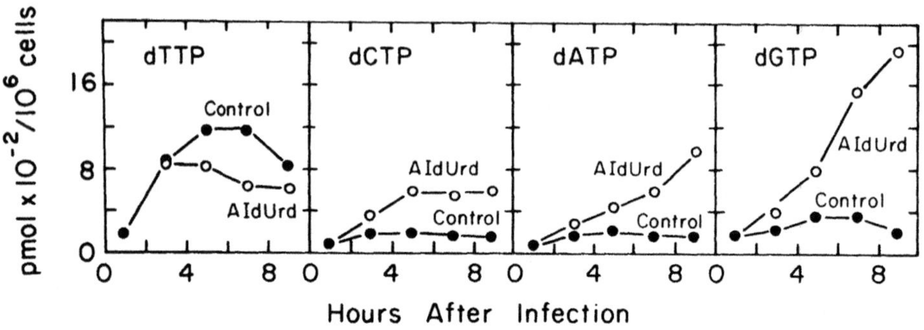

FIGURE 1. Effect of AIdUrd on dNTP pool sizes in HSV-1 infected Vero cells at various times after infection. Confluent Vero cells were infected with 10 plaque-forming units per cell of HSV-1 (Cl-101) and allowed to adsorb for 1 hour. AIdUrd (200 μM) was added immediately after removal of the virus inoculum, and both drug-treated (o) and control (•) cultures were removed at the indicated times. Cell extracts were prepared by the addition of 5 vol of 0.5 M perchloric acid to the cell pellet. The acid soluble fraction was neutralized with 4 M KOH and the salt precipitate was removed by centrifugation. The resulting supernatant was used for deoxyribonucleoside triphosphate determination using the method of Solter and Handschumacher (36) (Cheng & Prusoff, unpublished).

Although AIdUrd does not significantly inhibit viral DNA synthesis its incorporation brings about significant changes in the quality of the DNA. Fischer et al. (35) using alkaline and neutral sucrose gradients, found that the presence of AIdUMP in the DNA results in double- and single-stranded breaks in the viral DNA molecules. The number of breaks in the DNA increased proportionately to the concentrations of AIdUrd in the medium (Fig. 2,3).

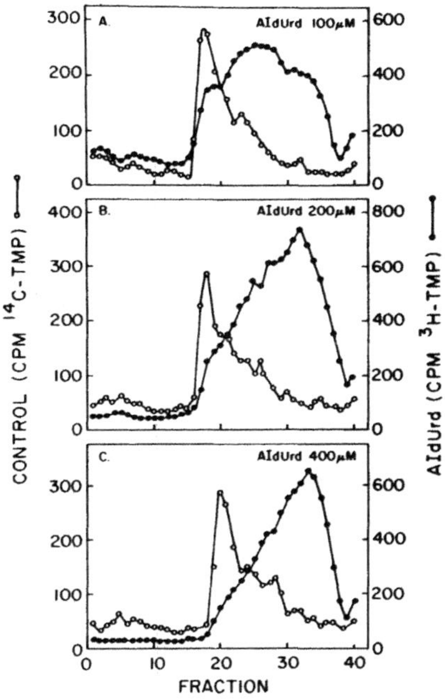

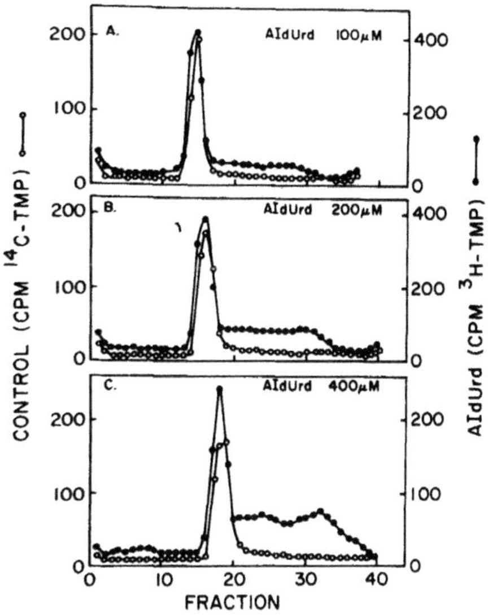

FIGURE 2. The sedimentation of HSV-1 DNA substituted with AIdUrd in alkaline sucrose gradients. To HSV-1-infected Vero cells either 0.25 ml buffer or 0.25 ml AIdUrd was added to achieve a final concentration of (A) 100 μM (B) 200 μM or (C) 400 μM in a total volume of 10 ml of medium. At 3.5 h postinfection control cultures were labeled with 10 μM [^{14}C]thymidine (0.5 μCi/ml) and the AIdUrd-treated cells were labeled with 0.4 μM [^{3}H]thymidine (10 μCi/ml). The cells were harvested at 20 h postinfection and samples from the control (o——o) and drug-treated (•——•) cultures were cosedimented in alkaline sucrose gradients. The direction of sedimentation is from right to left.

FIGURE 3. The sedimentation of AIdUrd-substituted HSV-1 DNA in neutral sucrose gradients. Control HSV-1 DNA ([^{14}C]thymidine labeled; 0.5 μCi/ml, 10 μM) was cosedimented with HSV-1 DNA ([^{3}H]thymidine labeled; 10 μCi/ml; 0.4 μM) obtained from HSV-1 infected Vero cells exposed to AIdUrd at concentrations of (A) 100 μM (B) 200 μM and (C) 400 μM. •——•, control; o——o, AIdUrd. The direction of sedimentation is from the right to the left.

The reason for these breaks is still unclear. It was originally thought that they were the result of the fragility of the phosphoramidate bond. It now seems unlikely that either the presence of the phosphoramidate alone or the presence of iodine alone at the 5 position is responsible for the breaks. Studies with 5'-amino-2',5'-dideoxythymidine (Otto and Prusoff, unpublished), which also forms a phosphoramidate bond in the DNA, or with IUdR (Figure 4) indicate no increased breakage of the viral DNA (34). It is possible that the combination of the 5-iodo and the 5'-amino groups brings about a less stable conformation for the DNA. In this regard it is of interest to note that the conformation of AIdUrd in the crystalline solid state and in solution has been examined by Birnbaum et al. (37) using X-ray crystallography and the Bruker HX-270 NMR. The

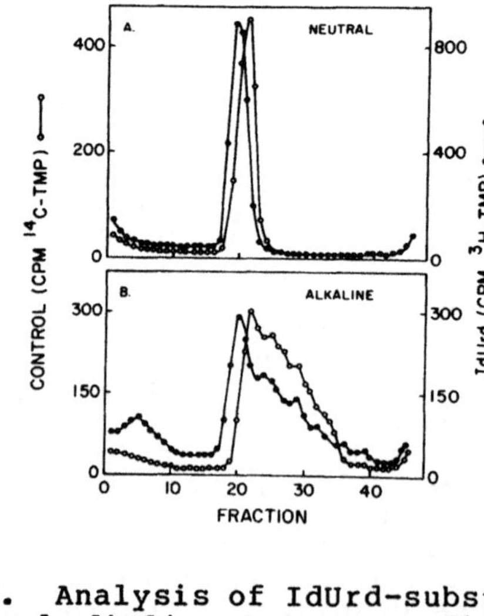

FIGURE 4. Analysis of IdUrd-substituted HSV-1 DNA in neutral and alkaline sucrose gradients. HSV-1 infected Vero cells were labelled with 10 μM [^{14}C]thymidine (0.5 μCi/ml) or exposed to 20 uM IdUrd and labelled with 0.2 μM [^{3}H]thymidine 5 μCi/ml. Samples obtained from the control and IdUrd treated cultures were cosedimented in both (A) a neutral sucrose gradient and (B) an alkaline sucrose gradient. (From Fischer et al., 35).

structure is zwitterionic with a protonated 5'-NH_2 group and an unusual negative charge at the N(3) position. Unlike the usual conformations of ribose and deoxyribose in nucleosides and nucleotides where the conformation is C(2')-endo or C(3')-endo (38), the deoxyribose in AIdUrd has an unusual O(1')-endo pucker. Further studies may reveal how this conformation might affect the structure of the viral DNA.

Analysis of the DNA in these progeny virions by restriction nucleases and electrophoresis revealed no change in the cleavage patterns produced by EcoRI endonuclease. There was however a slower rate of cleavage of the DNA indicating that the substitution of AIdUrd or IdUrd at levels greater than 15% for DNA thymidine either interfered with the recognition by the nuclease or the rate of cleavage at the EcoRI restriction sites.

The effects of the incorporation of AIdUrd or IdUrd into the DNA on herpes-viral gene expression have also been investigated. As might be expected, the most pronounced effects are observed in those processes which rely on newly synthesized viral DNA, that is, late RNA and protein synthesis. Since early transcription does not depend on newly synthesized viral DNA, it is unaffected during exposure to AIdUrd or IdUrd (Otto and Prusoff, unpublished). Although total accumulation of [^{3}H]uridine into RNA in HSV-1 infected Vero cells is not significantly affected, the amount of HSV-1 specific poly A^+ RNA begins to decline with respect to untreated infected cells as early as eight hours after infection. The reduced levels of HSV-1 specific poly A^+ RNA is more pronounced at 18-24 hr after infection since poly A^+ RNA in control cells continues to rise whereas that in the AIdUrd-treated cells remains essentially unchanged. At the same time there is a small increase in HSV-1 specific poly A^- RNA at 8 hr after infection (Table 4) which appears more pronounced at 18 hrs after infection because the control poly A$^-$ RNA is markedly decreased at this latter time.

Table 4. HSV-1 Specific RNA [^{3}H] Uridine (cpm/10^6 cells)

Analogue	8 hr		18 hr	
	Poly A$^+$	Poly A$^-$	Poly A$^+$	Poly A$^-$
None	22,600	12,800	35,000	6,420
AIdUrd	11,200	15,300	12,700	16,200
IdUrd	12,400	18,000	21,500	26,200

These changes in late HSV-1-specific RNA probably reflect the consequences of substitution of AIdUrd or IdUrd for thymidine in the newly synthesized DNA. The relative increase of poly A$^-$ along with the decrease in total HSV-1 specific RNA sequences suggests an inhibition of transcription and/or polyadenylation as well as a possible stabilization of late viral transcripts.

If such changes in HSV-1-specific poly A$^+$ RNA reflect a decrease in translatable late mRNA and no effect on early mRNAs then there should also be a reduction in the synthesis of only the late HSV-1 specific proteins. Such a reduction is seen in late (β and γ) but not early (α) proteins (39). Analysis of HSV-1 proteins by SDS gel electrophoresis and autoradiography revealed a wide range of inhibition of β and γ proteins at a concentrations of AIdUrd or IdUrd which inhibited the yield of infectious virus 99%. Although most β and γ proteins were inhibited (14-56%), two (ICP 35 and 39) remained unchanged and one (ICP 36) was stimulated with AIdUrd (Table 5). The stimulated protein was identified by its kinetics of synthesis, precipitation by antibody and enzymatic activity to be the herpes-specific pyrimidine deoxyribonucleoside kinase (Figure 5,6) (40).

Table 5. Effect of AIdUrd or IdUrd on Proteins Synthesized by HSV-1 (Percent of proteins synthesized as related to untreated infected cells)

ICP #	Kinetic Class	AIdUrd	IdUrd
4		99 ± 3	95 ± 5
0	α	100 ± 2	96 ± 6
27		100 ± 2	100 ± 2
22 h		100 ± 2	100 ± 2
34 h		102 ± 2	101 ± 2
6		76 ± 3	46 ± 14
8		85 ± 3	79 ± 2
26		70 ± 5	76 ± 3
29/30	β	86 ± 2	83 ± 3
36		123 ± 6	106 ± 3
39		100 ± 2	100 ± 3
40/41		66 ± 4	75 ± 6
5		65 ± 2	55 ± 8
10		82 ± 3	80 ± 4
11/12		55 ± 5	89 ± 6
20	γ	85 ± 7	83 ± 7
23		58 ± 3	64 ± 4
32		46 ± 6	35 ± 4
35 h		100 ± 2	100 ± 2
37		79 ± 6	83 ± 4
43		77 ± 3	68 ± 6
44		69 ± 4	58 ± 6

Vero cell monolayers were infected at 20 pfu/cell and after adsorption for 1 hour AIdUrd (800 μM) was added to the cells. Proteins were labelled at various times after infection with [^{35}S] methionine. Labelled proteins were analyzed by SDS gel electrophoresis and autoradiography. Densitometer scans of the autoradiographs were used to quantitate individual bands. Numbers shown are computed from 3-5 autoradiographs and are presented as percent of bands from untreated HSV-1 infected Vero cells. h indicates band identified as host protein by Honess and Roizman (41).

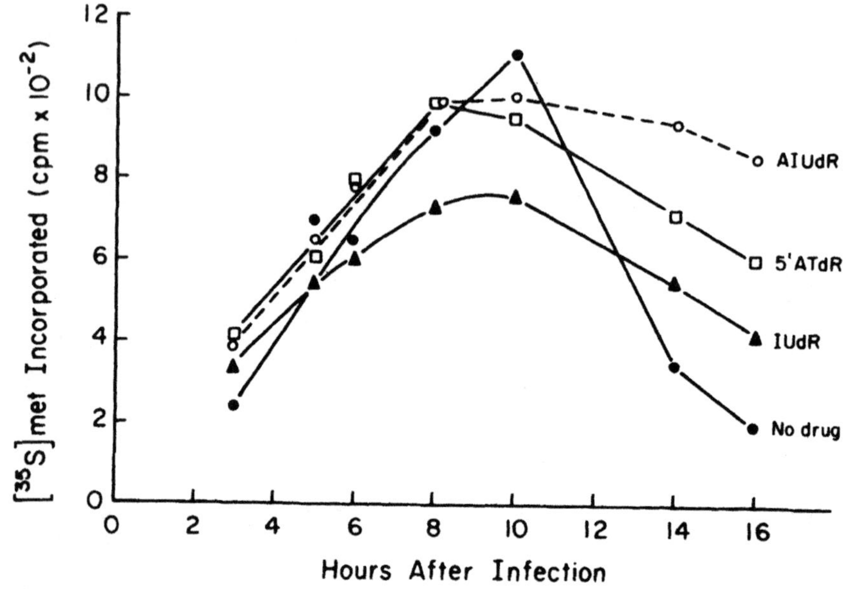

FIGURE 5. Time course of synthesis of ICP 36 (thymidine kinase) in the presence of nucleoside analogs. Incorporation of [^{35}S]methionine into ICP 36 in 30 minutes at different times after infection. Vero cells were infected at 20 pfu/cell and one hour later exposed to 800 μM AIdUrd, 25 μM IdUrd, or no drug. Infected cells were exposed to [^{35}S]methionine for 30 minutes at indicated times and lysed with SDS sample buffer. Lysates were analyzed by SDS-PAGE and autoradiography as in Figure 1. ICP 36 was located and excised and the gel slice counted.

Exposure of HSV-1 infected cells to AIdUrd or IdUrd, in addition to disrupting the synthesis of certain proteins, resulted in progeny virions with altered protein patterns (Fig. 7).

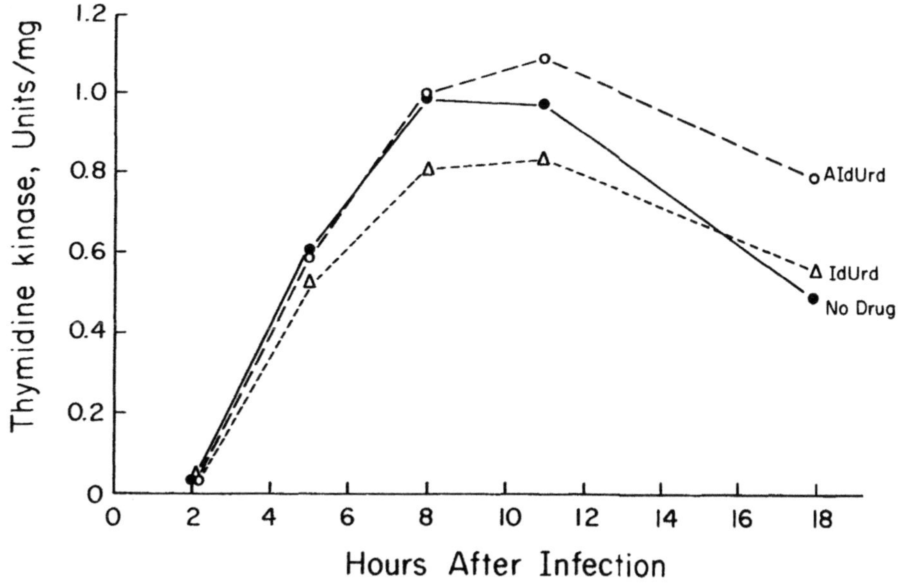

FIGURE 6. Induction of HSV-1 thymidine kinase in the presence of nucleoside analogues. Culture flasks of LMTK$^-$ cells were infected at 10 pfu/cell and one hour later drugs were added. At the indicated times after infection duplicate samples were harvested. Cells were scraped into phosphate-buffered saline (PBS) and washed two times by low speed centrifugation and resuspended in PBS. The cell pellet was lysed in 25 mM HEPES pH 7.5, 3 mM $MgCl_2$ and 4 mM dithiothreitol (DTT) by sonication. Cell debris was removed by centrifugation and the resulting supernatant was assayed for thymidine kinase activity in a reaction mix containing 2 mM ATP, 2 mM DTT, 100 μM dThd, 1.5 μCi [^{14}C]dThd, 50 mM NaF, and 50 mM HEPES, pH 7.5. Proteins were determined by the Lowry method.

These altered virions have a reduced ability to produce plaques, altered ability to code for new proteins, increased particle to pfu ratio, and the ability to interfere with the replication of standard HSV-1 virus (42). A single passage in the presence

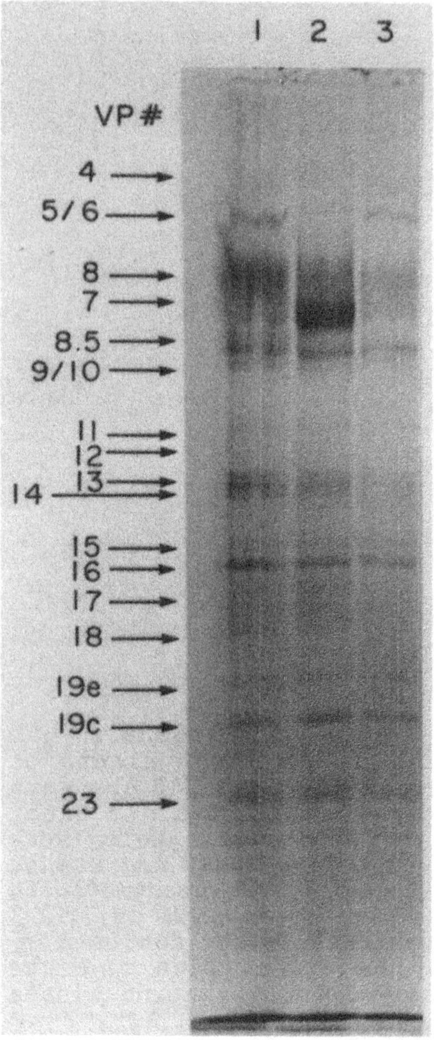

FIGURE 7. Autoradiograms of electrophoretically separated [^{35}S] methionine labelled virion proteins. Cells were infected, drug treated and labelled as described in Table IV. Virions were harvested, purified and particle number determined. Equal numbers of particles were analyzed by SDS-PAGE. Virion proteins are designated by number. Lane 1 - virions from untreated cells, lane 2 - virions from AIdUrd-treated cells, lane 3-virions from IdUrd-treated cells.

of AIdUrd resulted in the production of an interfering factor (presumably defective virion particles) which purifies with infectious progeny HSV-1 on sucrose gradients. SDS-PAGE and autoradiogrpahy revealed that these progeny HSV-1, whether alone (20 pfu/cell) or mixed with control HSV-1 (5 pfu progeny/cell: 20 pfu control/cell) were deficient in inducing HSV-1 proteins in the absence of drugs when compared to control HSV-1 at 20 pfu/cell (Fig. 8).

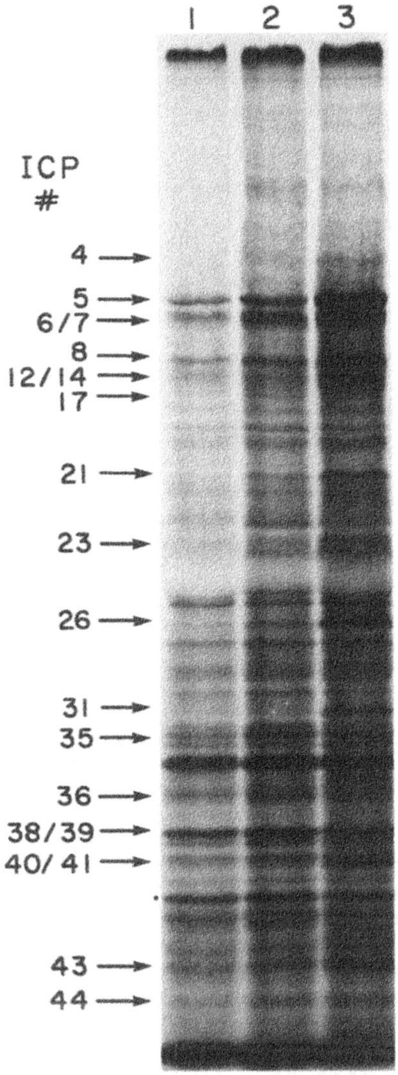

FIGURE 8. Effect of mixing progeny virions with standard HSV-1 on synthesis of virus-induced proteins. Autoradiograph of electrophoretically separated [^{35}S] methionine-labelled proteins from Vero cells infected with mixtures of purified progeny virus and standard HSV-1. Vero cells were infected with a mixture of standard HSV-1 (20 pfu/cell) and progeny virus (5 pfu/cell) from AIdUrd-treated cells (Lane 1), IdUrd-treated cells (Lane 2), or untreated cells (Lane 3). Proteins were labelled with [^{35}S]methionine from 2-18 hours after infection and analyzed as in Figure 7.

Infection of Vero cells with a mixture of wild type HSV-1 and progeny HSV-1 from AIdUrd-treated or IdUrd-treated cells resulted in a progressive interference in the replicative ability of control HSV-1 which was proportional to the input titer of the drug-substituted-HSV-1 progeny virions (Fig. 9).

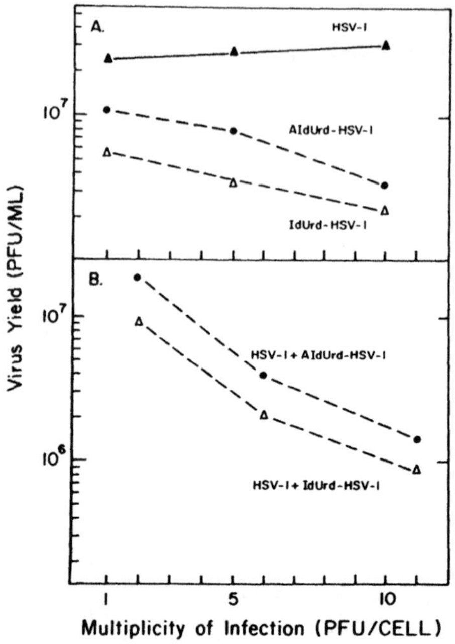

FIGURE 9. Production of infectious virus by passage of progeny virions.
A. Yield of infectious virus at various multiplicities of infection with sucrose gradient-purified progeny virus. Confluent monolayers of Vero cells were infected with the indicated progeny virus preparation at 1, 5 or 10 pfu/cell. Infectious virus from these passages was harvested by freezing and thawing and the titers determined by plaque assay on Vero cells.
B. Yield of infectious virus with mixtures of standard HSV-1 and progeny virus preparations. Standard HSV-1 was kept constant at 1 pfu/cell and mixed with progeny virus at 1, 5, or 10 pfu/cell. Infectious virus was harvested and titered as in A.

DISCUSSION OF THE BIOCHEMICAL EFFECTS

The lethal activity of AIdUrd or IdUrd depends upon two essential steps: first the initial phosphorylation of the nucleoside and second the incorporation of the nucleoside triphosphate into viral DNA. Both of these critical events depend upon the activity of herpes viral enzymes, the pyrimidine deoxyribonucleoside kinase and the DNA polymerase. IdUrd in addition is phosphorylated by cellular thymidine kinase. The results of these activities is the substitution of the analogues for thymidine in the DNA. We have described a number of sequelae to this substitution, all of which when taken together, present a reasonable explanation as to why infectious virus production is inhibited. The question which remains partly unanswered is how this substitution brings about the changes which we and others have documented.

If our recent results are considered in light of some earlier experiments, a possible explanation seems to present itself. Schwartz and Kirsten (43), amongst others (43) have reported that in rat embryo cells, BrdUrd, relative to deoxythymidine, was preferentially incorporated into intermediate repetitive sequences. Such sequences are interspersed contiguous with single copy sequences and are believed to have a regulatory role. Studies are in progress in our laboratory to determine the distribution of IdUrd and/or AIdUrd in the herpes viral DNA, in particular to identify any "hot spots" of incorporation. The significance of these studies relates to the observation from a number of laboratories (44) that indicate that IdUrd or BrdUrd substitution in DNA increases the binding of regulatory proteins to the DNA. Lin and Riggs (45) showed the substitution for thymidine resulted in increased binding of the lac repressor to the lac operator and in Syrian hamster melanoma cells, BrdUrd substitution, brought about increased binding of nonhistone chromosomal proteins to the DNA (45). With regard to this, preliminary experiments in our laboratory (Otto and Prusoff, unpublished) indicate an increase in certain proteins tightly bound to the drug substituted viral DNA.

Our observations regarding the change in viral-specific RNA and protein synthesis could be explained in light of increased binding of proteins to the DNA. Tight binding of proteins to DNA is consistent with the gene not being transcribed. For example a regulatory protein could be bound too tightly or the RNA polymerase, once bound at a promoter site, might not proceed down the DNA chain. Since herpes viral DNA contains several promoter sites, the expression of the different genes could be affected differently by IdUrd or AIdUrd substitution with each promoter having a different affinity for the RNA polymerase. Such an idea appears to have merit since there is direct evidence of BrdUrd substitution affecting promoter sites in fd bacteriophage DNA.

Hofer and Koster (47) used a cell-free system to synthesize BrdUrd-substituted or normal DNA which they fragmented with HpaII restriction endonuclease. They then compared the binding of RNA polymerase in the presence of nucleoside triphosphates to the promoter region in the various fragments of substituted and normal DNA. BrdUrd substitution affected the association constant of RNA polymerase for each fragment differently. The fold increase in association constant compared to unsubstituted fragments were 1, 2, 5 and 6 for four fragments studied. Even though the base sequences of the promoter regions are known, Hofer and Koster (46) could not pinpoint the critical BrdUrd substituted site(s) that altered polymerase binding.

One thing that is clear is that with the major strides made in the field of nucleic acid research such as DNA sequencing we should, before long, get a much more sophisticated understanding of how IdUrd inhibits viral replication.

CONCLUSIONS

1. AIdUrd, and IdUrd are phosphorylated by the herpes thymidine kinase to AIdUDP and IdUDP and then presumably by cellular enzymes to AIdUTP and IdUTP respectively.
2. Both are incorporated into herpes progeny DNA by herpes DNA polymerase in place of thymidine.
3. Incorporation into DNA results in:
   a) formation with AIdUrd of phosphoramidate bonds in DNA was well as single- and double-stranded breaks. IdUrd produces normal phosphodiester bonds in DNA with no single- or double stranded breaks.
   b) inhibition of HSV-specific late RNA transcription and an incease in poly A^- HSV-specific RNA.
   c) reduction in the synthesis of most late HSV proteins.
   d) overproduction of HSV pyrimidine deoxyribonucleoside kinase synthesis with AIdUrd and no significant change with IdUrd.
4. In addition, progeny virions:
   a) have abnormal protein composition
   b) are less infective
   c) show only a slight reduction in total particles resulting in a greater particle/pfu ratio.
   d) have the ability to interfere with infection by standard virus.

ACKNOWLEDGEMENTS

The research performed in this laboratory was supported by United States Public Health Service Grant, CA-05262, from the National Cancer Institute.

REFERENCES

1. Prusoff, W.H. Biochim. Biophys. Acta. 32:295-296, 1959.
2. Herrmann, E.C., Jr. Proc. Soc. Exp. Biol. Med. 107:142-145, 1961.
3. Prusoff, W.H. and Goz, B. In: Antineoplastic Immunosuppressive Agents Vol. 2 (Eds. A.C. Sartorelli and D.G. Johns) Springer, Berlin, 1975 pp. 272-347.
4. Kaufman, H.E., Marola, E. and Dohlman, C. Arch. Opthalmol. 68:235-239, 1960.
5. Goz, B. and Prusoff, W.H. J. Biol. Chem. 243:4750-4756, 1968.
6. Aamodt, L. and Goz, B. Biochem. Pharmacol. 19:2400-2403, 1970.
7. Goz, B. and Prusoff, W.H. Ann. N.Y. Acad. Sci. 173:379-389, 1970.
8. Kan-Mitchell, J. and Prusoff, W.H. Biochem. Pharmacol. 28:1819-1829, 1979.
9. Pennington, T.H. J. Virol. 18:1131-1133, 1976.
10. Chang, P.K. and Welch, A.D. Biochem. Pharmacol. 8:327-328, 1961.
11. Cooper, G.M. Proc. Natl. Acad. Sci. USA 70:3788-3792, 1973.
12. Camiener, G.W. and Smith, G.S. Biochem. Pharmacol. 14:1405-1416, 1965.
13. Greer, S., Schildkraut, I., Zimmerman, T. and Kaufman, H. Ann. N.Y. Acad. Sci. 255:359-365, 1975.
14. Lin, T.S., Neenan, J.P., Cheng, Y.-C., Prusoff, W.H. and Ward, D.C. J. Med. Chem. 19:495-498, 1976.
15. Lin, T.S. and Prusoff, W.H. J. Med. Chem. 21:109-112, 1978.
16. Horwitz, J.P., Thompson, A.J., Urbanski, J.A. and Chua, J. J. Org. Chem. 27:3045, 1962.
17. Cheng, Y.C., Goz, B., Neenan, J.P., Ward, D.C. and Prusoff, W.H. J. Virol. 15:1284-1285, 1975.
18. Iltis, I.P., Lin, T.S., Prusoff, W.H. and Rapp, F. Antimicrob. Agents Chemotherapy 16:92-97, 1979.
19. Henderson, E.E., Long, W.K. and Ribecky, R. Antimicrob. Agents Chemotherapy 15:101-110, 1979.

20. Prusoff, W.H., Ward, D.C., Lin, T.S., Chen, M.S., Shiau, G.T., Chai, C., Lentz, E., Capizzi, R., Idriss, J., Ruddle, N.H., Black, F.L., Albert, D., Bhatt, P.N., Hsiung, G.C., Strickland, S. and Cheng, Y.C. Ann. N.Y. Acad. Sci. 284:335-341, 1977.
21. Albert, D.M., Lahav, M., Bhatt, P.N., Reid, T.W., Ward, R.E., Cykiert, R.C., Lin, T.S., Ward, D.C. and Prusoff, W.H. Invest. Opthalmol. 15:470-478, 1976.
22. Puliafito, C.A., Robinson, N.L., Albert, D.M., Pavan-Langston, D., Lin, T.S., Ward, D.C. and Prusoff, W.H. Proc. Soc. Exp. Biol. Med. 156:92-96, 1977.
23. Park, N.H., Pavan-Langston, D., Hettinger, M.E., McLean, S.L., Albert, D.M., Lin, T.S. and Prusoff, W.H. J. Infect. Diseases 141:575-579, 1980.
24. Sim, I.S., Stebbing, N. and Carey, N.H. Antiviral Res., 1:393-404, 1981.
25. Kan-Mitchell, J., Mitchell, M.S., Lin, T.S. and Prusoff, W.H. Cancer Research 40:3491-3494, 1980.
26. Jamieson, A.T. and Subak-Sharpe, J.H. J. Gen. Virol. 24:481-492, 1974.
27. Chen, M.S. and Prusoff, W.H. J. Biol. Chem. 254:10449-10452, 1979.
28. Lee, L.-S. and Cheng, Y.-C. Biochemistry 15:3686-3690, 1976
29. Cheng, Y.-C. Biochim. Biophys. Acta 452:370-381, 1976.
30. Ostrander, M. and Cheng, Y.-C. Biochim. Biophys. Acta 609:232-235, 1980.
31. Cheng, Y.-C., Goz, B. and Prusoff, W.H. Biochim. Biophys. Acta 390:253-263, 1975.
32. Chen, M.S., Ward, D.C. and Prusoff, W.H. J. Biol. Chem. 251:4833-4838, 1976.
33. Chen, M.S., Ward, D.C. and Prusoff, W.H. J. Biol. Chem. 251:4839-4842, 1976.
34. Munyon, W., Kraiselburd, E., Davis, D. and Mann, J. J. Virol. 7:813-820, 1971.
35. Fischer, P.H., Chen, M.S. and Prusoff, W.H. Biochim. Biophys. Acta 606:236-245, 1980.
36. Solter, A.W. and Handschumacher, R.E. Biochim. Biophys Acta 174:585-590, 1969.
37. Birnbaum, G.I., Lin, T.S., Shiau, G.T. and Prusoff, W.H. J. Am. Chem. Soc. 101:3353-3358, 1979.
38. Sundaralingam, M. Biopolymers 7:821-860, 1969.
39. Otto, M.J., Lee, J.J. and Prusoff, W.H. Antiviral Res. in press, 1982.
40. Mancini, W.R., Otto, M.J. and Prusoff, W.H. Abstracts Ann. Meet. Amer. Soc. Micro. S43:242, 1982.
41. Honess, R.W. and Roizman, B. Virology 14:18-19, 1974.

42. Otto, M.J. and Prusoff, W.H. Abstracts Ann. Meet. Amer. Soc. Microb. S44:243, 1982.
43. Schwartz, S.A. and W.H. Kirsten. Proc. Natl. Acad. Sci. USA 71:3570-3574, 1974.
44. Goz, B. Pharmacol. Rev. 29:249-272, 1978.
45. Lin, S.-Y. and Riggs, A.D. Proc. Natl. Acad. Sci. USA 69:2574-2576, 1976.
46. Bick, M.D. and Devine, E.A. Nucl. Acids Res. 4:3687-3700, 1977.
47. Hofer, B. and Koster, H. Nucl. Acids Res. 8:6143-6162, 1980.

3

ANTIVIRAL PROPERTIES OF ARANUCLEOSIDES: METABOLIC CONSIDERATIONS

THOMAS W. NORTH
Department of Biochemistry and Pharmacology, Tufts University Schools of Veterinary Medicine and Medicine, Boston, MA 02111, USA.

INTRODUCTION

Several D-arabinosyl nucleosides have been shown to possess significant antiviral and antitumor activities in experimental systems and two of these, 9-β-D-arabinofuranosyladenine (araA) and 1-β-D-arabinofuranosylcytosine (araC), have shown antiviral activity in humans. AraA is now in widespread use for treatment of human herpetic infections. AraC has proven too toxic for use in viral chemotherapy but has proven effective in the treatment of certain leukemias. Both of these aranucleosides have also been useful as biochemical tools to study DNA synthesis. The antiviral properties of these and other aranucleosides have been recently reviewed (1,2). Structures of the aranucleosides are shown in Figure 1.

Following the discovery of 1-β-D-arabinofuranosylthymine (araT) and 1-β-D-arabinofuranosyluracil (araU) in the sponge *Cryptotethya crypta* (3), chemical syntheses of araC (4) and of araA (5) were reported. AraA was later produced as a fermentation product of *Streptomyces antibioticus* by Parke, Davis and Co. The corresponding nucleotides of araC and araA (5' mono-, di-, and triphosphates) were synthesized by Cohen and his colleagues (6-8). The availability of these nucleotides made possible extensive study of the metabolism of these aranucleosides and their modes of action. The chemistry, biochemistry, and molecular biology of these compounds have been extensively reviewed (7,9,10).

There are many similarities between araA and araC both in their metabolism and in their modes of action. Both are inhibitors of DNA synthesis and are also incorporated into DNA, and these activities require formation of the 5'-triphosphates, araCTP and araATP (9). These two aranucleotides are competitive inhibitors of DNA polymerase (8,11), and at least part of the antiviral selectivity of araA is due to a selective inhibition of the herpes simplex virus (HSV) DNA polymerase by araATP (12,13). One difference, however, between these two aranucleosides is that araATP inhibits ribonucleotide reductase, but araCTP does not (14). It has been suggested that araA may self-potentiate its activity by

Y. Becker (ed.). Antiviral Drugs and Interferon: The Molecular Basis of Their Activity.

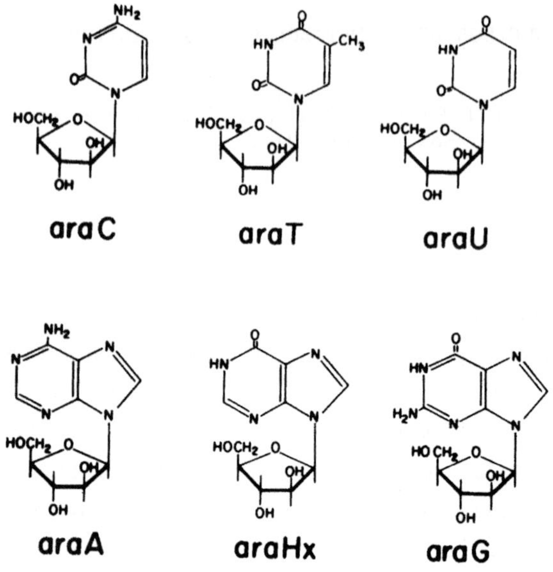

Figure 1. Structures of the aranucleosides.

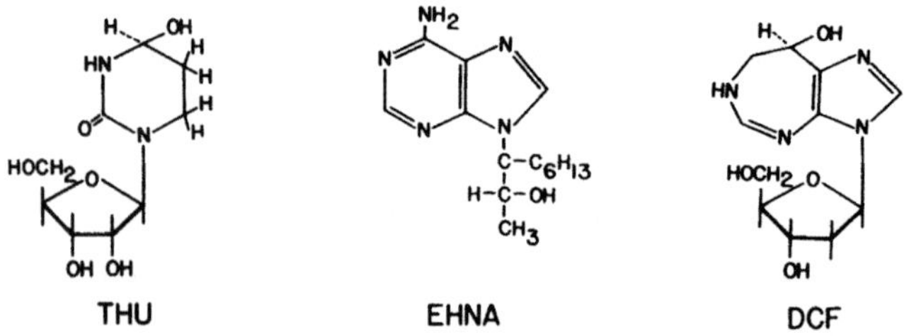

Figure 2. Structures of the inhibitor of cytidine deaminase, tetrahydrouridine (THU), and of the inhibitors of adenosine deaminase, erythro-9-(2-hydroxy-3-nonyl)adenine (EHNA) and 2'-deoxycoformycin (DCF).

blocking provision of the competing nucleotide, dATP (15).

Another feature common to araA and araC is that each may be enzymatically deaminated to a less toxic form. In this manner, araA is converted to 9-β-D-arabinofuranosylhypoxanthine (araHX) by adenosine deaminase (ADA) and araC is converted to araU by cytidine deaminase (CDA) (1,9). The therapeutic use of these analogs is greatly limited by these deaminase activities present in serum and tissues. As discussed below, the means to inhibit these deaminases are available and they promise to improve the therapeutic properties of araA and araC.

Despite the similarities of araA and araC, a striking difference is that araA is a selective inhibitor of HSV replication but araC is not. In fact, araC is more cytotoxic to uninfected cells than it is inhibitory to HSV replication. One explanation that was put forth to explain this difference is a failure of araCTP to inhibit the HSV DNA polymerase (12). However, Reinke et al. (13) have reported that araATP is only slightly more potent than araCTP as an inhibitor of the HSV DNA polymerase and both are more potent inhibitors of the viral polymerase than of cellular DNA polymerases. These data suggest that there may be other sites or other factors that contribute to the antiviral selectivity of araA. Because araATP and araCTP compete with dATP and dCTP, respectively, the activities of aranucleosides and their selectivities may be affected by factors that alter levels of either the deoxyribonucleoside 5'-triphosphates (dNTPs) or the aranucleotides. In recent years inhibitors of ADA and CDA have been used to increase conversion of araA and araC to their corresponding nucleotides (see below). To further evaluate metabolic consequences of the aranucleosides I have begun to examine their effects upon DNA precursor metabolism. These experiments were designed to determine whether the effects of araATP upon ribonucleotide reductase contribute to its biological activity and whether any of the antiviral properties of araA and araC (particularly their differences) can be explained by effects on DNA precursor metabolism. It is hoped that a full understanding of the metabolic effects produced by aranucleosides will allow improvement of their therapeutic efficacy.

MATERIALS AND METHODS

Cells and virus

HeLa F cells and the Miyama strain of HSV-1 were grown and maintained in Joklik-modified Minimum Essential Medium supplemented with 10% horse serum under conditions previously described (16). In experiments where concentrations of intracellular dNTPs were to be measured cells were grown in media supplemented with dialyzed horse serum. Mycoplasma assays were routinely performed as described previously (17) or by the procedure of Schneider et al. (18), and stocks of cells and virus were consistently free of mycoplasma contamination. Experiments to measure production of progeny virus following infection of cells by HSV, in the presence and absence of the indicated compounds, were also performed as previously described (16).

Other materials

AraA and araC were purchased from Sigma Chemical Co., EHNA was provided by H. Schaeffer and G.B. Elion (Wellcome Research Laboratories) and THU was provided by G. Neil (The Upjohn Co.). All other nucleosides and nucleotides were purchased from Sigma Chemical Co. All radioisotopes were obtained from New England Nuclear Corp. *E. coli* DNA polymerase I was obtained from Bethesda Research Labs, Inc. All other chemicals were reagent grade.

Measurement of dNTPs

Levels of deoxyribonucleoside 5'-triphosphates were measured with the enzymatic assay previously described (19). Extracts from uninfected HeLa cells used for these measurements were prepared by the two-step procedure described in that report. This protocol consists of an initial extraction of cells with 60% methanol followed by treatment of the extracted material with 0.5 N perchloric acid (19). However, I found that dTTP and dCTP were not stable to extraction from HSV-infected cells with these procedures. These nucleotides were degraded to their respective monophosphates during the initial extraction with 60% methanol. This is most likely due to activity of the deoxypyrimidine triphosphatase described by Wohlrab and Francke (20). When the two-step procedure is reversed so that infected cells are extracted first with 0.5 N perchloric acid, the extracted material taken to dryness, and the resulting residue treated with 60% methanol, none of the dNTPs are degraded. Therefore, this modification of the extraction procedure has been used for preparation of samples from HSV-infected cells. All other aspects of the dNTP assays were identical to the previously reported methods (19). The enzymatic assay for dNTPs can be used for analysis of samples containing araATP or araCTP because these aranucleotides do not inhibit DNA polymerase I (11), and neither araATP nor araCTP are substrates for or inhibitions of the polymerase under my reaction conditions (data not shown).

All other procedures were as described in previous reports (16,17) or are described in legends to figures and tables.

RESULTS

AraA and inhibitors of ADA

Early metabolic studies of araA and araC demonstrated that deamination limits the availability of these aranucleosides for conversion to their respective nucleotides. Workers at the Upjohn Co. showed that most of the araC administered to animals appears rapidly in the urine as araU. This led to a search for inhibitors of cytidine deaminase which resulted in the synthesis of tetrahydrouridine (THU) (21). THU is a competitive inhibitor of CDA with a K_i of 1.3 x 10^{-7}M (22). Similarly, massive deamination of araA to araHx was shown to occur in cultured cells and in animals (23,24). Subsequently, several inhibitors of ADA have become available. The first of these was erythro-9-(2-hydroxy-3-nonyl)adenine (EHNA), synthesized by Schaeffer and Schwender (25). EHNA is a competitive inhibitor of ADA with a K_i of 2-4 x 10^{-9} M (26). Another widely studied inhibitor of ADA is 2'-deoxycoformycin (DCF) which was isolated from cultures of <u>Streptomyces antibioticus</u> by researchers at Parke, Davis and Co. (27). DCF is a tight binding inhibitor of ADA with a K_i of 2.5-15 x 10^{-12}M (26). Structures of these inhibitors of CDA and ADA are shown in Figure 2.

Studies of these deaminase inhibitors in combination with araA or araC in cultured cells demonstrated that the antiproliferative activity of araA or araC can be greatly increased when the appropriate deaminase is inhibited. These results were of particular concern for studies of the antiherpes activity of araA, because araA had been shown to selectively inhibit the replication of HSV. Plunkett and Cohen (28) were the first to report a potentiation of the antiproliferative activity of araA against cultured mouse cells by combination with EHNA. This prompted evaluation of this combination against replication of HSV.

Potentiation of the antiherpes activity of araA by EHNA was first shown by myself and Cohen (16). Results similar to those previously reported are shown in Figure 3. These data show the effect of araA, alone or in combination with 10^{-6}M EHNA, upon production of progeny virus following infection of HeLa cells with HSV. The concentration of EHNA (10^{-6}M) used in these experiments has little or no effect upon HSV replication (dashed line), although higher concentrations of EHNA reduce HSV DNA synthesis (see below). The data in Figure 3 show that araA is much more inhibitory to HSV replication in the presence of 10^{-6}M EHNA than in the absence of the inhibitor of ADA. For example, combination of EHNA with 10^{-5}M araA increases the antiherpes activity 60-fold relative to 10^{-5}M araA alone. EHNA was shown to similarly potentiate the antiherpes activity of another adenosine analog, 3'-deoxyadenosine (cordycepin) (16).

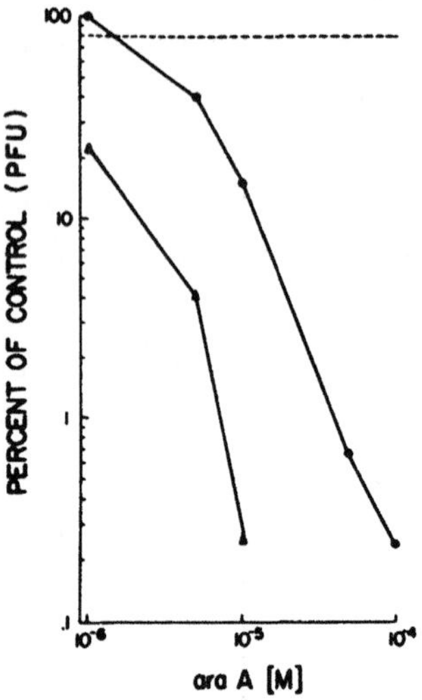

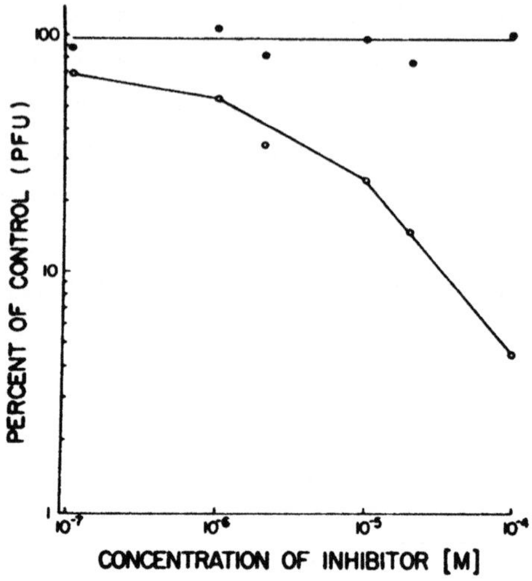

Figure 3 (left). Effect of araA upon production of progeny virus in HSV-infected HeLa cells in the presence (▲) or absence (●) of 10^{-6}M EHNA. The dashed line indicates activity of 10^{-6}M EHNA alone.

Figure 4 (right). Effects of the two inhibitors of ADA, EHNA (O) and DCF (●), upon production of progeny virus in HSV-infected HeLa cells.

Table 1. Effect of EHNA on dNTP Pools

	dNTP levels (pmoles/10^6 cells)[a]			
	dATP	dTTP	dGTP	dCTP
A. HSV-infected cells[b]				
Control	14.0	169.6	34.3	41.2
10^{-5}M EHNA	11.0	74.1	19.0	36.6
5 x 10^{-5}M EHNA	6.4	43.7	9.2	21.2
B. Uninfected HeLa Cells				
Control	19.2	42.6	8.1	41.0
10^{-5}M EHNA	20.0	25.2	6.9	48.6
5 x 10^{-5}M EHNA	17.3	19.7	11.7	31.5

[a]Assays for dNTPs are described in Methods.

[b]Cells were infected with HSV at an input moi of 10 PFU/cell and harvested 6 hr after infection for analysis of dNTPs as described in Methods.

We have also compared the combination of araA and EHNA to araA alone in treatment of HSV infections of mice. In these studies EHNA was shown to enhance the antiherpes activity of araA. The combination of araA plus EHNA was shown to decrease mortality and increase mean survival time of mice inoculated intraperitoneally with 6.4 LD_{50} of HSV-1 (29). Thus, EHNA has now been shown to increase the biological activity of araA in several experimental systems and has been shown to enhance the antiviral activity of araA in cultured cells and in animals.

Results similar to these with EHNA have been obtained in studies of DCF in combination with araA. DCF increases the antiherpes activity of araA in cultured cells and in animals (30,31). However, there are several reasons that EHNA may be better suited than DCF for use in combination with araA. Lapi and Cohen (32) reported that the ability of EHNA to inhibit ADA in cultured cells is readily reversible while DCF is not. This is a particularly important concern since long-term inhibition of ADA may impair the immune response. Deficiency of ADA is associated with severe-combined immunodeficiency disease in humans (33). Studies in mice have further demonstrated that inhibition of ADA by DCF is of much longer duration than inhibition by EHNA; there is also a lag of several hours following injection of DCF before inhibition of ADA is achieved, whereas EHNA produces immediate inhibition of ADA (34). Moreover, as discussed elsewhere, DCF seems to be much more toxic than EHNA to animals (1,2).

Another reason that makes EHNA more attractive than DCF for use in combination with araA in treatment of herpetic disorders is that EHNA alone displays some antiherpes activity while DCF does not (16). As shown in Figure 4, production of progeny virus in HSV-infected cells is inhibited by EHNA. Equal concentrations of the more potent inhibitor of ADA, DCF, do not. This activity is somewhat selective. A concentration of EHNA, which inhibits production of progeny virus by 75% (10^{-5}M) does not inhibit cell growth or decrease viability of cultured mouse or human cells (16,28). This selectivity was shown to be due to a specific inhibition of HSV DNA synthesis by EHNA (16). Another inhibitor of ADA which is similar in structure to EHNA, (S)-9-(2,3-dihydroxypropyl)adenine, also inhibits replication of HSV and several other viruses in cultured cells (35). It has not been determined whether this compound is also a specific inhibitor of HSV DNA synthesis.

Although EHNA significantly reduces production of HSV in infected cells, HSV replication is not completely blocked at concentrations of EHNA up to 10^{-4}M. At concentrations of EHNA above 2 x 10^{-5}M cell growth is impaired. This incomplete block of HSV replication probably explains the failure of EHNA to protect mice against severe HSV infections (29). Nevertheless, the fact that EHNA exhibits some antiviral activity when used alone may increase its efficacy when used in combination with araA more than by the simple inhibition of ADA.

The mechanism by which EHNA blocks HSV DNA synthesis is of interest for another reason. It seems to represent a new class of antiherpes compounds that act in a way other than as a substrate for the HSV thymidine kinase or as an inhibitor of the HSV DNA polymerase. If derivatives of EHNA with greater antiviral potency can be obtained they may be useful antiviral drugs, particularly if they fail to inhibit ADA.

Several lines of evidence suggest that the inhibition of HSV by EHNA is not due solely to inhibition of ADA. First of all, the K_i of EHNA for inhibition of ADA is 2-4 x 10^{-9}M (26) whereas inhibition of HSV replication by 50% requires 4-8 x 10^{-6}M (Figure 4). Secondly, the more potent inhibitor of ADA, DCF, does not inhibit replication of HSV (Figure 4). And finally, I have recently found that both stereoisomers of EHNA that were synthesized by Bastian _et al._ (36) have equal anti-HSV activity; these two stereoisomers are 250-fold different in their potencies as inhibitors of ADA (36). Both stereoisomers of EHNA also inhibit HSV DNA synthesis to the same extent as the racemic mixture (manuscript in preparation). These data demonstrate that inhibition of ADA is not by itself inhibitory to replication of HSV. However, since ADA is inhibited by concentrations of EHNA necessary to block HSV replication, it cannot yet be determined whether concomminant inhibition of ADA is required for the antiviral activity of this compound.

EHNA does not appear to act at any of the sites previously employed in antiherpes chemotherapy, namely, thymidine kinase or DNA polymerase (see reference 1). I have found no effect of EHNA upon activity of the HSV DNA polymerase, measured in extracts from infected cells (data not shown). Moreover, no phosphorylated derivatives of EHNA have been detected in metabolic studies and EHNA is not phosphorylated by the HSV thymidine kinase (J. Fyfe, personal communication). Thus, it is not yet clear how EHNA produces an inhibition of viral DNA synthesis. However, in recent experiments I have found that EHNA may exert its activity through an effect on DNA precursor metabolism. Effects of EHNA upon levels of deoxyribonucleoside 5'-triphosphates in HSV-infected and uninfected cells are shown in Table I. Treatment of HSV-infected cells with EHNA leads to decreases in levels of all four dNTPs, measured at 6 hr post-infection. These concentrations of EHNA have much less effect on dNTP pools in uninfected cells that are treated with EHNA in the same manner.

Levels of dNTPs have been reported to increase after infection of cells with HSV (37). Results in Table I show that this is particularly true for dTTP and dGTP, which increase in concentration several-fold after infection. In this experiment there was no increase in dCTP concentration, although in other experiments it sometimes increases about 1.5-fold. Interestingly, levels of dATP decrease somewhat after infection. In several experiments dATP was consistently found to decrease by 25-50% following infection of cells by HSV. Treatment of HSV-infected cells with EHNA seems to block the expansion of dTTP and dGTP pools. If the increased dNTP levels are required for a maximal rate of DNA synthesis, this effect of EHNA on dNTP levels may explain its ability to reduce the rate of viral DNA synthesis.

AraC and an inhibitor of CDA

There has been much less interest in evaluation of the combination of araC and inhibitors of CDA due to the failure of araC to selectively inhibit replication of herpesviruses. Nevertheless, potentiation of araC activity by tetrahydrouridine has been demonstrated against cultured cells and in treatment of tumors in animals. This combination has been particularly useful in systems with a high ratio of CDA to cytidine kinase activity. Although very little information is available concerning the effect of this combination against replication of HSV, the results of several studies have prompted a more careful look at this interaction. First of all, exogenously supplied deoxycytidine is poorly incorporated into HSV DNA unless THU is present (17). And secondly, the induction of deoxycytidine deaminase (dCDA) by HSV was reported (38). These results suggested that the antiviral activity of araC might be decreased due to massive deamination in HSV-infected cells. However, in experiments done by myself and Mathews (17) we found that THU co-administration produced only a slight increase in the ability of araC to block HSV replication in cultured cells. Results of these experiments are shown in Figure 5. This failure of THU to potentiate the antiviral activity of araC is in sharp contrast to its ability to facilitate conversion of deoxycytidine to dCTP and its subsequent incorporation into viral DNA.

Recently, the induction of dCDA by HSV has been questioned. Charron and Langlier have reported that they could detect induction of this activity only in infections of cells with virus stocks that were contaminated with mycoplasma (39). Our stocks of cells and virus are negative for mycoplasma contamination by two criteria, as described in Methods. In agreement with Charron and

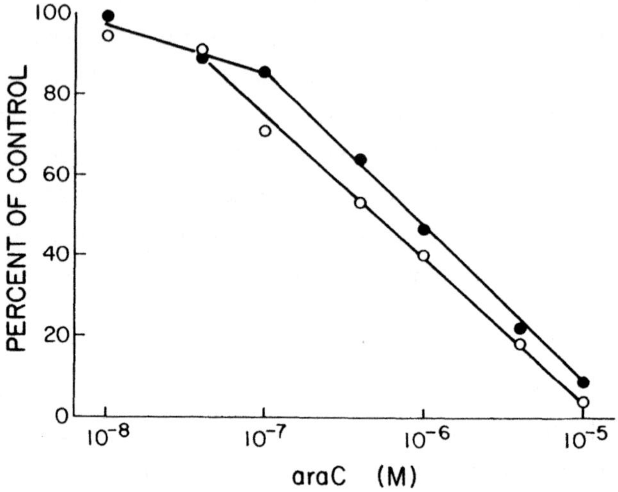

Figure 5. Effect of araC upon production of progeny virus in HSV-infected HeLa cells in the presence (O) or absence (●) of 10^{-5} M THU. THU alone has no antiviral activity.

Table 2. Deoxycytidine Deaminase Activities[a]

Extract	Specific Activity[b]	% Control
HeLa cells	12.6	100
HSV-infected HeLa cells[c]	5.3	42

[a]Assays were performed as described by Chan (38).

[b]Specific activity is expressed in nmoles/min per mg of protein.

[c]Cells were infected with HSV (moi = 10) for 6 hr.

Langlier, John Case has determined in my laboratory that dCDA is not increased after infection of cells by HSV. In fact, by 6 hr post-infection, the specific activity of this enzyme decreases to less than half of the activity present in uninfected cells (Table 2). This decrease in dCDA activity was seen with two strains of HSV-1 (Miyama and KOS) following infection of two cell lines (HeLa and Vero).

The failure of HSV to increase dCDA activity leaves us at a loss to explain the need for THU in order to incorporate exogenously supplied deoxycytidine into HSV DNA. Nevertheless, we have consistently observed this phenomenum with both strains of HSV and both lines of cultured cells indicated above. A possibility that I am currently evaluating is that THU is phosphorylated to the corresponding 5'-monophosphate by the viral thymidine kinase. Accordingly, THU-MP might facilitate conversion of dC to dCTP by blocking deamination of dCMP via dCMP deaminase.

Effects of aranucleosides upon DNA precursor levels

Rates of phosphorylation and deamination control conversion of aranucleosides to their active nucleotides. However, these pathways are not the only metabolic considerations that are important in determining their biological activities. AraATP and araCTP also compete with the natural DNA precursors, dATP and dCTP, respectively, for inhibition of DNA polymerase and for incorporation into DNA. Accordingly, factors that alter levels of competing nucleotide will alter the biological activities of araA or araC. It has been demonstrated that conditions which decrease intracellular dCTP concentration, such as provision of thymidine, increase the activity of araC in cultured cells (40) and in animals (41). This and similar synergistic combinations may be developed based upon the use of nucleosides to modulate dNTP levels.

The abilities of araADP and araATP to inhibit ribonucleoside diphosphate reductase (rNDP reductase) (14) suggests that araA treatment of cells might lead to a "self-potentiation" of its biological activity if such treatment produces a decrease in dATP concentration (15). Although inhibition of rNDP reductase by 50% requires concentrations of araATP (30-70 μM) that are much higher than necessary to block DNA polymerases (1 μM), these concentrations are well below those found in araA-treated cells (42). AraCTP does not inhibit rNDP reductase (14) and, accordingly, would not be expected to decrease intracellular levels of DNA precursors. In the experiments described below I have examined effects of araA and araC upon DNA precursor levels to determine whether these effects correlate with the abilities of the corresponding aranucleotides to inhibit rNDP reductase.

Results presented in Figure 6 show dNTP levels following treatment of uninfected HeLa cells with 10^{-4}M araA in combination with 10^{-6}M EHNA. This treatment inhibits DNA synthesis by >99%. As described in Methods, the presence of araATP in extracts from these cells does not interfere with the assay for dNTPs. Following treatment with araA + EHNA, there is a rapid decrease in levels of all four dNTPs to levels less than 40% of those in untreated controls. Maximal depression of levels of all four occurs by 6 hr after treatment. At later times, levels of all four dNTPs are restored and by 24 hr they are elevated to 1.5-2.0 times the levels in untreated controls. The decrease in dNTP levels at early times after treatment with araA may be important because this is the time at which inhibition of DNA synthesis is established. It is interesting that dATP levels are the most severely depressed of the four; dATP falls to 12% of the control by 6 hr. This decrease in dATP should enhance the activity of araATP formed under these conditions. Although the reason dNTP pools are restored to levels greater than in controls is not known, it is not unexpected because any dNTP synthesized _de novo_ or by salvage pathways would not be utilized in the absence of DNA synthesis.

There are many factors that control levels of dNTPs and inhibition of rNDP reductase by araATP is only one of the possible ways in which araA treatment may reduce intracellular concentrations of dNTPs. However, if such an effect upon the reductase is responsible for these results it is expected that araC will not produce a similar response because its nucleotides fail to inhibit rNDP reductase. Indeed, I have found that treatment of cells with araC (10^{-4}M araC in combination with 10^{-5}M THU) leads to an increase in levels of all four dNTPs. All four DNA precursors are elevated to 1.2-1.5 times the control levels by 3 hr, and continue to increase through the 24 hr period in which these determinations were made (data not shown). These data are consistent with an inhibition of rNDP reductase as the basis for depression of dNTP levels by araA.

Since araA treatment of uninfected cells results in a temporary depression of dNTP levels, I have examined effects of araA and araC upon levels of DNA precursors in HSV-infected HeLa cells. The major objectives of this study were to determine whether effects on DNA precursor metabolism might contribute to the antiviral selectivity of araA, or explain the failure of araC to show antiviral selectivity. HSV induces the appearance of an rNDP reductase activity which is different from the cellular enzyme in several properties; namely, the viral

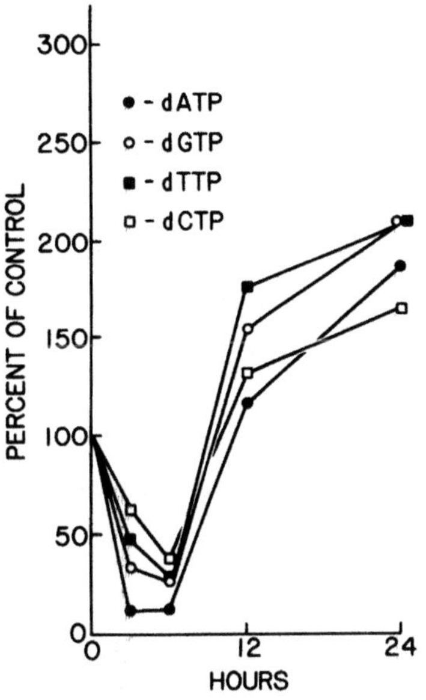

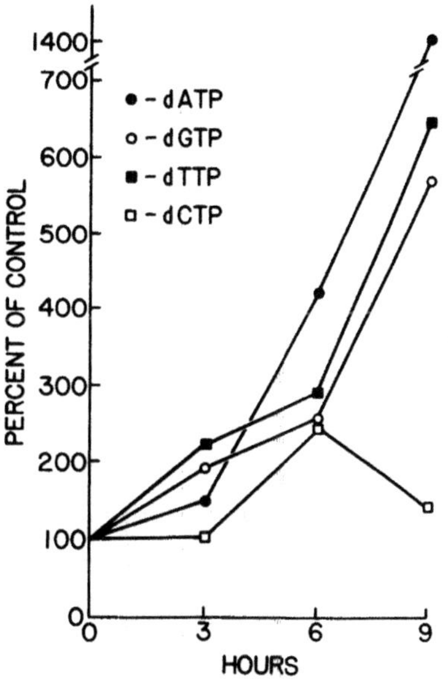

Figure 6 (left). Effect of 10^{-4}M araA (in the presence of 10^{-6}M EHNA) upon levels of deoxyribonucleoside 5'-triphosphates in uninfected HeLa cells. Inhibitors were added at time zero and samples taken at the indicated times were analyzed for dNTPs as described in Methods.

Figure 7 (right). Effect of 10^{-4}M araA (in the presence of 10^{-6}M EHNA) upon levels of dNTPs in HSV-infected HeLa cells. Cells were infected with an input moi of 10 PFU/cell. Inhibitors were added after a 1 hr adsorption period (time zero) and samples taken at indicated times were analyzed for dNTPs as described in Methods.

activity is insensitive to inhibition by dTTP and dATP, and shows no requirement for ATP (43,44). If the araA-induced depression of dNTP pools is due to effects of araATP upon rNDP reductase, the alteration in this activity following HSV-infection might change the response of dNTP levels to araA.

Results presented in Figure 7 show the effect of araA (10^{-4}M in combination with 10^{-6}M EHNA) upon dNTP levels in HSV-infected HeLa cells. In contrast to the decrease in concentrations of these four dNTPs seen in uninfected cells, the four DNA precursors are elevated at all times examined in HSV-infected cells. By 6 hr all four dNTPs are elevated to more than double the levels found in untreated, HSV-infected controls. By 9 hr all pools except dCTP are expanded several-fold. Interestingly, dATP is elevated to the greatest extent some 14 times the level found in the absence of araA. This increase is much greater than is seen in uninfected cells even after 24 hr (Figure 6). It is apparent that the effect of araA upon DNA precursor metabolism in HSV-infected cells is much different from that in uninfected cells. It has recently been reported that the HSV-induced rNDP reductase is not inhibited by araATP (45). The failure of araA to depress dNTP levels in HSV-infected cells is consistent with this. I have also found in preliminary experiments that araC does not depress levels of dNTPs in HSV-infected cells (data not shown).

In all of these experiments, dNTP levels were decreased by aranucleosides only under conditions where the corresponding aranucleotide is known to inhibit rNDP reductase. This is consistent with the "self-potentiation" concept whereby araATP activity is increased by depletion of the competing nucleotide (14,15). However, further work will be necessary to ascertain that the target of this effect is rNDP reductase and not some other enzyme of DNA precursor biosynthesis.

DISCUSSION

AraA has been approved for use in treatment of human herpetic disorders and is currently in widespread use for treatment of herpetic keratitis and herpetic encephalitis. However, the efficacy of araA has been limited by its low solubility (2mM in H_2O at room temperature) and its rapid deamination to the less active form, araHx. This problem has been of particular concern in treatment of herpetic encephalitis where infusions of required doses sometimes lead to problems from fluid overload (46).

It is now well established that the antiviral activity of araA can be greatly increased by combination with an inhibitor of adenosine deaminase. The potentiation of araA activity by inhibitors of ADA has been demonstrated in cell culture systems and in animals. This combination allows achievement of antiviral activity with 10 to 100-fold lower concentrations of araA than previously obtained with araA alone. If comparable results are obtained in humans, this combination should eliminate the problem of fluid overload in treatment of herpetic encephalitis.

Both EHNA and DCF have been shown to potentiate the antiviral activity of araA in cultured cells and in animals. However, as discussed above, there are several reasons to believe that EHNA may be the better suited of the two for use in combination with araA. EHNA produces a more rapid and controllable inhibition of ADA. Inhibition of ADA by EHNA is readily reversible, whereas inhibition of this enzyme by DCF is not. Moreover, DCF appears to be more toxic to animals than EHNA. In animal studies the combination of araA plus EHNA produced no detectable adverse effects in mice at therapeutic doses (29). Despite these promising results, I am unaware of any human clinical trials of this combination.

In contrast to araA, araC has shown no selective antiviral activity, even when used in combination with THU, an inhibitor of cytidine deaminase. This is surprising, particularly in view of the need for THU to allow conversion of exogenously supplied deoxycytidine to dCTP and its incorporation into DNA in HSV-infected cells (17), and the reported induction of a new deoxycytidine deaminase following infection of cells with HSV (38). However, as reported above, the specific activity of dCDA in HeLa cells decreases after HSV-infection to less than half of the activity in uninfected cells. Thus, araC appears to be metabolized differently than deoxycytidine but the lack of antiviral selectivity displayed by araC cannot be explained by its massive deamination.

In addition to the competition between deaminases and kinases which controls formation of the active species, araATP and araCTP, these aranucleotides must compete with the natural DNA precursors (dATP and dCTP, respectively) in order to exert their biological activity. Accordingly, alterations of DNA precursor pools will effect activity of aranucleosides. This has been demonstrated by the combination of araC with thymidine. This combination is synergistic due to the ability of thymidine to reduce intracellular concentration of dCTP. In order to predict synergistic combinations of inhibition to use with araA or araC, we must also know how the aranucleosides affect dNTP pools. This is particularly important with araA because araATP inhibits ribonucleoside diphosphate reductase (14) and it has been suggested that this action might "self-potentiate" the ability of araA to inhibit DNA synthesis (15). Results presented here show that treatment of cells with araA (plus 10^{-6}M EHNA) leads to a rapid decrease in concentrations of all four dNTPs. The most dramatic decrease is in dATP concentration which falls to 12% of its level prior to treatment. This should produce a higher ratio of araATP/dATP, assuming that araATP formation is not impeded. These data provide the first in vivo evidence for the "self-potentiation" of araA by its effect on DNA precursor provision.

In contrast to the situation in uninfected cells, araA fails to decrease dNTP levels in HSV-infected cells. In fact, araA-treatment of HSV-infected cells leads to expansion of all four dNTP pools, relative to untreated controls. Thus, it is apparent that the "self-potentiation" which may be occurring in uninfected cells does not contribute to antiviral selectivity. Recently it has been reported that HSV induces an rNDP reductase which is resistant to araATP (45). Therefore, the ability of araA to decrease dNTP levels correlates with the ability of araATP to inhibit rNDP reductase in these two cases. Moreover, araCTP, which does not inhibit rNDP reductase (14), fails to decrease dNTP pools. Although these data are not enough to draw a firm conclusion, there exists a tentative relationship between ability of aranucleosides to decrease dNTP provision and the ability of their corresponding nucleotides to inhibit rNDP reductase.

In recent years several nucleoside analogs have been developed which have much greater antiherpes selectivity than araA. Most of these are nucleosides that are converted to toxic nucleotides only by the HSV-encoded thymidine kinase. The selectivity arises from a failure to phosphorylate these analogs in uninfected cells. The more promising of these compounds include 9-(2-hydroxyethoxymethyl)guanine (acyclovir) (47), which has been approved for use in humans in the U.S.A., (E)-5-(2-bromovinyl)-2'-deoxyuridine (BVUdR) (48), and 2'-fluoro-5-iodo-araC (49). All of these compounds have also been reported to be more potent than araA. However, all comparisons have been done in studies where araA was used in the absence of inhibitors of ADA. In the presence of EHNA the potency of araA is comparable to at least some of these other agents.

Although the antiviral selectivities of these substrates for the viral thymidine kinase are unquestionably higher than the selectivity of araA, these other compounds suffer from another problem. The viral thymidine kinase is not essential for HSV replication and, therefore, drug-resistant strains arise rapidly through loss of this activity. For example, HSV strains resistant to acyclovir were obtained early in clinical trials whereas clinical resistance to araA has never been reported despite extensive trials (M. Hirsch, personal communication). Although it has been reported that HSV mutants deficient in thymidine kinase are less pathogenic in animals than wild-type virus, it is naive to think that all acyclovir-resistant mutants will be less pathogenic in humans. HSV mutants have already been isolated with an altered enzyme that phosphorylates thymidine but not acyclovir (50). Moreover, some acyclovir-resistant mutants have an altered HSV DNA polymerase, and some of these are cross-resistant to araA (51). Thus, care must be taken in using these nucleosides, particularly in cases of recurrent herpes where repeated treatment may be necessary. AraA has a wide margin of safety and is less likely to lead to development of resistant virus.

It must be cautioned, however, that there are serious problems with araA that limit its scope. As discussed by Cohen (52), the most serious limitation is its potential mutagenicity. This will certainly limit its use to life-threatening situations (herpetic keratitis and herpetic-encephalitis) and preclude its use from common, recurrent disorders such as herpes genitalis and herpes labialis. Thus, with all the problems of resistance and mutagenicity it appears that none of the current antivirals is really safe enough for treatment of the non-threatening forms of recurrent herpes.

The study of nucleosides has taught us a lot about strategies that may be useful in further attacks on the problem of viral chemotherapy. There is need for compounds with the degree of selectivity of acyclovir, BVUdR, and related compounds, but which allow a minimal risk for development of resistance. In addition, it is imperative that such an agent be neither mutagenic nor carcinogenic. To achieve all of these goals we may have to turn away from enzymes of nucleic acid and nucleotide metabolism since most of the inhibitors of these processes are potential mutagens. Herpesviruses have more than 50 genes and it is likely that some more of them will be exploited in chemotherapeutic strategies in the years to come. Rational approaches to this problem have been outlined (53).

ACKNOWLEDGEMENTS

I would like to thank Drs. Seymour S. Cohen and Christopher K. Mathews for all their help and encouragement during all stages of this work. I also thank Lorraine O'Connor for excellent technical assistance and Rachel Runfola for preparation of the manuscript. This work was supported in part by a grant from the Massachusetts Division of the American Cancer Society.

REFERENCES

1. North, T.W. and Cohen, S.S. 1979. Pharmac. Therap. 4:81-108.
2. North, T.W. and Cohen, S.S. 1982, in press. In: The International Encyclopedia of Pharmacology and Therapeutics, Shugar, D. (ed.).
3. Bergmann, W. and Feeney, R. 1951. J. Org. Chem. 16:981-987.
4. Walwick, E.R., Robert, W.K. and Dekker, C.A. 1959. Proc. Chem. Soc. p. 84.
5. Lee, W.W., Benitez, A., Goodman, L. and Baker, B.R. 1960. J. Am. Chem. Soc. 82:2648-2649.
6. Cardeilhac. P.T. and Cohen, S.S. 1964. Cancer Res. 24:1595-1603.
7. Cohen, S.S. 1966. Prog. Nucleic Acid Res. Mol. Biol. 5:1-88.
8. Furth, J.J. and Cohen, S.S. 1967. Cancer Res. 27:1528-1533.
9. Cohen, S.S. 1976. Med. Biol. 54:299-326.
10. Suhadolnik, R.J. 1979. Nucleosides as Biological Probes, John Wiley and Sons, New York.
11. Furth, J.J. and Cohen, S.S. 1968. Cancer Res. 28:2061-2067.
12. Müller, W.E.G., Zahn, R.K., Bittlingmaier, K. and Falke, D. 1977. Ann. N.Y. Acad. Sci. 284:34-48.
13. Reinke, C.M., Drach, J.C., Shipman, C. Jr. and Weissbach, A. 1978. In: Oncogenesis and Herpesviruses III (Part 2), pp 999-1005, DeThe, G., Henle, W. and Rapp, F. (eds.) IARC, Lyon, France.
14. Moore, E.C. and Cohen, S.S. 1967. J. Biol. Chem. 242:2116-2118.
15. Chang, C.-H. and Cheng, Y.-C. 1980. Cancer Res. 40:3555-3558.
16. North, T.W. and Cohen, S.S. 1978. Proc. Nat. Acad. Sci. USA, 75:4684-4688.
17. North, T.W. and Mathews, C.K. 1981. J. Virol. 37:987-993.
18. Schneider, E.L., Stanbridge, E.J. and Epstein, C.J. 1974. Exp. Cell Res. 84:311-318.
19. North, T.W., Bestwick, R.K. and Mathews, C.K. 1980. J. Biol. Chem. 255:6640-6645.
20. Wohlrab, F. and Francke, B. 1980. Proc. Nat. Acad. Sci. USA 77:1872-1876.
21. Hanze, A.R. 1960. J. Am. Chem. Soc. 89:6720-6725.
22. Stollar, R.G., Drake, J.C. and Chabner, B.A. 1975. Proc. Am. Assoc. Cancer Res. 16:88.
23. LePage, G.A. 1970. Can. J. Biochem. 48:75-78.
24. Plunkett, W. and Cohen, S.S. 1975. Cancer Res. 35:415-422.
25. Schaeffer, H.J. and Schwender, D.F. 1974. J. Med. Chem. 17:6-8.
26. Agarwal, R.P., Spector, T. and Parks, R.E. Jr. 1977. Biochem. Pharmac. 26:359-367.
27. Woo, P.W.K., Dion, H.W., Lange, S.M., Dahl, L.F. and Durham, L.T. 1974. J. Heterocyclic Chem. 11:641-643.
28. Plunkett, W. and Cohen, S.S. 1975. Cancer Res. 35:1547-1554.
29. Shannon, W.N., Arnett, G., Shabel, F.M. Jr., North, T.W. and Cohen, S.S. 1980. Antimicrob. Agents Chemother. 18:598-603.
30. Bryson, Y., Connor, J.D., Sweetman, L., Carey, S., Stucky, M.A. and Buchanan, R. 1974. Antimicrob. Agents Chemother. 6:98-101.
31. Sloan, B.J., Kielty, B.J. and Miller, F.A. 1977. Ann. N.Y. Acad. Sci. 284:60-80.
32. Lapi, L. and Cohen, S.S. 1977. Biochem. Pharmac. 26:71-76.
33. Meuwissen, H.J., Pickering, R.J., Pollara, B.P. and Porter, I.H. 1975. Combined Immunodeficiency Disease and Adenosine Deaminase Deficiency, Academic Press, N.Y.

34. Plunkett, W., Alexander, L., Chubb, S. and Loo, T.L. 1979. Biochem. Pharmac. 28:201-206.
35. DeClercq, E., Descamps, J., Desomer, P. and Holy, A. 1978. Science 200:563-565.
36. Bastian, G., Bessodes, M., Panzica, R.P., Abushanab, E., Chen, S.-F., Stoeckler, J.D. and Parks, R.E. Jr. 1981. J. Med. Chem. 24:1383-1385.
37. Cheng, Y.-C., Goz, B. and Prusoff, W.H. 1975. Biochim. Biophys. Acta 390:253-263.
38. Chan, T. 1977. Proc. Nat. Acad. Sci. USA 74:1734-1738.
39. Charron, J. and Langlier, Y. 1981. J. gen. Virol. 57:245-250.
40. Grant, S., Lehman, C. and Cadman, E. 1980. Cancer Res. 40:1525-1531.
41. Danhauser, L.L. and Rustum, Y.M. 1980. Cancer Res. 40:1274-1280.
42. Plunkett, W., Chubb, S., Alexander, L. and Montgomery, J.A. 1980. Cancer Res. 40:2349-2355.
43. Ponce de Leon, M.R., Eisenberg, R.J. and Cohen, G.H. 1977. J. gen Virol. 36:163-173.
44. Huszar, D. and Bacchetti, S. 1981. J. Virol. 37:580-588.
45. Langlier, Y. and Buttin, G. 1981. J. gen. Virol. 57:21-31.
46. Ho, S.U. and Harter, D.N. 1982. J. Am. Med. Assoc. 247:337.
47. Elion, G.B., Furman, P.A., Fyfe, J.A., DeMiranda, P. Beauchamp, L. and Schaeffer, H.J. 1977. Proc. Nat. Acad. Sci. USA 74:5716-5720.
48. DeClercq, E., Descamps, J., DeSomer, P., Barr, P.J., Jones, A.S. and Walker, R.T. 1979. Proc. Nat. Acad. Sci. USA 76:2947-2951.
49. Watanabe, K.A., Reichman, U., Hirota, K., Lopez, C. and Fox, J.J. 1979. J. Med. Chem. 22:21-24.
50. Darby, G., Field, H.J. and Salisbury, S.A. 1981. Nature 289:81-83.
51. Coen, D.M., Furman, P.A., Gelep, P.T. and Schaffer, P.A. 1982. J. Virol. 41:909-918.
52. Cohen, S.S. 1979. In: Nucleoside Analogues p. 225-245, Walker, R.T., DeClercq, E. and Eckstein, F. (eds.) Plenum Pub. Co., New York.
53. Cohen, S.S. 1979. Science 205:964-971.

4

HERPES SIMPLEX VIRUS THYMIDINE KINASE—DEPENDENT ANTIVIRAL AGENTS

YUNG-CHI CHENG
Department of Pharmacology and Medicine, the University of North Carolina, Chapel Hill, North Carolina, USA

Herpes simplex virus type 1 (HSV-1), type 2 (HSV-2) and varicella zoster virus (VZV) are capable of inducing their specified thymidine kinase (TK) in virus-infected cells. These virus-specified thymidine kinases differ from human cytosol and mitochondrial thymidine kinases not only in their physical and immunological properties (Table 1) but also in their kinetic behaviors (1-8).

Table 1. Some properties of various thymidine kinases

	Human cytosol	mito-condrial	HSV-1 (KOS)	HSV-2 (333)	VZV
Electrophoretic mobility (R_f)	0.15	0.6	0.70	0.45	0.40
Activation energy (Kcal/mole)	15.2	10.9	17.2	24.5	N.D.
Reaction mechanism	Sequential	Ping-pong	Sequential	Sequential	Sequential
Inactivation by antibodies raised against					
1) HSV-1	--	--	+	slightly	N.D.
2) HSV-2	--	--	slightly	+	N.D.
Substrate behavior					
TdR (Km, μM)	2.6	5.2	0.6	0.4	0.4
CdR (Km, μM)	no substrate	22	25	88	180
ATP 1) (Km, μM)	220	100	30	140	16
2) 1/v vs 1/ATP	not linear	linear	not linear	not linear	not linear
a/v vs a/ATP^2	linear	not linear	linear	linear	linear
Inhibitor Behavior					
dTTP (K_I μM), non-Competitive w/TdR	0.6	--	120	20	5
Competitive w/TdR	--	5.7	--	--	--

PHYSIOLOGICAL ROLES OF VIRAL THYMIDINE KINASES

Thymidine kinase is a salvage enzyme responsible for securing enough thymidine nucleotide for DNA synthesis. Under normal circumstances, cultured cells could utilize the *de novo* pathway via thymidylate synthetase to synthesize thymidine nucleotide for DNA synthesis. Thus, thymidine kinase activity is not critical for cell growth. Cell mutants deficient in the activity of this enzyme could grow well under regular culture conditions. Upon HSV or VZV infection, a new species of thymidine kinase is induced in infected cells. Virus-specified kinase is coded by the virus gene, at least in the case of HSV-1 and HSV-2. HSV mutants that cannot induce virus-specified thymidine kinase in infected cells have been isolated. These TK^- mutants could replicate efficiently in TK^- host cells in cell culture. This indicated that virus-specified thymidine kinase does not appear to be critical for virus replication in cell culture, whereas the importance of this activity *in vivo* should not be discounted. Two lines of evidence suggested the possible importance of this enzyme in the pathological process of HSV *in vivo*. The TK^- strains of HSV are much less lethal and establish latency at a lower frequency than that of parental strains of HSV in animals. Most fresh clinical HSV isolates so far are able to induce viral thymidine kinase in infected cells. Furthermore, the latent virus was identified to be in neuron cells and thus neuron cells in general do not have the machinery for synthesizing thymidine nucleotide *de novo*, thus the salvaging of thymidine extracellularly through the action of thymidine kinase could be critical for the latent virus reactivation process providing these processes require viral DNA synthesis in the neurons. This is not clear at the present.

STRATEGIES IN DEVELOPING SELECTIVE ANTIVIRAL AGENTS BASED ON VIRUS-SPECIFIED THYMIDINE KINASE

1) Alternative substrate approach (9,10)

This approach postulated in 1976 was based on the concept that virus-induced thymidine kinase may have different substrate specificity from that of the host cellular thymidine kinases. Therefore, it will be possible to find nucleoside analogs which could be preferentially recognized by the viral thymidine kinase as a substrate but not by the host enzyme. Thus, these analogs will be selectively phosphorylated in cells having viral thymidine kinase. Once phosphorylated to the monophosphate, they could be trapped in these cells and exert action by either further phosphorylation or as the monophosphate metabolites. Since normal cells do not have viral thymidine kinase, these nucleoside analogs could not be phosphorylated. Therefore, no toxicity should be observed (10). Several newly discovered selective antiviral nucleoside analogs were found to exert their selective antiviral activity by mechanisms related to this approach. A simple schematic description is shown below:

Uninfected cells

Infected Cells and transformed cells

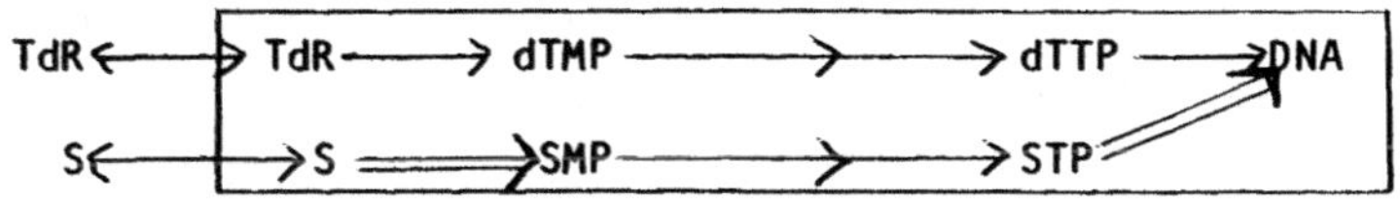

S: nucleoside analogs⟶ host enzyme ⟹ virus enzyme

Since most of the selective antiviral nucleosides based on this mechanism of selectivity developed so far were not absolutely selective as alternative nucleoside substrates for viral thymidine kinase, they could also act as substrate, although poorly, for host enzymes. Cytotoxicity at high concentrations of these compounds was observed. In order to avoid such potential toxicity without sacrificing antiviral activity, the following strategies are proposed.

2) Selective protection approach (11,12)

This is based on the following rationales: (a) since there are differences in the ratios of the binding affinity of thymidine and the antiviral thymidine analog between the viral and cellular enzymes, low concentrations of thymidine may selectively protect normal cells against such compounds by preferentially inhibiting analog phosphorylation in normal cells without affecting to a great extent the phosphorylation of the analogs in cells containing the viral TK. E5-(2-bromovinyl)deoxyuridine (BVdU) cytotoxicity was found to be prevented by thymidine at a concentration that did not affect its antiviral activity. (b) Sometimes, the enzyme responsible for the phosphorylation of the antiviral nucleoside in uninfected cells was found not to be cellular thymidine kinase but another nucleoside kinase. Thus, phosphorylation of the antiviral nucleoside could be prevented by the nucleoside which utilizes the same enzyme in uninfected cells. Therefore, the nucleoside could reverse the cytotoxicity of the compound without affecting its antiviral action. For example, acyclovir (ACV) is likely to be phosphorylated by the guanosine or deoxyguanosine phosphorylating enzyme in normal cells. Thus, its cytoxicity could be reversed by guanosine or deoxyguanosine at low concentrations without affecting its antiviral activity. 2'-F 5-Iodo araC (FIAC) is phosphorylated by deoxycytidine kinase in normal cells. Thus, its cytoxicity could be reversed by deoxycytidine providing the cells do not have cytidine deaminase.

3) Selective inhibition of host thymidine kinase approach (11)

This approach could only apply to thymidine analogs that have both antiviral and cytotoxic activity. The concept was developed independently by Dr. Prusoff's and our laboratories and involves use of an inhibitor which selectively inhibits the host thymidine kinase without affecting the viral thymidine kinase. 5'-NH_2 IUdR (iododeoxyuridine) is a potent inhibitor of cellular cytosol thymidine kinase and is a substrate of HSV-1-induced thymidine kinase. This compound, in combination with IUdR, which could serve as the substrate of both cellular and virus-induced thymidine kinase, could prevent IUdR cytotoxicity and potentiate the antiviral activity of IUdR. Combination of 5'-NH_2 IUdR or 5'-NH_2 TdR with other antiviral thymidine analogs can be expected to have a similar effect.

4) Selective inhibition of virus thymidine kinase approach

This is based on the assumption that the viral thymidine kinase plays an important role in virus replication or the establishment of virus latency *in vivo*. Selective inhibitors of viral thymidine kinase should prevent these processes *in vivo* and achieve an antivirus effect. Compounds of this nature are not anticipated to have activity in cell culture since the viral thymidine kinase is not critical for virus replication. No such compound has been discovered yet.

All of the previously listed strategies are based on the differences of the thymidine binding sites among host enzymes and viral thymidine kinase. The structure of those binding sites was explored using various thymidine analogs. This will be discussed in the following section.

STRUCTURE-ACTIVITY RELATIONSHIP OF VARIOUS THYMIDINE KINASES

The interaction of highly purified human cellular and viral-specified thymidine kinases, with thymidine analogs with the modification at the 5,2' and 3' positions of the thymidine molecule and deoxycytidine analogs with the modification at the 5 and/or 2' positions of the deoxycytidine molecule, were examined: the relative binding affinity with reference to thymidine of some of these analogs is summarized in Table 2 (3,5,10,12-14). It should be pointed out that the Km of thymidine for viral TK is lower than that for host TK.

Table 2. Relative binding affinity of thymidine analog and thymidine to various types of TK

Group at 5 position of thymidine	Enzyme Source				
	Human cytosol	mitochondrial K_I/K_m	HSV-1 (KOS)	HSV-2 (333)	VZV
$-CH_3$	1	1	1	1	1
$-CF_3$			0.7	1	-
-I	1	2	1	1	1
-CHO	-	-	9	-	4
$-CH=CH_2$	10	0.3	1	1	1
$-CH_2CH_3$	30	60	1	1	0.5
$-CH(CH_3)_2$	60	30	2	4	-
-CH=C(Cl)(H)	60	0.06	0.5	5	-
-CH=C(Br)(H)	60	0.16	0.5	10	0.14
-CH=C(H)(Br)	60	7	4	14	0
-CH=C(I)(H)	60	0.20	0.5	6	-
$-CH=C(CH_3)(H)$	60	0.16	0.4	15	0.5
$-(CH_2)_2CH_3$	30	30	1	4	1
$-(CH_2)_3CH_3$	30	30	3	10	-
$-(CH_2)_4CH_3$	30	30	15	60	-

Table 2 (continued)

2' up position of thymidine					
-H	1	1	1	1	-
-F	40	20	1	6	-
-OH	40	20	5	100	-
3' position of thymidine					
-H	1	1	1	1	-
-F	4	2	4	40	-

Several generalizations could be made:

1) All the virus-specified enzymes could tolerate more structural variations of thymidine or deoxycytidine than the cellular enzymes.

2) Human cytosol and mitochondrial thymidine kinases interact differently with the thymidine molecule. Substitution of the methyl group by a conjugated vinyl group at the 5 position could facilitate its binding to the mitochondrial enzyme. This could be partly responsible for the more potent cytotoxicity of these analogs than either 5-ethyl-or 5-propyl-deoxyuridine.

3) Different types of virus-specified thymidine kinases interact differently with the thymidine molecule. The HSV-1 enzyme could tolerate substitution of a longer linear alkyl group at the 5 position of thymidine than the HSV-2 or VZV enzyme. The introduction of the E-5 bromovinyl group at the 5 position of thymidine could enhance its binding to HSV-1 and VZV but decrease its binding to the HSV-2 enzyme. The binding affinity correlates well with potency of these enzymes against different types of herpesviruses.

All of above data are based on a study utilizing enzyme derived from cells infected with the KOS strain of HSV-1 and the 333 strain of HSV-2. Several relevant facts should be noted: 1) There was no apparent difference in the virus enzyme when different host cells were used in the study; 2) different clinical strains of HSV-1 induced thymidine kinases with similar substrate specificities, but a greater variation was observed in the properties of the TK induced by different strains of HSV-2 isolated from patients (15) (Table 3); 3) under selective conditions, it is possible to isolate herpesvirus variants that have a different spectrum of substrate specificity from the parental strains (16).

Table 3. Effect of nucleoside analogs on viral thymidine kinase activity in HeLa cells infected with clinical isolates of HSV

Clinical isolates	Type	-	Addition (400 μM) IUrD	BVdUrd	PdUrd
			% activity		
1	HSV-1	100	69	13	32
2	HSV-1	100	53	26	31
3	HSV-1	100	55	14	31
4	HSV-1	100	64	9	29
5	HSV-2	100	128	129	80
6	HSV-2	100	118	127	66
7	HSV-2	100	99	72	39
8	HSV-2	100	91	75	40

The binding affinity is only one determinant of the compound's behavior as the substrate of a given enzyme; the catalytic turnover rate (Kcat) is the other determinant. In other words, not all the compounds once they are bound to a given enzyme could be converted to their product at the same rate. Some of the compound may bond poorly, although once it is bound, conversion to its product could be very efficient. The catalytic conversion rate of each compound in comparison with thymidine was studied by examining these compounds at 400 μM which is much higher than the K_I (similar to Km in this case) of many analogs; some of these results are shown in Table 4.

Some generalizations could be made:

1) There is no correlation between Kcat and binding affinity.

2) At a low concentration of substrate ($S < Km$), the binding affinity plays a determining role in the rate of phosphorylation other than Kcat, whereas at high concentration of substrate ($S > Km$) Kcat is the more important factor. Thus the amount of drug used is important in determining its selectivity.

3) Among the thymidine analogs examined, no analogs could serve as the substrate for the viral thymidine kinase only and not for the host thymidine kinase, although a preference for the former was found.

Table 4. Relative phosphorylation rates of antiviral nucleoside analogs (400 μM) by various kinds of thymidine kinase

Position of dThd		Velocity (%) relative to dThd			
5	2'-up	Human cytosol	Human mitochondrial	HSV-1	HSV-2
$-CH_3$	-H	100	100	100	100
$-(CH_2)_2CH_3$	-H	22	20	128	116
$-CH{=}C(Br)(H)$	-H	5	33	90	127
$-CH_2{=}C(Br)(H)$	-F	4	71	234	162
$-CH_2{=}C(Br)(H)$	-OH	0	176	71	27
$-CH_3$	-F	82	219	42	147

GENERAL ANTIVIRAL PROPERTIES OF THOSE ANALOGS REQUIRING VIRUS THYMIDINE KINASE TO EXERT THEIR SELECTIVE EFFECT

1) All of these analogs require phosphorylation at the 5' position in order to act; thus both the ability of the virus to induce TK and the behavior of the analog toward the virus-induced TK play an important role in deciding the sensitivity of the virus to a given analog. 5-propyl-dThd could not inhibit HSV-1 and HSV-2 mutants lacking the ability to induce the viral thymidine kinase in infected cells. Recently, in collaboration with Dr. Darby's group of Cambridge University, we demonstrated an HSV-1 mutant that lost its sensitivity to E-5-bromovinyl-dThd due to the alteration of virus-induced thymidine kinase in recognizing this analog as the substrate (16).

2) Most, if not all, of these anlaogs could be further phosphorylated to the nucleoside triphosphate in cells. As the triphosphate nucleoside, it could exert its effect on several potential enzymes; their mode of action could be quite different. The most important site is likely to be the virus-specified DNA polymerase. 5-propyl-dUTP or E-5-bromovinyl-UTP could be incorporated into DNA in place of dTTP and allow the DNA chain to be elongated. E-5-

bromovinyl-araUTP could be incorporated into DNA to replace dTTP while the incorporated E-5-bromovinyl-araU at the terminal end of the DNA growing chain could be excised by the viral DNA polymerase slowing down the DNA chain elongation. Acyclo GTP could be incorporated into DNA in place of dGTP. The acyclovir at the terminal of DNA could not be cleaved by exonuclease and thus served as a potent inhibitor of DNA polymerase by competing with DNA. Thus, these antiviral nucleosides may exert their selectivity through a similar mechanism but their mechanism of action could be quite different (17,18,19).

3) Some of these analogs were found to exert their action at the early phase of virus replication and had no effect on virus replication at its late phase (20). The antiviral action was irreversible with certain analogs as shown in Table 5, but not with acyclovir.

Table 5. Reversibility of antivirus activity of nucleoside analogs against HSV-1 and HSV-2 grown in culture

Compound	Reversibility (6 hr exposure started at 0 hr postinfection)
5-propyl-dUrd	no
E-5-bromovinyl-dUrd	no
2'-F araT	no
2'-F 5-Iodo araC	no
Acyclovir	yes

This should be taken into consideration for designing the proper protocol to utilize these compounds effectively clinically. It is advisable to keep an effective concentration of the compound in infected tissues for more than one generation of virus replication. This will allow virus in the late phase of replication at the time when the drug is applied to go through the early phase of replication by infecting other cells. Furthermore, the reversibility of the antiviral compound could also be overcome by long exposure (20,21,22). The reversibility of the antiviral action of these analogs, for a finite six-hour period of drug treatment, could be related to how these analogs are incorporated into DNA. ACV is incorporated only at the terminal region. Endonuclease, which can remove the terminal fragments of DNA containing ACV, could reverse the action of ACV; thus, the action of ACV was reversible once the compound was removed from the culture medium, whereas 5-propyl-dUrd, E-5-bromovinyl dUrd, and 2'-F araT were incorporated into the internucleotide region of DNA. Such mass incorporation of the drug would be difficult to remove, and its action was therefore quite irreversible.

4) Some of these analogs could inhibit cell proliferation at high concentration. This suggests that these nucleoside analogs could have adverse effects as long as they are phosphorylated. The rate of phosphorylation of these analogs could determine the sensitivity of the cells. Thus, these analogs could conceivably be used as anticancer agents, especially for the treatment of tumors that are associated with HSV, such as cervical carcinoma. Preliminary results from this laboratory indicate that some but not all cervical carcinoma tumor tissues have viral thymidine kinase activity.

5) Thymidine analogs with the modification at the 5' position are substrates of thymidine phosphorylase which could inactivate these compounds, whereas compounds with modifications at the 2' position are not (Table 6).

Table 6. Substrate specificity of human thymidine phosphorylase

Compound	Substrate activity
dThd	yes
araT	no
E-5-Bromovinyl-dUrd	yes
E-5-Bromovinyl-araU	no
IdUrd	yes

Thus, pharmacokinetically, the latter group of compounds is more favorable than the former group. FIAC could be deaminated by human deoxycytidine deaminase to a more cytotoxic agent, FIAU. Thus, it may have a unique place in the treatment of cancer, especially solid tumors which have high activity of deoxycytidine deaminase and thymidine kinase (23).

Future considerations:

Once these viral thymidine kinase-dependent agents are used clinically, variants of HSV that could be selected have either altered or lost their ability to induce virus-specified thymidine kinase and are resistant to these compounds. In order to avoid these virus variants, agents that exert their action independently of the virus-specified thymidine kinase should be used in combination with thymidine kinase-dependent antiviral nucleoside analogs of thymidine or deoxycytidine. The argument that the rise of a TK⁻ HSV variant may not be a potential problem clinically due to its poor ability to establish latency in neurons is not necessarily a valid one for the following reasons: 1) HSV TK⁻ variants may establish latency in different organs, which could lead to a different spectrum of diseases (this question has not been addressed adequately); 2) during the course of infection, a TK⁻ variant could arise from the replication of a TK⁺ variant that was reactivated from the latent form. These TK⁻ variants will not be inhibited by thymidine kinase-dependent agents and will continue to replicate. In immunodeficient patients, this could cause serious problems. Furthermore, HSV variants resistant to these analogs due to altered thymidine kinase were demonstrated and were found to have the ability to establish latency in neurons. This could add another complication to future use of these compounds. Therefore, it is advisable to use the highest tolerable dosage of the compound in the initial treatment of herpes infections in combination with agents that do not require thymidine kinase.

Most of the selective antiviral compounds developed so far were aimed at the inhibition of virus replication; virus in the latent stage is unlikely to be affected by these compounds. The question of whether we can wipe out the latent virus is going to be a continuous challenge for the next decade.

LITERATURE PUBLISHED FROM THIS LABORATORY RELATED TO THIS CHAPTER

1. Cheng, Yung-chi and Ostrander, M. J. Biol. Chem. 251:2605-2610, 1976.
2. Lee, L-S. and Cheng, Yung-chi. J. Biol. Chem. 251:2600-2604, 1976.
3. Lee, L-S. and Cheng, Yung-chi. Biochemistry 15:3686-3690, 1976.
4. Cheng, Yung-chi, Domin, B.A., Sharma, R.A. and Bobek, M. Antimicrobiol. Agents and Chemotherapy 10:119-122, 1976.
5. Cheng, Yung-chi. Biochim. Biophys. Acta 452:370-381, 1976.
6. Cheng, Yung-chi, Chadha, K. and Hugh, R. Infection and Immunity 16:486-492, 1977.
7. Wigler, M., Silverstein, S., Lee, L-S., Pellicer, A., Cheng, Yung-chi and Axel, R. Cell 11:223-232, 1977.
8. Cheng, Yung-chi, and Tsou, J. J. Virol. 31:172-177, 1979.

9. Cheng, Yung-chi, In: The Human Herpes Viruses (Eds. Nahmias, Dowdle and Schinazi), Elsevier North Holland, 1981 pp. 189-190.
10. Cheng, Yung-chi. Ann. N.Y. Acad. Sci. 284:594-598, 1977.
11. Cheng, Y-C. In: Proc. Antimetabolites in Biochemistry, Biology and Medicine (Eds. J. Skoda and Langen), Pergamon Press, Ltd., Oxford, 1979.
12. Cheng, Y.-C., Grill, S., Ruth, J. and Bergstrom, D. E. Antimicrob. Agents Chemother. 18:957-961, 1980.
13. Cheng, Y-C., Dutschman, G., DeClercq, E., Jones, A.S., Rahim, S.G., Verhelst, G. and Walker, R.T. Molec. Pharmacol. 20:230-233, 1981.
14. Cheng, Y.-C., Dutschman, G., Fox, J.J., Watanabe, K.A. and Machida, H. Antimicrob. Agents 20:420-423, 1981.
15. Cheng, Y-C., Schinazi, R.F., Dutschman, G.E., Tan, T.S., and Grill, S.P. J. Viol. Methods, 5: 209-219, 1982.
16. Larder, B.A., Derse, D., Cheng, Y-C., and Darby, G. J. Biol. Chem., 258: 2027-2033, 1983.
17. Derse, D. and Cheng, Y.-C. J. Biol. Chem. 256:8525-8530, 1981.
18. Ruth, J.L. and Cheng, Y.-C. Molecular Pharmacology 20:415-422, 1981.
19. Derse, D., Cheng, Y.-C., Furman, P.A., St. Clair, M.H. and Elion, G.B. J. Biol. Chem. 256, 11447-11451, 1981.
20. Cheng, Yung-chi, Grill, S. and Dutschman, G. Biochem. Pharmac. 28:3529-3532, 1979.
21. Cheng, Yung-chi, Grill, S., Derse, D., Chen, J.-Y., Caradonna, S.J. and Connor, K. Biochimica et Biophysica Acta 652:90-98, 1981.
22. Cheng, Yung-chi. In: Herpetic Eye Diseases. R. Sundmacher, ed., J.F. Bergmann, Verlag, Muchen 12(14):237-245, 1981.
23. Cheng, Y-C., Tane, R.S., Ruther, J.L., and Dutschman, G.E. Biochemical Pharmacology, 32, No. 4, pp. 726-729, 1983.

5

ACYCLOVIR

GERTRUDE B. ELION
Wellcome Research Laboratories, Burroughs Wellcome Co., Research Triangle Park, NC 27709, USA

SUMMARY

Acyclovir, an acyclic analogue of deoxyguanosine, is a potent inhibitor of the replication of herpes simplex virus, types 1 and 2, and varicella zoster virus, and has extremely low toxicity for uninfected host cells. This selectivity is due to the specific phosphorylation of acyclovir by a herpes virus-coded thymidine kinase. Acyclovir monophosphate is subsequently converted to a triphosphate which is a potent inhibitor of the DNA polymerases of herpesviruses. These viral polymerases also use acyclovir triphosphate as a substrate, and incorporate the analog into the viral DNA primer-template, causing chain termination and inactivation of the polymerase. Uninfected cells do not phosphorylate acyclovir to any significant extent and have DNA polymerases which are much less sensitive to acyclovir triphosphate than are the herpesvirus DNA polymerases.

Following extensive toxicologic, metabolic and pharmacokinetic studies, acyclovir has undergone randomized, controlled clinical trials with a variety of formulations, -topical, intravenous and oral. It has shown efficacy against herpes keratitis, initial and recurrent genital herpes, cutaneous and generalized herpes simplex virus infections in immunocompromised hosts, and herpes zoster.

INTRODUCTION

The search for antiviral agents of high selectivity for DNA viruses has been hampered for years by the toxicity of such compounds for the DNA of the host. A systematic program of synthesis and testing of purines and pyrimidines, and their analogs, as inhibitors of nucleic acid synthesis has been pursued by our laboratory for many years. The finding that 2,6-diaminopurine (DAP) was active against vaccinia virus (1) provided some encouragement that purines might be useful antiviral

Y. Becker (ed.), Antiviral Drugs and Interferon: The Molecular Basis of Their Activity.

agents. Unfortunately, diaminopurine was too cytotoxic for this purpose although it showed some promise as an antileukemic agent (2). The discovery of the antiviral activity of 9-β-D-arabinofuranosyladenine (ara-A) (3) led to a resurgence of our interest in purines and purine nucleosides as potential antiviral agents. In particular, 9-β-D-arabinofuranosyl-2,6-diaminopurine (ara-DAP) seemed like an attractive candidate, because of the previous findings with DAP, and because of the likelihood that ara-DAP would be less readily deaminated than ara-A by adenosine deaminase in vivo. These expectations were fulfilled. Not only was ara-DAP effective against vaccinia and herpes simplex viruses in vitro and in vivo, but was less cytotoxic than ara-A (4). Moreover, its deamination product, 9-β-D-arabinofuranosylguanine (ara-G) was as potent an antiviral agent as ara-DAP (4).

Studies by Schaeffer et al. (5) had shown that an acyclic side chain could mimic the ribose moiety of adenosine with respect to binding to adenosine deaminase. Based on this observation, a program was begun by Schaeffer in these laboratories to synthesize nucleoside analogs with acyclic side chains attached to a heterocyclic base and to examine the potential of such compounds as chemotherapeutic agents. Since ara-DAP and ara-G had shown efficacy against DNA viruses, their acyclic analogs were synthesized and tested (6). The 9-(2-hydroxyethoxymethyl) derivative of guanine (acyclovir, acycloguanosine, ACV, acyclo-Guo) proved to be extremely active against herpes simplex viruses, types 1 and 2, (HSV-1, HSV-2), surpassing in potency any of the previously known antiherpetic agents (6).

ANTIVIRAL SELECTIVITY

Acyclovir is highly selective for herpesviruses, including HSV-1, HSV-2, varicella zoster virus (VZV), and Epstein-Barr virus (EBV) (7-10), but, unlike the purine arabinosides, it is inactive against DNA viruses such as vaccinia virus and adenovirus. A surprising finding was its lack of toxicity for normal, uninfected cells. Thus, while the concentration of ACV for 50% inhibition (I_{50}) for HSV-1 and HSV-2 is in the range of 0.1-1 μM, depending on the strain and cell type (7,8), the I_{50} for Vero cells is 300 μM, and for WI38 cells the I_{50} = >3000 μM (6,9, 11-13). The human cytomegalovirus (HCMV) is much less sensitive than HSV-1, HSV-2 or VZV (Fig. 1).

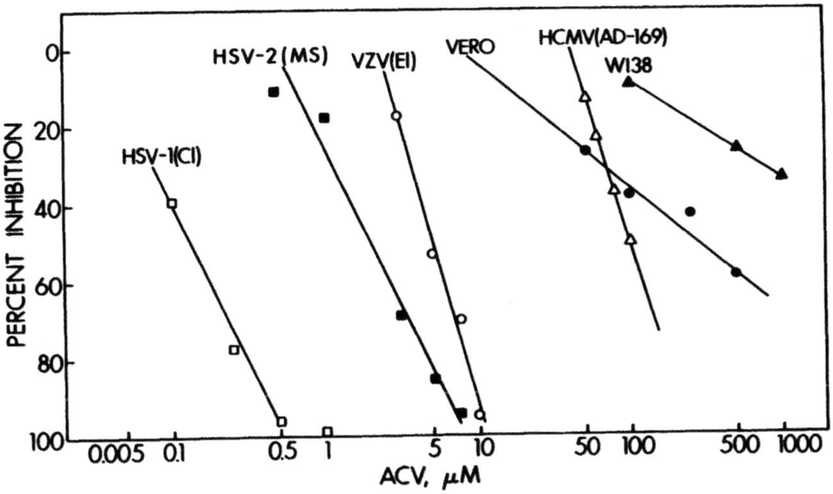

FIGURE 1. Dose response curves of various viruses and cell lines to acyclovir (13). (Reprinted with permission.)

MECHANISM OF ACTION

Activation

Enzyme studies _in vitro_ showed that acyclovir was not a substrate for several known mammalian purine nucleoside kinases, e.g. adenosine kinase, deoxyguanosine kinase, suggesting that it was not behaving like a purine nucleoside. In order to determine the basis of its selectivity against herpesviruses, radioactive acyclovir, labeled either at the C-8 of guanine with ^{14}C, or on the acyclic side chain with ^{3}H, was synthesized. Cell extracts were prepared from HSV-1-infected Vero cells following incubation with 500 μM ACV for 6 hr after adsorption of the virus. High pressure liquid chromatography (HPLC) of these extracts revealed that acyclo-Guo had been transformed to three new radioactive products in HSV-1-infected cells, but not in uninfected Vero cells (Fig. 2) (13). These products were identified as the mono-, di-, and triphosphates of acyclovir (acyclo-GMP, acyclo-GDP, acyclo-GTP) (11). It was concluded that viral infection had resulted in the induction of one or more virus-specified enzymes capable of phosphorylating acyclovir. It was anticipated that the first enzyme in this transformation would be a virus-coded purine nucleoside kinase. However, isolation and

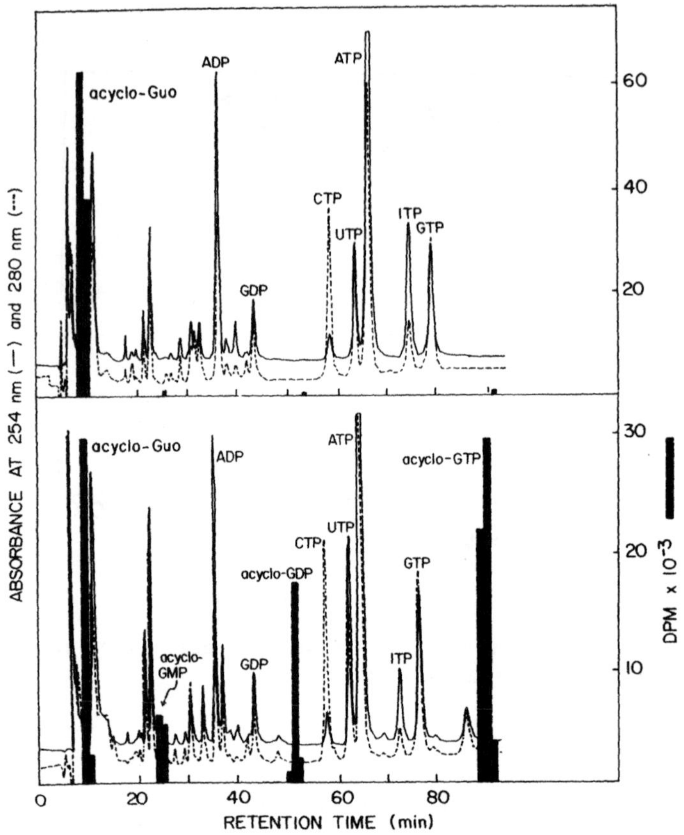

FIGURE 2. High pressure liquid chromatographic profiles of extracts of uninfected and HSV-1-infected Vero cells after 6 hr of incubation with ^{3}H-acyclovir. Top: uninfected cells incubated with 1 mM acyclovir; bottom: infected cells incubated with 0.5 mM acyclovir(13) (with permission).

purification of the phosphorylating activity from HSV-infected cells by affinity chromatography showed that this activity coincided with that of the HSV-specified thymidine kinase (HSV-TK). This enzyme also phosphorylated deoxycytidine, but did not use purine nucleosides such as ara-G or ara-A as substrates (14). Further evidence that it was, indeed, the HSV-TK which was converting acyclo-Guo to acyclo-GMP came from studies showing that thymidine or deoxycytidine could compete with acyclo-Guo and prevent its phosphorylation, and that a temperature-sensitive mutant of HSV-1 (tsA1) which did not express an HSV-TK at 39°, did not phosphorylate acyclo-Guo (11,14). Subsequently, it was shown that other TK^- mutants of HSV likewise were not sensitive to inhibition by ACV (15). Moreover, if mouse fibroblasts (L cells) which were TK^-, and insensitive to the effects of acyclovir, were biochemically transformed with herpesvirus genetic information containing the HSV-TK, these cells were capable of phosphorylating ACV and their growth and DNA synthesis were inhibited (16,17).

The identity of the enzyme responsible for the conversion of acyclo-GMP to acyclo-GDP was found to be a cellular enzyme, guanylate kinase, which is present in the same amount in uninfected cells and in HSV-infected cells (18). It is interesting that, although ACV behaves like a pyrimidine deoxynucleoside with respect to HSV-TK, the product of the phosphorylation, acyclo-GMP, is not a substrate for the HSV-specified thymidylate kinase, but behaves like an analogue of guanylate. In this respect, acyclo-GMP is different from the 5-bromovinyldeoxyuridylate, which is a substrate for the HSV-1 thymidylate kinase (19). The final step in the activation of ACV, acyclo-GDP to acyclo-GTP, is likewise catalyzed by cellular enzymes (Fig. 3). While a number of kinases and

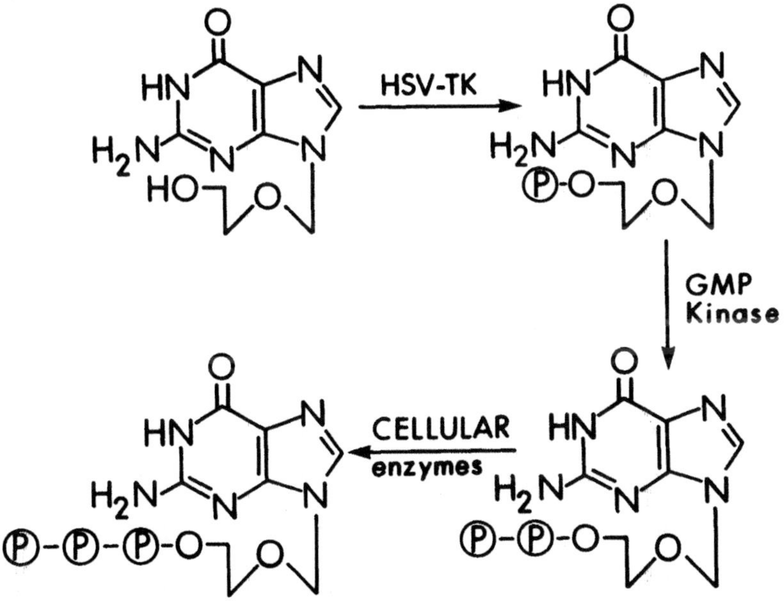

FIGURE 3. Enzymatic conversion of acyclovir to its mono-, di-, and triphosphate derivatives (13). (Reprinted with permission).

phosphotransferases are capable of carrying out this reaction, kinetic studies and enzyme concentrations suggest that phosphoglycerate kinase is probably responsible for most of that conversion (20).

Effects of Acyclo-GTP on DNA Polymerases

The inhibitory effect of acyclo-GTP on the replication of DNA viruses was suspected to be due to competition with dGTP for the viral DNA polymerases. Therefore, a number of viral and cellular DNA polymerases were isolated and purified, and the kinetics of their reaction with acyclo-GTP was examined in cell-free systems (21). The K_i values for acyclo-GTP were different for the various DNA polymerases, but were consistently lower for herpesvirus-specified DNA polymerases than for the cellular polymerases (Table 1) (12,21).

Table 1. Effect of acyclovir on viral and cellular replication, acyclo-GTP formation and apparent K_i values for DNA polymerases

Virus or Cell	I_{50} µM	ACV µM	Acyclo-GTP µM	K_i, DNA polymerase µM
HSV-1 (H29)	0.1	0.5 100	0.80 14.2	0.08
HSV-2 (MS)	1.4	1.0	0.70	0.45
HSV-2 (333)	1.4	1.0 100	0.40 74.0	0.56
VZV (Benj)	3.6			0.024*
EBV	7.0	250	0.06	0.015
Vaccinia	200	100	0.20	3.7
HCMV (AD-169)	115	100	0.30	0.25
L-929	50	100	0.78	2.8
Vero	300	100	0.28	2.2
WI38	>3000	100	0.04	1.86

*Template = Poly dC. oligo $(dG)_{12-18}$ (K. Biron, personal communication)

The inhibition of viral DNA synthesis is dependent both on the concentration of acyclo-GTP formed and on the K_i of the individual DNA polymerase. The amount of acyclo-GTP formed in normal and in virus-infected cells varies greatly (Table 1), depending on whether or not a virus-specified thymidine kinase is formed and whether that kinase has the appropriate specificity. Uninfected mammalian cells can phosphorylate acyclovir only to a very small extent but the enzyme responsible for this is not the normal cellular thymidine kinase (22). Purified cellular TK from Vero cells can not catalyze this reaction. Moreover, the phosphorylation of acyclovir by uninfected cell extracts is not prevented by thymidine. The exact nature of the cellular kinase or phosphotransferase capable of this conversion is still unidentified. Herpesviruses such as HCMV, EBV and murine cytomegalovirus (MCMV) do not code for specific viral thymidine kinases, and cells infected with these viruses make very low amounts of acyclo-GTP (12,23,24). Vaccinia virus does code for a specific viral TK but acyclovir is not a substrate for this particular enzyme. On the other hand, HSV-1, HSV-2, VZV and the simian delta herpesvirus specify TKs which phosphorylate acyclovir well (9,14,25).

Besides the nature of the cell and the virus, other factors which affect the amount of acyclo-GTP formed are the amount of acyclovir in the medium and the time after infection (11,26). In HSV-1-infected Vero cells the maximum intracellular accumulation of acyclo-GTP occurred 8 hours after viral infection. The amount of acyclo-GTP formed in uninfected Vero cells and in HSV-1 (H29)-infected cells varies with ACV concentration (Fig. 4) (26). However, the large difference in the amounts formed in infected and uninfected cells is maintained over the entire concentration range.

To answer the question, as to how long acyclo-GTP persists in the cell after acyclovir is removed from the surrounding medium, HSV-1 (H29)-infected Vero cells were exposed to 100 μM 8-[^{14}C]acyclovir for 4 hr, washed, and placed in a medium containing either no drug or 10 μM acyclovir. Aliquots of these cells were extracted and examined for acyclo-GTP content by HPLC at various time intervals (Fig. 5) (26). Intracellular acyclo-GTP levels decreased rapidly in the control medium for the first 1.5 hr, then more slowly; at the end of 6 hr the concentration leveled off at approximately 15 pmol/10^6 cells (≅ 3 μM). In

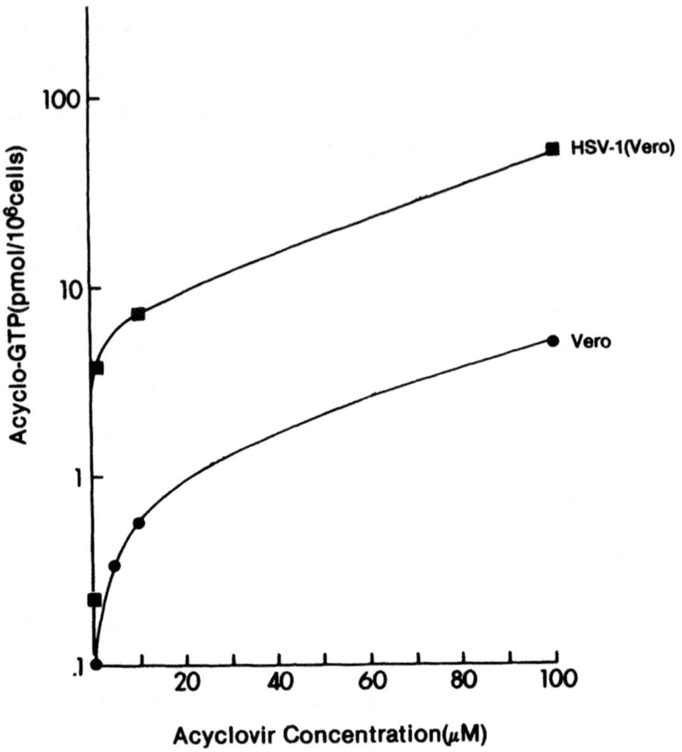

FIGURE 4. Effect of concentration of acyclovir on amount of acyclo-GTP formed in uninfected and HSV-1-infected Vero cells, following 7 hr of incubation with drug.

the presence of a medium containing 10 μM acyclovir, the acyclo-GTP concentration fell more slowly and reached a plateau after 4 hr at about 40 pmol/10^6 cells (8 μM). Disappearance of acyclo-GTP from the cells is due to gradual dephosphorylation and leakage of acyclovir back into the medium. The plateau undoubtedly represents the equilibration of the synthesis and degradation of the triphosphate. The persistence of acyclo-GTP in virus infected cells is of importance in the clinical situation, since levels of acyclovir in the plasma and tissues will vary with intermittent administration.

Inhibition of the viral DNA polymerase by acyclo-GTP results in an increase in the dGTP and dATP pools in the HSV-infected cells (30). Since dGTP competes with acyclo-GTP for the polymerase, the ratio of acyclo-GTP to dGTP would influence the extent of enzyme inhibition in a particular cell line.

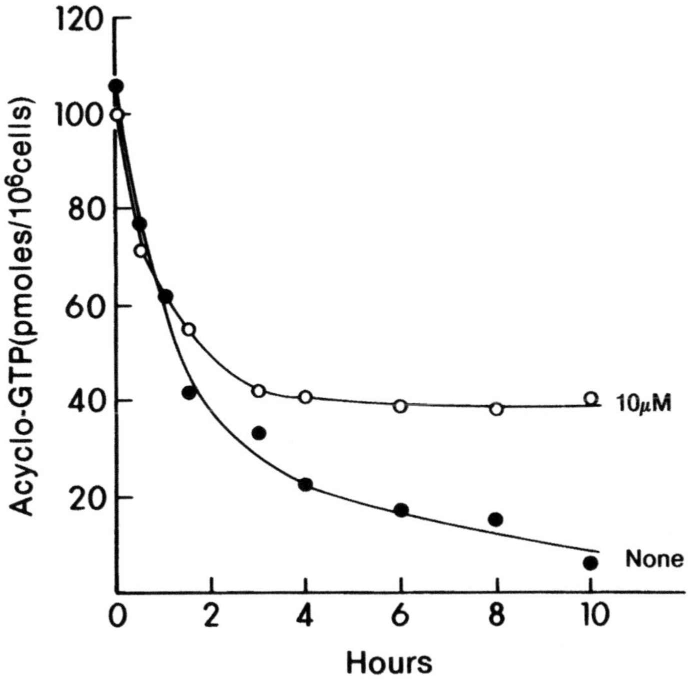

FIGURE 5. Persistence of acyclo-GTP in preloaded HSV-1-infected Vero cells after incubation in medium containing zero or 10 μM acyclovir.

For the DNA polymerases specified by HSV-1 and HSV-2, the initial antagonism between acyclo-GTP and dGTP was competitive early in the enzymic reaction at low concentrations of inhibitor. At higher concentrations of acyclo-GTP there was a gradual slowing of the polymerization reaction and inactivation of the viral DNA polymerase (21). This was associated with the ability of the viral DNA polymerase to use acyclo-GTP as a substrate and to incorporate acyclo-GMP into the DNA template (16,21,27). This resulted in termination of the growing DNA chain, since acyclovir does not have a 3'-hydroxyl group on the acyclic side chain on which to continue the elongation. Moreover, the acyclo-GMP residues could not be excised by the polymerase-associated 3',5'-exonuclease. In addition to the chain termination of the template which occurs with acyclo-GTP and the HSV-1 and HSV-2 DNA polymerases, there is also an inactivation of the viral DNA polymerase during the process (21,27). On the other hand, acyclo-GTP, even at high concentrations, is a very poor substrate for cellular DNA polymerase α (21). Further evidence that ACV can be incorporated into growing chains of viral DNA, resulting in very low molecular weight fragments of viral DNA, was obtained in HSV-infected resting cells in which cellular DNA synthesis was not occurring (28).

For some viral DNA polymerases, e.g. the EBV polymerase, the antagonism between acyclo-GTP and dGTP appears to be strictly competitive, with no inactivation of the polymerase (29). Presumably acyclo-GTP does not act as a substrate for this polymerase. However, the EBV polymerase is extremely sensitive to acyclo-GTP, with a $K_i = 0.015$ μM.

Resistance

As with most infectious agents, it is not difficult to develop drug-resistant strains in the laboratory. For drugs like ACV which require activation by a virus-induced enzyme such as HSV-TK, the loss of this enzyme results in drug resistance. A number of TK^- mutants have been isolated in vitro which are resistant to ACV, arabinosylthymine, and bromovinyldeoxyuridine (31-33). However, these TK^- mutants have been less neurovirulent in vivo (34) and have a very poor ability to establish latency (35,36). Indeed, Klein et al. (35) have described the use of a HSV-TK^- mutant to protect mice against infection with the parental virus. Several mutants with an altered TK specificity have also been described (37); these have been isolated under very specialized laboratory conditions and are rare.

Another type of drug-resistant mutant is one with an altered DNA polymerase (32,33,38,39). Such mutants are generally less sensitive to phosphonoacetate (PAA), phosphonoformate, ara-A triphosphate, and acyclo-GTP. However, lack of cross resistance between acyclo-GTP and PAA has also been described (40).

PHARMACOLOGY

Metabolic Disposition and Pharmacokinetics

In mice, rats, and dogs given ACV by various routes of administration (i.p., s.c., i.v., p.o.) ACV showed excellent metabolic stability and was excreted in the urine essentially unchanged (6,41). The drug was widely distributed is all body tissues, although the levels in brain were lower than in other tissues (6,41,42). Two urinary metabolites of ACV have been detected. The 9-carboxymethoxymethylguanine, in which the terminal hydroxymethyl group has been converted to a carboxyl group, accounts for approximately 10% of the total dose of acyclovir in man (43); in rats, mice and dogs the amount is considerably less (6,41). Another metabolite, 8-hydroxy-9-(2-hydroxyethoxymethyl)guanine is formed in monkeys, rabbits and guinea pigs (44) but is a very minor metabolite (less than 0.2% of the dose) in man (43). During chronic toxicity experiments in mice, rats, and dogs, plasma levels of acyclovir were the same after long-term (1-2 years) administration of drug as after short-term dosing, indicating that there was no induction of drug metabolism (41).

Extensive pharmacokinetic studies have been conducted in dogs after i.v. bolus injection (45) and in man following intravenous administration by one hr infusion (46-48) continuous infusion (49), and bolus injection (50). The drug has been assayed by a sensitive radioimmunoassay (51). In man the overall plasma half-life during the elimination phase is approximately 3 hr. Since elimination of the drug is mainly by urinary excretion, patients with poor renal function show a longer plasma half-life and require less frequent administration of the drug (52,53). Plasma protein binding is low. Proportionality between dose and peak plasma levels are seen after single doses by 1 hr infusion or at steady state after multiple dosing at 8 hr intervals (Table 2) (54). Acyclovir levels in cerebrospinal fluid are approximately 50% of corresponding plasma levels.

Table 2. Steady state plasma levels of acyclovir after intravenous one-hour infusions every eight hours

Dose mg/kg	Mean plasma concentration (μM ± SD) C^{ss}_{max}	C^{ss}_{min}
2.5	22.5 ± 3.4	1.5 ± 0.9
5.0	43.2 ± 4.3	2.9 ± 1.0
10.0	88.9 ± 5.5	9.7 ± 3.5
15.0	105 ± 21	8.8 ± 0.5

The extent of the oral absorption of acyclovir is species-dependent (42). The dog and the mouse absorb the drug better than man or the rat. The poorest oral absorption was found in rhesus monkeys. In man approximately 20% of a 200 mg dose of acyclovir is absorbed and peak plasma levels are found 1.5 to 1.75 hr after oral administration (55). Mean peak plasma levels of 2.5 μM were attained when 200 mg was given at 4 hr intervals; with 400 mg oral doses the peak levels were 5.4 μM (56). The drug in the saliva was present at about 13% of plasma levels, and vaginal secretions contained 0.5-3.6 μM acyclovir following 200 mg every 4 hr (55).

Toxicology

Numerous and extensive toxicologic studies of ACV in animals, by various routes of administration showed a very low order of toxicity (57). In mice and rats, oral doses up to 450 mg/kg/day for 2 years produced no significant toxicity and no carcinogenicity. Dogs tolerated oral doses of 60 mg/kg/day (in 3 divided doses) for a year. Secondary toxicity, in the form of precipitation of acyclovir crystals in the lower nephron of the kidney, occurred after multi-day rapid intravenous bolus injections of 20 mg/kg/day in rats and 25 mg/kg b.i.d. in dogs, and were related to the relatively low solubility of the drug in the urine. Impairment of renal function in the dog was transient. Dermal and ophthalmic ointment formulations were well tolerated by rabbits and guinea pigs. Acyclovir showed no sensitization potential in guinea pigs. In tests measuring the immune response in mice, e.g. antibody response to sheep erythrocytes, delayed hypersensitivity, graft vs. host reaction, ACV showed no inhibitory effects at i.p. doses up to 100 mg/kg/day (58).

In 11 different in vitro mutagenicity assays, 9 were negative and 2 showed positive effects at extremely high drug concentrations, e.g. 1-10 mM. In 2 distinct cell transformation assays, acyclovir was moderately positive at the highest dosage tested in one system and completely negative in the other. Acyclovir had no effect on reproductive processes or prenatal, perinatal or postnatal development in mice, rats or rabbits. It showed no oncogenic potential in lifetime carcinogenicity bioassays in rats and mice (57).

EFFICACY IN ANIMAL MODELS

Acyclovir has shown efficacy as an antiherpetic agent in a variety of animal models. Topical application of an ophthalmic ointment has been effective in acute herpetic keratitis in rabbits (6,59-61). Cutaneous herpes simplex virus infections in guinea pigs have responded well to topical treatment with ACV (6,62,63). The same has been true for genital infections caused by HSV-2 in mice and guinea pigs (64) and oral HSV-1 infections in mice (65). Systemic administration (i.p., p.o., s.c.) has reduced mortality in mice with herpes encephalitis (6,63,66) and in rabbits with B virus infections (67). ACV was also active against mouse cytomegalovirus infections (22) and in monkeys infected with the simian varicella virus (68). Although early administration of the drug has sometimes prevented the establishment of latency in the spinal ganglia of mice (61,70-72), the treatment of established latent infections has generally failed to abolish latency.

CLINICAL EFFICACY

Clinical studies have been conducted with various formulations of ACV by a large number of investigators. The ophthalmic ointment has been found effective in double-blind, controlled trials in the treatment of herpes simplex keratitis both for dendritic and geographic ulcerations (73-78). In these trials, it was equal to or better than idoxuridine or vidarabine and equal to trifluorothymidine in efficacy.

Topical application of a water-soluble ointment, 5% ACV in polyethylene glycol, has reduced the period of viral shedding in patients with both initial and recurrent genital herpes infections (79,80). In initial genital herpes it reduced the duration of pain and shortened the time of healing. Among men, but not among women with recurrent disease, topical ACV produced a significant reduction in the mean time to crusting and healing of lesions. In general, recurrent episodes of genital herpes in nonimmunocompromised patients have a shorter healing time than do initial episodes. This may account for the more modest clinical benefit observed in recurrent disease.

The application of acyclovir ointment was of clinical benefit in immunocomprised patients with cutaneous herpes infections (81). The period of viral shedding was substantially reduced and pain experienced by drug recipients resolved at an accelerated rate. However, nonimmunosuppressed patients with recurrent herpes labialis showed only a decreased period of viral shedding but no significant change in other parameters (82).

Intravenously administered acyclovir has produced substantial clinical benefit in initial genital herpes infections (83), cutaneous HSV infections in immunocompromised patients (84-87), and nonimmunocompromised patients treated early in the course of herpes zoster infections (88). It has also been used with success prophylactically in bone marrow transplant patients to prevent HSV infections in the first two weeks after transplantation (89). In immuncompromised patients with CMV infections results have been less satisfactory, as might be expected from the low sensitivity of CMV to acyclovir (90,91). However, some benefit has been noted in those not terminally ill. Controlled clinical trials with the intravenously administered drug are currently in progress in herpes encephalitis, neonatal herpes, herpes zoster, cutaneous VZV infections in immunocompromised hosts and severe mononucleosis.

Very encouraging clinical results have been reported with oral ACV in initial and recurrent genital herpes (55,87,92). Further studies to determine efficacy in the prophylaxis for recurrent genital disease are in progress. Oral drug is also being evaluated currently in the treatment of herpes zoster, herpes labialis, proctitis in homosexual males, and herpes virus-induced erythema multiforme. Although the absorption of ACV by the oral route is somewhat limited, the plasma and tissue concentrations achieved by oral dosing is, in general, well above the level needed to inhibit the virus in vitro (55).

Acyclovir has been an extremely well tolerated drug in man (93). The principal adverse reaction has been increases in serum creatinine and BUN, particularly in patients given high dose bolus intravenous injections without adequate hydration. These effects have been almost always transient. Inflammation at the injection site due to extravasation of the solution into subcutaneous tissue has also occurred. There have been a few cases of dizziness, disorientation, lethargy and coma reported in severely ill patients.

The development of drug resistance has not been a clinical problem to date. A few cases have been reported where the HSV isolated from immunocompromised patients has been less sensitive to ACV after treatment than before (94,95). In a large number of treated cases, however, no such changes in sensitivity have occurred (96). In all of the clinical strains examined to date, decreased sensitivity to ACV has been due to a reduction or loss in HSV-TK. Since clinical isolates of HSV are generally quite heterogeneous (96,97), more work is required to determine to what extent resistance may reside in only a small portion of the clones isolated from clinical material.

At this time, acyclovir appears to be unable to eliminate latent virus in the sensory ganglia in animals or to prevent recurrences in man. However, it remains to be determined whether repeated courses of treatment with ACV over a prolonged period will eventually alter the pattern of recurrences or affect latency. In the meantime, early diagnosis and treatment will play an important part in the ability of the clinician to reduce morbidity and mortality from herpesvirus infections.

REFERENCES

1. Thompson, R.L., Price, M.L., Minton, S.A., Jr., Elion, G.B. and Hitchings, G.H. J. Immunol. 65:529-534, 1950.
2. Burchenal, J.H., Karnofsky, D.A., Kingsley-Pillers, E.M., Southam, C.M., Myers, W.P.L., Escher, G.C., Craver, L.F., Dargeon, H.W. and Rhoads, C.P. Cancer 4:549-569, 1951.
3. Schabel, F.M. Chemother. 13:321-338, 1968.
4. Elion, G.B., Rideout, J.L., de Miranda, P., Collins, P. and Bauer, D.J. Ann. N.Y. Acad. Sci. 255:468-480, 1975.
5. Schaeffer, H.J., Gurwara, S., Vince, R. and Bittner, S. J. Med. Chem. 14:367-369, 1971.
6. Schaeffer, H.J., Beauchamp, L., de Miranda, P., Elion, G.B., Bauer, D.J. and Collins, P. Nature 272:583-585, 1978.
7. Collins, P. and Bauer, D.J. J. Antimicrob. Chemother. 5:431-436, 1979.
8. Crumpacker, C.S., Schnipper, L.E., Zaia, J.A. and Levin, M.J. Antimicrob. Agents Chemother. 15:642-645, 1979.
9. Biron, K.K. and Elion, G.B. Antimicrob. Agents Chemother. 18:443-447, 1980.
10. Colby, B.M., Shaw, J.E., Elion, G.B. and Pagano, J.S. J. Virol. 34:560-568, 1980.
11. Elion, G.B., Furman, P.A., Fyfe, J.A., de Miranda, P., Beauchamp, L. and Schaeffer, H.J. Proc. Natl. Acad. Sci. USA 74:5716-5720, 1977.
12. St. Clair, M.H., Furman, P.A., Lubbers, C.M. and Elion, G.B. Antimicrob. Agents Chemother. 18:741-745, 1980.
13. Elion, G.B. Am. J. Med. 73(1A):7-13, 1982.
14. Fyfe, J.A.,Keller, P.M., Furman P.A., Miller, R.L. and Elion, G.B. J. Biol. Chem. 253:8721-8727, 1978.
15. Coen, D.M. and Schaffer, P.A. Proc. Natl. Acad. Sci. USA 77:2265-2269, 1980.

16. Furman, P.A., McGuirt, P.V., Keller, P.M., Fyfe, J.A. and Elion, G.B. Virology 12:420-430, 1980.
17. Elion, G.B. Adv. Enzyme Regul. 18:53-66, 1980.
18. Miller, W.H. and Miller, R.L. J. Biol. Chem. 255:7204-7207, 1980.
19. Fyfe, J.A. Molec. Pharmacol. 21:432-437, 1982.
20. Miller, W.H. and Miller, R.L. Biochem. Pharmacol. in press, 1982.
21. Furman, P.A., St. Clair, M.H., Fyfe, J.A., Rideout, J.L., Keller, P.M. and Elion, G.B. J. Virol. 32:72-77, 1979.
22. Keller, P.M., Fyfe, J.A., Beauchamp, L., Lubbers, C.M., Furman, P.A., Schaeffer, H.J. and Elion, G.B. Biochem. Pharmacol. 30:3071-3077, 1981.
23. Burns, W.H., Wingard, J.R., Sanford, G.R., Bender, W.J. and Saral, R. Am. J. Med. 73(1A):118-124, 1982.
24. Colby, B.M., Furman, P.A., Shaw, J.E., Elion, G.B. and Pagano, J.S. J. Virol. 38:606-611, 1981.
25. Fyfe, J.A., Biron, K.K., McKee, S.A., Kelly, C.M., Elion, G.B. and Soike, K.F. Am. J. Med. 73 (1A):58-61, 1982.
26. Furman, P.A., de Miranda, P., St. Clair, M.H. and Elion, G.B. Antimicrob. Agents Chemother. 20:518-524, 1981.
27. Derse, D., Cheng, Y.-C., Furman, P.A., St. Clair, M.H. and Elion, G.B. J. Biol. Chem. 256:11447-11451, 1981.
28. McGuirt, P.V. and Furman, P.A. Am. J. Med. 73(1A):67-71, 1982.
29. Datta, A.K., Colby, B.M., Shaw, J.E. and Pagano, J.S. Proc. Natl. Acad. Sci. USA 77:5163-5166, 1980.
30. Furman, P.A., Lambe, C.U. and Nelson, D.J. Am. J. Med. 73(1A): 14-17, 1982.
31. Field, H.J., Darby, G.K. and Wildy, P. J. Gen. Virol. 49:115-124, 1980.
32. Coen, D.M. and Schaffer, P.A. Proc. Natl. Acad. Sci. USA 77:2265-2269, 1980.
33. Crumpacker, C.S., Chartrand, P., Subak-Sharpe, J.H. and Wilkie, N.M. Virology 105:171-184, 1980.
34. Field, H.J. and Darby, G. Antimicrob. Agents Chemother. 17:209-216, 1980.
35. Klein, R.J., DeStefano, E., Brady, E. and Friedman-Kien, A.E. Arch. Virol. 65:237-246, 1980.
36. Tenser, R.B., Miller, R.L. and Rapp, F. Science 205:915-917, 1979.
37. Field, H.J., Larder, B.A. and Darby, G. Am. J. Med. 73(1A):369-371, 1982.
38. Furman, P.A., Coen, D.M., St. Clair, M.H. and Schaffer, P.A. J. Virol. 40:936-941, 1981.
39. Coen, D.M., Furman, P.A., Gelep, P.T. and Schaffer, P.A. J. Virol. 41:909-918, 1982.
40. Coen, D.M., Schaffer, P.A., Furman, P.A., Keller, P.M. and St. Clair, M.H. Am. J. Med. 73(1A):351-359, 1982.
41. de Miranda, P., Krasny, H.C., Page, D.A. and Elion, G.B. J. Pharmacol. Exp. Ther. 219:309-315, 1981.
42. de Miranda, P., Krasny, H.C., Page, D.A. and Elion, G.B. Am. J. Med. 73(1A):31-35, 1982.
43. de Miranda, P., Good, S.S., Laskin, O.L., Krasny, H.C., Connor, J.D. and Lietman, P.S. (1981): Clin. Pharmacol. Ther. 30:662-672, 1981.
44. Good, S.S. and de Miranda, P. Am. J. Med. 73(1A):91-95, 1982.
45. Krasny, H.C., de Miranda, P., Blum, M.R. and Elion, G.B. J. Pharmacol. Exp. Ther. 216:281-288, 1981.

46. de Miranda, P., Whitley, R.J., Blum, M.R., Keeney, R.E., Barton, N., Cocchetto, D.M., Good, S., Hemstreet, B.P., Kirk, L.E., Page, D.A. and Elion, G.B. Clin. Pharmacol. Ther. 26:718-728, 1979.
47. Spector, S.A., Connor, J.D., Hintz, M., Quinn, R.P., Blum, M.R. and Keeney, R.E. Antimicrob. Agents Chemother. 19:608-612, 1981.
48. Laskin, O., Longstreth, J.A., Saral, R., de Miranda, P., Keeney, R.E. and Lietman, P.S. Antimicrob. Agents Chemother. 21:393-398, 1982.
49. Spector, S.A., Hintz, M., Wyborny, C., Connor, J.D., Keeney, R.E. and Liao, S. Am. J. Med. 73(1A):275-281, 1982.
50. Brigden, D., Bye, A., Fowle, A.S.E. and Rogers, H. J. Antimicrob. Chemother. 7:399-404, 1981.
51. Quinn, R.P., de Miranda, P., Gerald, L. and Good, S.S. Analyt. Biochem. 98:319-328, 1979.
52. Laskin, O.L., Longstreth, J.A., Whelton, A., Rocco, L., Lietman, P.S., Krasny, H.C. and Keeney, R.E. Clin. Pharmacol. Ther. 31: 594-601, 1982.
53. Blum, M.R., Liao, S.H. and de Miranda, P. Am. J. Med. 73 (1A): 186-192, 1982.
54. Elion, G.B. In: Herpesvirus. Clinical, Pharmacological and Basic Aspects (Eds.: H. Shiota, Y.-C. Cheng, and W.H. Prusoff) Excerpta Medica, Amsterdam-Oxford-Princeton, 1982, pp. 229-235.
55. VanDyke, R.B., Connor, J.D., Wyborny, C., Hintz, M. and Keeney, R.E. Am. J. Med. 73(1A):172-175, 1982.
56. de Miranda, P. Program and Abstracts, 22nd Interscience Conf. Antimicrob. Agents Chemother., Miami Beach, October 4-6, 1982, p. 140.
57. Tucker, W.E., Jr. Am. J. Med. 73(1A):27-31, 1982.
58. Quinn, R.P., Wolberg, G., Medzihradsky, J. and Elion, G.B. Am. J. Med. 73(1A):62-66, 1982.
59. Bauer, D.J., Collins, P., Tucker, W.E. and Macklin, A.W. Br. J. Ophthal. 63:429-435, 1979.
60. Kaufman, H.E., Varnell, E.D., Centifano, Y.M. and Rheinstrom, S.D. Antimicrob. Agents Chemother. 14:842-845, 1978.
61. Trousdale, M.D., Dunkel, E.C. and Nesburn, A.B. Invest. Ophthalmol. Vis. Sci. 19:1336-1341, 1980.
62. Park, N.-H., Pavan-Langston, D., McLean, S.L. and Lass, J.H. Arch. Dermatol. 116:672-675, 1980.
63. Collins, P. and Oliver, N.M. Am. J. Med. 73(1A):96-99, 1982.
64. Kern, E.R. Am. J. Med. 73(1A):100-108, 1982.
65. Park, N.-H., Pavan-Langston, D. and Hettinger, M.E. J. Dent. Res. 59:2080-2086, 1980.
66. Kern, E.R., Richards, J.T., Glasgow, L.A., Overall, J.C., Jr. and de Miranda, P. Am. J. Med. 73(1A):125-131, 1982.
67. Boulter, E.A., Thornton, B., Bauer, D.J. and Bye, A. Br. Med. J. 280:681-683, 1980.
68. Soike, K.F. and Gerone, P.J. Am. J. Med. 73(1A):112-117, 1982.
69. Field, H.J., Bell, S.E., Elion, G.B., Nash, A.A. and Wildy, P. Antimicrob. Agents Chemother. 15:554-561, 1979.
70. Park, N.-H, Pavan-Langston, D. and McLean, S.L. J. Infect. Dis. 140:802-806, 1979.
71. Klein, R.J., Friedman-Kien, A.E. and DeStefano, E. Antimicrob. Agents Chemother. 15:723-729, 1979.
72. Klein, R.J. Am. J. Med. 73(1A):138-142, 1982.

73. Jones, B.R., Coster, D.J., Fison, P.N., Thompson, G.M., Cobo, L.M. and Falcon, M.G. Lancet 1:243-244, 1979.
74. Collum, L.M.T. Br. J. Ophthalmol. 64:766-769, 1980.
75. McGill, J. and Tormey, P. Am. J. Med. 73(1A):286-289, 1982.
76. Shiota, H. Am. J. Med. 73(1A):307-310, 1982.
77. LaLau, C., Oosterhuis, J.A., Versteeg, J. Van Rij, G., RenardelDeLavalette, J.G.C., Craandijk, A. and Lamers, W.P.M.A. Br. J. Ophthalmol. 66:506-508, 1982.
78. Coster, D.J., Wilhelmus, K.R., Michaud, R. and Jones, B.R. Br. J. Ophthalmol. 64:763-765, 1980.
79. Corey, K., Benedetti, J.K., Critchlow, C.W., et al. Am. J. Med. 73(1A):326-334, 1982.
80. Corey, L., Nahmias, A.J., Guinan, M.E., Benedetti, J.K., Critchlow, C.W. and Holmes, K.K. New Engl. J. Med. 306:1313-1319, 1982.
81. Whitley, R., Barton, N., Collins, E., Whelchel, J. and Diethelm, A.G. Am. J. Med. 73(1A):236-240, 1982.
82. Spruance, S.L., Schnipper, L.E., Overall, J.C., Jr., Kern, E.R., Wester, B., Modlin, J., Wenerstrom, G., Burton, C., Arndt, K.A., Chiu, G.L. and Crumpacker, C.S. J. Infect. Dis. 146:85-90, 1982.
83. Mindel, A., Adler, M.W., Sutherland, S. and Fiddian, A.P. Am. J. Med. 73(1A):347-350, 1982.
84. Mitchell, C.D., Bean, B., Gentry, S.R., Groth, K.E., Boen, J.R. and Balfour, H.H., Jr. Lancet 1:1389-1392, 1981.
85. Meyers, J.D., Wade, J.C., Mitchell, C.D., Saral, R., Lietman, P.S., Durack, D.T., Levin, M.J., Segreti, A.C. and Balfour, H.H. Am. J. Med. 73(1A):229-235, 1982.
86. Spector, S.A., Hintz, M., Wyborny, C., Connor, J.D., Keeney, R.E. and Liao, S. Am. J. Med. 73(1A):275-281, 1982.
87. Straus, S.E., Smith, H.A., Brickman, C., de Miranda, P., McLaren, C. and Keeney, R.E. Ann. Intern. Med. 96:270-277, 1982.
88. Esmann, V., Ipsen, J., Peterslund, N.A., Seyer-Hansen, K., Schonheyder, H. and Juhl, H. Am. J. Med. 73(1A):320-325, 1982.
89. Saral, R., Burns, W.H., Laskin, O.L., Santos, G.W. and Lietman, P.S. New Engl. J. Med. 305:63-67, 1982.
90. Balfour, H.H., Bean, B., Mitchell, C.D., Sachs, G.W., Boen, J.R. and Edelman, C.K. Am. J. Med. 73(1A):241-248, 1982.
91. Wade, J.C., McGuffin, R.W., Springmeyer, S.C., Connor, J.D. and Meyers, J.D. Am. J. Med. 73(1A):249-256, 1982.
92. Fiddian, A.P., Halsos, A.M., Kinge, B.R., Nilsen, A.E. and Wilstrom, K. Am. J. Med. 73(1A):335-337, 1982.
93. Keeney, R.E., Kirk, L.E. and Brigden, D. Am. J. Med. 73(1A):176-181, 1982.
94. Sibrack, C.D., McLaren, C. and Barry, D.W. Am. J. Med. 73(1A): 372-375, 1982.
95. Schnipper, L.E., Crumpacker, C.S., Marlowe, S.I., Kowalsky, B.A., Hershey, B.J. and Levin, M.J. Am. J. Med. 73(1A):387-392, 1982.
96. Collins, P., Appleyard, G. and Oliver, N.M. Am. J. Med. 73(1A): 380-382, 1982.
97. McLaren, C., Sibrack, C.D. and Barry, D.W. Am. J. Med. 73(1A): 376-379, 1982.

6

BVDU [(E)-5-(2-BROMOVINYL)-2'-DEOXYURIDINE]

ERIK DE CLERCQ
Rega Institute for Medical Research, Katholieke Universiteit Leuven, B-3000 Leuven, Belgium

SUMMARY

BVDU [(E)-5-(2-bromovinyl)-2'-deoxyuridine] is one of the most efficient antiherpes agents yet discovered. In cell culture, it inhibits the replication of several herpesviruses, i.e. herpes simplex virus type 1 (HSV_1), varicella-zoster virus (VZV), pseudorabies virus (PRV) and bovid herpesvirus type 1 (BHV_1), at a concentration of about 0.01 μg/ml, while not affecting normal cell metabolism at a concentration up to 30-100 μg/ml. BVDU is specifically phosphorylated by the virus-induced deoxythymidine (dThd) kinase and the target for its antiviral action may be the viral DNA synthesis or viral DNA itself. BVDU has demonstrated high efficacy in several animal model infections, i.e. HSV_1 keratitis in rabbits, HSV_1 encephalitis in mice, mucocutaneous HSV_1 lesions in mice and guinea pigs, and simian varicella virus infection in monkeys. Preliminary clinical studies indicate that BVDU may have great potentials in the topical and systemic (i.e. oral) treatment of HSV_1 and VZV infections in humans.

INTRODUCTION

The field of antiviral chemotherapy has recently been revolutionized by the discovery of several new nucleoside analogues which are highly selective and potent inhibitors of herpesvirus replication and which hold great promise for the systemic treatment of herpes simplex virus (HSV) and varicella-zoster virus (VZV) infections in humans. Foremost among these are ACV [acyclovir, acycloguanosine, 9-(2-hydroxyethoxymethyl)guanine], FIAC [1-(2-fluoro-2-deoxy-β-D-arabinofuranosyl)-5-iodocytosine] and BVDU [(E)-5-(2-bromovinyl)-2'-deoxyuridine]. The spectrum of antiviral activity of BVDU, its selectivity as an antiviral agent, its molecular mechanism of action and structure-function relationship, and, finally, the efficacy of BVDU in animal model infections and humans have been the subject of extensive studies and these studies are briefly reviewed in the present report. For a more comparative description of the potency of BVDU, its mechanism of action and therapeutic potentials relative to other antiviral compounds such as ACV and FIAC the reader is referred to previous publications (1-4).

SPECTRUM OF ANTIVIRAL ACTIVITY

BVDU is particularly effective against HSV_1, VZV, PRV, BHV_1 and SVV (Table 1). It inhibits the replication of these viruses at a concentration of approximately 0.01 μg/ml (5-10). Many of the herpesviruses which appear to be highly susceptible to inhibition by BVDU, i.e. VZV, PRV, BHV_1, SVV, HVS and NPV, are not inhibited by ACV, or only at concentrations that are 100- or 1000-fold higher than those required for BVDU (7-12). BVDU is at an average 5 to 10 times more effective than ACV against HSV_1, as has been ascertained in primary rabbit kidney cells (6) and a number of cell lines of either murine, simian or human origin (13). However, ACV is about 25 times more effective than BVDU against HSV_2 (6). In fact, HSV_2 is 100- to 200-fold less susceptible to inhibition by BVDU than HSV_1, and the differential sensitivity of HSV_1 and HSV_2 to BVDU could be advocated as a useful marker test for the differentiation between type 1 and type 2 strains in clinical isolates, as originally proposed by De Clercq _et al._ (6) and later confirmed by Mayo (14). BVDU has also proven useful in identifying the virus type that was recovered from latently HSV_1-infected cell cultures following superinfection with temperature-sensitive HSV_2 mutants (15).

Table 1. Spectrum of antiviral activity of BVDU

Virus	Minimal inhibitory concentration (μg/ml)	References
HSV_1 (herpes simplex virus-1)	0.007-0.01	(5,6)
HSV_2 (herpes simplex virus-2)	1-2	(6)
VZV (varicella-zoster virus)	0.001-0.01	(7)
PRV (pseudorabies virus)	0.006-0.06	(8)
BHV_1 (bovid herpesvirus-1)	0.01	(9)
SVV (simian varicella virus)	0.01	(10)
HVS (herpesvirus saimiri)	ND*	(11)
NPV (nuclear polyhedrosis virus)	0.5	(12)

*Not determined but significantly lower than 10 μg/ml.

BVDU is virtually inactive against dTK^- (deoxythymidine (dThd) kinase deficient) mutants of HSV_1 or VZV (6,7). This indicates that the compound must be phosphorylated by the virus-induced dThd kinase to exert its antiviral effects. dTK^- HSV mutants are also resistant to other nucleoside analogues, i.e. ACV, which require phosphorylation by viral dTK for their antiviral activity, and to the extent that drug resistance of HSV or VZV mutants is based upon the absence of viral dTK activity, BVDU-resistant mutants should also be cross-resistant to other nucleoside analogues, and vice versa (16,17).

BVDU is only weakly active against vaccinia virus (5,6), inactive against adenovirus (8), vesicular stomatitis virus (18) and other RNA viruses (8). The activity of BVDU against cytomegalovirus (CMV) and Epstein-Barr virus (EBV) has not been fully assessed, but, considering the mode of action of the compound against HSV and VZV (which is essentially based upon an interaction with the viral dThd kinase and DNA polymerase) there are no reasons to expect a significant activity against either CMV or EBV.

Herpesviruses are notorious for establishing a latent infection once the acute primary infection has subsided. For HSV and VZV the latent infection occurs at the neurones of the sensory **ganglia** . BVDU and ACV both block the reactivation of latent HSV_1 and the multiplication of reactivated latent virus, yet, neither drug completely eliminates the latent virus from the sensory ganglia (19,20).

SELECTIVITY AS AN ANTIVIRAL AGENT

BVDU does not affect the growth or metabolism of normal (uninfected) cells unless it is used at concentrations that are approximately 3,000- to 10,000-fold higher than the concentrations required for inhibition of HSV_1 or VZV replication. This relative lack of cytotoxicity has been evidenced by several parameters such as cell counts (following a 3-day incubation period in the presence of the compound) (5), microscopic appearance of the cells (6), incorporation of (methyl-^{3}H)dThd and (2-^{14}C or 1',2'-^{3}H)dUrd into host cell DNA (5,6) and cell colony formation (21). For example, at 200 μg/ml BVDU did not reduce the replicative ability of human diploid fibroblasts by more than 30% (21). BVDU appears to be more cytotoxic for tumor cells : its ID_{50} (50% inhibitory dose) for the growth of murine leukemia L1210 cells is 27 μg/ml (22), and for the growth of human lymphoblast cells (Namalva and Raji) ID_{50} values of 95 and 36 μg/ml have been noted (23). Quite surprisingly, BVDU turned out to be 20 times more cytotoxic for a dTK^- L1210 cell line than for the parent dTK^+ cell line (23). This increased toxicity of BVDU for dTK^- L1210 cells may reflect an alteration in the substrate specificity of the dThd kinase of the mutant cell line.

BVDU is non-mutagenic in the Ames mutagenicity test, as has been ascertained with five bacterial strains in the presence and absence of liver microsomal fraction (24). While 5-iododeoxyuridine (IDU) and 5-trifluorothymidine (TFT) impair normal morphogenesis of the chick embryo at a dose of 0.1 μg/embryo (1 mg/kg) and 0.01 μg/embryo (0.1 mg/kg), respectively, BVDU does not show any embryo toxicity, even at 100 μg/embryo (1000 mg/kg) (25). BVDU does not affect the lymphocyte blastogenic responses to mitogens (i.e. phytohemagglutinin) and specific antigens (i.e. tetanus toxoid) at concentrations up to 20 μg/ml (26). Nor does BVDU lead to chromosomal aberrations (as monitored by the rate of sister chromatid exchanges) in human fibroblasts or lymphocytes, unless it is added to the cells at a concentration (50 μg/ml) that is 5000-fold higher than the minimal antiviral concentration (27). TFT, however, increased the rate of sister chromatid exchange at a concentration that coincided with its minimal antiviral concentration (0.5 μg/ml) (27). Unlike IDU, BVDU does not lead to the release of retrovirus (RNA tumor virus) particles from carrier cell lines, i.e. murine BALB/3T3 cells (28). This suggest that BVDU is not incorporated into BALB/3T3 cell DNA or incorporated to an extent that does not lead to the expression of retroviral genes. The possibility that BVDU might be incorporated into DNA of BALB/3T3 cells was directly examined with a radiolabeled analogue of BVDU, namely (E)-5-(2-(^{125}I)iodovinyl)-2'-deoxyuridine ((^{125}I)IVDU), and under conditions where (^{125}I)IDU was effectively incorporated into BALB/3T3 cell DNA, no such incorporation was observed with (^{125}I)IVDU (28).

MECHANISM OF ACTION AT MOLECULAR LEVEL

The selectivity of BVDU as an antiherpes agent depends essentially on two virus-induced enzymes, dThd kinase and DNA polymerase (Fig. 1). The virus-induced dThd kinase would convert BVDU to its 5'-monophosphate (BVdUMP), and some virus (i.e. HSV_1)-induced dThd kinases may convert BVdUMP further onto its 5'-diphosphate (BVdUDP). Such phosphorylation would not occur in the uninfected cell. Indeed, the Ki for BVDU of HSV_1-, HSV_2- and VZV-induced dThd kinase is 0.24, 4.24 and 0.07 μM respectively, as compared to > 150 μM for the Ki of cytosol dThd kinase (29). Thus, BVDU has a much higher affinity for the virus-encoded dThd kinase than for the cytosol dThd kinase. Moreover, BVDU has a markedly higher affinity for the HSV_1 dThd kinase than for the HSV_2 dThd kinase, which seems to correlate with its differential inhibitory activity towards these viruses.

When added to cell cultures, i.e. primary rabbit kidney cells, which have been infected with HSV_1 or HSV_2, (^{125}I)IVDU is promptly converted to its 5'-monophosphate form (30). Such phosphorylation is not observed in uninfected cells, or cells which have been infected with a dTK$^-$ mutant of HSV_1. HSV_1-infected cells, but not HSV_2-infected cells, are capable of phosphorylating (^{125}I)IVDU to its 5'-di- and 5'-triphosphate (30), which would indicate that HSV_1, but not HSV_2, induces a dTMP kinase (associated with dThd kinase) that converts the 5-(2-halogenovinyl)-dUrd 5'-monophosphates to their 5'-diphosphate forms. In fact, Fyfe (31) has shown that HSV_1 and VZV induce a dThd (dTMP) kinase which is capable of catalyzing the phosphorylation of BVdUMP to BVdUDP. HSV_2-infected cells lack such dTMP kinase activity. Obviously, the differential phosphorylation of BVDU in HSV_1- and HSV_2-infected cells bears on the relative sensitivities of HSV_1 and HSV_2 to the antiviral effects of BVDU.

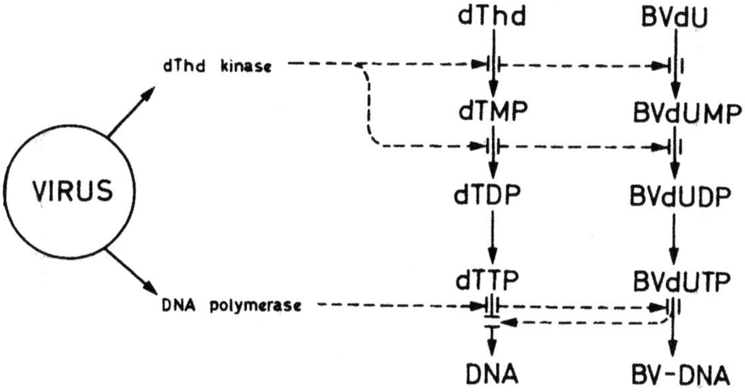

FIGURE 1. Mechanism of antiviral action of BVDU. This mode of action has been demonstrated for HSV_1 but may also be applicable to other herpesviruses such as VZV, PRV, BHV_1 and SVV.

The dThd kinase appears to play an important role in the selectivity of BVDU as an antiherpes agent, as has been ascertained by using HSV_1 x HSV_2 intertypic recombinants and dTK$^-$ L cells (32). The HSV_1 x HSV_2 recombinant proved susceptible to the inhibitory activity of BVDU if it contained the 0.295 - 0.385 coordinates of HSV_1, that is the region where the dTK gene is located. While not susceptible to 1.0 μM BVDU in dTK$^-$ L cells, HSV_2 was readily inhibited by the compound if tested in dTK$^-$ L cells which had been transformed by the HSV_1 dTK gene.

The target for the antiviral action of BVDU may be the viral DNA synthesis or viral DNA itself. Indeed, the 5'-triphosphate of BVDU (BVdUTP) inhibits the HSV_1 DNA polymerase to a considerably greater extent than the cellular DNA polymerases α, β and γ (33). BVdUTP would inhibit HSV_2 DNA polymerase to the same extent as HSV_1 DNA polymerase (32) which corroborates the hypothesis that differential susceptibility of HSV_1 and HSV_2 towards inhibition by BVDU resides at the dThd (dTMP) kinase rather than the DNA polymerase level.

When added to intact cells, i.e. primary rabbit kidney cells, BVDU inhibits DNA synthesis of HSV_1-infected cells (as monitored by [methyl-^{3}H]dThd incorporation) at a concentration that is approximately 3,000-fold lower than the concentration required to inhibit DNA synthesis of uninfected cells (34). Since an inhibitory effect on dThd incorporation into DNA may be mediated by an inhibition of any step in dThd metabolism (i.e. dThd kinase, dTMP kinase, ...) and does not necessarily reflect an inhibition of DNA synthesis *per se*, additional experiments were performed in which [^{32}P]ortophosphate rather than [methyl-^{3}H]-dThd served as the radiolabel for monitoring DNA synthesis. Again, it was found that BVDU inhibited DNA synthesis of HSV_1-infected cells at a 3,000-fold lower concentration than DNA synthesis of uninfected cells (35). BVDU, ACV and FIAC all showed high selectivity when their effects on HSV_1 DNA synthesis were compared with their effects on cellular DNA synthesis in uninfected cells (36).

The 5'-triphosphates of BVDU and IVDU can substitute for dTTP as substrates for both viral and cellular DNA polymerases (37,38), and these substitutions have little, if any, effect on the primer-template activity of the resulting DNA (39). BVDU can also be incorporated into DNA of intact cells. However, this incorporation would primarily be restricted to the DNA of herpesvirus-infected cells (38). It remains to be established to what extent this incorporation contributes to the antiviral activity of BVDU.

There appears to be a perfect correlation between the susceptibility of different HSV_1 and HSV_2 strains to BVDU and their ability to express deoxypyrimidine triphosphatase (dPyTPase) activity : HSV_1 strains (which are highly sensitive to BVDU) induce dPyTPase activity, whereas HSV_2 strains (which are much less sensitive to BVDU) fail to express the enzyme (40). The molecular basis for this correlation has not yet been determined.

Since the mechanism of action of BVDU is essentially based upon the viral dThd kinase and viral DNA polymerase, BVDU-resistant virus strains may arise by mutations within either the dThd kinase or DNA polymerase locus. That both loci confer drug resistance has been clearly demonstrated with ACV (41,42), and there is little doubt that they also confer resistance to BVDU. The region specifying BVDU resistance within the DNA polymerase locus has actually been defined: it maps within coordinates 0.410 and 0.428, that is partially overlapping with the map limits of ACV resistance (coordinates 0.402 - 0.418) (43).

STRUCTURE-FUNCTION RELATIONSHIP

The critical determinant in the antiviral activity of BVDU is undoubtedly the 2-bromovinyl substituent at C_5 of the uracil ring (Fig. 2). The 2-bromovinyl substituent should occur in the E ("entgegen") configuration, since the Z ("zusammen") isomer of BVDU is much less active against HSV_1 than the E isomer (44). This is accompanied by a decreased affinity for the HSV_1-induced dThd kinase (Ki = 2.47 μM for the Z isomer, as compared to 0.24 μM for the E isomer (29)). The bromine of BVDU can be replaced by an iodine or chlorine without significant loss of either antiviral potency or selectivity. The resulting E-5-(2-chlorovinyl)- and E-5-(2-iodovinyl)-2'-deoxyuridines behave in all aspects like BVDU (6,45). (E)-5-(1-Propenyl)- and (E)-5-(3,3,3-trifluoro-1-propenyl)-2'-deoxyuridine can also be regarded as close analogues of BVDU (46). They are almost as selective in their antiherpes activity as BVDU (47); their minimal inhibitory concentration for HSV_1 is in the range of 0.01 - 0.1 μg/ml, and, akin to BVDU, they also discriminate between HSV_1 and HSV_2 (46). Substitution of cytosine for uracil is also compatible with selective antiviral activity, since (E)-5-(2-bromovinyl)-2'-deoxycytidine (BVDC) and (E)-5-(2-iodovinyl)-2'-deoxycytidine (IVDC) are equally specific and almost as potent inhibitors of HSV_1 replication as their dUrd counterparts (48). Again, BVDC and IVDC are much less inhibitory to HSV_1 than to HSV_2.

In attempts to develop, if at all possible, a compound with superior anti-HSV_2 activity, several BVDU analogues have been synthesized with modifications at the sugar moiety, i.e. arabinofuranose, 2-deoxyxylofuranose, 3-amino-2,3-dideoxyribose, 3-azido-2,3-dideoxyribose, 3-chloro-2,3-dideoxyribose, 5-amino-2,5-dideoxyribose, 5-azido-2,5-dideoxyribose, 5-chloro-2,5-dideoxyribose, 5-bromo-2,5-dideoxyribose, and 5-iodo-2,5-dideoxyribose (49). Of these sugar-modified derivatives of BVDU only four compounds (1-β-D-arabinofuranosyl-(E)-5-(2-bromovinyl)uracil) (BVaraU), (E)-5-(2-bromovinyl)-3'-amino-2',3'-dideoxyuridine (3'-NH_2-BVDU), (E)-5-(2-bromovinyl)-3'-azido-2',3'-dideoxyuridine (3'-N_3-BVDU) and (E)-5-(2-bromovinyl)-5'-iodo-2',5'-dideoxyuridine (5'-I-BVDU) display an appreciable anti-HSV_1 activity (50). They suppress HSV_1 replication in primary rabbit kidney cell cultures at a minimal inhibitory concentration of 0.1, 0.1, 1 and 1 μg/ml, respectively. In human diploid fibroblasts, BVaraU is as active as inhibitor of HSV_1 replication as BVDU (51), but in other cell lines, i.e. monkey kidney (Vero, BSC-1), it is definitely less active (13,52).

Neither of the four sugar-modified derivatives of BVDU offers significant activity against HSV_2, and, since BVaraU and 3'-NH_2-BVDU inhibits HSV_2 replication at a 300-fold higher concentration than HSV_1 replication, they could both be advocated as useful markers for the identification of HSV_1 and HSV_2 strains (50). As compared to BVDU, 3'-NH_2-BVDU is about 4 times less active, and BVaraU about 7 times more active as inhibitor of VZV replication (53). This brings the minimal inhibitory concentration of BVaraU for VZV to 1 ng/ml (54).

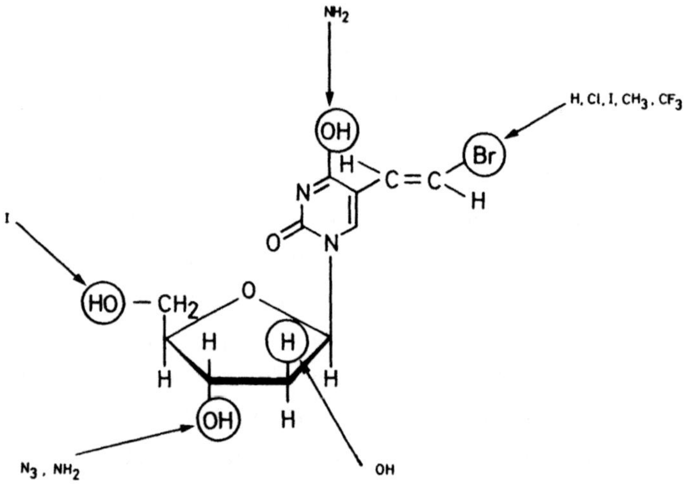

FIGURE 2. Structural analogues of BVDU that have proven effective as antiherpes agents. Full and short names of BVDU derivatives resulting from substitution of

H for Br : 5-vinyl-2'-deoxyuridine (VDU)
Cl for Br : (E)-5-(2-chlorovinyl)-2'-deoxyuridine (CVDU)
I for Br : (E)-5-(2-iodovinyl)-2'-deoxyuridine (IVDU)
CH_3 for Br : (E)-5-(1-propenyl)-2'-deoxyuridine (PeDU)
CF_3 for Br : (E)-5-(3,3,3-trifluoro-1-propenyl)-2'-deoxyuridine (TFPeDU)
NH_2 for OH at C4 : (E)-5-(2-bromovinyl)-2'-deoxycytidine (BVDC)
OH for H at C'2 : 1-β-D-arabinofuranosyl-(E)-5-(2-bromovinyl)uracil (BVaraU)
N_3 for OH at C'3 : (E)-5-(2-bromovinyl)-3'-azido-2',3'-dideoxyuridine (3'-N_3-BVDU)
NH_2 for OH at C'3: (E)-5-(2-bromovinyl)-3'-amino-2',3'-dideoxyuridine (3'-NH_2-BVDU)
I for OH at C'5 : (E)-5-(2-bromovinyl)-5'-iodo-2',5'-dideoxyuridine (5'-I-BVDU)

All structural analogues of BVDU depicted in Fig. 2 can be considered as selective inhibitors of HSV_1 replication, since they inhibit normal host cell metabolism as well as HSV_2 and vaccinia virus replication at concentrations which are greater by several orders of magnitude than those required to inhibit HSV_1 replication (Table 2). The least selective in its anti-HSV_1 activity is

5-vinyl-2'-deoxyuridine (VDU), which inhibits HSV_2 and vaccinia virus replication at a concentration which is only 5- to 20-fold higher than the minimal inhibitory concentration for HSV_1. In fact, VDU emerged as the most potent anti-HSV_2 agent among all 5-substituted 2'-deoxyuridines that have so far been evaluated for their antiviral properties (47).

Table 2. Antiviral potency and selectivity of BVDU and its structural analogues

Compound*	Minimal inhibitory concentration (μg/ml)				References
	HSV_1	HSV_2	Vaccinia virus	Cell metabolism†	
BVDU	0.008	1	7	20	(6,47)
VDU	0.018	0.1	0.4	7	(6,47)
CVDU	0.02	2	85	70	(45,47)
IVDU	0.012	2	10	20	(6,47)
PeDU	0.07	22	225	40	(46,47)
TFPeDU	0.07	65	350	150	(46,47)
BVDC	0.07	10	200	100	(48)
BVaraU	0.1	35	> 200	100	(50)
3'-N_3-BVDU	1	50	200	150	(50)
3'-NH_2-BVDU	0.1	25	> 200	200	(50)
5'-I-BVDU	1	10	70	50	(50)

*The full names of the compounds are indicated in the legend to Fig. 2.
†As determined by [1',2'-^3H]dUrd incorporation into host cell DNA.

To some extent, the mechanism of antiviral action of the sugar-modified derivatives of BVDU may be similar to that of BVDU itself; this means, dependent on two virus-induced enzymes, the dThd kinase and the DNA polymerase. Yet, important differences have been noted in the action of the 5'-triphosphates of BVDU, BVaraUTP and 3'-NH_2-BVdUTP. While BVdUTP is recognized by DNA polymerase as an alternate substrate and is incorporated into DNA probably within the interior of the DNA chain, BVaraUTP does not serve as an efficient alternate substrate, although it can be incorporated at the 3'-end of the DNA chain, thereby acting as a chain terminator (39,55). Unlike BVdUTP (33), BVaraUTP and 3'-NH_2-BVdUTP do not inhibit HSV_1 DNA polymerase to a greater extent than cellular DNA polymerase α (53,55). One may postulate, therefore, that the selectivity of BVaraU and 3'-NH_2-BVDU as antiherpes agents depends mainly on a specific phosphorylation by the viral dThd kinase. This would be consistent with the high affinity of BVaraU for HSV_1 dThd kinase (Ki = 0.94 μM, as compared to > 100 μM for the cytosol dThd kinase) (56).

EFFICACY IN ANIMAL MODEL INFECTIONS

BVDU has proven to be efficacious against HSV_1 infections in various animal model systems (Table 3); i.e. in athymic-nude (nu/nu) mice, where both topical application of BVDU (at 1% in a water-soluble ointment) (5,57) and systemic administration (at 60 mg/kg/day, in the drinking water) (58) suppressed the development of cutaneous HSV_1 lesions and mortality associated therewith. When tested under similar conditions, interferon failed to alter the course of the infection (58). When administered intraperitoneally, BVDU protected hairless (hr/hr) mice against a lethal HSV_1 skin infection (59), and, when administered subcutaneously or orally, it prevented the establishment of latent HSV_1 infections in either hairless or BALB/c mice (60,61). However, latent HSV_1 infections once established were intractable and were not amenable to ACV therapy (60,61). In mice inoculated intracerebrally with HSV_1, BVDU effected a significant reduction of the mortality rate if (oral, intraperitoneal or subcutaneous) treatment was initiated shortly after virus infection, i.e. on day 0 or 2 (or day 4, if BVDU was administered subcutaneously), at a dosage of 80 mg/kg/day or higher (62). In guinea pigs, BVDU suppressed the development of HSV_1 skin lesions, when administered either orally or topically (32). However, HSV_2 infections did not appear to respond to systemic or topical BVDU treatment in either mice (32,59) or guinea pigs (32).

Table 3. Efficacy of BVDU in animal model infections

- Topical and systemic (oral, intraperitoneal) treatment of cutaneous HSV_1 lesions, and mortality therewith associated, in athymic-nude mice (5,57,58) and hairless mice (59).
- Topical and systemic (subcutaneous) treatment of orofacial HSV_1 infection in hairless mice (60).
- Systemic (oral) treatment of HSV_1 ear infection in BALB/c mice (61).
- Systemic (oral, intraperitoneal, subcutaneous) treatment of HSV_1 encephalitis in NMRI mice (62).
- Topical and systemic (oral) treatment of cutaneous HSV_1 lesions in guinea pigs (32).
- Topical treatment of epithelial and stromal HSV_1 keratitis, and topical and systemic (oral) treatment of HSV_1 uveitis in rabbits (63-67).
- Systemic (oral, intravenous, intramuscular) treatment of simian varicella virus infection in African green monkeys (68).
- Systemic (oral) treatment of pseudorabies virus infection in pigs (69).

When applied as 0.1% eye drops or 0.5% ointment, BVDU was superior to IDU in promoting the healing of experimental HSV_1 keratitis in rabbits (63,64). The efficacy of BVDU in the topical treatment of herpetic keratitis has also been demonstrated with 0.25% eye drops (65). BVDU (at either 0.1 or 0.5% eye drops) is also effective in the local treatment of stromal HSV_1 keratitis, and in this regard it is superior to 1% TFT drops if the treatment is started one day after infection (66). BVDU has also proven efficacious in the topical and systemic (oral) treatment of HSV_1 uveitis in rabbits, where a significantly faster healing of the iritis was noted with 0.5% BVDU eye drops than with 1% TFT eye drops (67).

BVDU has been shown to completely eliminate all manifestations of disease, including viremia, rash and anorexia, when administered orally at 15 mg/kg/day to African green monkeys infected with simian varicella virus (68). In this simian varicella model BVDU was equally effective when administered by the oral, intramuscular or intravenous route. Even if the dosage was lowered to 1 mg/kg/day or if treatment was deferred as long as 6 days after virus infection BVDU conferred an inhibitory effect on the disease (68). Preliminary findings with oral BVDU therapy in pigs indicate that the drug would also be effective in curtailing the symptoms (fever, weight loss, mortality) associated with a pseudorabies virus infection (69).

BVDU is very well absorbed when given orally, as demonstrated in several animal species, including mice (70), rats, rabbits, monkeys, and humans (71). Pharmacokinetic studies have indicated that 84% and 90% of a single oral dose of 10 mg BVDU/kg is absorbed in monkeys and rats, respectively (32). The fact that effective blood drug levels can be sustained for several hours after oral administration of BVDU (70,71) suggests that oral administration may be the route of choice for the application of BVDU in the systemic treatment of HSV and VZV infections.

EFFICACY IN HUMANS

Prompted by the efficacy of BVDU in animal model systems and its apparent freedom of toxicity when applied at therapeutically effective doses, initial clinical studies have been undertaken to assess the usefulness of BVDU in the topical and systemic (oral) treatment of HSV_1 and VZV infections in humans (Table 4). The drug proved highly efficacious in the topical treatment (0.1% eye drops) of patients with either dendritic corneal ulcers, geographic corneal ulcers or stromal keratitis (72,73). Most of these patients had first been treated with topical IDU or ara-A (adenine arabinoside), albeit unsuccessfully, before BVDU treatment was started. They all responded promptly to BVDU therapy. Following BVDU treatment, corticosteroid medication could be ceased in several patients with deep stromal keratitis who had become corticosteroid-dependent for a long time (72). No toxic side effects, whether local or systemic, were observed in any of the patients treated with BVDU.

Table 4. Clinical conditions in which BVDU has shown efficacy

Virus	Disease	Administration	Dosage	References
HSV_1	Keratitis	Topical (eye drops)	0.1%	(72,73)
HSV_1	Mucocutaneous herpes	Oral (capsules)	7.5-15 mg/kg/day	(74)
VZV	Ophthalmic zoster	Topical (eye drops)	0.1%	
		Oral (capsules)	7.5-15 mg/kg/day	(73,75,76)
VZV	Varicella-zoster in leukemic children	Oral (capsules)	7.5-15 mg/kg/day	(77)
VZV	Localized or disseminated zoster in cancer patients	Oral (capsules)	7.5-15 mg/kg/day	(71,78)

When administered orally at 7.5-15 mg/kg/day to cancer patients who developed a severe mucocutaneous herpes simplex infection (74), or (localized or disseminated) herpes zoster infection (71,78) as the consequence of an intensive anticancer chemotherapy, BVDU caused a rapid healing of the herpes lesions: no new lesions formed and existing lesions regressed within the first few days after the start of BVDU treatment, and in no case it appeared necessary to treat for longer than 5 days. Similar promising results have been obtained with combined (oral or topical) BVDU treatment of patients suffering from ophthalmic zoster (73,75,76). Equally successful are the results that have been noted following oral BVDU treatment of leukemic children with a severe varicella or zoster infection (77) : they all recovered rapidly from their varicella-zoster episode. Neither adults nor children showed any evidence of drug toxicity for bone marrow, liver, kidney or any other organ.

The encouraging results that have so far been obtained with BVDU in the topical and systemic treatment of HSV_1 and VZV infections in humans only concern uncontrolled clinical trials. Obviously, these studies should be extended to double-blind controlled trials before the therapeutic efficacy of BVDU could be fully assessed.

The clinical utility of BVDU, as well as that of ACV and FIAC, is hampered by two limitations. These are the emergence of drug-resistant virus strains (79,80) and the failure to eradicate the virus during its latent phase (81). It should be emphasized, however, that, while the virus may become resistant to the drug, clinical improvement may still ensue as the consequence of a diminished pathogenicity of the mutated virus. That BVDU (like ACV and FIAC) is unable to eradicate a latent virus infection would not be surprising since the two enzymes (dThd kinase and DNA polymerase) interacting with BVDU are not expressed during this latent infection. However, if given early enough during the acute manifestation of the primary or recurrent infection, BVDU may reduce the number of neurones that become colonized (or re-colonized) by the virus, and thereby lessen the frequency and severity of further recurrences.

REFERENCES

1. De Clercq, E. Acta microbiol. Acad. Sci. hung. 28:289-306, 1981.
2. De Clercq, E. Trends in Pharmacological Sciences, in press, 1982.
3. De Clercq, E. Biochem. J., in press, 1982.
4. De Clercq, E., Whitley, R.J., Helgstrand, E., Keeney, R.E., Crumpacker, C., Young, C.W., Kern, E.R. and Galasso, G.J. In: Current Chemotherapy and Immunotherapy, American Society for Microbiology, Washington, D.C., in press, 1982.
5. De Clercq, E., Descamps, J., De Somer, P., Barr, P.J., Jones, A.S. and Walker, R.T. Proc. Natl. Acad. Sci. USA 76:2947-2951, 1979.
6. De Clercq, E., Descamps, J., Verhelst, G., Walker, R.T., Jones, A.S., Torrence, P.F. and Shugar, D. J. Infect. Dis. 141:563-574, 1980.
7. De Clercq, E., Descamps, J., Ogata, M. and Shigeta, S. Antimicrob. Agents Chemother. 21:33-38, 1982.
8. Reefschläger, J., Bärwolff, D., Engelmann, P., Langen, P. and Rosenthal, H.A. Antiviral Res., in press, 1982.
9. Weinmaster, G.A., Misra, V., McGuire, R., Babiuk, L.A. and De Clercq, E. Virology 118:191-201, 1982.
10. Soike, K.F. Personal communication, 1981.
11. Honess, R.W., O'Hare, P. and Young, D. J. gen. Virol. 58:237-249, 1982.
12. Wang, X., Lescott, T., De Clercq, E. and Kelly, D.C. Submitted for publication, 1982.
13. De Clercq, E. Antimicrob. Agents Chemother., in press, 1982.
14. Mayo, D.R. J. Clin. Microbiol. 15:733-736, 1982.
15. Wigdahl, B.L., Isom, H.C., De Clercq, E. and Rapp, F. Virology 116:468-479, 1982.
16. Field, H., McMillan, A. and Darby, G. J. Infect. Dis. 143:281-285, 1981.
17. Veerisetty, V. and Gentry, G.A. Virology 114:576-579, 1981.
18. De Clercq, E., Descamps, J., Maudgal, P.C., Missotten, L., Leyten, R., Verhelst, G., Jones, A.S., Walker, R.T., Busson, R., Vanderhaeghe, H. and De Somer, P. In: Developments in Antiviral Therapy (Eds. L.H. Collier and J. Oxford), Academic Press, London, New York, Toronto, Sidney, San Francisco, 1980, pp 21-42.
19. Park, N.-H., Pavan-Langston, D. and De Clercq, E. Amer. J. Med., in press, 1982.
20. Pavan-Langston, D., Park, N.-H., De Clercq, E., Albert, D.M., Lin, T.S. and Prusoff, W.H. Submitted for publication, 1982.

21. Dhariwal, M.A.H. M.Sc. Thesis, University of Birmingham, England, 1980.
22. De Clercq, E., Balzarini, J., Torrence, P.F., Mertes, M.P., Schmidt, C.L., Shugar, D., Barr, P.J., Jones, A.S., Verhelst, G. and Walker, R.T. Molec. Pharmacol. 19:321-330, 1981.
23. Balzarini, J., De Clercq, E., Torrence, P.F., Mertes, M.P., Park, J.S., Schmidt, C.L., Shugar, D., Barr, P.J., Jones, A.S., Verhelst, G. and Walker, R.T. Biochem. Pharmacol. 31:1089-1095, 1982.
24. Richold, M. and Jones, E. Personal communication, 1981.
25. Jelinek, R. and De Clercq, E. In preparation, 1982.
26. Marmer, D.J., Steele, R.W. and De Clercq, E. In: Current Chemotherapy and Immunotherapy, American Society for Microbiology, Washington, D.C., in press, 1982.
27. Cassiman,J.J., De Clercq, E., Jones, A.S., Walker, R.T. and Van den Berghe, H. Brit. Med. J. 283:817-818, 1981.
28. De Clercq, E., Heremans, H., Descamps, J., Verhelst, G., De Ley, M. and Billiau, A. Molec. Pharmacol. 19:122-129, 1981.
29. Cheng, Y.-C., Dutschman, G., De Clercq, E., Jones, A.S., Rahim, S.G., Verhelst, G. and Walker, R.T. Molec. Pharmacol. 20:230-233, 1981.
30. Descamps, J. and De Clercq, E. J. Biol. Chem. 256:5973-5976, 1981.
31. Fyfe, J.A. Molec. Pharmacol. 21:000-000, 1982.
32. Sim, I.S. Personal communication, 1981.
33. Allaudeen, H.S., Kozarich, J.W., Bertino, J.R. and De Clercq, E. Proc. Natl. Acad. Sci. USA 78:2698-2702, 1981.
34. De Clercq, E. and Descamps, J. In: Herpetische Augenerkrankungen (Ed. R. Sundmacher), J.F. Bergmann Verlag, München, 1981, pp 329-338.
35. De Clercq, E. and Fukui, T. Unpublished data, 1981.
36. Larsson, A. and Öberg, B. Antimicrob. Agents Chemother. 19:927-929, 1981.
37. Sagi, J., Szabolcs, A., Szemzo, A. and Ötvös, L. Nucleic Acids Res. 9:6985-6994, 1981.
38. Allaudeen, H.S., Chen, M.S., Lee, J.J., De Clercq, E. and Prusoff, W.H. J. Biol. Chem. 257:603-606, 1982.
39. Ruth, J.L. and Cheng, Y.-C. Molec. Pharmacol. 20:415-422, 1981.
40. Wohlrab, F., Mayo, D.R., Hsiung, G.D. and Francke, B. In: International Workshop on Herpesviruses, held in Bologna (Italy) on 27-31 July 1981 (Eds. A.S. Kaplan, M. La Placa, F. Rapp and B. Roizman), Esculapio Publ. Co., Bologna, 1981, p 181.
41. Coen, D.M. and Schaffer, P.A. Proc. Natl. Acad. Sci. USA 77:2265-2269, 1980.
42. Schnipper, L.E. and Crumpacker, C.S. Proc. Natl. Acad. Sci. USA 77:2270-2273, 1980.
43. Crumpacker, C. and Kowalsky, P. In: International Workshop on Herpesviruses, held in Bologna (Italy) on 27-31 July 1981 (Eds. A.S. Kaplan, M. La Placa, F. Rapp and B. Roizman), Esculapio Publ. Co., Bologna, 1981, p 50.
44. Jones, A.S., Rahim, S.G., Walker, R.T. and De Clercq, E. J. Med. Chem. 24:759-760, 1981.
45. De Clercq, E., Descamps, J., Verhelst, G., Jones, A.S. and Walker, R.T. In: Current Chemotherapy and Infectious Disease, American Society of Microbiology, Washington, D.C., 1980, pp 1372-1374.
46. De Clercq, E., Verhelst, G., Descamps, J. and Bergstrom, D.E. Acta microbiol. Acad. Sci. hung. 28:307-312, 1981.
47. De Clercq, E. Meth. and Find. Exptl. Clin. Pharmacol. 2:253-267, 1980.
48. De Clercq, E., Balzarini, J., Descamps, J., Huang, G.-F., Torrence, P.F., Bergstrom, D.E., Jones, A.S., Serafinowski, P., Verhelst, G. and Walker, R.T. Molec. Pharmacol. 21:217-223, 1982.
49. Busson, R., Colla, L., Vanderhaeghe, H. and De Clercq, E. Nucleic Acids Res. Symp. Series No 9:49-52, 1981.

50. De Clercq, E., Busson, R., Colla, L., Descamps, J., Balzarini, J. and Vanderhaeghe, H. In: Current Chemotherapy and Immunotherapy, American Society for Microbiology, Washington, D.C., in press, 1982.
51. Machida, H., Sakata, S., Kuninaka, A. and Yoshino, H. Antimicrob. Agents Chemother. 20:47-52, 1981.
52. Stening, G., Gotthammar, B., Larsson, A., Alenius, S., Johansson, N.G. and Öberg, B. Antiviral Res. 1:213-223, 1981.
53. De Clercq, E., Descamps, J., Fukui, T. and Allaudeen, H.S. Submitted for publication, 1982.
54. Machida, H., Kuninaka, A. and Yoshino, H. Antimicrob. Agents Chemother. 21: 358-361, 1982.
55. Descamps, J., Sehgal, R.K., De Clercq, E. and Allaudeen, H.S. Submitted for publication, 1982.
56. Cheng, Y.-C., Dutschman, G., Fox, J.J., Watanabe, K.A. and Machida, H. Antimicrob. Agents Chemother. 20:420-423, 1981.
57. Descamps, J., De Clercq, E., Barr, P.J., Jones, A.S., Walker, R.T., Torrence, P.F. and Shugar, D. Antimicrob. Agents Chemother. 16:680-682, 1979.
58. De Clercq, E., Zhang, Z.-X., Descamps, J. and Huygen, K. J. Infect. Dis. 143:846-852, 1981.
59. De Clercq, E., Zhang, Z.-X. J. Infect. Dis. 145:130, 1982.
60. Park, N.-H., Pavan-Langston, D., Boisjoly, H.M. and De Clercq, E. J. Infect Dis., in press, 1982.
61. Field, H.J. and De Clercq, E. J. gen. Virol. 56:259-265, 1981.
62. De Clercq, E., Zhang, Z.-X. and Sim, I.S. Submitted for publication, 1982.
63. Maudgal, P.C., De Clercq, E., Descamps, J. and Missotten, L. Bull. Soc. belge Ophtal. 186:109-118, 1979.
64. Maudgal, P.C., De Clercq, E., Descamps, J., Missotten, L., De Somer, P., Busson, R., Vanderhaeghe, H., Verhelst, G., Walker, R.T. and Jones, A.S. Antimicrob. Agents Chemother. 17:8-12, 1980.
65. Hettinger, M.E., Pavan-Langston, D., Park, N.-H., Albert, D.M., De Clercq, E. and Lin, T.-S. Arch. Ophthalmol. 99:1618-1621, 1981.
66. Maudgal, P.C., De Clercq, E., Descamps, J., Missotten, L., and Wijnhoven, J. Arch. Ophthalmol. 100:653-656, 1982.
67. Maudgal, P.C., Uyttebroeck, W., De Clercq, E. and Missotten, L. Arch. Ophthalmol., in press, 1982.
68. Soike, K.F., Gibson, S. and Gerone, P. J. Antiviral Res. 1:325-337, 1981.
69. Biront, P. and De Clercq, E. Unpublished data, 1981.
70. De Clercq, E., Descamps, J., De Somer, P., Barr, P.J., Jones, A.S. and Walker, R.T. Antimicrob. Agents Chemother. 16:234-236, 1979.
71. De Clercq, E., Degreef, H., Wildiers, J., De Jonge, G., Drochmans, A., Descamps, J. and De Somer, P. Brit. Med. J. 281:1178, 1980.
72. Maudgal, P.C., Missotten, L., De Clercq, E., Descamps, J. and De Meuter, E. Albrecht von Graefes Arch. Klin. Ophthalmol. 216:261-268, 1981.
73. Maudgal, P.C., De Clercq, E. and Missotten, L. Unpublished data, 1982.
74. Tricot, G., Van Hoof, A., Verwilghen, R. and De Clercq, E. Unpublished data, 1982.
75. Maudgal, P.C., Missotten, L., De Clercq, E. and Descamps, J. Bull. Soc. belge Ophtal. 190:71-86, 1980.
76. Maudgal, P.C., Dralands, L., Lamberts, L., De Clercq, E., Descamps, J. and Missotten, L. Bull. Soc. belge Ophtal. 193:49-56, 1981.
77. Benoit, Y., Delbeke, M.-J. and De Clercq, E. Unpublished data, 1982.
78. Wildiers, J., Drochmans, A. and De Clercq, E. Unpublished data, 1982.
79. Crumpacker, C.S., Schnipper, L.E., Marlowe, S.I. Kowalsky, P.N., Hershey, B.J. and Levin, M.J. New Engl. J. Med. 306:343-346, 1982.

80. Burns, W.H., Saral, R., Santos, G.W., Laskin, O.L., Lietman, P.S., McLaren, C. and Barry, D.W. Lancet i:421-423, 1982.
81. Field, H.J. and Wildy, P. Brit. Med. J. 282:1821-1822, 1981.

7

2'-FLUORO-5-IODO-1-β-D-ARABINOFURANOSYLCYTOSINE [FIAC]: SYNTHESIS AND MODE OF ANTI-HERPESVIRUS ACTIVITY

Carlos Lopez, Ting-Chao Chou, Kyoichi A. Watanabe and Jack J. Fox.
Sloan-Kettering Institute for Cancer Research, Memorial Sloan-Kettering Cancer Center, New York, N.Y. 10021, U.S.A.

SUMMARY

1-(2'-Deoxy-2'-fluoro-β-D-arabinofuranosyl)-5-iodocytosine [FIAC or 2'-fluoro-5-iodo-ara-C] is a potent and selective new anti-herpesvirus drug. The anti-herpes simplex virus type 1 (HSV-1) activity depends, at least in part, on the phosphorylation of FIAC by the virus-specified pyrimidine nucleoside kinase. FIAC is phosphorylated poorly by cellular kinases which probably accounts for its low level of toxicity. FIAC is incorporated preferentially into viral DNA but only short length chains appear to be generated. FIAC exhibits selective activity against cytomegalovirus (CMV) even though this virus does not express a virus-specified nucleoside kinase for its replication. Comparative anti-viral data of FIAC with acycloguanosine, ara-A, ara-C and IUDR are given.

INTRODUCTION

Human herpes viruses [herpes simplex virus - types 1 and 2 (HSV-1, -2), herpes zoster virus (HZV), cytomegalovirus (CMV) and Epstein-Barr virus (EBV)] are important pathogens (1). Members of this group of viruses are especially problematic for individuals with primary or secondary immunodeficiencies. Furthermore, evidence has been accumulated which indicates that three of these viruses may cause cancer in man (2).

Replication of herpesviruses in permissive cells results in the expression of several virus-specified enzymes (3). Although these enzymes function to catalyze reactions similar to those mediated by cellular enzymes, the viral-specified enzymes are clearly different in molecular weight, charge, and substrate specificity. Because some of these viral eyzymes have much less stringent requirements with respect to the substrates which they will accept, they have been widely discussed as targets for selective antiviral chemotherapy (4). The enzymes which have received the most attention are the pyrimidine nucleoside kinase (specified by HSV-1, HSV-2 and HZV) and the DNA polymerase (specified by all herpesviruses) (5-10).

2'-Fluoro-5-iodo-ara-C (FIAC) was synthesized as one of a series of 2'-deoxy-2'-fluoro-substituted arabinosyl-cytosines and -uracils (11) (see Fig.1) and was found to have potent anti-HSV-1, HSV-2 and HZV activity with minimal cytotoxicity for replicating (uninfected) cells (12). In addition, FIAC was found to have potent anti-CMV activity (13). FIAC was selected for further studies because of its potent activity and because it had the greatest in vitro therapeutic index (ID_{50}/ED_{50}) (12). Studies in mice inoculated with HSV-1 showed that FIAC was effective in protecting them from lethal infections (14). More recent studies have shown that FIAC is more effective than arabinosyl adenine in the treatment of herpes zoster infections in immunosuppressed cancer patients (15).

In this report, we present the method of synthesis of FIAC as well as the results of studies designed to reveal the mode of action against HSV-1.

MATERIALS AND METHODS

Synthesis of FIAC

Based upon reports from our laboratory (16), and others, on pyrimidine nucleoside transformations, it is clear that the direct introduction of a fluoro function in the 2'-"up" (arabino) position would be most difficult, if not impossible, due to neighboring group participation by the 2-carbonyl function of the pyrimidine moiety in any displacement reactions on the sugar portion. We therefore developed a 9-step synthesis (17) of the appropriately protected 2'-deoxy-2'-fluoro-arabinofuranosyl bromide (2, Fig.1) from readily available di-isopropylidene-D-allofuranose (1). The synthesis of FIAC was then accomplished (18) by condensation of this key sugar intermediate (2) with trimethylsilylated 5-iodocytosine to afford the 3',5'-acylated FIAC (3e) which was deprotected to FIAC (4e) with methanolic ammonia. Condensation of 2 with other trimethylated 5-substituted -uracils or -cytosines afforded a host of 2'-fluoro-ara-cytosine nucleosides [4a → f] and their corresponding uracil nucleosides [5a → f] (11,18). 2-^{14}C labelled FIAC was also synthesized by condensation of 2-^{14}C 5-iodocytosine with 2 followed by de-acylation (19).

FIGURE 1. Synthesis of FIAC and related compounds

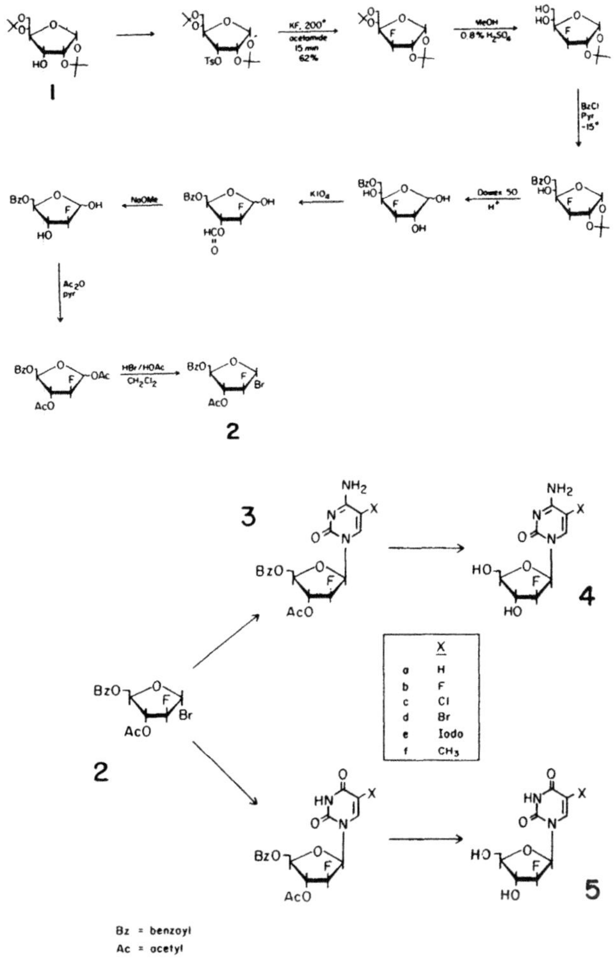

Antiviral Activity: Most studies were carried out with HSV-1 (Strain 2931) (12), isolated from a patient with a cold sore. Other strains of HSV-1 included HSV (1023), HSV (1061), B 2006 and HSV (F), which were obtained from Dr. Bernard Roizman, University of Chicago. CMV strains AD 169 and Davis were obtained from the American Type Culture Collection, Rockville, Maryland. The strains of HSV-1 were propagated by passaging at low multiplicity of infection (MOI) in Vero cell monolayers. Quantitation of virus preparation was by the plaque assay on Vero cells under methylcellulose overlays (12). CMV was grown in human foreskin fibroblast monolayers and assayed on these cells under a methylcellulose overlay.

Antiviral activity was determined using the plaque reduction assay (12). Briefly, monolayers of permissive cells were prepared in wells of Linbro FB-16-24-TC plates (Linbro Chemical Co., Inc., New Haven, Conn.) in 1 ml medium. Monolayers were inoculated with 20-30 plaque forming units (PFU) of virus and, after a 2-hour absorption period, were overlaid with maintenance medium containing 1% methylcellulose. Antiviral drugs were incorporated into the maintenance medium at several different concentrations. When the plaques were fully developed, they were counted and the concentration of drug required to reduce plaque formation by 50% was calculated (12).

Kinase assays were done using the disc assay method as described earlier (20). All samples were incubated with and without the cytidine deaminase inhibitor tetrahydrouridine (2 mg/assay), for evaluation of the effect of deamination upon the activities of the kinases.

Suppression of viral DNA synthesis was evaluated by the method of Drach and Shipman (21). Briefly, monolayers of Vero cells were mock infected or infected with HSV-1 at one plaque per cell and were treated or untreated with various concentrations (0.01 μM to 100 μM) of FIAC. Medium containing [H^3]-thymidine was added and the cells were incubated for 12 hours at 37°C. After washing the cells thoroughly, the DNA was precipitated with trichloroacetic acid (TCA). Following isopycnic centrifugation in CsCl gradients, the amount of radioactive label in cellular versus viral DNA was determined.

DNA was also extracted from HSV-1 infected or uninfected Vero cells using a modified phenol procedure (22). Briefly, the cells were homogenized in equal volumes of 88% aqueous phenol and 8% ammonium acetate. The homogenate was centrifuged and the upper aqueous layer was removed and dialyzed to remove excess phenol. DNA was hydrolyzed in 0.3 M KOH at 37°C for 18 hours and dialyzed to remove RNA.

RESULTS

The anti-HSV-1 activity of FIAC was compared to that of acycloguanosine ACG), arabinosylcytosine (Ara-C), iododeoxyuridine (IUDR) and arabinosyladenine (Ara-A) (Table 1). Clearly, FIAC demonstrated the greatest *in vitro* activity while ACG was almost as potent. Table 1 also details the cytotoxicity of each of these drugs for normal replication of Vero cells.

FIAC and ACG were much more active against HSV-1 replication than against normal cell division. Ara-C and Ara-A, on the other hand, demonstrated poor selectivity.

TABLE 1. Anti-HSV-1 Activity and Cytotoxicity of FIAC Compared To That Of Acycloguanosine (ACG), Arabinosyladenine (Ara A), Arabinosyl Cytosine (Ara C) and Iododeoxyuridine (IUDR).

	FIAC	ACG	ARA-C	IUDR	ARA-A
*ED_{50} (μM)	0.01	0.05	0.2	3	25
**ID_{50} (μM)	6.7	42	0.1	ND	60

*ED_{50} = effective dose to reduce HSV-1 plaque formation by 50%.
**ID_{50} = inhibitory dose to reduce normal cell replication by 50%.

Inhibition studies using thymidine (TdR) and deoxycytidine (dCTD) showed that FIAC's antiviral activity could be inhibited by TdR but not by dCTD, while the cytotoxicity could be blocked by dCTD but not by TdR (12). These early results showed that FIAC's antiviral effect was due to a mechanism which was clearly different from the mechanism responsible for the cytotoxicity.

Earlier studies showed that FIAC was not active against a strain of HSV-1 which failed to express the virus-specified thymidine kinase (12). Studies were therefore carried out to determine whether FIAC could be selectively phosphorylated by the viral versus the cellular kinase. Table 2 demonstrates the results of assays in which normal cellular kinases were compared to enzymes extracted from uninfected or HSV-infected Vero cells. FIAC was phosphorylated as much as 9,000 times better by the virus-specified enzyme than by enzymes obtained from normal mouse spleen and thymus. Furthermore, the HSV-1 specified enzyme phosphorylated FIAC 1,200 times better than did the enzyme from dividing Vero cells.

As noted above, FIAC failed to inhibit the replication of HSV-1 (B 2006), a strain of virus which completely lacks the genetic information for the viral nucleoside kinase. We have now evaluated the antiviral activity of FIAC using a series of HSV-1 mutants selected because of charges in the viral kinase or DNA polymerase. Table 3 compares the results with FIAC to those with some other well known anti-herpesvirus drugs. The first mutant, TK^-(B 2006) clearly lacks viral kinase and, consequently, is not inhibited by FIAC, ACG or Ara-T, drugs whose selectivity depends on the fact that they are substrates for the viral but not the cellular kinase. PAA, in contrast, only depends on the viral DNA polymerase for its selectivity and thus demonstrates normal activity against TK^-(B 2006). The second mutant was selected by growing HSV-1 in 100 μm PAA and is thus resistant (PAA^R). Although the PAA^R mutant was no more resistant to Ara-T or FIAC than the wild type virus, this mutant was resistant to ACG, suggesting that a change in the viral DNA polymerase altered selectivity of ACG but not Ara-T or FIAC. The latter suggests that these drugs may have different mechanisms of antiviral activity. Mutant HSV (1061) was resistant to both Ara T-(TK^-) and PAA (PAA^R) but was still very susceptible to FIAC. In fact, this mutant probably has an altered kinase unable to use Ara-T but still able to phosphorylate FIAC.

TABLE 2. Phosphorylation of FIAC by Homogenates of Mouse Tissue and Uninfected and HSV-1-Infected Vero Cells.

Tissue	Total FIAC phosphates (nmoles/ mg/hr)	FIAC phosphates in % of total cmp eluted from HPLC					
		FIACMP	FIAUMP	FIACDP	FIAUDP	FIACTP	FIAUTP
Normal mouse							
Liver	0.12	0.7	1.0	ND*	ND	0.1	ND
Spleen	0.09	0.5	0.9	ND	ND	0.3	T
Thymus	0.09	0.3	0.3	0.2	ND	2.3	0.1
Bone marrow	0.25	0.3	0.2	0.3	ND	0.7	0.1
Vero cell cytosol							
Control	0.07	0.3	ND	ND	ND	T	T
HSV-1-infected	88.86	88.4	ND	ND	ND	ND	ND

*ND = not detectable

The studies of Allaudeen et al., (23) indicated that the triphosphate of FIAC (FIACTP) was selectively utilized by HSV-1 DNA polymerase but not by DNA polymerase α from uninfected cells. These workers enzymatically generated the FIACTP and then showed that it competitively inhibited the incorporation of dCTP into acid precipitable material. When dCTP was omitted from the reaction mixture for DNA synthesis catalyzed by HSV-1 DNA polymerase, addition of FIACTP enhanced the reaction, indicating that FIACTP was an alternate substrate for DNA synthesis. A similar experiment with DNA polymerase α from uninfected cells showed that FIACTP was not used by this enzyme. Thus, the selectivity of FIAC may be due, in part, to the selective utilization of the FIACTP by viral DNA polymerase.

TABLE 3. Comparative Antiviral Activity of FIAC, Acycloguanosine (ACG), Arabinosyl Thymine (ARA-T) and Phosphonoacetic Acid (PAA).

	Mutants (ED_{50} μM)			
DRUGS	TK^- (B2006)	PAA^R	1061	WILD
FIAC	>100	0.032	0.063	0.01
ACG	>100	7.3	8.5	0.05
ARA-T	>100	2.8	34	1.7
PAA	25	>100	>100	25.6

Utilizing the procedure of Drach and Shipman (21), we found that synthesis of viral DNA, as gauged by [^{3}H]-thymidine incorporation into acid precipitable counts, was preferentially suppressed in HSV-1 infected Vero cells. Thus, as little as 0.01 μM of FIAC inhibited viral DNA synthesis by 90% while, in the same cells, synthesis of cellular DNA was inhibited only by concentrations greater than 1 μM. The therapeutic index (calculated as concentration of FIAC required to yield 50% reduction of cellular DNA synthesis divided by the concentration of FIAC required to produce 50% reduction of HSV-1 DNA synthesis) was calculated to be about 500.

The studies of Chou, et al., (24) showed that [2-^{14}C]FIAC was preferentially incorporated into acid precipitable material in HSV-1-infected cells much better than in uninfected Vero cells. The incorporation into infected cell material could be competitively inhibited by TdR but not by dCTD, while the incorporation into uninfected cell material could be inhibited by dCTD but not by TdR.

Of special interest has been the recent finding that incorporation of [2-^{14}C]FIAC into viral material depended greatly upon whether DNA was extracted with phenol or was precipitated out with acid. Table 4 shows a typical result of an experiment comparing trichloroacetic acid (TCA) precipitation with phenol extraction. Phenol extraction of virus-infected cells resulted in a 70-fold increase in incorporation of radiolabelled FIAC. A comparison of uninfected cells extracted by these procedures revealed only a miner difference. These results and more recent studies suggest that in HSV-1-infected cells, FIAC is incorporated into viral DNA but that the pieces of DNA are very small - so small that they are not precipitated by acid. The incorporated FIAC could terminate chain elongation or make the chain fragile. The end result is that full pieces of 1 x 10^6 MW DNA are not made and infectious virus is thus not packaged and generated by the cells.

TABLE 4. Incorporation of [2-^{14}C]FIAC into DNA Fractions.[a]

Method	Vero Cells	HSV-Vero Cells
	(pmole/10^6 cells/45 min)	
TCA Procedure	1.0	1.8
Phenol Procedure	2.5	127.3

a [2-^{14}C]FIAC at 10 μM, 37°C

Unlike the other herpesviruses, CMV does not encode its own pyrimidine nucleoside kinase. Rather, this virus induces an elevation in the expression of cellular enzyme (25). Nevertheless, we found (13) that CMV is very sensitive to the antiviral activity of FIAC. The ED_{50} for strain AD 169 was 0.05 μM and for the Davis strain was 0.1 μM. Both of these drug concentrations were well below the toxic dose for the uninfected tissue culture cells used in this plaque reduction assay (ID_{50} = 5 μM).

DISCUSSION

FIAC is a potent new anti-HSV drug which demonstrates minimal cytotoxicity. It has demonstrated great activity against the herpesviruses which carry the genetic information for their own pyrimidine kinase (HSV-1, HSV-2 and HZV) as well as against CMV, a herpesvirus which does not have its own pyrimidine kinase and induces high levels of cellular enzyme (25). In recent clinical studies, FIAC was found to be significantly better than Ara-A in a double blind randomized trial against HZV infections in immunosuppressed cancer patients (15). The side effects of FIAC were few and minor. FIAC is, therefore, an excellent candidate for further clinical development as an anti-herpesvirus drug.

The studies of its mode of action indicate that FIAC's potent antiviral activity, yet minimal cytotoxicity, is dependent at least in part on the fact that it is preferentially phosphorylated by the viral pyrimidine kinase, but very poorly by the cellular enzyme. Thus, FIAC is readily converted to the 5'-phosphate in HSV-1 infected cells but poorly in uninfected cells. In these experiments, very little of the FIAC was found to be converted to the di- and tri-phosphates (16). This may be because very little is converted to these intermediates or because they are rapidly incorporated into DNA. If the latter is true, the enzymes responsible have not been determined to be virus encoded or not.

The studies of Allaudeen et al., (20) indicate that FIACTP is utilized by viral DNA polymerase in the place of dCTP. These studies demonstrated a preference by viral enzyme for FIACTP over the natural substrates. Since these investigators utilized acid precipitation rather than phenol extraction to determine incorporation, they may not have detected incorporation into small molecular weight DNA (19). In fact, the affinity of viral DNA polymerase for FIACTP may be even greater.

Once incorporated into viral DNA, the pieces do not appear to be able to form the normal 1×10^6 MW pieces which make up the viral genome. Since only large pieces can be encapsidated, only empty particles can be formed.

The antiviral activity against HSV-1 appears to depend on phosphorylation of FIAC by the viral kinase. However, CMV is a herpesvirus which causes an increase in cellular kinase but does not specify its own enzyme. Yet, CMV is still sensitive to FIAC's antiviral activity (ED_{50} = 0.1 - 0.5 μM). FIAC is far more active against CMV than against mutant TK^- (B 2006) which also lacks viral kinase activity and must depend on the cellular enzyme. This is an enigma upon which, it is hoped, future studies will shed light.

ACKNOWLEDGMENT

These investigations were supported in part by funds from the National Cancer Institute, U.S. Department of Health and Human Services, Grants No. CA-08748, 18856 and 18601. C. Lopez is a recipient of a Faculty Research Award (No. 193) from the American Cancer Society.

REFERENCES

1. The Herpesviruses (A. Kaplan, Ed.), Academic Press, New York, NY 1973.
2. Oncogenic Herpesviruses (F. Rapp, Ed.), CRC Press, Boca Raton, Florida 1980.
3. Kit, S., Viral-associated and induced enzymes. Pharmac. Ther., 4: 501-565, 1979.
4. Prusoff, W. H. and Ward, D. C., Nucleoside analogs with antiviral activity. Biochem. Pharmacol., 25:1233-1239, 1976.
5. Kit, S., Dubbs, D. R. and Anken, M. Altered properties of thymidine kinase after infection of mouse fibroblast cells with herpes simplex virus. J. Virol., 1:238-240, 1967.
6. Ogino, J., Otsuka, T. and Takahashi, M. Induction of deoxypyrimidine kinase activity in human embryonic lung cells infected with varicella zoster virus. J. Virol., 21:1232-1235, 1977.
7. Aron, G. M., Purifoy, J. M. and Schaffer, P. A. DNA synthesis and DNA polymerase activity of herpes simplex virus type 1 temperature-sensitive mutants. J. Virol., 16:498-507, 1975.
8. Huang, E. S. Human cytomegalovirus. IV. Virus-induced DNA polymerase. J. Virol., 16:298-310, 1975.
9. Mar, E. C., Huang, V. C. and Huang, E. S. Purification and characterization of varicella zoster virus induced DNA polymerase. J. Virol., 26:249-256, 1978.
10. Ooka, T., Lenoir G. and Daillie, J. Characterization of an Epstein-Barr virus induced DNA polymerase. J. Virol., 29:1-10, 1979.
11. Watanabe, K. A., Reichman, U., Hirota, K., Lopez, C. and Fox, J. J. Nucleosides. 110. Synthesis and antiherpes virus activity of some 2'-fluoro-2'-deoxyarabinofuranosyl-pyrimidine nucleosides. J. Med. Chem., 22:21-24, 1979.
12. Lopez, C., Watanabe, K. A. and Fox, J. J. 2'-Fluoro-5-iodo-aracytosine, a potent and selective anti-herpesvirus agent. Antimicrob. Agents Chemother., 17:803-806, 1980.
13. Lopez, C., Livelli, T., Watanabe, K. A., Reichman, U. and Fox, J. J. 2'-Fluoro-5-iodo-aracytosine: A potent anti-herpesvirus nucleoside with minimal toxicity to normal cells. Proc. Am. Assoc. Cancer Res., 20:183, 1979.
14. Fox, J. J., Lopez, C. and Watanabe, K. A. 2'-Fluoro-arabinosyl pyrimidine nucleosides: Chemistry, antiviral and anticancer activities, in Proceedings of the VIIth International Symposium on Medicinal Chemistry, Pergamon Press, 1980.
15. Leyland-Jones, B., Donnelly, H., Myskowski, P., Donner, A., Groshen, S. and Fox, J. J. FIAC, a potent antiviral agent: Efficacy in Immuno-suppressed patients with Varicella Zoster infection, (submitted for publication).
16. Fox, J. J. Pyrimidine nucleoside transformations via anhydronucleosides. Pure Applied Chem., 18:223-255, 1969, and references therein.
17. Reichman, U., Watanabe, K. A. and Fox, J. J. A practical synthesis of 2-deoxy-2-fluoro-D-arabinofuranose derivatives. Carbohydr. Res., 42:233-240, 1975.
18. Watanabe, K. A., Su, T-L., Klein, R. S., Chu, C. K., Matsuda, A., Chun, M. W., Lopez, C. and Fox, J. J. Nucleosides. 123. Synthesis of antiviral nucleosides: 5-Substituted 1-(2-deoxy-2-halogeno-β-D-arabinofuranosyl)-cytosines and -uracils. Some structure-activity relationships. J. Med. Chem., 26, 1983, In Press.

19. Chou, T-C., Feinberg, A., Grant, A. J., Vidal, P., Reichman, U., Watanabe, K. A., Fox, J. J. and Philips, F. S. Pharmacological Disposition and Metabolic fate of 2'-fluoro-5-iodo-arabinofuranosyl-cytosine in mice and rats. Cancer Res., 41:3336-3342, 1981.
20. Kreis, W., Damin, L., Colacino, J. and Lopez, C. In vitro metabolism of 1-β-D-arabinofuranosylcytosine and 1-β-2'-fluoro-arabino-5-iodocytosine in normal and herpes simplex type 1 virus infected cells. Biochem. Pharmacol., 31:767-773, 1982.
21. Drach, J. C. and Shipman, C. The selective inhibition of viral DNA synthesis by chemotherapeutic agents: an indication of clinical usefulness? Ann. N.Y. Acad. Sci., 284:396-409, 1977.
22. Grant, A. J., Feinberg, A., Chou, T-C., Watanabe, K. A., Fox, J. J. and Philips, F. S. Incorporation of metabolites of 2'-fluoro-5-iodo-1-β-D-arabinofuranosylcytosine into deoxyribonucleic acid of neoplastic and normal mammalian tissues. Biochem. Pharmacol., 31:1103-1108, 1982.
23. Allaudeen, H. S., Descamps, J., Sehgal, R. K. and Fox, J. J. Selective inhibition of DNA replication in herpes simplex virus infected cells by 1-(2'-deoxy-2'-fluoro-β-D-arabinofuranosyl)-5-iodocytosine. J. Biol. Chem., 257:11879-11882, 1982.
24. Chou, T-C., Lopez, C., Colacino, J. and Philips, F. S. Metabolism and effects of 2'-fluoro-5-iodo-1-β-D-arabinofuranosylcytosine (FIAC) in human leukemia and virus infected Vero cells. Proc. Am. Assoc. Cancer Res., 21:269, 1980.
25. Estes, J. E. and Huang, E. S. Stimulation of cellular thymidine kinases by human cytomegalovirus. J. Virol., 24:13-21, 1977.

8

HERPESVIRUS-SPECIFIED DNA POLYMERASE AS THE TARGET FOR DEVELOPING SELECTIVE ANTIHERPESVIRUS AGENTS

YUNG-CHI CHENG, DAVID DERSE, KARL FRANK & KEN BASTOW
Department of Pharmacology, University of North Carolina at Chapel Hill, Chapel Hill, NC

There are five types of herpesvirus associated with a variety of human diseases. These are herpes simplex virus type 1 (HSV-1) and type 2 (HSV-2), varicella zoster virus (VZV), cytomegalovirus (CMV) and Epstein-Barr virus (EBV). A biochemical characteristic shared by these viruses is the induction of a virus-specified DNA polymerase in infected cells. It is known that this enzyme must be functional for productive infection by HSV-1 and HSV-2 and it is likely that the corresponding DNA polymerases specified by CMV, EBV and VZV are also essential for virus growth. Furthermore, these virus-specified DNA polymerases share some common kinetic properties which differ from DNA polymerases α and β. These include sensitivity to pyrophosphate analogs and behavior toward salt when activity is measured in vitro. These observations suggest that compounds could be developed which selectively inhibit DNA polymerases induced by all types of herpesviruses without affecting cellular DNA polymerases. Such compounds could potentially have a broad spectrum of antiherpetic effect with limited toxicity to the host. Many laboratories, including ours, have been searching for such compounds. Our approach was initiated by comparing the properties of the HSV-1 and HSV-2 DNA polymerases to those of human α and β DNA polymerases, with the aim of finding unique characteristics of the virus replicative enzymes.

Strategies for the development of antiherpes compounds based on these studies could be devised and will be discussed in the first part of this article. One major drawback to the approach of developing selective antivirus agents targeted to virus-specified enzymes is the development of virus variants which induce enzymes with altered kinetic properties. The development of viruses with decreased sensitivity to phosphonoformic acid (PFA), a compound targeted against the virus DNA polymerase, has been reported previously. We recently isolated a group of PFA-resistant variants of HSV-1. The development of these variants during virus growth, their sensitivities to other antiviral agents that exert their action by inhibiting virus DNA synthesis, and the properties of highly-purified variant DNA polymerases, is discussed in the second

Y. Becker (ed.), Antiviral Drugs and Interferon: The Molecular Basis of Their Activity.

part of this article. The final part examines the possible development of agents with a broad spectrum of activity against all human herpesviruses. The DNA polymerases induced by these viruses as a group appear to have a higher binding affinity for deoxynucleoside triphosphates compared to host cell DNA polymerases. If a deoxynucleoside analog were equally phosphorylated to its triphosphate in both virus-infected and uninfected cells it should, a priori, have a greater effect on virus growth than on normal cell proliferation. Also, the rapid rate of virus DNA synthesis and mode of virus growth (i.e. hundreds of progeny templates accumulated per infected cell), could be an important biological factor contributing to a differential effect of such agents. To test this hypothesis we have examined 2'-F 5-iodo arabinosyl cytosine (FIAC), 2'-F arabinosyl thymine (FMAU) and E-5-Bromovinyl deoxyuridine (BVdU) as inhibitors of CMV and EBV. These drugs exhibit cytotoxicity at higher concentrations than those active against HSV owing to their selective phosphorylation by a second enzyme induced in HSV-infected cells, the deoxythymidine kinase. There is no counterpart of this virus-specified enzyme induced in cells infected with either CMV or EBV. Preliminary results obtained in collaboration with Dr. E. Huang and Dr. J. Pagano will be discussed.

Properties of HSV-1 and HSV-2 DNA Polymerases and Strategies for Developing Antiviral Agents Based on those Unique Properties of Viral DNA Polymerase

HSV-1 and HSV-2 induced DNA polymerases were purified from HSV-1 (KOS strain) and HSV-2 (333 strain)-infected human cells utilizing procedures described previously (1). Two polypeptides with molecules weights of 140,000 and 62,000 were copurified and coincided with the peak of polymerase activity. Based on the electrophoretic data and glycerol gradient centrifugation experiments, HSV DNA polymerase was found to have 3' to 5' exonuclease activity (1,2). It is unclear at present whether both polypeptides are required for DNA polymerase and associated exonuclease activity. The mechanism of virus DNA polymerase is shown as follows:

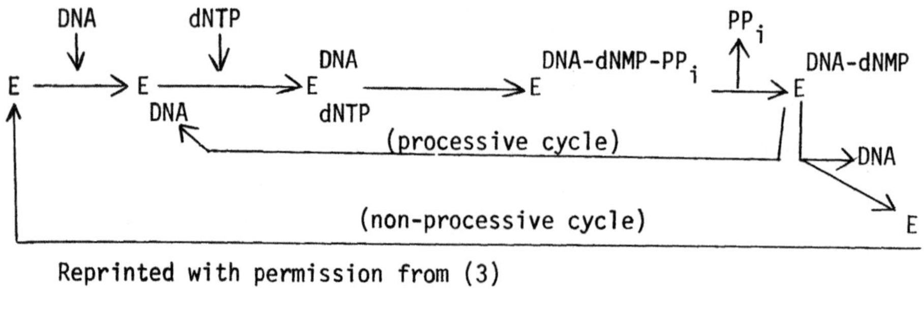

Reprinted with permission from (3)

The processivity of DNA polymerase was found to be 10; in other words, there are an average of 10 nucleotides added before the DNA-enzyme complex is dissociated. The binding of DNA could potentiate the binding of dNTP. The K_M of dNTP is about one order of magnitude less than that for human α and β DNA polymerase (4). This suggested more efficient recruitment of dNTP by virus DNA polymerase. An ionic strength of 0.25 is required for optimal activity of the virus enzyme, whereas the same ionic strength inhibits both α and β DNA polymerases (5). Physiological concentrations of polyamine could replace salt and provide optimal ionic strength requirements of the virus DNA polymerase reaction, suggesting the possible role of polyamine in virus DNA synthesis (1). The stimulation of activity by increased ionic strength or polyamine in the reaction is primarily due to an increase of V_{max}. The associated exonuclease could be inhibited by purine ribonucleoside monophosphate; the purine deoxyribonucleoside monophosphates are less inhibitory than their ribonucleotide counterparts (2,6). Neither human α or β DNA polymerase have associated exonuclease. There were no apparent differences between HSV-1 and HSV-2 DNA polymerases studied. Based on these observations, strategies for the development of antivirus compounds, targeted to the virus DNA polymerase or its associated exonuclease activity, could be devised as outlines in the following discussion.

1. Olignucleotide analog approach

This involves finding DNA that could selectively bind and sequester virus DNA polymerase. Consequently, virus DNA templates could not replicate due to a lack of functional polymerase and would eventually be broken down by DNase, which is also induced by herpes virus in infected cells. Recently, we observed that acyclovir (ACV)-terminated DNA could compete with activated DNA for binding to viral DNA polymerase. The K_I value for virus DNA polymerase is approximately 100 times less than that for human HeLa DNA polymerase (7). This observation suggested the possibility of finding an oligonucleotide analog which exhibits selectivity in binding to and inactivating virus DNA polymerase. The need for transport of such oligonucleotides through cell membranes in order for action to be exerted was a major concern for the potential use of this type of compound. Dr. T. Bardos and his colleagues have synthesized some thiolated polynucleotides. These polynucleotides were found to have cytotoxicity and were transported to the nuclei of cells being exposed (8). Dr. P.T'so and his colleagues have also synthesized some nonionic oligonucleotides which could be taken up by cells (9). Thus the transport of oligonucleotides through cell membranes may not be as critical a problem as had been previously conceived.

2. dNTP analog approach

This is to find an analog of dNTP that is metabolized more readily by virus DNA polymerase than by host DNA polymerase. Based on the apparent differences of K_M values of dNTP between the virus and host enzymes, development of such types of analogs with selectivity should be possible. Indeed, we have observed a number of analogs (Table 1) which, in nucleoside form (10,13), are potent antiherpes agents (7,14). The mode of interaction of these analogs with DNA polymerase could be as follows.

(1) Alternative substrate: The dNTP analog could be incorporated into DNA in place of the naturally occurring dNTP. The incorporation of such analogs could have several consequences with respect to elongation of the growing DNA chain:

a. No effect on the rate of DNA chain elongation (14).
   5-propyl dUTP and E-5-Bromovinyl dUTP (BVdUTP) are in this category.
b. Slowing the rate of DNA chain elongation (6).
   araATP could be incorporated and excised continuously, resulting in a diminished rate of DNA chain elongation. This type of compound could still be in internucleotide linkages, depending on the relative activities of DNA polymerase and exonuclease. E-5-Bromovinyl araUTP and FIACTP to some extent are also in this category.
c. Inhibition of further DNA elongation by termination.
   Acyclovir (ACV) could be incorporated into DNA, which then could not be further elongated. The presence of ACV monopnosphate at the terminus of DNA could not be recognized as a substrate for the DNA polymerase-associated exonuclease. Thus, ACV will stay at the primer-terminus (7).

(2) Inhibitor: This type of compound will not be able to be incorporated into DNA and act as an inhibitor of DNA polymerase. Compounds with a modification at the 5' phosphate group may behave like this.

Table 1. Inhibition constants of analog triphosphates for various types of DNA polymerase.

Compound	Competing Nucleotide	K_I (μM) HeLa alpha	HeLa beta	HSV-1	HSV-2
5-Propyl dUTP	dTTP	19.7	21	0.24	0.38
BVdUTP	dTTP	3.6	6.5	0.07	0.05
FIACTP	dCTP	1.3	5.2	0.03	0.04
ACVTP	dGTP	0.15	no inhibition	0.003	0.003

There are several points to consider regarding the approach of finding dNTP analogs with antivirus selectivity.

(1) dNTPs in general are not transported efficiently through cell membranes in an intact form, whereas nucleosides are. Thus, it will require the cell to have the biochemical machinery to phosphorylate nucleosides to their triphosphate derivatives in order for their action to be realized. Further discussion will be in the third part of this article.

(2) Once dNTP analogs are incorporated into DNA, and depending on how such analogs are incorporated, cells may have repair mechanisms to remove the incorporated analogs. This will result in a decrease in the antiviral effect of such agents. Compounds such as BVdU, 5-Propyl dU, FIAC or FMAU could be incorporated into internucleotide linkage in DNA; thus the repairing mechanism, if any, could not correct the incorporation without drastically changing the structure of DNA. The action of these analogs is irreversible. In contrast, ACV is incorporated into DNA termini. Although such termini are not substrates for removal by exonuclease, endonucleases may remove terminal oligonucleotides containing ACV, allowing reversal of antiviral effects. Indeed, this is what we observed in cell culture (Table 2). Thus, the position at which the analog is incorporated could play an important role in the reversibility of drug action. More studies will be needed to further establish this concept.

Table 2. Reversibility of antivirus action of nucleoside analogues against HSV-1 (KOS) growth in HeLa cells (24 hr).

Drug (μM)	Treatment Period (hrs)	Virus Titer (PFU/ml)	Inhibition
5-propyl dUrd (25)	0 - 24	1.3×10^6	Irreversible
	0 - 7	4.4×10^6	
BVdU (5)	0 - 24	1.5×10^5	Irreversible
	0 - 7	1.2×10^5	
FIAC (5)	0 - 24	1.3×10^5	Irreversible
	0 - 7	2×10^5	
ACV (20)	0 - 24	2.6×10^5	Reversible
	0 - 7	1.3×10^7	
	7 - 24	3.3×10^7	
---	---	6.4×10^7	

3. Pyrophosphate analog approach

This approach was taken by several laboratories and the most promising analogs discovered as antivirus compounds are phosphonoformate (PFA) and phosphonoacetate (PAA) (15,16). Both compounds were found to inhibit herpesvirus DNA polymerase more potently than host DNA polymerase. The action of phosphonoformate and phosphonoacetate on DNA polymerase is reversible, and its mode of action as an antivirus compound is also found to be reversible (17). A more detailed discussion of pyrophosphate binding behavior will be in the second part of this article.

4. Exonuclease inhibitor approach

The physiological role of the DNA polymerase-associated exonuclease is unclear at present. Based on the observation made with respect to E. coli DNA polymerase, it may play a critical role in proofreading of deoxynucleotide incorporation into DNA (18). Furthermore, it was demonstrated that the amount of araA incorporated will depend on the activity of associated exonuclease (6). Inhibition of this exonuclease could enhance the incorporation of araA into internucleotide linkages of DNA. It is conceivable that the inhibition of exonuclease could decrease the fidelity of DNA being replicated, which could affect the yield of infectious virus synthesized in cells. In addition, the inhibition of exonuclease may increase the activity of some antivirus nucleosides such as araA or 5-bromovinyl araA due to the enhancement of their incorporation into DNA. Among the many compounds examined, AMP, GMP, and IMP were found to be the most potent inhibitors of this activity. This suggests that analogs of purine ribonucleosides could be considered as potential antivirus compounds. Furthermore, compounds such as ribavirin, which increases the pool of IMP (19), may exert their antivirus activity to a certain extent through the inhibition of exonuclease. This should be further investigated.

Isolation of Virus Variants with Altered DNA Polymerase and Possible Implications for Effective Antivirus Chemotherapy

The strategies for drug development presented in the preceding section are based on known unique properties of the replicative enzymes of herpesviruses. Any virus which induces a DNA polymerase with altered kinetic properties could express a change in its sensitivity to such agents. PAA and PFA are potent antivirus compounds and variants which exhibit decreased sensitivity have been isolated and characterized by several groups (20,21). However, no investigators have analyzed the kinetic properties of highly purified variant DNA polymerases in detail. One might expect biochemical variation of this virus enzyme to be minimal since its activity is essential for virus growth and therefore its basic catalytic functions must be maintained. We have recently isolated a group of variants which can grow in a single high concentration of PFA and have analyzed the extent of kinetic variation found in the HSV DNA polymerase and also looked at its development during virus growth. It was found that the appearance of variants with decreased sensitivity to PFA was an intrinsic property of virus growth, since independent populations obtained from plaque--purified isolates of HSV-1 were heterogeneous with respect to PFA sensitivity, in spite of low passage. This heterogeneity was apparent from dose response curves obtained from examining PFA sensitivity using a plaque reduction assay and it was noted that the slope of this dose response curve was identical for the variant populations examined. This may indicate that the genetic mechanism responsible for biochemical variation at this enzyme's locus results from the progressive accumulation of spontaneous mutations.

We have compared the potency of several antiviral agents against individual virus in the population by using a specific biological unit of drug activity: that concentration which reduces virus yield by 90% (E.D.$_{90}$ units). Dose-response curves from plaque reduction assays replotted using E.D.$_{90}$ units revealed that compounds such as ACV, which require two virus-induced enzymes to exert their antiviral selectivity, exhibit a shallower dose response compared to PFA. This is expected if biochemical variation occurs independently in both enzyme targets during outgrowth of populations from cloned virus. Unfortunately, we do not know if variants with kinetically altered DNA polymerase occur with such ease during growth of herpes in humans. However, the routine use of combinations of antivirus agents with different modes of action or the highest tolerable dosage of antiviral agents would clearly be the most effective means of preventing this occurrence. The results from our systematic biochemical analysis of purified variant DNA polymerases (22) indicate: 1. Resistance to a given concentration of PFA may arise by different types of alterations in the active center of the virus DNA polymerase, since five variant enzymes were divided into two distinct groups with respect to their Mg^{++} optima and apparent affinities for PFA and PAA. 2. One variant enzyme exhibited a higher dNTP K_M value, lower processivity, and lower ratio of polymerase to exonuclease activity which may have resulted from another active site alteration unrelated to PFA resistance. 3. Two of the five variant enzymes exhibited cross-resistance to β-phenyl-PAA, whereas the other three had the same sensitivity as the parent enzyme. This suggests variants resistant to PFA may still be sensitive to other pyrophosphate analogs. Furthermore, generalizations about patterns of cross-resistance should not be made from the study of single virus isolates. 4. The K_I values of acycloGTP and ddGTP were more closely correlated with K_I values of PFA than to the K_M values of dNTPs. This may suggest these agents share a common underlying inhibitory mechanism, i.e. preventing release or translocation of DNA polymerase along the primer-template. This explanation for the observed resistance to ACV following selection of variants with pyrophosphate analogs is different from a suggestion made previously. 5. Affinities for PFA were 10 to 20 times lower with the variant DNA polymerases, while affinities for pyrophosphate (PPi) were only 2.5 to 3-fold lower, relative to the parent DNA polymerase. This suggests PFA and PPi bind to different, perhaps overlapping regions, in the active center of HSV-1 DNA polymerase. 6. All variant polymerases exhibited decreased sensitivities to the trphosphates of arabinosyl hypoxanthine (araHxTP) and arabinosyl adenine (araATP) in synthetic reactions. When tested by plaque reduction assay in the absence of a deaminase inhibitor variants showed some degree of resistance to araHX, but significant resistance to araA. However, in the presence of the deaminase inhibitor (deoxycoformycin), variants were no more sensitive than parental virus. This suggests that HSV DNA polymerase may be a site of only secondary importance for the mechanism of araA action and that the "araA-resistant" variants which have been reported may be, in reality, less susceptible to araHX, or a metabolite common to araA and araHX (23).

In conclusion, it appears that biochemical variation of the viral DNA polymerase is a common occurance when HSV is grown in culture. Such enzymes may confer a growth disadvantage upon virus variants and this possibility is under current investigation.

Potential Use of Anti-HSV Nucleoside Analogs for the Treatment of CMV or EBV-Associated Diseases

A common feature of all herpesviruses is their ability to induce their virus-specified DNA polymerase and a distinguishing feature is that only HSV-1, HSV-2 and VZV, but not CMV and EBV, could induce virus-specified thymidine kinase. Thus, development of anti-HSV and VZV compounds, which rely on phosphorylation by virus-specified thymidine kinase, may not have any value for the treatment of CMV or EBV associated diseases. Several anti-HSV nucleoside analogs such as BVdU, FIAC, FMAU and ACV were found to be phosphorylated more efficiently in HSV or VZV-infected cells due to high activity as well as the broader spectrum of substrate specificity of virus thymidine kinase, whereas they could still be phosphorylated in uninfected cells, although poorly. Since the antivirus and anticell activity was dependent on the amount of triphosphate metabolite of those analogs, which could exert adverse actions on DNA replication through the action of DNA polymerase inside cells, selectivity against HSV and VZV was observed. The question raised is whether these analogs, although poorly phosphorylated by the host enzyme, have selectivity against CMV and EBV. Since all herpes-virus induced DNA polymerases share a lot of common features and may have similar behaviors toward those analog triphosphates, which have better binding affinity toward HSV-1, HSV-2 and in some cases EBV DNA polymerase, it is conceivable that the amount of analog triphosphate required to inhibit EBV or CMV DNA replication inside cells is less than that required to inhibit DNA synthesis in uninfected cells. Furthermore, the replication cycle of virus DNA is much shorter than that of cellular DNA. Even those compounds having equal effect on virus and host DNA polymerization could have more pronounced effects on virus growth than cell proliferation. Thus, it is possible that these nucleoside analogs could have anti-CMV or EBV effects. In fact, ACV was found to have potent anti-EBV effects (24) and ACVTP was found to have potent binding affinity to EBV DNA polymerase (25). The mechanism of selectivity could be due to what we have suggested above. The effect of ACV on EBV replication appears to be reversible, which could be due to the action of ACV as chain terminator, which could be easily corrected by endonucleases, as previously discussed. Since BVdU, FIAC and FMAU have some cell toxicity (indicating some degree of phosphorylation by host enzymes) they could be incorporated into intra-nucleotide linkages of the DNA chain, which is difficult to repair. We examined the effect of those compounds on replication of EBV in collaboration with Dr. Lin and Dr. Pagano and of CMV in collaboration with Dr. Huang, Dr. Patel and Dr. Ma, of this institute. BVdU, FIAC and FMAU could inhibit both types of virus replication at concentrations not inhibitory to growth of human cells in culture (FIAC and FMAU results are consistent with unpublished results obtained by Dr. Lopez and Dr. Fox of Sloan-Kettering Institute). Furthermore, their inhibition of virus growth is irreversible. Those preliminary observations are consistent with our predictions and further experiments will be needed to support our hypothesis. In conclusion, the presence in infected cells of virus DNA polymerase, which has a higher binding affinity to dNTPs in general, and the shorter virus replication cycle provides a rationale for the potential use of anti-HSV nucleosides analogs, which have some cytotoxicity, for the treatment of CMV or EBV associated diseases.

LITERATURE CITED

1. Ostrander, M. and Cheng, Y.C. 1980. Biochim. Biophys. Acta. 609, 232.
2. Knopf, K.W. 1979. Eur. J. Bioch. 98, 231.
3. Leinbach, S.S., Reno, J.M., Lee, L.F., Isbell, A.F. and Boezi, J.A. 1976. Biochemistry 15, 426.
4. Muller, W.E.G., Zahn, R.K., Bitlingmaier, K. and Falke, D. 1977. N.Y. Acad. Sci. 284, 34.
5. Weissbach, A., Hong, S.L., Aucker, J. and Muller, R. 1973. J. Biol. Chem. 248, 6270.
6. Derse, D. and Cheng, Y.C. 1981. J. Biol. Chem. 256, 8525.
7. Derse, D., Cheng, Y.C., Furman, P.A., St. Clair, M.H. and Elion, G.B. 1981. J. Biol. Chem. 256, 11447.
8. Paffenholz, V., Le, H.V., Ho, Y. and Bardos, T.J. 1976. Cancer Res. 36, 1443.
9. Miller, P.S., McParland, K.B., Jayaraman, K. and Ts'o, P.O.P. 1981. Biochemistry. 20, 1874.
10. Elion, G., Furman, P., Fyfe, J., deMiranda, P., Beauchamp, L. and Shaeffer, H. 1977. Proc. Natl. Acad. Sci. U.S.A. 74, 7516.
11. DeClercq, E., Descamps, J., DeSomer, P., Barr, B., Jones, A. and Walker, R. 1979. Proc. Natl. Acad. Sci. U.S.A. 76, 2947.
12. Prusoff, W. and Fisher, P. 1979. in Nucleoside Analogs (R. Walker, E. DeClercq, and F. Eckstein, eds.). Plenum Press, New York. p 281-318.
13. Watanabe, K., Reichman, U., Hirota, K., Lopez, C. and Fox, J. 1979. J. Med. Chem. 22, 21.
14. Ruth, J.L. and Cheng, Y.C. 1981. Mol. Pharm. 20, 415.
15. Reno, J.T., Lee, L.F. and Boezi, J.A. 1978. Antimicrob. Agents Chemother. 13, 188.
16. Eriksson, B., Larsson, A., Helgstrand, E., Nils-Gunnar, J. and Oberg, B. 1980. Bioch. Biophys. Acta. 607, 53.
17. Cheng, Y.C., Grill, S., Derse, D., Chen, J.Y., Caradonna, S.J. and Connor, K. 1981. Biochim. Biophys. Acta. 652, 90.
18. Brutlag, D. and Kornberg, A. 1972. J. Biol. Chem. 247, 241.
19. Miller, J.P., Kigwana, L.J., Streeter, D.G., Robbins, R.K., Simon, L.N. and Roboz, J. 1977. Ann. N.Y. Acad. Sci. 284, 211.
20. Honess, R.W. and Watson, D.K. 1977. J. Biol. 21, 584.
21. Eriksson, B. and Oberg, B. 1979. Antimicrob. Agents Chemother. 15, 758.
22. Derse, D., Bastow, K. and Cheng, Y.C. 1982. J. Biol. Chem. 257, 10251.
23. Bastow, K., Derse, D., and Cheng, Y.C. 1982 submitted for publication Antimicrob. Ag. Chemother.
24. Datta, A.K., Colby, B.M., Shaw, J.E. and Pagano, J.S. 1980. Proc. Natl. Acad. Sci. U.S.A. 77, 5163.
25. Colby, B.M., Shaw, J.E., Elion, G. and Pagano, J.S. 1980. J. Virol. 34, 560.

9

PHOSPHONOFORMATE AND PHOSPHONOACETATE

B. ERIKSSON AND B. ÖBERG
Department of Antiviral Chemotherapy, Research and Development Laboratories, Astra Läkemedel AB, Södertälje, Sweden

ABSTRACT

Despite the similarities in chemical structure, phosphonoformate (foscarnet) and phosphonoacetate (PAA) are different in their inhibition of viral enzymes and inhibition of virus multiplication in cell culture. A major difference in dermal toxicity has been observed and will probably prevent the clinical use of PAA. Recent double-blind placebo-controlled clinical trials have shown that foscarnet, when used topically as a cream, has therapeutic effects on cutaneous (labial) and genital herpes infections.

INTRODUCTION

Today more antiviral agents are known against herpesviruses than against any other group of viruses. The majority of the antiherpes drugs are nucleoside analogs inhibiting herpesvirus DNA polymerase, often after first being phosphorylated by herpesvirus thymidine kinase. The efficacy of these compounds on cutaneous and mucocutaneous herpes infections in animals, including man, has not been as impressive as their inhibition of herpesvirus replication in cell cultures. The two pyrophosphate analogs phosphonoformic acid (foscarnet) and phosphonoacetic acid (PAA), on the other hand, are very effective therapeutic agents when used topically in animal herpes models, whereas their antiviral effects on herpesvirus multiplication in cell cultures are not too impressive. Foscarnet has furthermore shown therapeutic effects in the clinic. This makes pyrophosphate analogs interesting as herpesvirus drugs, both from a mechanistic and a therapeutic point of view. Reviews on PAA (1, 2) and foscarnet (3, 4) have dealt both with their mechanisms of action and with their therapeutic effects. Possible explanations for the relative efficacy of pyrophosphate analogs and nucleoside analogs in different systems have

Y. Becker (ed.), Antiviral Drugs and Interferon: The Molecular Basis of Their Activity.

recently been discussed (5). In this review we will compare recent studies on foscarnet and PAA, outline their different enzyme inhibitory properties, discuss their antiviral efficacy in cell cultures and in animal models, and finally summarize the clinical data available for foscarnet. The lack of data on PAA makes a comparison to foscarnet regarding toxicity and pharmacokinetics difficult but some results will be discussed.

EFFECTS ON HERPESVIRUS-INDUCED DNA POLYMERASE ACTIVITIES

All herpesviruses induce the synthesis of virus-specific DNA polymerases (6, 7). Phosphonoacetate (PAA) and foscarnet are effective inhibitors of all human herpesvirus DNA polymerase activities. Mao et al. (8, 9) first observed the inhibition of the partially purified DNA polymerase of herpes simplex virus (HSV) by PAA. Later, Helgstrand et al. (10) and Reno et al. (11) independently observed the efficacy of foscarnet as an inhibitor of herpesvirus DNA polymerases. Since the discovery of PAA and foscarnet, a large number of pyrophosphate analogs have been compared for their inhibitory effects on herpesvirus DNA polymerases (1, 2, 12 - 16). Tables 1 and 2 summarize the 50% inhibitory concentrations of the most interesting analogs on some herpesvirus DNA polymerases. For comparison, cellular DNA polymerase α and other virus nucleic acid polymerases of interest, such as hepatitis B virus DNA

Table 1. Inhibition of polymerase activities by pyrophosphate analogues.

Compound	Concentration (μM) giving 50% reduction of						
	HSV-1 DNA pol	HSV-2 DNA pol	HCMV DNA pol	Hepatitis B DNA pol	Influenza RNA pol	AMV RNA dep. DNA pol	Cellular DNA pol α
PP	>500	>500	>500	~500	450	500	>500
MDP	>500	>500	>500	>500	>500	>500	>500
CDP	10	27	6	~500	n.d.	200	100
MHDP	120	260	105	>500	n.d.	>500	>500
EHDP	>500	n.d.	>500	>500	n.d.	>500	>500
HP	150	12	150	97% at 500	n.d.	25	250
OA	115	400	130	>500	>500	>500	>500
PFA	0.3	0.5	0.3	20	30	8	40

Abbreviations: PP, pyrophosphate; MDP, methylenediphosphonate; CDP, carbonyldiphosphonate; MHDP, methanehydroxydiphosphonate; EHDP, ethane-1-hydroxy-1,1-diphosphonate; HP, hypophosphate; OA, oxalate; PFA, phosphonoformate, n.d. = not determined. The results are from refs. 14, 16, 19, 21, 28, and 29.

polymerase, reverse transcriptase and influenza virus RNA polymerase have been included. In this class of compounds, foscarnet and PAA are, to date, the two most potent inhibitors. Both compounds affect all tested herpesvirus DNA polymerases about equally. However, these enzyme activities are also inhibited by several other analogs, such as carbonyldiphosphonate (CDP), methanehydroxydiphosphonate (MHDP), hypophosphate (HP), and oxalate (OA), as shown in Table 1. CDP is about one order of magnitude more effective than are the other three inhibitors. In comparison to foscarnet, however, CDP is 20-50 times less effective. Hypophosphate (HP) preferentially inhibits HSV-2 DNA polymerase. This compound is, however, highly toxic to cells and animals (17, 18). The effects of PAA and analogs with a systematic variation of substituents on the methylene carbon are presented in Table 2. The different herpesvirus DNA polymerases are inhibited about equally by most of the analogs. However, some notable exceptions were found. The HSV-2 DNA polymerase is about 25 times less susceptible to 2-phosphonopropionate (M-PAA) than are HSV-1 and human cytomegalovirus (HCMV) DNA polymerases. Phosphonoacetates carrying large and bulky hydrophobic α-substituents (P-PAA, N-PAA, Ph-PAA) are preferential inhibitors of the HCMV DNA polymerase.

Table 2. Inhibition of polymerase activities by α-substituted phosphonoacetates.

Compound	Concentration (μM) giving 50% reduction of						
	HSV-1 DNA pol	HSV-2 DNA pol	HCMV DNA pol	Hepatitis B DNA pol	Influenza RNA pol	AMV RNA dep. DNA pol	Cellular DNA pol α
PAA	0.5	0.7	0.4	>500	300	>500	35
M-PAA	15	550	18	>500	>500	>500	>500
E-PAA	8	14	13	n.d.	n.d.	>500	>500
P-PAA	200	160	30	n.d.	n.d.	>500	>500
N-PAA	10	5	1.5	n.d.	n.d.	>500	160
Ph-PAA	100	120	18	n.d.	n.d.	>500	>500
H-PAA	2	3	3	n.d.	n.d.	700	100
H,M-PAA	>500	n.d.	>500	n.d.	n.d.	>500	>500

Abbreviations: PAA, phosphonoacetate; M-PAA, α-methylphosphonoacetate (or 2-phosphonopropionate); E-PAA, α-ethyl-PAA; P-PAA, α-n-propyl-PAA; N-PAA, α-nonyl-PAA; Ph-PAA, α-phenyl-PAA; H-PAA, α-hydroxy-PAA; H,M-PAA; α-hydroxy-α-methyl-PAA (or 2-hydroxy-2-phosphonopropionate), n.d. = not determined. The results are from refs. 14, 16, 19, 21, 28, and 29.

EFFECTS ON OTHER DNA AND RNA POLYMERASES

The isolated DNA polymerase activities from different herpesviruses are about equally inhibited by foscarnet and PAA. This parallelism was not observed when the effects of these two compounds were compared on other DNA and RNA polymerases. Human hepatitis B virus and woodchuck hepatitis virus DNA polymerases are effectively inhibited by foscarnet (20 μM reduces the activity by 50%) whereas PAA is not inhibitory (19, 20). The only example of an RNA dependent RNA polymerase activity which is strongly affected by foscarnet is influenza A virus RNA polymerase (21). Mn^{2+} used instead of Mg^{2+} as divalent cation in the reaction mixture, the concentration of foscarnet giving a 50% reduction of enzyme activity decreases from 30-60 μM to 0.1-0.4 μM depending on the influenza virus strain. Foscarnet has also been used as an inhibitor of vesicular stomatitis virus (VSV) RNA polymerase to confirm the sequential transcription of the viral genes N-NS-M-G (22). A high concentration of foscarnet (5 mM) has to be used to inhibit this enzyme activity by 90%. Neither influenza nor VSV RNA polymerases are inhibited by PAA. Likewise PAA does not inhibit the reverse transcriptase activities of Rous Sarcoma virus (RSV) or avian myeloblastosis virus (AMV) (9, 12). In contrast, foscarnet is an effective inhibitor of the reverse transcriptase activities of several retroviruses, such as Rauscher murine leukemia virus (RMuLV), Simian sarcoma virus (SSV), baboon endogeneous virus (BaEV), bovine leukemia virus (BLV), AMV, visna virus (VV), and Moloney murine leukemia virus. With $(rA)_n \cdot (dT)_{10}$ as template-primer, these enzyme activities are inhibited by at least 90% by 100 μM foscarnet (23 - 28).

Among the other pyrophosphate analogs, only HP is an effective inhibitor of both hepatitis B DNA polymerase and AMV reverse transcriptase activities, whereas CDP only slightly affects the latter enzyme (28, 29).

MECHANISM OF INHIBITION

Overby et al. (30) reported that the inhibition of HSV-1 replication by PAA appeared to be a result of an inhibition of the viral DNA synthesis. The molecular mechanism by which PAA exerts its anti-herpetic effect was later demonstrated by Mao et al. (9). PAA was observed to inhibit herpesvirus DNA polymerase activities in a non-competitive manner with respect to the varied dNTPs and in an uncompetitive manner with respect to the varied amounts of activated DNA with the dNTPs at a saturating level. Later Leinbach et al. (12) proposed a detailed mechanism by which PAA inhibits the DNA polymerase activity. By interacting at the pyrophosphate binding site, PAA was suggested to block the formation of the 3'-5'-phosphodiester bond between the primer and substrate and thus prevent further elongation. Enzyme kinetic data support this, so called, alternate product inhibitor mechanism although no successful experiment leading to direct evidence has been reported (2).

The mechanism of inhibition by foscarnet has been studied using a variety of herpesvirus DNA polymerases and has been found similar to that of PAA. The DNA polymerase of a PAA-resistant mutant of herpesvirus of turkeys was about two times less susceptible to pyrophosphate (31). We have confirmed this observation for a PAA-resistant mutant of HSV-1 and further found this to be true also for a foscarnet-resistant mutant (32). A cross-resistance of DNA polymerases of the latter two mutants to both compounds has also been observed (32). In a study of several cellular DNA polymerases, the effect of foscarnet was found to be analogous to that for PAA (33). All these observations support the idea that both compounds interact at a common presumed pyrophosphate binding site.

Detailed studies on the mechanism by which foscarnet inhibits various nucleic acid polymerases have been extended to enzymes which are not affected by PAA. Foscarnet is a non-competitive inhibitor of influenza RNA polymerase activity with respect to the four NTPs (21). In the case of vesicular stomatitis virus, however, foscarnet is a competitive inhibitor of RNA polymerase activity with respect to ATP, and appears to exert its inhibitory effect at the initiation step of RNA synthesis (22).

The mechanism of inhibition of AMV reverse transcriptase by foscarnet is non-competitive with respect to substrate and template (23, 26 - 28). The activity of this enzyme was found to be highly dependent on the template-primer used. Table 3 summarizes the apparent Michaelis-Menten (K_m) and inhibition (K_i) constants for foscarnet using four different template-primers.

Among the other analogs, carbonyldiphosphonate and α-hydroxyphosphonoacetate also cause linear non-competitive inhibition patterns. On the other hand, hypophosphate, imidodiphosphonate and pyrophosphate were observed to be substrate-dependent inhibitors of AMV reverse transcriptase. Thus two sets of pyrophosphate analogs exist with different modes of inhibition (28).

Table 3. Influence of template-primers on the apparent kinetic constants for the linear non-competitive inhibition of AMV reverse transcriptase activity by foscarnet. Data from Eriksson et al. (28).

Template-primer	Substrate	K_m (μM)	K_i(μM)
$(rA)_n \cdot (dT)_{12-18}$	dTTP	36	8
$(rC)_n \cdot (dG)_{12-18}$	dGTP	2.0	100
$(dC)_n \cdot (dG)_{12-18}$	dGTP	0.25	30
Activated DNA	All four dNTPs	4.0	5

INHIBITION OF VIRUS MULTIPLICATION

Table 4 presents the antiviral effects of foscarnet and PAA on a selection of viruses (for reviews see 1 - 4). The multiplication of all human herpesviruses are inhibited by either compound. Vaccinia virus replication is inhibited by PAA only, whereas several retroviruses are

Table 4. Inhibition of virus multiplication in cell culture by foscarnet and PAA.

Virus	Inhibition 100 μM of			
	Foscarnet		PAA	
	Effect	References	Effect	References
DNA viruses				
HSV-1 and HSV-2	+	10, 11, 35, 36	+	8, 30
VZV	+	40, B. Wahren, pers. comm.	+	1, 52
HCMV	+	53, 54	+	55 - 57
EBV	+	58	+	59 - 61
Adeno	-	10	-	1
Vaccinia	-	10	+	62
African Swine Fever	+	63	+	64
RNA viruses				
Polio	-	10	-	1
Influenza	-	10	-	10
Retroviruses				
AMV	+	23	-	23
Visna	+	24	-	24
MuLV	+	23, 65	-, +	24, 66

+, >50 % inhibition; -, <50 % inhibition

affected only by foscarnet. In order to inhibit HSV-1 replication, foscarnet has to be added to the infected cells within 8 hours post-infection (34) and within 3 hours to be maximally effective (35). The amount of virus inoculated into the culture will also influence the inhibitory effect (36).

A selective inhibition of HSV-1 DNA synthesis by foscarnet, as visualized by isopycnic separation of cellular and viral DNA in CsCl gradients, gives a 50% reduction of HSV-1 DNA synthesis at 50 μM concentration (37). Similar results were previously observed for PAA (30, 38). These results correlate well to the plaque reduction effects of the two compounds.

A cross-resistance to foscarnet and PAA was observed for HSV-1 mutants selected for resistance to either compound (32, 35). Mutants of HSV-1 with defective thymidine kinase or lacking thymidine kinase are susceptible to foscarnet in contrast to their resistance to thymidine analogs (39). The susceptibility of HSV-1 to foscarnet is not affected by an addition of nucleosides to the culture medium (34), and the enzymatic activity of thymidine kinase is not inhibited by foscarnet (A. Larsson, pers. comm.) indicating that this enzyme is not involved in the mechanism of inhibition by foscarnet.

The cellular toxicity for foscarnet is low. The DNA synthesis in HeLa cells or primary human lung fibroblasts is reduced to 50% by 1 mM of foscarnet in a 24 hour treatment and is easily reversed by drug removal (17, 18). The only pyrophosphate analog in Tables 1 and 2 with a cell toxic effect is hypophosphate, which reduces the cellular DNA synthesis to 50 % at 4 - 8 μM concentration.

THERAPEUTIC EFFECTS IN INFECTED ANIMALS

The therapeutic effects of foscarnet and PAA on viral infections in animals have been reviewed previously (1 - 4). In Table 5 we have listed recent results, mostly for HSV-1 and HSV-2 infections. No major differences in efficacy have been observed between foscarnet and PAA when used topically on herpes infections. In contrast to foscarnet, however, PAA causes a skin irritation which can be severe. This was also observed when PAA is given i.p. (40). The best therapeutic effects in the cutaneous and genital infections are obtained when the compounds are

Table 5. Efficacy of foscarnet and PAA on herpesvirus infections in animals

Disease	Animal	Virus	Treatment	Effects	References
Cutaneous herpes infection	Guinea pig	HSV-1	5-100 mM foscarnet (0.15-3 %) topical	Decreased lesions Decreased time to healing Decreased virus titres in skin	10, 41, 43, 44, 67, 68
			2 % PAA, topical	Decreased lesions Decreased time to healing Decreased virus titres	42, 43
			2 % PAA, cream	Decreased lesions	69
	Hairless mouse	HSV-1	3 % foscarnet, topical	Decreased lesions Decreased mortality	70
			2 % PAA, topical	Decreased lesions Prevention of latency	71
	Nude mouse	HSV-1	1 % foscarnet, topical	Decreased mortality Decreased lesions	72
			1 % PAA, topical	Decreased mortality Decreased lesions	72
Genital herpes infection	Guinea pig	HSV-2	3 % foscarnet, topical	Decreased lesions Decreased virus shedding Normal weight gain Prevention of paralysis	73
	Guinea pig	HSV-2	10 % foscarnet, topical	Decreased lesions Decreased virus shedding	46
			3 % foscarnet, ointment	Decreased lesions Decreased virus shedding Reduced cellular changes	74
	Mouse	HSV-2	10 % foscarnet, suspension, topical	Decreased virus shedding Decreased mortality	46
			5 % PAA, cream or 500 mg/kg/day, i.p.	Decreased virus shedding Decreased mortality	46
	Mouse	HSV-1 HSV-2	8 % foscarnet, suspension, topical	Decreased virus shedding Decreased mortality	75
		HSV-1 HSV-2	5 % PAA, cream, topical	Decreased virus shedding Decreased mortality	75

Disease	Animal	Virus	Treatment	Effects	References
	Hamster	HSV-2	2 % PAA, topical	Decreased mortality Decreased virus shedding	76
	Cebus monkey	HSV-2	0.2-5 % PAA, topical	No therapeutic effect, irritating	77
Herpes keratitis	Rabbit	HSV-1	3 % foscarnet, solution	Decreased corneal ulcers	45
			5 % foscarnet, ointment	Decreased corneal ulcers	45
			5 % PAA, ointment or drops	Decreased corneal ulcers	78
			5 % PAA, ointment	Decreased corneal ulcers Decreased virus shedding	79
Systemic herpesvirus infection	Mouse	HSV-2	500 mg/kg/day of foscarnet, i.p.	Increased survival time Decreased mortality	46
			500 mg/kg/day of PAA, i.p.	Decreased mortality	80
Herpes encephalitis	Mouse	HSV-1	800 mg/kg/day of foscarnet, i.p.	Increased survival time No effect on mortality	46
		HSV-2	500 mg/kg/day of PAA, i.p.	Increased survival time No effect on mortality	80
	Rat	HSV-1	2 mg/kg/day of PAA, inracerebro-ventricular or 500 mg/kg/day of PAA, i.p.	Increased survival time No effect on mortality	81
Cytomegalo-virus infection	Mouse	MCMV	500 mg/kg/day of foscarnet, i.p.	Decreased mortality	46
		MCMV	500 mg/kg/day of PAA, i.p.	Decreased virus titres Decreased mortality	57
Simian varicella	Patas monkey	Simian vari-cella	200 mg/kg/day of foscarnet, i.p.	Decreased rash and viremia	K F Soike, pers. comm.
			200 mg/kg/day of PAA, i.p.	Decreased clinical illness Dermatitis, increased transaminases	40

applied topically. In comparative studies, foscarnet has been more effective than any nucleoside analog as a therapeutic agent on the cutaneous herpes infection in guinea pigs (41 - 44). This is illustrated by the reduction in virus titre observed after topical treatment of cutaneous herpes infections in guinea pigs with foscarnet and acyclovir (Figure 1). The effects on herpes keratitis and herpes encephalitis have

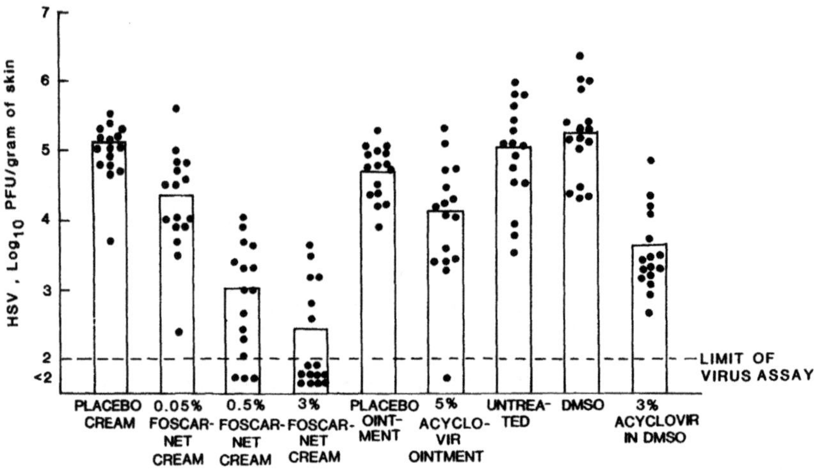

Figure 1. Effect of foscarnet and acyclovir on titres of HSV-1 in the skin of guinea pigs five days after inoculation. Treatment started 48 h post inoculation and consisted of four daily applications. Foscarnet cream is a water based cream and acyclovir ointment is a polyethylene formulation. The results are from (44) with permission.

been less pronounced than those observed with nucleoside analogs (45, 46). This probably is due to a rapid removal of the pyrophosphate analogs by the tear fluid, in contrast to the nucleoside analogs, which after phosphorylation probably are trapped in the cornea cells. Furthermore, foscarnet does not penetrate into the brain (3).

Foscarnet has not been tested on vaccinia infections, but judging from cell culture data (Table 3) it is likely that foscarnet will be less active than PAA, which has shown activity in vivo (47). Foscarnet has been tested for effect on chronic hepatitis infections in woodchucks. No effect was observed in this animal infection although serum levels sufficiently high to inhibit the viral DNA polymerase were obtained (48). This indicates that a functional viral DNA polymerase is not required during the chronic hepatitis infection. The severity of a myxoma virus infection in rabbits decreased after s.c. injections of 400 mg/kg/day of PAA (49), but no results have been reported for foscarnet in this infection. Pseudorabies virus infection in rabbits and pigeon herpesvirus infection in pigeons were treated with 100 mg/kg/day of foscarnet without any therapeutic effect (50).

The mechanism of action of foscarnet on a cutaneous HSV-1 infection in guinea pigs seems to be an inhibition of the viral DNA polymerase. This is indicated by the lack of therapeutic effect on an infection caused by a foscarnet-resistant mutant of HSV-1 coding for a foscarnet-resistant DNA polymerase (32, 43). Percutaneous infection of hairless mice with a PAA-resistant strain of HSV-1 likewise results in an infection not suppressed by PAA (51). Neither foscarnet nor PAA affects the latent infection of HSV-1 in mice (82, 83).

PHARMACOKINETICS AND METABOLISM

Foscarnet and PAA are not metabolized, and foscarnet is rapidly excreted in the urine (2, 3, 84). A comparison of the pharmacokinetic properties of foscarnet and PAA is not possible due to lack of data for PAA. A deposition in bone has been reported for PAA (2, 84). The small amount, less than 10% , of the bioavailable foscarnet which is bound to the inorganic matrix of bone is released to 50% within 3 weeks in mice (3, 4). The penetration through guinea pig skin is 2-4 % when a 3 % foscarnet cream is used (3). The serum half-life of foscarnet after single dose administration is 0.7 h for mouse, 1.2 h for cynomolgus monkey, 2 h for dog (19.6 kg) and 3.6 h for pig (78 kg) when given intravenously (3), and the half-life of PAA in mice has been reported to be 20-30 min (85).

TOXICITY

The acute toxicity (LD_{50}) of foscarnet in mice is 510 mg/kg when given intravenously and >4 g/kg when given orally (3). When given subcutaneously to mice, PAA has an LD_{50} of 1500 mg/kg (85). Toxic reactions to PAA has been reported for rabbits given 300 mg/kg/day intravenously (78) and monkeys given 500-1000 mg/kg intravenously (85). In patas monkeys 200 mg/kg/day of PAA given intramuscularly causes liver degeneration, ulcerative dermatitis, gingivitis and a darkening of the skin (40). In the same patas monkey model foscarnet did not give any skin toxic effects when given 200 mg/kg/day intramuscularly (K.F. Soike, pers. comm.).

A large difference between foscarnet and PAA in dermal toxicity has been observed in guinea pigs (41) and in cynomolgus monkeys, as illustrated in Figure 2. No dermal toxicity was observed for foscarnet, while PAA caused severe dermal degeneration. A dermal toxicity in cebus monkeys has also been reported for PAA when used topically as a 2% cream (77).

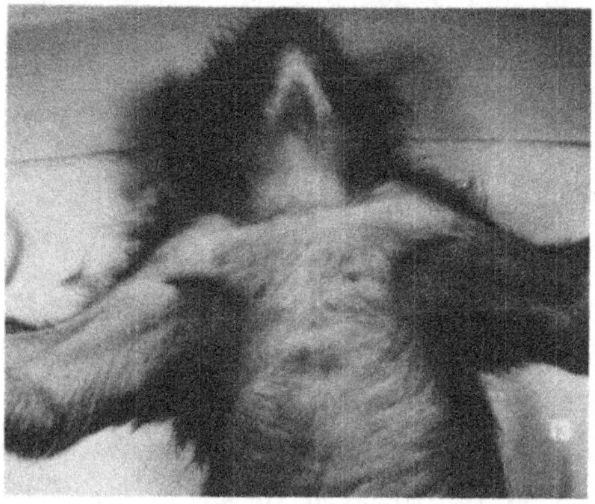

Figure 2. Cynomolgus monkey treated twice daily for 7 days on the upper arms with 2% foscarnet cream and an equimolar (1.78%) phosphonoacetate cream. The red and irritated part on the left arm was treated with phosphonoacetate. No irritation was seen on the right arm treated with foscarnet (Alenius and Öberg, unpublished data).

Extensive toxicity, fertility and teratogenicity studies have been carried out with foscarnet (3, 4, and unpublished results) showing a good safety margin and permitting human studies. A 3% foscarnet cream has been used clinically for labial herpes without any side effects, and 0.3 and 1% foscarnet cream have been used for the treatment of genital herpes without any side effects (86 - 89). The 3% foscarnet cream, however, can cause some skin irritation when used on genital lesions in man (Wallin, pers. comm.). Foscarnet has been given by infusion to a few patients with severe CMV infections and seems to be tolerated when the serum levels were kept at 100-200 μM for 5-10 days (J.-O. Lernestedt, pers. comm.).

CLINICAL STUDIES ON HERPES VIRUS INFECTIONS

Whereas the dermal toxicity of PAA has obviously prevented any clinical studies with this compound, several studies on labial and genital herpes are presently underway using foscarnet cream.

The design and results of the first completed clinical trial on the treatment of recurrent cutaneous herpes (mainly labial) using 3% foscarnet cream have been reported (86 - 88). The study was carried out double-blind and placebo controlled. After the first episode the patients were put on a cross-over study for another two episodes. During the cross-over study, medication was kept at home to ensure an early treatment institution. The results from this study show that early topical treatment with 3% foscarnet cream significantly shortens the vesicular period, decreases the development of new vesicles and results in a patient preference for the foscarnet cream as compared to placebo cream.

Results from a study on recurrent genital herpes infections using a double-blind placebo-controlled cross-over design have recently shown a considerably decreased period during which sores are present and also a significantly decreased healing time for the group treated with 0.3% foscarnet (89).

No resistence development has been reported so far in the foscarnet studies. The risk for a resistence development seems less for an inhibitor, such as foscarnet, acting on a viral DNA polymerase than for an inhibitor utilizing herpesvirus thymidine kinase, which is not an obligatory enzyme for herpesvirus replication. However, the use of any specific antiviral agent can possibly lead to resistance development. The importance of this will depend on the ability of the resistant virus to persist in a latent form in the host. Should resistance development be a problem for an antiviral drug there is a possibility to combine drugs with different modes of action.

ACKNOWLEDGEMENTS

We thank our colleagues at the Department of Antiviral Chemotherapy for valuable discussions, Mrs. M. Kropp for typing the manuscript and Mrs. J. de Paulis for linguistic corrections.

REFERENCES

1. Overby, L.R., Duff, R.G. and Mao, J.C.-H. Ann. N.Y. Acad. Sci. 284: 310-320, 1977.
2. Boezi, J.A. Pharmacol. Ther. 4: 231-243, 1979.
3. Helgstrand, E., Flodh, H., Lernestedt, J.-O., Lundström, J. and Öberg, B. In: Developments in Antiviral Chemotherapy p. 63-83, 1980, L.H. Collier and J. Oxford, (eds.). Academic Press, N.Y.
4. Öberg, B. Pharmacol. Ther. In press 1982.
5. Öberg, B., in Beecham Colloquier: Problems with antiviral therapy. Eds. C. Harris-Stuart and J. Oxford. In press.
6. Kit, S. Pharmacol. Ther. 4: 501-585, 1979.
7. Weissbach, A., Hong, S.-C.L., Aucker, J. and Muller, R. J. Biol. Chem. 248: 6270-6277, 1973.
8. Mao, J. C.-H., Robishaw, E.E. and Overby, L.R. J. Virol. 15: 1281-1283, 1975.
9. Mao, J. C.-H. and Robishaw, E.E. Biochemistry 14: 5475-5479, 1975.
10. Helgstrand, E., Eriksson, B., Johansson, N.-G., Lannerö, B., Larsson, A., Misiorny, A., Norén, J.O., Sjöberg, B., Stenberg, K., Stening, G. Stridh, S., Öberg, B., Alenius, S. and Philipson, L. Science 201: 819-821, 1978.
11. Reno, J.M., Lee, L.F. and Boezi, J.A. Antimicrob. Agents Chemother. 13: 188-192, 1978.
12. Leinbach, S.S., Reno, R.M., Lee, L.F., Isbell, A.F. and Boezi, J.A. Biochemistry 15: 426-430, 1976.
13. Herrin, T.R., Fairgrieve, J.S., Bower, R.R., Shipkowitz, N.L. and Mao, J.C.-H. J. Med. Chem. 20: 660-663, 1977.
14. Eriksson, B., Larsson, A., Helgstrand, E., Johansson, N.-G. and Öberg, B. Biochim. Biophys. Acta 607: 53-64, 1980.
15. Ostrander, M. and Cheng, Y.-C. Biochim. Biophys. Acta 609: 232-245, 1980.
16. Eriksson, B., Öberg, B. and Wahren, B. Biochim. Biophys. Acta 696: 115-123, 1982.
17. Stenberg, K. and Larsson, A. Antimicrob. Agents Chemother. 14: 727-730, 1978.
18. Stenberg, K. Biochem. Pharmacol. 30: 1005-1008, 1981.
19. Nordenfelt, E., Helgstrand, E. and Öberg. B. Acta path. microbiol. scand. Sect. B 87: 75-76, 1979.
20. Nordenfelt, E. and Werner, B. Acta path. microbiol. scand. Sect. B 88: 183-184, 1980
21. Stridh, S., Helgstrand, E., Lannerö, B., Misiorny, A., Stening, G. and Öberg, B. Arch. Virol. 61; 245-250, 1979.
22. Chanda, P.K. and Banerjee, A.K. Virology 107: 562-566, 1980.
23. Sundquist, B. and Öberg, B. J. Gen. Virol. 45: 273-281, 1979.
24. Sundquist, B. and Larner, E. J. Virol. 30: 847-851, 1979
25. Reno, J.M., Kung, H.-J. and Boezi, J.A. Fed. Proc. 39/6: 1846, 1980.
26. Modak, M.J., Srivastava, A. and Gillerman, E. Biochem. Biophys. Res. Commun. 96: 931-938, 1980.

27. Margalith, M., Falk, H. and Panet, A. Molecular and cellular Biochemistry 43: 97-103, 1982.
28. Eriksson, B., Stening, G. and Öberg, B. Antiviral Research 2: 81-95, 1982.
29. Nordenfelt, E., Öberg, B., Helgstrand, E. and Miller, E. Acta path. microbiol. scand Sect. B 88: 169-175, 1980.
30. Overby, L.R., Robishaw, E.E., Schleicher, J.B., Rueter, A., Shipkowitz, N.L. and Mao, J. C.-H. Antimicrob. Agents Chemother. 6, 360-365, 1974.
31. Lee, L.F., Nazerian, K., Witter, R.L., Leinbach, S.S. and Boezi, J.A. J. Natl. Cancer Inst. 60, 1141-1146, 1978.
32. Eriksson, B. and Öberg, B. Antimicrob. Agents Chemother. 15, 758-762, 1979.
33. Sabourin, C.L.K., Reno, J.M. and Boezi, J.A. Arch. Biochem. Biophys. 187: 96-101, 1978.
34. Schnürer, J. and Öberg, B. Arch. Virol. 68: 203-209, 1981.
35. Cheng, Y.-C., Grill, S., Derse, D., Chen, J.-Y., Caradonna, S.J. and Connor, K. Biochim. Biophys. Acta 652: 90-98, 1981.
36. Svennerholm, B., Vahlne, A. and Lycke, E. Proc. Soc. Exp. Biol. Med. 61: 115-118, 1979.
37. Larsson, A. and Öberg, B. Antiviral Research 1: 55-62, 1981.
38. Drach, J.C. and Shipman Jr., C. Ann. N.Y. Acad. Sci. 284, 396-409, 1977.
39. De Clercq, E., Descamps, J., Verhelst, G., Walker, R.T., Jones, A.S., Torrence, P.F. and Shugar, D. J. Infect. Dis. 141: 563-574, 1980.
40. Felsenfeld, A.D., Abee, C.R., Gerone, P.J., Soike, K.F. and Williams, S.R. Antimicrob. Agents Chemother. 14: 331-335, 1978.
41. Alenius, S., Dinter, Z. and Öberg, B. Antimicrob. Agents Chemother. 14: 408-413., 1978.
42. Alenius, S. and Öberg, B. Arch. Virol. 58: 277-288, 1978.
43. Alenius, S. Arch. Virol. 65: 149-156, 1980.
44. Alenius, S., Berg, M., Broberg, F., Eklind, K., Lindborg, B., and Öberg, B. J. Infect. Dis. 145: 569-573, 1982.
45. Alenius, S., Laurent, U. and Öberg, B. Acta Ophthalmol. 58: 167-173, 1980.
46. Kern, E.R., Glasgow, L.A., Overall, Jr., J.C., Reno, J.M. and Boezi, J.A. Antimicrob. Agents Chemother. 14: 817-823, 1978.
47. Friedman-Kien, A.E., Fondak, A.A. and Klein, R.J. J. Invest. Dermatol. 66: 99-102, 1976.
48. Nordenfelt, E., Widell, A., Löfgren,, B., Möller-Nielsen, C. and Öberg, B. Acta Path. Scand. In press.
49. Goodrich, J.M., Lee, K.W. and Hinze, H.C. Arch. Virol. 70, 157-164, 1981.
50. Vindevogel, H., Pastoret, P.P. and Aguilar-Sẽtien, A. J. Comp. Path. 92: 1-4, 1982.
51. Klein, R.J. and Friedman-Kien, A.E. Antimicrob. Agents Chemother. 7: 289-293, 1975.
52. May, D.C., Miller, R.L. and Rapp, F. Intervirology 8: 83-91, 1977.
53. Wahren, B. and Öberg, B. Intervirology 12: 335-339, 1979.
54. Wahren, B. and Oberg, B. Intervirology 14: 7-15, 1980.
55. Huang, E.-S. J. Virol. 16: 1560-1565, 1975.
56. Huang, E.-S., Huang, C.-H., Huong, S.-M. and Selgrade, M. Yale J. Biol. Med. 49: 93-98, 1976.
57. Overall, Jr., J.C., Kern, E.R. and Glasgow, L.A. J. Inf. Dis. 133: 237-244.

58. Margalith, M., Manor, D., Usieli, V. and Goldblum, N. Virology 102: 226-230, 1980.
59. Summers, W.C. and Klein, G. J. Virol. 18: 151-155, 1976.
60. Yajima, Y., Tanaka A., and Nonoyama, M. Virology 71: 352-354, 1976.
61. Nyormoi, O., Thorley-Lawson, D.A., Elkington, J. and Strominger, J.L. Proc. Natl. Acad. Sci. USA 73: 1745-1748.
62. Bolden, A., Aucker, J. and Weissbach, A. J. Virol. 16, 1584-1592, 1975.
63. Viñuela, E. cited in 3.
64. Gil-Fernandez, C., Paez, E., Vilas, P. and Gancedo, A.G. Chemother. 25: 162-169, 1979.
65. Varnier, O.E., Repetto, C.M., Raffanti, S.P. and Carloni, G. Microbiologica 5: 137-141, 1982.
66. Shannon, W.M. Ann. N.Y. Acad. Sci. 284: 472-507, 1977.
67. Helgstrand, E., Öberg, B. and Alenius, S. Adv. Ophthal. 38: 276-280, 1979.
68. Öberg, B., Alenius, S., Eriksson, B., Helgstrand, E., Lundberg, C. and Lundström, J. Excerpta Medica 571: 175-178, 1982.
69. Mc Carty, J.R. and Jarratt, M.T. J. Am. Acad. Derm. 1: 244-248, 1979.
70. Klein, R.J., De Stefano, E., Brady, E. and Friedman-Kien, A.E. Antimicrob. Agents Chemother. 16: 266-270, 1979.
71. Klein, R.J. Infection and Immunity 16: 842-848, 1977.
72. Descamps, J., De Clercq, E., Barr, P.J., Jones, A.S., Walker, R.T., Torrence, P.F. and Shugar, D. Antimicrob. Agents Chemother. 16: 680-682, 1979.
73. Alenius, S. and Nordlinder, H. Arch. Virol. 60: 197-206, 1979.
74. Mayo, D.R., Lucia, H.L. and Hsiung, G.D. Intervirology. In press 1982.
75. Kern, E.R., Richards, J.T., Overall Jr., J.C. and Glasgow, L.A. Antiviral Research 1: 225-235, 1981.
76. Renis, H.E. Antimicrob. Agents Chemother. 11, 701-707, 1977.
77. Palmer, A.E., London, W.T. and Sever, J.L. Antimicrob. Agents Chemother. 12: 510-512, 1977.
78. Meyer, R.F., Varnell, E.D. and Kaufman, H.E. Antimicrob. Agents Chemother 9: 308-311, 1976.
79. Gordon, Y.J., Lahav, M., Photiou, S. and Becker, Y. Brit. J. Ophthalmol. 61: 506-509, 1977.
80. Kern, E.R., Glasgow, L.A., Overall Jr., J.C., Reno, J.M. and Boezi, J.A. Antimicrob. Agents Chemother. 14: 817-823, 1978.
81. Aoki, F.Y., Trépanier, G. and Lussier, G. J. Inf. Dis. 139: 158-165, 1979.
82. Svennerholm, B., Vahlne, A. and Lycke, E. Arch. Virol. 69: 43-48, 1981.
83. Wohlenberg, C.R., Walz, A.M. and Notkins, A.L. Inf. Imm. 13, 1519-1521, 1976.
84. Bopp, B.A., Estep, C.B. and Andersson, D.J. Fed. Proc. 36: 939, 1977.
85. Roboz, J., Suzuki, R., Bekesi, G. and Hunt, R. Biochem. Mass Spectrometry 4: 291-296, 1977.
86. Wallin, J., Lernestedt, J.-O. and Lycke, E. In: Current Chemother. and Inf. Dis. 2: 1361-1362, 1981, J.D. Nelson and C. Grassi (eds) American Society for Microbiology, Washington, D.C.
87. Wallin, J., Lernestedt, J.-O. and Lycke, E. 12th International Congress of Chemotherapy. Current Chemoth. and Inf. Dis. II: 1068-1069, 1982.
88. Wallin, J., Lernestedt, J.-O. and Lycke, E. Excerpta Medica 571: 179-182, 1982.
89. Wallin, J. Lernestedt, J.-O. and Lycke, E. Twenty-second Interscience conference on Antimicrobial Agents and Chemotherapy. ASM. Session 52, 1982.

10

DISTAMYCIN AND DERIVATIVES: THE PYRROLE AMIDINE ANTIVIRAL ANTIBIOTICS

MEIR BIALER
Pharmacy Department, School of Pharmacy, Hebrew University, Jerusalem
91120 ISRAEL

Distamycin A(I) is a basic oligopeptide antibiotic isolated from the fermentation medium of Streptomyces distallicus (1). Its chemical structure was established by acidic and basic chemical degradations and was confirmed by a total synthesis (2,3,4). As determined by chemical analysis, distamycin A contains three 1-methylpyrrole-2-carboxamide residues linked to formamido and amidinopropyl groups. Distamycin A has a distinctive ultraviolet spectrum (2) (λmax = 237 nm $\varepsilon = 3 \cdot 10^4$ and λmax = 303 nm $\varepsilon = 3 \cdot 710^4$) and is optically nonactive.

Distamycin A possesses several interesting biological activities:

1) It has a significant effect on ascites tumors such as Erlich and S 180 in mice; at a daily dose of 50 mg/Kg, the mean survival time of the mice increased by more than 70 percent. The antibiotic also caused a marked decrease in the growth of solid tumors (5).
2) Distamycin A inhibits the growth of phages T_1, T_2 or T_4 in infected Escherichia coli K_{12}. At concentrations from 1 to 100 μg/ml (up to $2 \cdot 10^{-4}$ M), distamycin A has no influence on the growth of the host bacteria but prevents their infection with phages (6,7).
3) As an antibacterial agent distamycin A has no effect on the growth of E. coli or Salmonellae (8-11) when these bacteria are inoculated into liquid media in which the nutrients can be utilized by constitutively

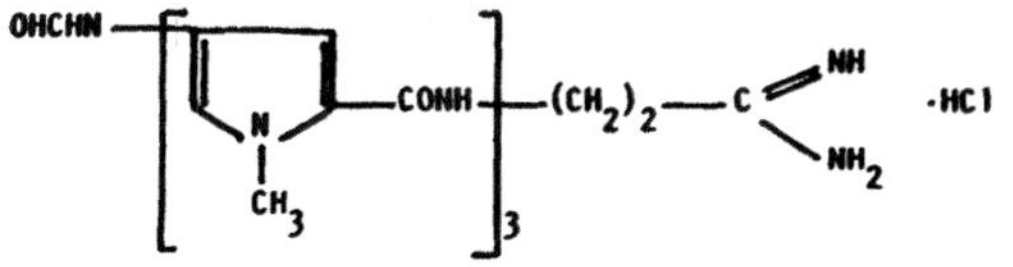

distamycin A (I)

synthesized bacterial enzymes. On the other hand, when the test bacteria need the *ad hoc* synthesis of enzymes , growth is decreased or prevented by the antibiotic.

4) The most important antibiotic effect of distamycin A is against DNA viruses and retroviroses, as shown in Table I. It does not inhibit the multiplication of RNA viruses such as Newcastle disease virus (NDV) and poliovirus (12,13).

TABLE 1. Antiviral activity of distamycin A

Virus group	Virus	References
DNA viruses		
Poxviridae	Vaccinia virus	12, 14-16
	Shope fibroma virus (SFV)	17, 18
Herpesviridae	Herpes simplex virus (HSV)	14, 16, 18-22
	Herpes zoster virus (HZV)	33, 34
	Epstein-Barr virus (EBV)	23, 24
Adenovirdae	Adenoviruses*	14, 18
Retroviruses (Oncovirinae)		
Sarcoma viruses	Rous sarcoma virus (RSV)	25
	Murine sarcoma virus (MSV)	26
Leukemia viruses	Friend leukemia virus (FLV)	27, 28
	Feline leukemia virus (FeLV)	27, 28

* Limited inhibition

In man distamycin ointments have been used successfully for the topical treatment of chickenpox, herpes zoster and eruptions resulting from smallpox vaccination (29-34).

The toxicity of distamycin A depends on the route of administration. The LD_{50} of distamycin administered I.V. to mice is 75 mg/Kg and 500 mg/Kg when given intraperitoneally (I.P.). Mice tolerate daily doses of 50-75 mg/Kg subcutaneously while rats are sensitive (5).

Only a limited amount of information was available until 1979 on the structural requirements for the antiviral activity of distamycin A. Replacement of the formamide side chain by nitro, amino, acetamide and cyclopentylpropionamide groups led to reduction of activity against vaccinia virus (13, 35-37). These side chain derivatives of distamycin A were synthesized by the same first total synthesis of distamycin (4) except in the last step where instead of a formylation, the appropriate acylation was conducted.

Substitution of the formyl moiety by N-formylglycinamide and N-formyl-alanylamide decreased the antiviral activity against FLV (38). These experiments support the fact that the formyl group must be linked to the 4-amino-1 methylpyrrole-2-carboxylic acid residue in order to achieve its full antiviral activity in biological and biochemical systems.

Relacement of the β-aminopropionamidine side chain by γ-aminobutyroamidine or a p-aminobenzamidine side chain causes a sharp reduction in activity against vaccinia virus (35,36) while an acetamidine side chain causes retention of activity against vaccinia virus and reduction of cytotoxicity but considerable reduction of activity against phage T_2 (39).

One of the reasons for lack of synthetic work on the structure-activity relationship (SAR) of distamycin A was due to the fact that the first published synthesis of distamycin A (4) does not allow for facile preparation of analogs and each alkyl amidine side chain homolog has to be prepared by a separate total synthesis. In 1978 we described a new total synthesis of distamycin A (40). The essential novel feature is that it employs as a key intermediate the tripyrrolic acid (II) which can readily be used for synthesis of side chain analogs. This method (40) was used for the synthesis of three types of side-chain analogs of distamycin A (41).

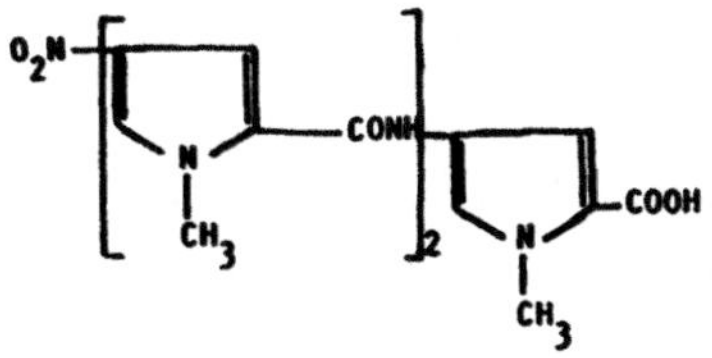

(II)

The three types were: a) analogs in which the β-aminopropionamidine side chain of the parent drug was replaced by aminoacetamidine or by the β-amino-β-methyl propionamidine group, b) analogs with an aminobenzamidine side chain, and c) analogs in which the β-aminopropioamidine side chain was replaced by a β-aminoalkylonitrile moiety. All these derivatives were tested for cytotoxicity, inhibition of HSV replication in cultured cells, effects on the synthesis of HSV DNA in isolated nuclei in vitro, as well as on DNA synthesis by the purified HSV DNA polymerase (41). None of these thirteen analogs of distamycin A was more active than the parent drug. Distamycin A was the most active compound in all three antiviral tests, as well as the most toxic. However, several compounds, in particular the aromatic analogs (with substitution in position ortho and meta of the benzene ring),showed no toxicity under the experimental conditions used but were still active in the three antiviral tests (41).

Isomers and analogs of distamycin A,in which the peptide bonds were moved to positions 2 and 5,were synthesized (3, 42). This was done to determine if the location of the peptide in the pyrrole ring is critical for the activity of this series. The appropriate distamycin A isomer was less active antivirally (against HSV) and nontoxic. The new pyrrole derivative, N-methyl-5-nitropyrrole-2-carboxamido-β-propionamidine hydrochloride, was nontoxic and was almost as active antivirally as distamycin A (42).

When distamycin A (which is a tripeptide) is converted into an analogous recurring tetra- or pentapeptide (4),the activity against vaccinia virus (13,44), HSV (20), SFV (17),RSV(25,26), MSV (27,45) and FLV (27) is increased. At the same time the cytotoxicity is somewhat reduced indicating that the antiviral and cytotoxic activities are separable. None of these distamycin homologs demonstrates antiviral activity against RNA viruses (13). These distamycin homologs inhibit the proliferation of mouse lymphoma cells to a greater extent than the parent drug (46). All these tetra- and pentapeptide homologs of distamycin A were synthesized according to first total synthesis of distamycin (43).

By using this method, homologs of distamycin A with one and three pyrrole rings, in which the N-methyl group was substituted by the N-propyl group, were synthesized (47,48,111). A few derivatives of distamycin A containing N-propylpyrrole residue have been synthesized and examined with respect to DNA-binding properties (108-113).

Recently the synthesis of a distamycin A analog in which each of the three pyrrole rings is fully methylated was described (49). This structure modification resulted in pyrrole rings, which are extraordinarily electron rich. A new synthetic approach to the development of this compound was required. The key reaction involved development of a general method for the synthesis of 3-aminopyrroles for the formation of an amide bond between a pyrrole-2-carboxylic acid and these 3-aminopyrroles. Neither of the mono-, di-, nor tri-pyrrolic permethyl analogs of distamycin A were effective antimalarials, and none showed anticancer activity (49).

In 1981 a new total synthesis for distamycin A was published (115). The starting compounds for this total synthesis were 4-[[(tert-butyloxy) carbonyl] amino]-1-methylpyrrole-2-carboxylic acid and its corresponding formyl derivative 4-(formaido)-1-methylpyrrole-2-carboxylic acid (115). This new total synthesis is characterized by direct application of a stable versatile derivative of the unstable monomer 4-amino-1-methylpyrrole-2-carboxylic acid,as well as the preformed amino amidine side chain.

Among the many compounds which possess antiviral activity there are a number of compounds with a chemical structure related to distamycin A. These compounds are included in a group named pyrrole amidine antiviral

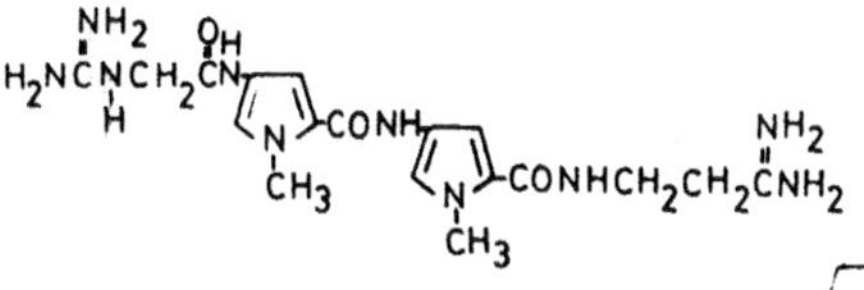

CONGOCIDINE (NETROPSIN)

(III)

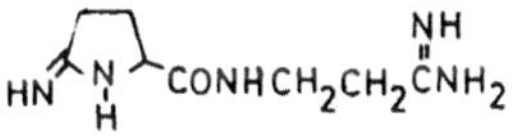

NOFORMYCIN IV

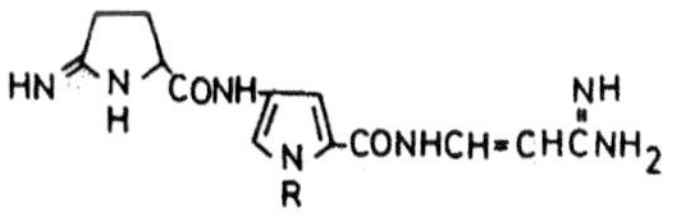

KIKUMYCIN (VI) R=H or CH_3

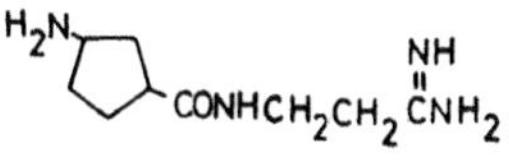

(VII)

AMIDINOMYCIN

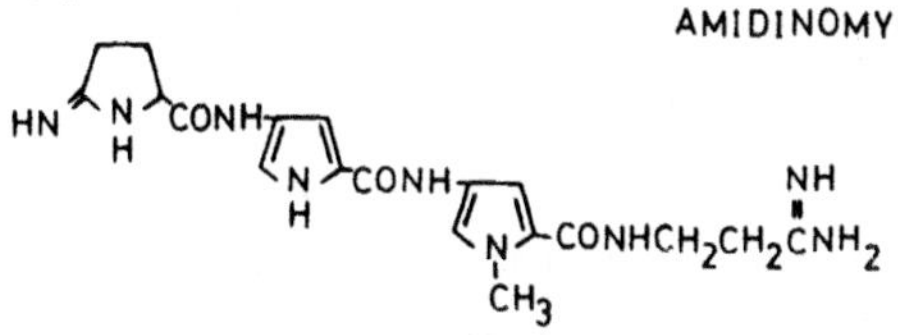

ANTHELVENCIN (TENTATIVE) (V)

antibiotics (50-55). The members of this group are: congocidine (netropsin) (III), noformycin (IV), anthelvencin (V), kikumycin (VI) and amidinomycin (VII).

Congocidine (III) is a basic oligopeptide isolated from the fermentation medium of Streptomyces chromagens and Streptomyces umbifacirens (56,57). Its chemical structure was established by a series of degradations and by a total synthesis (58,59). An antibiotic named netropsin was isolated from the fermentation medium of Streptomyces netropsin (60), and different chemical structures were assigned to it (61-64). In 1964, the correct chemical structure of netropsin was established by proving its identity to congocidine (65). Like distamycin A congocidine exhibits UV absorption spectra with two characteristic maxima (65). Both antibiotics are optically inactive. In aqueous solution at 4°, congocidine is stable for 3 days and distamycin A for approximately 24 hr. After this time, there is a loss of the absorbance that amounts to 10% of the initial value after 8 days for congocidine and 18% for distamycin A after 6 days (54).

Congocidine has very interesting antibacterial, antiparasitic and antiviral activities. As an antibacterial agent,congocidine inhibits gram-positive and gram-negative bacteria in contrast to distamycin A which inhibits only gram-positive bacteria (66,67). Unlike distamycin A, congocidine shows antiparasitic activity against Trypanosoma congolense (the source of its name) and other parasites (56,57,65). As an antiviral drug, congocidine inhibits the multiplication of DNA viruses, such as vaccinia virus (68), HSV (20) and SFV (17) and retroviruses, such as RSV (26), murine leukemogenic Rauscher virus (69) and FeLV (70).

A very limited amount of information was available until 1980 on the structural requirements for the antiviral or biological activity of congocidine. In 1967 derivatives of congocidine were synthesized in which the N-methyl pyrrole ring was substituted by meta- or para-benzene, and triazine diamidine derivatives (71-73); but there was no report about the antiviral or biological activity of these derivatives.

Replacement of the N-methylpyrrole ring in congocidine by analog moieties such as thiophene, pyridine and benzene leads to reduction of the antiviral and antiparasitic activities (74). Replacement of the guanidinoacetyl side chain by an amine moiety causes a reduction in the binding of congocidine to DNA (75).

Congocidine and distamycin A are very similar in their chemical structures. They differ mainly in the number (n) of pyrrole rings. For congocidine, n = 2, for distamycin A, n = 3. When distamycin (tripeptide) is converted into tetra- or penta-peptide analogs, the antiviral activity against some viruses is increased (17,20,25,43-45). At the same time, the cytotoxicity is somewhat reduced, indicating that the antiviral and cytotoxic activities are separable.

Similar mono- and dipeptide homologs of distamycin A proved to be more toxic and less effective than the parent drug (27,43,44). The fact that the dipeptide congocidine demonstrates antiviral activity led us to challenge the effect of "n" on the antiviral and antiparasitic activity in the congocidine series (76). Additionally, four tripyrrole side chain derivatives of congocidine were prepared and tested. Some of these congocidine derivative compounds (IV-VII) are as follows:

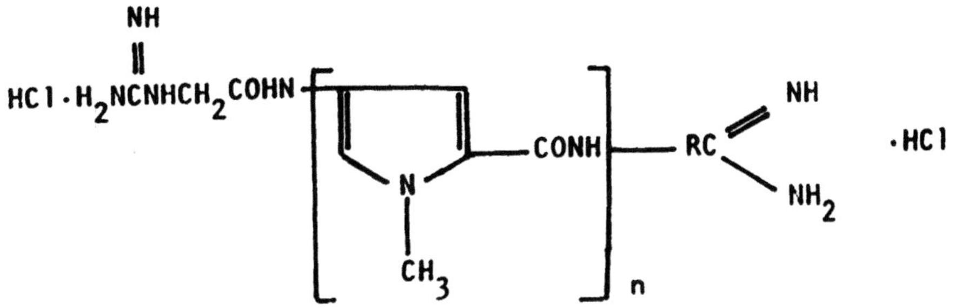

(VIII)	n = 3	R = CH_2
(IX)	n = 3	R = CH_2CH_2
(X)	n = 3	R = $CH(CH_3)CH_2$
(XI)	n = 1	R = CH_2CH_2

All these tripyrrole derivatives were synthesized by using our total synthesis of distamycin A with a small modification at the last step for the synthesis of congocidine (40,76). These compounds were tested for cytotoxicity, inhibition of replication in cultured cells, for their effects on the synthesis of HSV DNA in isolated nuclei in vitro, and on DNA synthesis by the purified HSV DNA polymerase. All the synthesized tripyrrole derivatives of congocidine were less cytotoxic and more active than congocidine in all the three antiviral tests. These compounds were as active and less cytotoxic than distamycin A (78).

These congocidine analogs were tested against the parasites Trypanosoma congolense and Leishmania tropica (77,78). The tripyrrole derivative β-[N-methyl-4-[N-methyl-4-[N-methyl-4-(guanidinoaietamido)-pyrrole-2-carboxamido]pyrrole-2-carboxamido]pyrrole-2-carboxamido]-butyramidine dihyrochloride, was less toxic and more active than congocidine. Distamycin A and all its other tested derivatives were totally nonactive antiparasitically. The guanidinoacetyl moiety appears to be a structural requirement for antiparasitic activity in the congocidine series (77,78). Isomers of mono- and tripyrrole homologs of congocidine containing 2,5 disubstituted pyrroles were less active antivirally than congocidine or distamycin A (42).

Noformycin (IV) was isolated from a culture of Nocardia formica and was identified as the active constituent of the microorganism (79-82). This material has quite a wide range of antimicrobial activity but it also appears to possess considerable toxicity. Of particular interest is its in vivo activity in mice against swine influenza and poliomyelitis (79-82). Noformycin was also reported to exhibit in vitro activity against equine rhinovirus, human rhinovirus type 2 and parainfluenza type 3 (83). Noformycin is the only member of the pyrrole amidine-antiviral antibiotics which demonstrates antiviral activity against RNA viruses. There is only one report on SAR study in which 24 derivatives of noformycin were synthesized. All these derivatives were less active antivirally than the parent drug when tested against human rhinovirus type 2 (83).

Anthelvencin A and B (V) have been isolated from culture filtrates of Streptomyces venezuelae. Anthelvencin inhibits a broad spectrum of microorganisms in vitro and is effective in mice against several gram-negative bacteria. In mice, the LD_{50} was 177 mg/Kg given I.P. and > 500 mg/Kg given orally. Nematode infections in mice and pigs are effectively controlled by oral administration of anthelvencin (84).

Kikumycin A and B (VI) were isolated from the culture filtrate of Streptomyces phaeochomagens R-719. Kikumycin has a broad antibacterial spectrum, though the minimum inhibitory concentration (MIC) is in the order of 10 μg/ml (85). The chemical structure identification of kikumycins

A and B was done by means of low and high mass spectrometry (86,87).

Amidinomycin (myxoviromycin) (VII) has been obtained from cultures of Streptomyces flavochromagens (88-90). It inhibits the growth of spore-bearing bacteria such as B. subtilis, B. megathenium and B. anthracis (91). A report was published in 1969 on a SAR study of amidinomycin (92).

During the past several years, evidence has been presented that distamycin A and congocidine bind tightly to double helical (dA-dT)-rich DNA regions, but poorly to (dG-dC)-rich domains, to RNA and to single-stranded DNA (50,52,54). Distamycin A, as well as congocidine, form a complex with DNA which causes poisoning of the template for the DNA polymerase activity (95).

The distamycin A-DNA complex is characterized by the following observations:

1) A bathochromic shift of the UV maximal absorbance at 303 nm. The red shift is from 303 nm to 326 nm or 340 nm (94,95).

2) A melting temperature increase of DNA with sharpening of the transition (96).

3) The typical circular dichroism (CD) spectrum of the DNA-distamycin complex exhibits very large amplitudes with increasing oligopeptide concentrations (45). Binding of distamycin A to DNA containing partially A-type structures is drastically diminished as indicated by a reduced CD amplitude at 330 nm (95).

4) Like other drugs which bind to DNA, distamycin A displaces methyl green from its complex with DNA (97,98). At similar molar concentrations of methyl green bound to DNA and of free distamycin A, the displacement reaction is of the second order with time and has a constant rate.

5) The fact that the antibiotic binds very strongly to DNA is shown by the fact that 0.1 M NaCl, 0.01 M Mg acetate or 6 M urea do not reverse the DNA-induced bathochromic shift in the absorption spectrum of distamycin (97). Extremely high $NaClO_4$ concentrations of 2 M LiCl plus 8 M urea, which are known as hydrophobic bond-breaking agents, probably change the DNA conformation and this allows an effective dissociation of the oligopeptide complex (96,99).

6) Distamycin A causes large shifts in the thermal denaturation profile (melting curve) of calf thymus DNA to higher temperature (90).

7) An essential difference between distamycin A and congocidine was found in the viscosity behavior complex with DNA contrary to congocidine: distamycin A

causes a lowering of η with increasing molar ratio up to 0.08, while at higher oligopeptide concentration, η starts to increase (93,96).

Distamycin A is bound to DNA without intercalation (101). Congocidine and distamycin A are bound to the (dA,dT)-rich region of DNA through hydrogen bonding and van der Waals forces. A second type of binding to G-C pairs in DNA also occurs (101-107).

The results of various physico-chemical investigations of the interaction of distamycin A, congocidine and their analogs enable one to propose a scheme for recognition of the DNA A-T-rich regions by the distamycin-type antibiotics (48, 108-110). According to this scheme the oligopyrrole carboxamide backbone of the distamycin molecule is localized in the minor grove of DNA and forms a helix isogeometrical to that of the B-form. The AT-specific interaction is due to hydrogen bond formation between the nitrogen atoms of the amide groups of the antibiotic molecules, the O_2 atom of thymines and the N_3 atom of adenines of the same DNA strain (109, 110). By using the proton nuclear magnetic resonance (H-NMR) technique it was established that the preferential conformation of the pyrrole carboxamide backbone of the distamycin A molecule in solution is very close to its possible conformation in the distamycin-DNA complex (110-114). This evidence was confirmed by an X-ray study of a dipyrrolic distamycin A analog (114).

CONCLUSIONS

Distamycin A and congocidine are the two major compounds of the groups named pyrrole amidine antiviral antibiotics. Both possess antiviral activity with the same mechanism of action against DNA and retroviruses. Congocidine has a wider antibacterial spectrum than distamycin A and it possesses antiparasitic activity which distamycin A does not have.

In SAR studies of the two compounds, replacement of the guanidinoacetyl moiety cancelled the critical need for the β-aminopropionamidine side chain as one of the structural requirements for the antiviral activity in the distamycin series. On the other hand, in the antiparasitic tests the guadino-acetyl moiety cannot be replaced by a formamido moiety. Moreover, increasing the pyrrole rings from one to three alone does not improve the antiparasitic activity though it improves the therapeutic index. Addition of a methyl group to the β-aminopropionamidine side chain increases the antiparasitic activity and further improves the therapeutic index.

REFERENCES

1. Arcamone, F., Biozioli, F., Canevazzi, G. and Grein, A. German Patent 1,027, 667, 1958; through Chem. Abstr. 55:2012, 1961. (Soc. Pharm. Italia: Brit. 872, 734 (12.7.61); U.S. 3, 190, 801 (22.6.1965); Germ. 1.039, 198 (18.9.1958).).
2. Arcamone, F., Penco, S., Orezzi, P.G., Nicollela, V. and Prelli, A. Nature 203: 1064-1066, 1964.
3. Arcamone, F., Orezzi, P.G., Barbieri, W., Nicolella, V. and Penco, S. Gazz. Chim. Ital. 97: 1097-1109, 1967.
4. Penco, P., Redaelli, S. and Arcamone, F. Gazz. Chim. Ital. 97: 1110-1115, 1967.
5. DiMarco, A., Gaetani, M., Orezzi, P., Scotti, T. and Arcamone, F. Cancer Chemother. 18: 15, 1962.
6. DiMarco, A., Ghione, M., Sanfilippo, A. and Morvillo, E. Experientia, 19: 134-136, 1963.
7. DiMarco, A., Ghione, M., Migliacci, A., Morvillo, E. and Sanfilippo, A. Giorn. Microbiol. 11: 87-100, 1963.
8. Sanfilippo, A., Monvillo, E. and Ghione, M. J. Gen. Microbiol. 43: 369-376, 1966.
9. Holldorf, A.W., Friebe, B. and Stoler, M. Zenter, Bakt. Parasitenk 42: 265-269, 1970.
10. Sanfilippo, A., Ann. N.Y. Acad. Sci. 182: 322-329, 1971.
11. Arcamone, A., Migliacci, A., Morvillo, E., Nicolella, V., Sanfilippo, A. and Schippcasse, G. Farmaco. Ed. Sci., 30: 859-869, 1975.
12. Verini, M.A. and Ghione, M. Chemotherapia 9: 144-157, 1964.
13. Verini, M.A., Casazza, A.M. and Fioretti, A. Personal communication from Farmitalia Co., Milano, Italy.
14. Werner, G.H., Ganter, P. and de Datuld, Y. Chemotherapia 9: 65-71, 1964.
15. Casazza, A.M. and Ghione, M. Chemotherapia 9: 80-87, 1964.
16. Casazza, A.M., Fioretti, A., Ghione, M., Soldati, M. and Verini, M. Antimicrob. Ag. Chemother. 593-598, 1966.
17. Becker, Y., Asher, Y. and Zakay-Rones, Z. Antimicrob. Ag. Chemother. 483-488, 1972.
18. de Ratuld, Y. and Werner, G.H. Antimicrob. Anticancer Chemother. 2: 14-16, 1979.
19. Casazza, A.M., Fioretti, A., Sanfilippo, A. and Verini, M.A. Antimicrob. Antineoplast. Chemother. Vol. I/1: 357-359, 1971.
20. Becker, Y., Neuman, S. and Levitt-Hadar, J. Isr. J. Med. Sci. 8: 1225-1230, 1972.
21. Verini, M.A., Fioretti, A., Casazza, A.M. Giorn. Mal. Infet. Parassit. 24: 790-794, 1972.
22. Verini, M.A., Diamico, G., Sanfilippo, A. and Arcamone, F. Farmaco Ed. Sci. 31: 705-713, 1976.
23. Becker, Y. and Weinberg, A. Isr. J. Med. Sci. 8: 75-78, 1972.
24. Brandner, G. and Koch, E., Z. Naturforsch. 31c: 214, 1976.
25. Kotler, M. and Becker, Y. Nature 234: 212-214, 1971.
26. Kotler, M. and Becker, Y. FEBS Lett. 21: 222-226, 1972.
27. Chandra, P., Zuinno, F., Gotz, A., Wacker, A., Gericke, P., Di Marco, A., Casazza, A.M. and Gilvani, F., FEBS Lett. 21: 154-158, 1972.
28. Chandra, P., DiMarco, A., Zunino, F., Casazza, A.M., Gericke, D., Giuliani, F., Soranzo, C., Thorbeck, R., Gotz, A., Arcamone, F. and Ghione, M. Naturwissenschaften 59: 448-455, 1972.
29. Bassetti, D., Giorn. Mal. Infett. Paras. 20: 827-832, 1968.
30. Bassetti, D., Giorn. Mal. Infett. Paras. 21: 849-855, 1969.
31. Scarzella, M., Miverva Pediat. 22: 1444-1448, 1970.

32. Bassetti, D., Antimicrob. Antineoplast. Chemother. Vol. I/1: 297-298, 1971.
33. Toscano, F. and Paschetta, A. Antimicrob. Antineoplast. Chemother. Vol. I/1: 299-300, 1971.
34. Bassetti, D. Minerva Med. 64: 2077-2079, 1973.
35. Chandra, P., Gotz, A., Wacker, A., Verini, M.A., Casazza, A.M., Fioretti, A., Arcamone, F. and Ghione, M. FEBS Lett. 19: 327-330, 1972.
36. Arcamone, F., Penco, S. and Delle Monache, F. Gazz. Chim. Ital. 99: 620-631, 1969.
37. Chandra, P., Gotz, A., Wacker, A., Zunino, F., Di Marco, A., Verini, M., M.A., Casazza, A.M., Fioretti, A., Arcamone, F. and Ghione, M. Hoppe-Seyler's Z. Physiol. Chem. 353: 393-398, 1972.
38. Chandra, P., Hausmann, S. and Metz, A. Pharm. Res. Comm. 6: 311-317, 1974.
39. Arcamone, F., Med. Chem. Spec. Contrib. Int. Symp., 3rd, 1972, p 29, 1973.
40. Bialer, M., Yagen, B. and Mechoulam, R. Tetrahedron 34: 2389-2391, 1978.
41. Bialer, M., Yagen, B., Mechoulam, R. and Becker, Y., J. Med. Chem. 22: 1296-1301, 1979.
42. Bialer, M., Yagen, B., Mechoulam, R. and Becker, Y., J. Pharm. Sci., 69: 1334-1338, 1980.
43. Arcamone, F., Nicolella, V., Penco, S. and Redaelli, S. Gazz Chim. Ital. 99: 632-640, 1969.
44. Chandra, P., Gotz, A., Wacker, A., Verini, M.A., Casazza, A.M., Fioretti, A., Arcamone, F. and Ghione, M. FEBS Lett. 16: 249-252, 1971.
45. Casazza, A.M., Gambarucci, C., Giuliani, F. and Di Marco, A. Antimicrob. Antineoplast. Chemother. Vol. I/1: 375-376, 1971.
46, Werner, E.G., Obermier, J., Maidhof, A. and Zahn, R.K. Chem. Biol. Inter. 8: 183-192, 1974.
47. Grokhovsky, S.L., Zhuze, S. and Gottikh, B.P. Bioorg. Khim. 1: 1616-1623, 1975.
48. Kolchinski, A.M., Mirzabekov, A.D., Zasedateelev, Ac., Gurski, G.Y., Grokhovsky, S.L., Zhuze, A.L. and Gottikh, B.P. Molekulzarnaya Biologiya 9: 19-27, 1975.
49. Gender, P.L. and Rapaport, H., J. Med. Chem. 24: 33-38, 1981.
50. Hahn, F.E. In: Antibiotics III (Eds. J.W. Corcoran and F.E. Hahn), Springer-Verlag, Berlin-Heidelberg-New York, 1975, 79-101.
51. Chandra, P. Topics in Current Chemistry 52: 99-139, 1974.
52. Chandra, P. Prog. Mol. Subcell. Biol. 4: 167-226, 1976.
53. Becker, Y. Israel J. Med. Sci. 11: 1135-1148, 1975.
54. Zimmer, C. Prog. Nucl. Acids Res. 15: 285-318, 1975.
55. Chirigos, M.A. and Papas, T.S. Advanc. Pharm. Chemother. 12: 89-124, 1975.
56. Cosar, C.R. and Ninet, L. C.R. Hebd. Seances Acad. Sci. 234: 1498-1499, 1952.
57. Depis, C. and Ninet, L. Abstr. Int. Microbiol. (Rome) 6th, 162: 241-242, 1953.
58. Julia, M. and Preau Joseph, N., C.R. Hebd. Seances Acad. Sci. 243: 961-964, 1956.
59. Julia, M., Preau Joseph, N. C.R. Hebd. Seances Acad. Sci. 257: 1115-1118, 1963.
60. Finaly, A.C., Hochstein, F.A., Sobin, B.A. and Murphy, F.X. J. Amer. Chem. Soc. 73: 341-343, 1951.
61. van Tamelen, E.G., White, D.W., Kogon, I.C. and Powell, A.D.G. J. Amer. Chem. Soc. 78: 2157-2159, 1956.
62. Waller, C.W., Wolf, C.F., Stein, W.J. and Hutchings, B.L. J. Amer. Chem. Soc., 79: 1265-1266, 1957.

63. Weiss, M.J., Webb, J.S. and Smith, J.M. J. Amer. Chem. Soc. 79: 1266, 1957.
64. van Tamelen, E.G.and Powell, A.P.G. Chem. and Ind. 365, 1957.
65. Nakamura, S., Vonehara, H. and Umezana, H. J. Antibiotics (Ser. A.) 14: 220-221, 1964.
66. Thrum, H., Haupt, I., Bradler, G., Zimmer, C. and Reinert, K.G. Antimicrob. Antineoplast. Chemother. Vol. I/2: 819-822, 1971.
67. Schuhmann, E., Hapt, I., Thrum, H. and Tabubeneck, U. Z. Allg. Mikrobiologie 14: 321-327, 1974.
68. Schabel, M., Laster, W.R., Brockman, R.W. and Skipper, H.E. Proc. Soc. Biol. Med. 83: 1-3, 1953.
69. Chirigos, M.A., Ruscher, R.J.,Kamel, I.A., Fanning, G.R. and Goldin, A. Cancer Res. 23: 1646-1650, 1963.
70. Sugiura, K. Gann. 50: 251-264, 1959.
71. Julia, M. and Preau Joseph, N. Bull. Soc. Chim. France 4348-4356, 1967.
72. Julia, M. and Gombert, R. Bull..Soc. Chim. France 369-375, 1968.
73. Julia, M. and Gombert, R. Bull. Soc. Chim. France 376-382, 1968.
74. Jones, D.O. and Wooldridge, K.R.H. J. Chem. Soc. (c) 550-554, 1968.
75. Zimmer, C., Luck, G., Thurm., H. and Pitra, C. Eur. J. Biochem. 26: 81-89, 1972.
76. Bialer, M., Yagen, B., Mechoulam, R. and Becker, Y. J. Med. Chem. 23: 1144-1148, 1980.
77. Bialer, M., El-On, J., Yagen, B. and Mechoulam, R. J. Pharm. Sci. 70: 822-824, 1981.
78. Bialer, M., Yagen, B., Mechoulam, R., Becker, Y. and El-On, J. Curr. Chemother. Immunother. 2: 1048-1050, 1982.
79. Harris, D.A. and Woodruff, H.B. Antibiot. Annun. 609-621, 1954.
80. Furusawa, E., Cutting, W. and Pursy, A. Chemotherapia, 8: 95-108, 1964.
81. Furusawa, E., Cutting, W., Buckley, P. and Furasawa, S. Proc. Soc. Exp. Biol. Med. 116: 938-944, 1964.
82. Diana, G.D., J. Med. Chem. 16: 857-859, 1973.
83. Diana, G.D., Salvador, U.J., Zalay, E.S. and Pancic, F. J. Med. Chem. 16: 1050, 1973.
84. Probst, G.W., Hoehn, M.N. and Woods, B.L. Antimicrob. Ag. Chemother. 789-795, 1973.
85. Kikuchi, M., Kamagui, K. Ishida, N., Ito, Y., Yamaguchi, T., Funumai and Okuda, T. J. Antibiotics (Ser.A.)., 18: 243-250, 1965.
86. Takaishi, T., Sugawara, Y. and Suzuki, M. Tetra. Lett. 1873-1876, 1972.
87. Takaishi, T., Suzuki, M. and Tatemazu, A. Or. Mass. Spect. 9: 635-648, 1974.
88. Nukamura, S., Ishida, N. and Umezawa, H. J. Antibiotics (Ser.A.) 14: 163, 1961.
89. Nakamura, S. Chem. Pharm. Bull. 9: 641-645, 1961.
90. Katsube, J. Chem. Pharm. Bull. 16: 232-238, 1968.
91. Takasawa, S., Utahara, R., Osato, Y., Maeda, K. and Umezawa, H. J. Antibiotics 21:
92. Paul, H., Berger, H., Booden, H. and Hilgetag, G. Arch. Pharm. 301: 512-519, 1969.
93. Chandra, P., Zimmer, C. and Thurm, H. FEBS Lett. 7: 90-94, 1970.
94. Krey, A.K. and Hahn, F.E., FEBS Lett. 10: 175-179, 1970.
95. Zimmer, C. and Luck, G. FEBS Lett. 10: 339-342, 1970.
96. Zimmer, C., Reinert, K.E., Luck, G., Loeber, G. and Thurm, H. J. Mol. Biol. 58: 329-
97. Krey, A.K. and Hahn, F.E. Proc. Europ. Bioph. Congr., 1st, 1: 223-225, 1971
98. Zuino, F. and Di Marco, A. Biochem. Pharmacol. 21: 867-873, 1972.

99. Luck, G. and Zimmer, C. Stud. Biophys. 558: 241-244, 1976.
100. Krey, A.K., Allison, K.G. and Hahn, F.E., FEBS Lett. 29: 58-62, 1973.
101. Becker, Y. Monographs in Virology, Vol. II, 1976, S. Karger, Basel.
102. Hahn, F.E., Krey, A.K. and Allison, R.G. Prog. Chemother. Proc. Int. Congr. Chemother, 8th 1: 398-401, 1974.
103. Reinert, K.E., J. Mol. Biol. 72: 593-607, 1972.
104. Wartell,R.M., Larsen, J.E. and Wells, R.D. J. Biol. Chem. 244: 6719-6731, 1974.
105. Wartell, R.M., Larsen, J.E. and Wells, R.D. J. Biol. Chem. 250, 2698-2703, 1975.
106. Luck, G., Triebel, H., Waring, M. and Zimmer, C. Nucl. Acids Res. 1: 503, 1974.
107. Dattagunta, N., Hogan, M. and Crotchers, P.M., Biochemistry 19: 5998-6005, 1980.
108. Krylov, A.S., Grokhovsky, S.L., Zasedatelev, A.S., Zhuze, A.L., Gurski, G.V. and Gottikh, B.P. Dokl. Akad. Nawk. S.S.S.R. 239: 732-735, 1978.
109. Zasedatelev, A.S., Zhuze, A.L., Zimmer, Ch., Grokhovsky, S.L., Tumayan, V.G., Gurski, G.V. and Gottikh, B.P. Dokl. Acad. Mauk. S.S.S.R. 231: 1006-1009, 1976.
110. Gurski, G.V., Tumanyan, V.G., Zasedatelev, A.S., Zhuze, A.L., Grohovsky, S.L. and Gottikh, B.P., in Nucleic Acid-Protein Recognition (Vogel, H.J. ed.) pp. 189-217, Academic Press, New York, 1977.
111. Turchin, K.F., Grokhovsky, S.L., Zhuze, A.L. and Gottikh, B.P. Bioorg. Khim. 4: 1065-1077, 1978.
112. Grokhovsky, S.L., Zhuze, A.L. and Gottikh, B.P. Bioorg. Khim. 4: 1078-1086, 1978.
113. Grokhovsky, S.L., Zhauze, A.L., Gottikh, B.P., Bioorg. Khim. 4: 1078-1085, 1978.
114. Gurskaya, G.V., Grokhovsky, S.V., Zhuze, A.L. and Gottikh, B.P. Biochim. Biophys. Acta 563: 336-342, 1979.
115. Grehn, L. and Ragnarsson, U. J. Org. Chem. 46: 3492-3497, 1981.

11

Clinical Insights into the Metabolism of Arabinosyl Adenine and Interferon as Derived by Microbiological Assays

A. Martin Lerner, M.D.
From the Department of Medicine, Wayne State University School of Medicine, Detroit, Michigan, USA

The past decade has witnessed the initiation of effective antiviral chemotherapy for disseminated varicella-zoster in immunosuppressed persons (1), herpes simplex virus keratitis (2), and herpes simplex virus encephalitis (3) among other conditions. Further advances in the treatment of a wide spectrum of viral diseases continue led by increments in basic understanding of the specific molecular biology of virally infected cells. My laboratory has emphasized the development of microbiologic assays to help guide the application of the treatment of virus infections. The basic tenet is that virus infection occurs within living cells, and that properly oriented in vitro microbiological assays in appropriate tissue cultures most closely simulate natural infection; therefore, the information derived might potently augment insights from chemical tests. Chemical (usually by high pressure liquid chromatography) or radioimmunoassay tests are vital to establishing firm in vivo pharmacokinetics of antiviral agents. Here, I outline a series of selected studies illustrating the special value of microbiological assays.

Experiments

No. 1. Assays of idoxuridine in serum, urine and cerebrospinal fluid of patients with suspected diagnoses of herpes simplex

Y. Becker (ed.), Antiviral Drugs and Interferon: The Molecular Basis of Their Activity.

Table 1.
Concentrations of IDU in Serum, Urine, and CSF

Patient	Day of treatment with IDU	IDU Serum	IDU Urine	IDU CSF
			mg/ml	
A. Received slow intravenous infusions of IDU, 4 mg/min				
1	1		55	
	+6*		≦2.5	
2	-4	≦2.5		
	4	≦2.5		
	5	≦2.5		
	+1	≦2.5		
	+17	≦2.5		
3	-4	≦2.5		
	1	≦2.5		
	2	≦2.5		
	3	≦2.5		
	4			≦2.5
	+5	≦2.5		
4	-1	≦2.5		
	1	≦2.5	≦2.5	
	2	≦2.5		
	3	33	≦2.5	≦2.5
	4	N.D.		≦2.5
	+2	≦2.5		≦2.5
5	-1	≦2.5	≦2.5	
	1	≦2.5	N.D.	833**
	2	≦2.5		
	3	≦2.5	45	≦2.5
	4	N.D.	≦2.5	
	+3	≦2.5		
6	-1	≦2.5	N.D.	
	3	N.D.	80	
	4	N.D.	N.D.	≦2.5
	5	≦2.5	16, ≦2.5***	
	+4	≦2.5	≦2.5	
7	-2	≦2.5		
	-1			≦2.5
	1	9		
	3	≦2.5		
	4	≦2.5	64, ≦2.5	
8	-5	≦2.5		
	1	≦2.5		
	4	≦2.5		
	5	≦2.5		
	+1	≦2.5		

Table 1. (continued)
Concentrations of IDU in Serum, Urine, and CSF

Patient	Day of treatment with IDU	IDU Serum	Urine	CSF
			mg/ml	
B. Received rapid intravenous infusion of IDU, 50 mg/min				
9	-1*	≦2.5		
	1 (15 min)****	10		
	1 (30 min)	10		
	1 (45 min)	20	1040	
	1 (60 min)	36		
	+1		≦2.5	

*+, day after cessation of intravenous IDU; -, day before beginning intravenous IDU; N.D., not done.
**CSF was obtained for assay immediately after beginning a 3 g infusion of IDU.
***Represents collection of two random samples of urine during day.
****Time after start of intravenous infusion of 3 g of IDU, delivered over 60 min.

Reprinted with permission from (4).

virus encephalitis (Table 1). A reproducible microbiologic assay in baby hamster kidney cells (BHK-21) of microgram quantities of 5-iodo-2'-deoxyuridine (idoxuridine, IDU) in serum, urine or cerebrospinal fluid was developed (4). By allowing virus to absorb to tissue cultures prior to application of the test body fluid, the antiviral assay was not interfered with by type-specific antibody. By using non-primate tissue cultures, the assay did not measure interferon. During slow intravenous infusions of idoxuridine (4 mg/min) in patients with suspected diagnoses of herpes simplex virus encephalitis, the rate of inactivation presumably by de-iodinization to inactive compound and/or removal of drug exceeded its administration. During rapid infusions of IDU (50 mg/min), however, significant quantities of the drug were found in serum, urine and cerebrospinal fluid. We used these assays to confirm that IDU was not significantly bound to serum proteins. Neither was IDU deiodinated in fresh serum or urine in vitro to its inactive products (iodouracil, uracil, iodide). It was rapidly excreted in the urine. Inactivation of IDU occurred in tissues. Further, minimal inhibitory concentrations (MIC) of IDU for fresh isolates of herpes simplex virus (Type 1 or 2, HSV-1, HSV-2) were determined. HSV-1 formed microplaques in baby hamster kidney cells and was sensitive to 2.5-10 μg/0.4 ml. HSV-2 macroplaques required 25-50 μg/0.4 ml. The latter characteristic was a biologic marker suggesting type-specificity of HSV isolates.

These transitory therapeutic drug levels were challenging but ultimately predicted the failure of efficacy and excessive toxicity of idoxuridine. This study was the first non-chemical measure of antiviral drug in human body fluids.

No. 2. Synergy of 9-β-D-arabinofuranosyladenine (adenine arabinoside, arabinosyl adenine, ara-A, vidarabine) and human interferon against herpes simplex virus, type 1 (5).

MICs of IDU, ara-A, 9-β-D arabinofuranosyl hypoxanthine (arabinosyl hypoxanthine, ara-H) and human leukocyte interferon (IFN-) were determined in an in vitro Vero tissue culture system with the Ket. strain of HSV-1. These values were 1.1 μg/ml, 22.7 μg/ml, 34.1 μg/ml and 91 μ/ml, respectively. The action of combinations of these agents against the Ket. strain was studied. An additive effect was noted when ara-A was combined with ara-H, but this effect was not observed when ara-A was combined with IDU. Human IFN- and ara-A were synergistic (Table 2).

This was the first demonstration of in vitro synergy of antiviral drugs in combination. Today, we know that the species and

Table 2. Effect of the combination of 9-β-arabinofuranosyladenine (ara-A) and human leukocyte interferon (IF) on activity against herpes simplex virus, type 1, (strain Ket.) [HSV-1 (Ket.)].

Virus control*	Ara-A (ug/ml)	IF (units/ml)	pfu	No. pfu of HSV-1 (Ket.) at 50% reduction in control virus at MIC	Total of fractional MICs of ara-A plus IF yielding one MIC of antiviral activity
A	...	...	72	36	
	34	...	13		
	22	...	21		
	11	...	55		
B	...	...	70	35	
	...	182	33		
	...	91	32		
	...	46	45		
C	...	...	67	34	
	11	91	11		
	11	46	32		0.50 ara-A + 0.50 IF
D	...	...	67	34	
	5.5	46	27		0.25 ara-A + 0.50 IF
	5.5	23	33		0.25 ara-A + 0.25 IF

NOTE. The MICs of ara-A and IF were found to be 22.7 μg/ml and 91 units/ml, respectively.

*The same sample of stock virus was used throughout. It was thawed and kept at 4 C during the experiment. At the beginning of each segment of the assays (about every 10 min.), stock virus was assayed again. (Reprinted with permisssion from (5)).

Table 3. Reduction in pfu of herpes simplex virus type 1 (HSV-1) by increasing amounts of an inhibitor of adenosine deaminase (DI; Parke, Davis preparation CL-67310465, lot no 95).

Inhibitor (μg/0.4 ml)	HSV-1, Ket. (pfu)*
0	56
8	56
17	53
25	45
50	38
100	29
166	26

*Assayed in Vero Monkey kidney cell tissue cultures. (Reprinted with permission from (6)).

even organ in tissue cultures may determine the presence or absence of synergy.

No. 3. Antiviral activity of a preparation of an adenosine deaminase inhibitor (6).

A unique seven-membered heterocyclic inhibitor of adenosine deaminase was examined for purity and activity by microbiologic assays. One preparation of the compound inhibited replication of HSV in the absence of ara-A. In this capacity, the MIC of the deaminase inhibitor for HSV-1, with 50% reduction of plaque-forming units as the end point, was 37.7 μg/ml. This activity compared favorably with the inhibitory activity of ara-H (34.1 μg/ml). Another preparation of deaminase inhibitor lacked antiviral activity (Table 3). The nature of the possible contaminant of the preparation in question is unknown. Coformycin, another inhibitor of adenosine deaminase, had no antiviral activity in the absence of ara-A.

Thus, a simple microbiological assay detected an apparent contamination of an inhibitor of adenosine deaminase.

No. 4. Differential sensitivity of herpes simplex virus types 1 and 2 to human interferon: antiviral effects of interferon plus ara-A in human cells (7).

The standard assay for the MIC of antiviral compounds adds the preparation to be tested after absorption of virus. When the MICs to human IFN for several strains of HSV-2 were tested in this fashion, values were 5-10x greater for HSV-2 than it was for 2 strains of HSV-1. This differential susceptibility of HSV types to IFN was found whether tests were done with a liquid overlay and microtiter plates or with agarose overlays and appeared to be a

Table 4. MICs of human interferon for herpes simplex virus types 1 and 2, as tested by several methods.

Type (strain) of herpes simplex virus	MICs of interferon (units/ml) by indicated method*			
	Interferon added after virus**		Interferon added 24 hr before virus	
	Microtiter	Agarose	Microtiter	Agarose
Type 1 (Ket)	1,880	1,364	60	109
Type 2 (MS)	>15,000	>5,455	60	14

*Each test was repeated at least once; values listed are from representative experiments

**This method is the standard antiviral assay in our laboratory.

Reprinted with permission from (7).

distinguishing biological marker (Table 4). When the MIC of IFN was tested by microtiter or agarose methods and IFN was allowed to incubate for 24 hr before virus was added, values for HSV-1 and HSV-2 were similar and much smaller in magnitude.

This experiment is striking. One possible clinical interpretation of this experiment is that as a prophylactic agent IFN is equally effective versus HSV-1 and HSV-2. However, in a therapeutic test (e.g., after virus absorption), which is always the case during recurrent disease, IFN would be generally less effective especially in the case of HSV-2 unless much higher quantities of IFN were available. This experiment preceded studies of the natural course of recurrent infection with HSV-1 and HSV-2 (8). More patients with HSV-2 infection have frequent recurrent disease and asymptomatic primary infection is uncommon. On the other hand, with HSV-1 infection, recurrences are often less frequent and subclinical primary infection is common.

No 5. Inhibitory and lethal concentrations of 9-β-D-arabinofuranosyladenine (ara-A) and its hypoxanthine-derivative (ara-H) versus herpes simplex virus, type 1 (9).

MICs of ara-A and ara-H at end points of 50 percent (MIC_{50}) and 100 percent (MIC_{100}) reduction to challenges of approximately 50 pfu of HSV-1 were determined in Vero renal tissue cultures. Adenosine deaminase is universally present in tissue cultures and serum. These same tests were repeated in the presence of a potent inhibitor of adenosine deaminase, covidarabine (co-ara-A). Addition of co-ara-A to assays of MIC_{50} or MIC_{100} for ara-A ensures standard reproducible results which can be compared in different laboratories. After incubations of HSV-1 in infected cultures for 96 hrs, 35^0, with concentrations of ara-A or ara-H at the MIC_{100} and over, cells were scraped and sonicated. Supernates were then re-inoculated into Vero flasks free of antiviral agents to determine minimum lethal concentrations (MLCs). Standard values (μg/ml) for ara-A with co-ara-A are 11.3 (MIC_{50}), 17.0 (MIC_{100}) and 34.0 (MLC) but were 68.1 (MIC_{50}), 170.4 (MIC_{100}) and 375 (MLC) for ara-H (Table 5). These data confirm that as a virustatic agent (MIC_{100}) ara-A is 10 times more active than ara-H. Ara-A and ara-H have virucidal potentials which require approximately 2 times the respective MIC_{100}. At the MLCs for ara-A and ara-H normal cellular protein and DNA synthesis as measured by ^{14}C-amino acid or ^{14}C-thymidine uptake was unimpaired.

This was the first demonstration that a virus-infected cell could be "cured" of infection by an antiviral agent.

No. 6. Antiherpesvirus activity in human sera and urines after administration of ara-A (10).

The ineffectiveness of idoxuridine and cytosine arabinoside as antiviral agents in patients appeared to correlate with

Table 5. Inhibitory and lethal concentrations of adenine arabinoside versus herpes simplex virus, type 1 (HSV-1, Ket)

Ara-A (µg/ml)	Co-ara-A* (2.2 µg/ml)	HSV-1 (p.f.u.)	Reduction in p.f.u. (%)	HSV-1 (pfu) after Subculture
A. 0	0	62	0	TNTC**
22.7	0	34	45	TNTC
79.5(MIC_{50})***	0	1	98	TNTC
90.9(MIC_{100})	0	0	100	TNTC
102.2	0	0	100	TNTC
113.6	0	0	100	18
125.0(MLC)	0	0	100	0
136.3	0	0	100	0
B. 0	+$	65	0	ND$
2.8	+	45	31	ND
5.6	+	38	42	ND
11.3(MIC_{50})	+	16	75	ND
17.0(MIC_{100})	+	0	100	TNTC
22.7	+	0	100	34
34.0(MLC)	+	0	100	0
45.4	+	0	100	0

*Co-Ara-A inhibitor of adenosine deaminase.
**TNTC, too numerous to count.
***The MIC_{50} is probably closer to 30 ug/ml.
$Present, ND, not done.

Reprinted with permission from (9).

pharmacokinetics showing transient measureable therapeuric levels in body fluids as well as excessive clinical toxicity. In human trials, ara-A was shown to be of benefit in the treatment of disseminated varicella-zoster in immunosuppressed persons and in HSV encephalitis (1, 3). However, measures of ara-A and ara-H in sera of patients treated with the recommended dosages of ara-A (10-20 mg/kg/day, intravenously) were below the MIC for each agent individually. This therapeutic paradox for ara-A was pharmacokinetics (high pressure liquid chromatography) predicting clinical ineffectiveness with paradoxical proved clinical usefulness (10).

We attempted to assay total ara-A plus ara-H in the body fluids of patients treated with ara-A in a microbiological assay expressed as "ara-A equivalents". The latter assay compared favorably in sensitivity with high pressure liquid chromatography. Measured serum values with this microbiological assay were sustained and above the MIC to HSV-1. The pharmacokinetics expressed by the microbiologic assay were consistent with the clinical effectiveness obtained in controlled clinical trials. Inhibition of cytopathic effects of HSV-1 in microtiter wells of rabbit kidney tissue cultures (RK-13) varied directly with the concentrations of ara-A or ara-H, and inversely with residual HSV-1 (Figure 1).

Of course, other explanations for this apparent paradox are possible including partial effects in combination with interferon and antibody, but the concept of "ara-A equivalents" and _in vitro_ and _in vivo_ synergy is also very attractive.

_No. 7_. Interferon in biopsy and autopsy specimens of brain. Its presence in herpes simplex virus encephalitis (11).

Quantitative virus and IFN titers were assayed in temporal lobe brain biopsy specimens (6 patients); in various regions of the brain at autopsy (5 of 6 patients); and in serums and CSF during courses of HSV-1 encephalitis. Until the 10th day of neurological disease, IFN (geometric mean titer, 25 u/ml) was present in each virus-positive brain biopsy specimen. The geometric mean HSV-1 titer at brain biopsy was 328 TCD_{50}/g. On the 6th day of disease, CSF from the sole survivor contained 160 u/ml, but 5 other samples of CSF contained no detectable IFN. No serums from any of the 6 patients contained IFN. At autopsies performed on the 6th to 19th day of neurologic disease, 13 of 35 of the regional areas of the

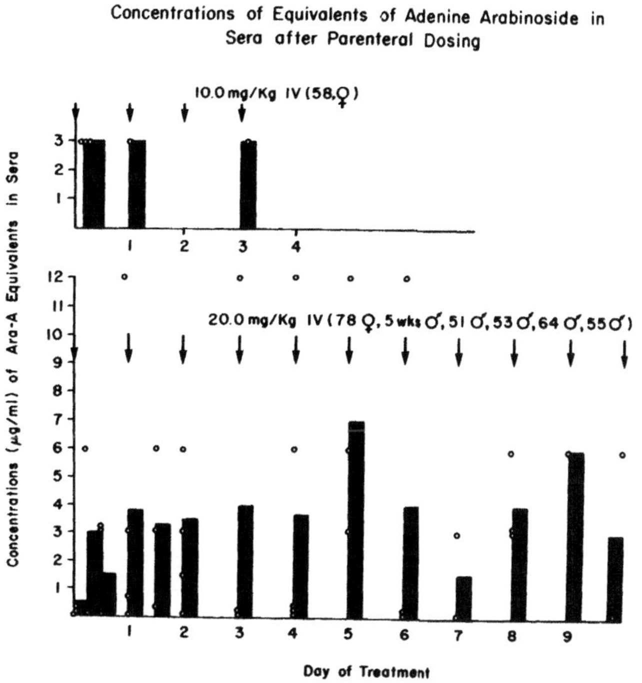

Figure 1. Concentrations of equivalents of ara-A in sera after parenteral dosing. AVA as ara-A equivalents in sera of patients receiving intravenous ara-A daily is shown. Covidarabine was not added to these specimens. The dots represent individual determinations, bars are arithmetic means, and patients' ages (yr.) are shown in parentheses. Urine values are concentrations per total collections over period of 4 hr. (Reprinted with permission from (10)).

brain retained HSV, but only a single specimen from a temporal lobe had measurable IFN. Early in the course of HSV encephalitis, substantial IFN titers are present in areas of virus multiplication (Figure 2). Local brain IFN in these cases was insufficient alone to

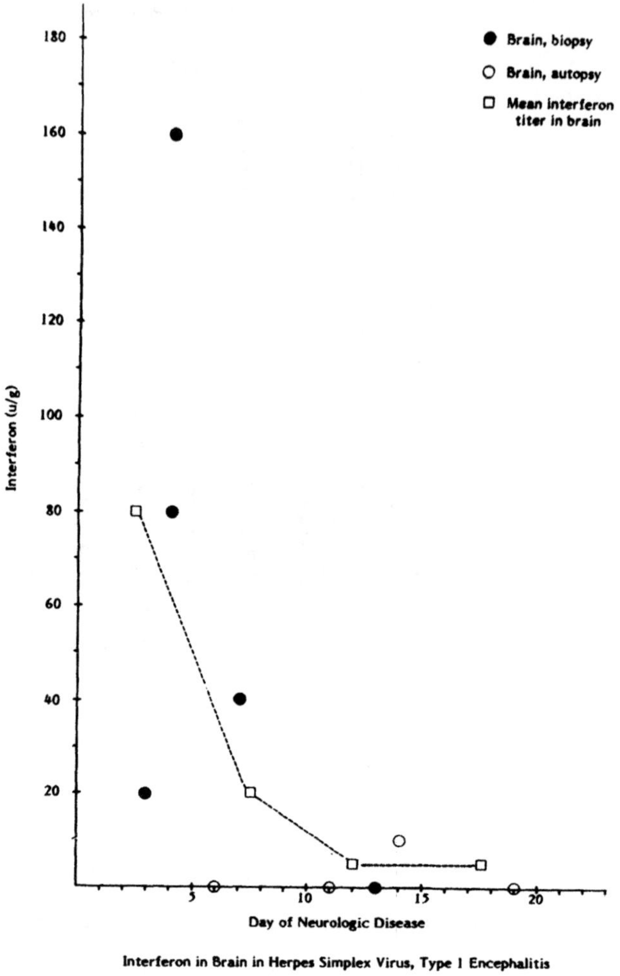

Figure 2. Interferon levels (units/g) in temporal lobes of patients with herpes simplex virus, type 1, encephalitis during course of disease are shown. Results of individual assays at biopsy (closed circle) and at autopsy (open circle) plus their means (open square) are noted. (Reprinted with permission from (11)).

inhibit progression of disease but could possibly be important in combination with antiviral chemotherapy.

It is clear that from the earliest time of detection of clinical HSV encephalitis that locally produced IFN is present in the brain. In clinically detected cases, IFN titers are insufficient to inhibit virus multiplication.

No. 8. Pharmacokinetics of interferon in blood, CSF and brain.(12).

Polyriboinosinic polyribocytidylic acid complexed with poly-L-lysine carboxymethyl (poly ICLC) was given intravenously to New Zealand rabits to trace the transfer of serum IFN to CSF and brain. After intravenous infusions of poly ICLC, peak serum IFN titers (C_{max}) were at 4 hrs, but IFN, CSF and brain C_{max} titers occurred at 8 hrs. At 4 hrs, after poly ICLC, CSF IFN was 0.6-2.7% serum titers, but at 8 hrs, CSF titers reached 11.7-51.1% of simultaneous serum titers. Autopsies (8 hrs) showed that poly ICLC-induced IFN was uniformly distributed to all areas of the brain and spinal cord. Brain IFN titers were independent of CSF values, and varied from 5.9%-44.1% of the simultaneous serum titer. These studies indicate that serum IFN does reach the brain in substantial quantities. This insight offers a potential for the use of exogenous IFN in the treatment of diseases affecting nervous tissue in the central and peripheral nervous system.

Discussion and Summary

These several experiments address basic clinically important problems to physicians attempting to use antiviral drugs rationally in the care of patients with serious life threatening virus diseases. The microbiological assays described seem to have regularly yielded important new insights. When given intravenously, the antiviral nucleoside idoxuridine gave only transient significant antiviral levels that ultimately proved to be ineffectual in patients. Interferon with ara-A was shown to be synergistic in vitro in primate tissue cultures in their combined anti-HSV effects suggesting a possibly useful mode of therapy in clinical trials. A biologic assay was able to detect a previously undetected small contamination of an unidentified antiviral agent in a preparation of an adenosine deaminase inhibitor. "Therapeutic" rather than "prophylactic" tests for the MIC of human IFN indicated that HSV-2 was significantly less susceptible to this antiviral agent than HSV-1. These experiments might be interpreted to correlate with the more severe disease associated with HSV-2 in comparison to HSV-1.

Ara-A and ara-H were demonstrated to be capable of "curing" an infected tissue culture of HSV infection without significantly inhibiting cellular protein or DNA synthesis. When the nucleoside ara-A is given intravenously to patients, ara-A is rapidly deaminated to ara-H. Ara-H has intrinsic anti-HSV activity and sustained anti-HSV levels due to combined ara-A (small amounts) and ara-H (larger amounts) were found in serum. Such combined antiviral activity were not expressed by measuring ara-A and ara-H singly in these same sera by high pressure liquid chromatography, but combined synergistic activity can be measured by microbiological assays of "ara-A equivalents".

In patients with HSV encephalitis, IFN was found from the earliest time after infection only in specific areas of the human brain containing virus. The quantitative IFN values at these sites in each case were below the MICs to HSV correlating with productive virus multiplication. However, exogenous serum IFN potentially could augment these natural values. IFN reaches the brain in substantial amounts from serum.

References

1. Whitley, R.J., Ch'ien, L.T., Dolin, R., Galasso, G.J., Alford Jr., C.A. and Collaborative Study Group. 1976. Adenine arabinoside therapy of herpes zoster in the immunosuppressed. N. Engl. J. Med. 294:1193-1199.
2. Kaufman, H.E. 1962. Clinical cure of herpes simplex keratitis by 5-iodo-2'-deoxyuridine. Proc. Soc. Exp. Biol. Med. 109:251-252.
3. Whitley, R.J., Soong, S.J., Dolin, R., Galasso, G.J., Ch'ien, L.T., Alford Jr., C.A. and the Collaborative Study Group. 1977. Adenine arabinoside therapy of biopsy-proved herpes simplex encephalitis. N. Engl. J. Med. 297:289-294.
4. Lerner, A.M. and Bailey, E.J. 1972. Concentrations of idoxuridine in serum, urine, and cerebrospinal fluid of patients with suspected diagnosis of <u>Herpesvirus hominis</u> encephalitis. J. Clin. Invest. 51:45-49.
5. Lerner, A.M. and Bailey, E.J. 1974. Synergy of 9-β-D-arabinofuranosyladenine and human interferon against herpes simplex virus, type 1. J. Infect. Dis. 130:549-552.
6. Williams, B.B. and Lerner, A.M. 1975. Antiviral activity of an adenosine deaminase inhibitor: decreased replication of herpes simplex virus. J. Infect. Dis. 131:673-677.

7. Lerner, A.M. and Bailey, E.J. 1976. Differential sensitivity of herpes simplex virus types 1 and 2 to human interferon: antiviral effects of interferon plus 9- -D-arabinofuranosyladenine. J. Infect. Dis. 134:400-404.

8. Corey, L., Adams, H.G., Brown, Z.A. and Holmes, K.K. Genital herpes simplex virus infections: clinical manifestations, course and complications. Ann. Int. Med. (in press).

9. Williams, B.B., Bailey, E.J. and Lerner, A.M. 1977. Inhibitory and lethal concentrations of 9-β-D-arabinofuranosyladenine and its hypoxanthine derivative versus herpes simplex virus, type 1. J. Lab. Clin. Med. 89:687-691.

10. Champney, K.J., Lauter, C.B., Bailey, E.J. and Lerner, A.M. 1978. Antiherpes virus activity in human sera and urines after administration of adenine arabinoside. *In vitro* and *in vivo* synergy of adenine arabinoside and arabinosylhypoxanthine in combination. J. Clin. Invest. 62:1142-1153.

11. Legaspi, R.C., Gatmaitan, B., Bailey, E.J. and Lerner, A.M. 1980. Interferon in biopsy and autopsy specimens of brain. Its presence in herpes simple virus encephalitis. Arch. Neurol. 37:76-79.

12. Williams, B.B., Paul, R.T. and Lerner, A.M. Pharmacokinetics of interferon in blood, CSF and brain using modified polyriboinosinic polyribocytidylic acid and amphotericin B. J. Infect. Dis. (in press).

12

MODULATION OF RETROVIRUS DNA POLYMERASE ACTIVITY BY POLYNUCLEOTIDES AND THEIR ANALOGS

P. CHANDRA, I. DEMIRHAN, U. EBENER, B. KORNHUBER AND D. GERICKE

Center of Biological Chemistry, Laboratory of Molecular Biology (P.C., I.D. and U.E.), Department of Pediatric Oncology (B.K.), University Medical School, Frankfurt 70, and Hoechst AG (D.G.), Frankfurt, W. Germany.

(Dedicated to Professor Klaus Munk on his 60th birthday)

Perhaps not as fundamental as the discovery of the DNA double helix, but most important to our understanding of the life cycle of retroviruses, has been the discovery of an RNA-dependent DNA polymerase (1,2), the reverse transcriptase. This was followed by another exciting discovery, that human leukemic cells possess an enzymic activity that biochemically resembles the DNA-polymerase activity associated with retroviruses (3). The latter discovery gave rise to the idea that retroviruses may be involved in human cancer. Immunological probing of reverse transcriptase has further contributed to our knowledge of the expression of retroviral information in human tumors (4-17). However, the recent discovery of human T-cell leukemia-lymphoma virus (HTLV) in fresh and in cultured T-cells from a lymph node biopsy of a patient with cutaneous T-cell lymphoma is direct evidence that retroviruses are associated with human cancer, or at least with this particular type of leukemia (18,19).

In view of the importance of reverse transcription in cancer, and as a biological process, attempts have been made in our laboratory to develop inhibitors of DNA synthesis in RNA tumor viruses. Depending upon their mode of action, the inhibitors of viral DNA polymerase can be divided into several classes (20-24). Of all these, the most meaningful approach is to design compounds that bind to the viral enzyme. In searching for this type of inhibitor, the enzymes chosen for study are important. A number of studies claim specificity (or selectivity) of a compound by comparing an inhibitory response to a bacterial DNA polymerase with the viral reverse

Y. Becker (ed.), Antiviral Drugs and Interferon: The Molecular Basis of Their Activity.

transcriptase. Such a comparison has no relevance to the selective nature of the compound. The most important approach to demonstrate selectivity of a compound is to compare its inhibitory effect against various eukaryotic cellular DNA polymerases, such as α-, ß-, and γ-DNA polymerases.

Our efforts to develop compounds that inhibit viral DNA polymerases by interacting directly with the enzyme led to the discovery of polycytidylic acid analog, containing 5-mercapto-substituted cytosine bases, a partially thiolated polycytidylic acid (25,26). This compound, abbreviated as MPC (mercapto-polycytidylic acid) was found to inhibit the oncornaviral DNA polymerase in a very specific manner (20-22, 25-29). In addition to studies with MPC, we have recently evaluated the effects of several polyadenylic acid analogs on various DNA polymerases. In the following pages we would like to describe the mode of action of polynucleotide analogs on viral reverse transcriptase, and discuss the biological and clinical implications of these studies. In order to document the specificity of action of these analogs, we have compared their activities against purified cellular DNA polymerases α, ß and γ and terminal deoxyribonucleotidyl transferase (TdT). This enzyme, though a cellular enzyme, is unique in the sense that, so far, only thymus tissue has been shown to contain it. In the strictest sense TdT is not a DNA polymerase as it does not require a template to direct the DNA synthesis; nevertheless, it does catalyze the incorporation of deoxyribonucleotides into DNA and thus we consider it here as one of the DNA synthetic enzymes. Like reverse transcriptase the TdT also merits considerable attention as a biochemical marker for certain types of human leukemias (30).

1. Studies with 5-mercapto-polycytidylic acid (MPC)

The inhibition of DNA polymerases from RNA tumor viruses by MPC was described earlier (25-29). Partially thiolated polycytidylic acid preparations, MPC I-III (containing 1.7%, 3.5% and 8.6% 5-mercaptocytidylate units, respectively) inhibited the DNA polymerase activity of Friend leukemia virus (FLV) in the endogenous reaction as well as in the presence of exogenous template-primers, such as $(rA)_n \cdot (dT)_{12}$ and $(rC)_n \cdot (dG)_{12}$. The inhibitory activities were directly related to the percent of thiolation. A maximum inhibition was observed with preparations containing 15-17% of the thiolated cytosine bases. In these experiments, non-thiolated samples of polycytidylic acid showed no inhibition of the reverse transcriptase reaction. Polyuridylic acid thiolated by the same procedure (25) was also found to inhibit the reverse transcriptase reaction (31); however, the non-thiolated samples of polyuridylic acid were also found to inhibit this reaction strongly. Besides, the inhibitory effect of thiolated samples of polyuridylic acid was not strictly related to their degree of thiolation. The inhibitory effect of non-thiolated polyuridylic acid is, presumably, due to hydrogen bonding between poly-U and the added template, $(rA)_n$. The inhibition of the endogenous reaction by non-thiolated poly-U is explainable by the fact, that genomic RNA contains large stretches of polyadenylic acid at 3'-terminal. For this reason, it was difficult to designate the role of thiolation on the inhibitory effect of MPC samples.

The mode of action of viral DNA synthesis by MPC was investigated by product analysis of the DNA polymerase reaction (endogenous) in the absence or in the presenc of MPC, as described by Kotler and Becker (32). The reaction mixtures were dissolved with Na-dodecyl sulfate (1%, wt/wt, final concentration), loaded on a hydroxylapatite column (1g, Bio-Rad Lab., Munich), eluted with a Na-phosphate gradient (0.05 - 0.4 M), collected into about 40 tubes (total volume approx. 100 ml), and the TCA-insoluble activity collected in GF/C filters (Whatman) and counted in a liquid scintillation counter.

Analysis of the endogenous products of the detergent-disrupted virions exhibits 3 DNA species: single-stranded DNA (ss-DNA), RNA-DNA hybrids (hy-DNA) and the double-stranded DNA (ds-DNA). As follows from Fig. 1, in the presence of MPC (open circles) there is an overall inhibition of ^{3}H-dTMP incorporation, indicating that the formation of all three species is blocked. This is to be expected since the inhibitor binds to the enzyme. This has been confirmed by the ultracentrifugation studies in which the binding of ^{35}S-labeled MPC to a purified FLV-DNA polymerase was investigated (Fig. 2).

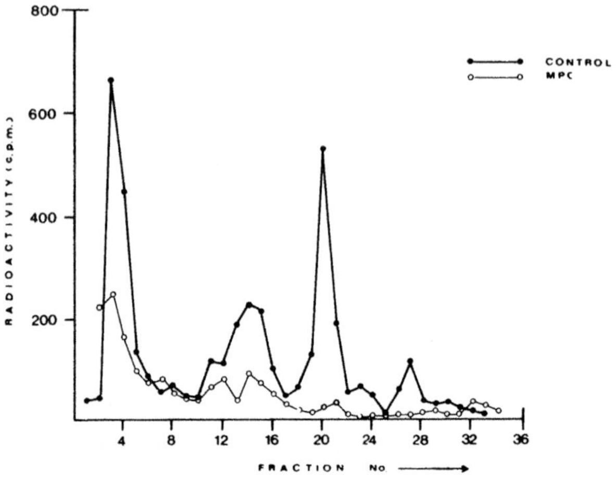

FIGURE 1: Analysis of the DNA species synthesized by FLV DNA polymerase by elution from hydroxylapatite column. Experimental details are described in the text. The first species to be eluted from the column contained ss-DNA, the second contained hy-DNA and finally, the ds-DNA, eluted in the last peak. The concentration of MPC in the reaction mixture was 20 μg (26). (Reprinted with permission).

In view of the fact that all of the oncornaviral DNA polymerases examined so far, do require a primer-template—like double stranded structure for the initiation of DNA synthesis, it is no surprise that single stranded synthetic polynucleotides (unprimed templates) can act as inhibitors of the polymerization reaction. This, presumably, is due to hydrogen bonding of the base sequences between the added polymer and the functional template. Thus, the specificity of inhibition by such polymers is not limited to the viral enzyme system only. On the other hand, minor modification in the chemical structure of synthetic polynucleotides might be useful to develop inhibitors that interact directly with the enzyme but fail to be transcribed, i.e. they function as a "dead template" for the enzyme. The data from our laboratory have shown that the partially thiolated polycytidylic acid is functioning as a "dead template" in the DNA polymerase system of FLV (26). The results of these studies can be summarized as follows: 1. The incorporation of ^{3}H-dGMP into DNA by the viral enzyme is stimulated to about 9-fold (compared to endogenous value) in the presence of $(rC)_n \cdot (dG)_{12}$. However, under similar conditions a hybrid of MPC.$(dG)_{12}$ failed to stimulate the incorporation of ^{3}H-dGMP into DNA; 2. In the presence of MPC.$(dG)_{12}$, the increasing concentrations of $(rC)_n \cdot (dG)_{12}$ in the reaction mixture have no effect on the activity of the enzyme; however, at higher enzyme concentrations the stimulatory effect of $(rC)_n \cdot (dG)_{12}$ gradually reappears. These data indicate that the viral enzyme has higher affinity towards MPC than to its optimal template $(rC)_n$. The presence of zinc in reverse transcriptase makes it attractive to suggest that the mercapto group may undergo an interaction with zinc to form a stable complex with the enzyme.

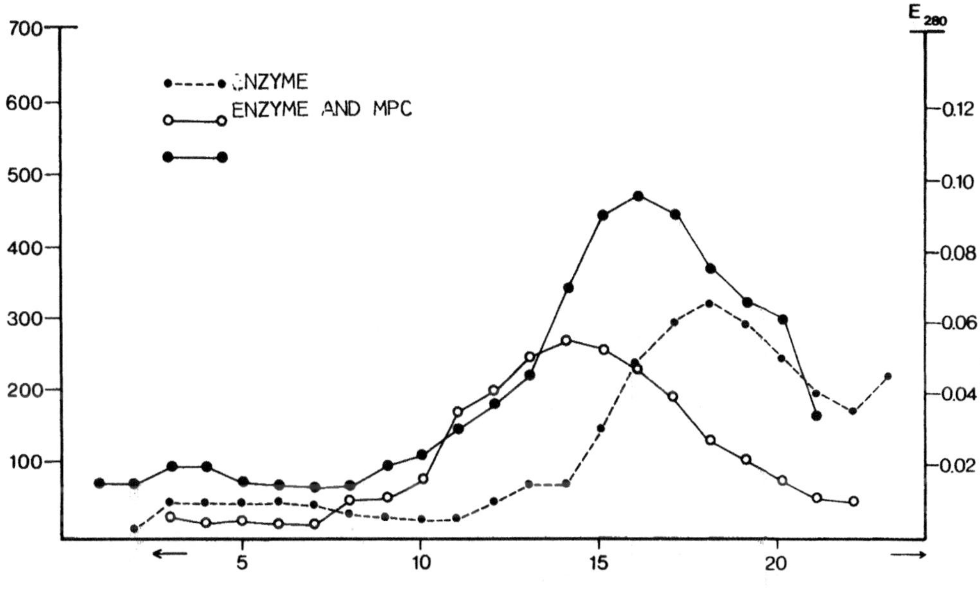

FIGURE 2: Binding of ^{35}S-labeled MPC to a purified FLV-DNA polymerase. The specimens in 0.1 Tris/HCl buffer (pH 7.4) were layered on linear gradients of 10-40% sucrose (RNase-free) in the same buffer and spun at 35.000 rpm (swingout rotor) at 4C for 20 hr. The gradients were dripped from below (bottom :left), fractions collected and after dilution, were analyzed for their radioactivity (——) or absorption at 280 nm (----). The MPC sample contained 10.1% thiolated cytosine bases (sp. act. 141 c.p.m./μg). Left ordinate depicts radioactivity (c.p.m./fraction); right ordinate depicts absorption (E_{280}). Reprinted with permission from (34).

In order to determine the selectivity of MPC action, cellular DNA polymerases (α , β and γ) were purified from normal and FLV-infected mouse spleen. The inhibitory effects of MPC on the cellular polymerases, and on the activity of reverse transcriptase, purified from FLV-infected mouse spleen were compared (Fig. 3). As follows from results, none of the cellular polymerases, from normal as well as from FLV-infected mouse spleen, was significantly inhibited by MPC. On the other hand, $(rA)_n \cdot (dT)_{12}$ and $(rC)_n \cdot (dG)_{12}$-catalyzed activities of reverse transcriptase were highly sensitive to MPC action.

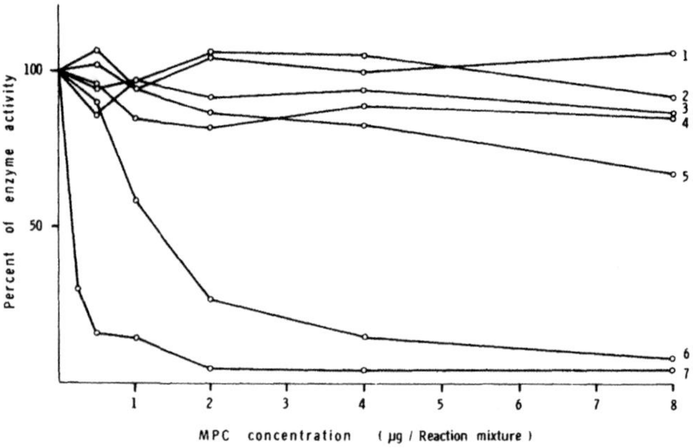

FIGURE 3: Effect of MPC on the activity of cellular DNA polymerases and reverse transcriptase from normal and FLV-infected mouse spleen.Cellular DNA polymerases from normal mouse spleen are depicted by curves 1(α), 2(ß) and 4(γ). Cellular DNA polymerases from FLV-infected spleen are depicted by curves 3(α) and 5(ß). Reverse transcriptase catalyzed by $(rC)_n \cdot (dG)_{12}$ and $(rA)_n \cdot (dT)_{12}$ is depicted by curves 6 and 7 respectively. Reprinted with permission from (29).

The studies on selectivity of MPC action were substantiated using a purified preparation of terminal deoxynucleotidyl transferase (TdT) from calf thymus. The comparative effect of MPC on the $(dA)_n$-catalyzed reaction of TdT,and on the $(rC)_n \cdot (dG)_{12}$ dependent activity of reverse transcriptase is shown in Fig.4. The non-thiolated samples of polycytidylic acid (PC) were in both cases without any inhibitory effect; on the contrary, at lower concentrations a significant stimulation of the reactions was observed. The thiolated samples (SH= 18.2%) of polycytidylic acid (MPC) inhibited both the enzymic reactions;however, the reverse transcriptase reaction was almost 15 times more sensitive to the action of MPC than the TdT reaction.

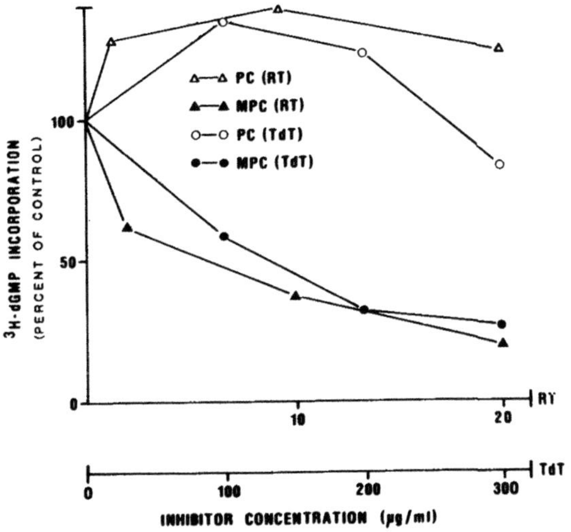

FIGURE 4: Effect of 5-mercapto-polycytidylic acid (SH= 18.2%) on the activities of retroviral reverse transcriptase and on calf thymus terminal deoxynucleotidyl transferase.

Of all the enzymatic studies carried out in our laboratory so far, MPC has proved to be relatively specific towards reverse transcriptase. The kinetic data have revealed that the MPC inhibition is of a non-competitive nature.

The protective effect of MPC on leukemogenesis in mice by FLV has been reported earlier (29). In these experiments, the effect of MPC was investigated by preincubating the virus-containing suspensions with MPC. These studies have been extended using MPC in-vitro and in-vivo.

The animals (mice,Groppel strain) were divided into four groups of five each (donors): 1) Group 1 was injected with a viral suspension (citrate plasma from FLV-infected animals, dose LD_{90}) preincubated with Tris/HCl buffer, pH 7.6 for 30 min at 37 °C; 2) Group 2 was injected with the viral suspension, as in 1, but preincubated with MPC (200 μg per 0.2 ml of suspension) at 37 °C for 30 min. These animals received in addition, on day 5 and day 9 (post-infection) 50 μg of MPC,injected intraperitoneally; 3) Group 3 was treated similarly to group 1, except that the viral suspensions were preincubated for 2 hr; 4) Group 4 was treated in a similar manner as group 2, except that the viral suspensions were preincubated for 2 hr at 37 °C. On the 10th day, animals were sacrificed and spleen extracts were prepared, as described elsewhere (33). The spleen extract from each mouse was then analyzed individually, with respect to their leukemogenic potentiality. Each "donor" spleen specimen was reinjected to a different "recipient" mouse (20 in total), and the leukemogenesis was followed, as shown in Table 1.

Table 1

ASSAY FOR LEUKEMOGENIC POTENTIAL OF SPLEEN EXTRACTS FROM FLV-INFECTED MICE AFTER THEIR IN-VITRO/VIVO TREATMENT WITH MPC.

TREATMENT OF DONOR MICE	LEUKEMOGENESIS IN RECIPIENT MICE AFTER INFECTION WITH SPLEEN EXTRACT(1)		
	NO. OF POSITIVE / TOTAL NO. OF MICE	MEAN SURVIVAL TIME (DAYS)	MEAN SPLEEN WEIGHT(g)
Virus Suspension(2) (0.2 ml) + Tris buffer (37°C,30 min)	5/5	47	2. 41
Virus Suspension + 200 μg of MPC (37 °C, 30 min.) + 50 μg MPC,i.p.(day 5 & 9)	2/5	123(3)	1. 05 (1.78, 2.10, 0.52,0.41,0.44)
Virus Suspension + Tris buffer (37 °C, 2hr)	5/5	52.2	1. 80
Virus Suspension + 200 μg of MPC (37 °C, 2 hr) + 50 μg MPC,i.p.(day 5 & 9)	1/5	110 4 (123)3) 1 (97)	0. 38 (0.74,0.29,0.34, 0.22, 0.31)

(1) Cell-free spleen extracts were prepared (33) from spleens of individual donors on the 10th day after being challenged with the virus, or other treatments as shown; (2) Citrate plasma from FLV-infected mice was used as a source of virus; (3) the experiment was terminated on day 123, and all animals were sacrificed on this day. Therefore, the term "mean survival" does not apply to these animals.
Reprinted with permission from (22).

All animals in groups 1 and 3 developed splenomegaly and died between 40-60 days, whereas, in the MPC-treated groups, of the 10 animals only 3 showed signs of splenomegally. In group 2, 2 animals had splenomegally but, in spite of that, all animals survived till the 123rd day, at which time our experiment was terminated. Similarly, in the last group animals survived till the 123rd day; one died on the 97th day. The spleen weights, shown in the last column, also exhibit large differences between the MPC-treated group, and the control group. In another study we have analyzed the effect of MPC on normal mice of the same strain. We failed to observe any effect of MPC on the spleen weights of non-infected mice.

In another study we have analyzed the in-vivo effect of MPC on leucocytes of mice infected with the active Friend virus (34). Within 12-24 hrs after MPC injection (50 µg/mouse) a dramatic fall in the leucocyte count of animals infected with FLV was observed; MPC failed to reduce the leucocyte number in mice not infected with the virus. It is interesting that, in one of the infected animals MPC failed to suppress the leucocyte number; on the contrary, there was a gradual increase in leucocyte number. This animal died on the 13th day of MPC treatment. Unfortunately, we were not able to analyze the spleen of this animal. It is therefore difficult to interpret the reasons for failure of the MPC effect in this animal.

Based on the biochemical selectivity of MPC action and its biological effects, clinical trials of MPC in the treatment of childhood leukemia were initiated. We began these trials with cases at the terminal phase of the disease (35).

Clinical data of patients submitted to MPC trials are shown in Table 2. Of the 18 cases treated with MPC, 13 were in the terminal phase of the disease. These patients were resistant to all previous chemotherapeutic regimes which involved drugs, such as prednisone, vincristin, daunorubicin, L-asparaginase, Ara-C, 6-mercaptopurine, methotrexate, cyclophosphamide and actinomycin D.

MPC used in our clinical trials contained 15% of thiolated cytosine bases. The lyophilized product (MPC) was dissolved in 0.1 M Tris/HCl buffer, pH 7.6, and diluted with 0.9% NaCl before use. This solution was sterilized by passing through a membrane filter (Millipore GmbH, Neu-Isenburg,Germany). It was kept at 4°C and used immediately, or within the next five days; solutions older than 5 days were reprecipitated, purified on the column and resterilized. In our clinical trials, MPC (sterile) was given intravenously at a dose 0.5 mg per kg body weight. The injections were given once a week.

Table 2. Clinical data of patients with acute lymphoblastic leukemia (ALL) and acute myeloblastic leukemia (AML) submitted to MPC trials.

No.	Init.	Age	Sex	Diag.	Stage	Results
1	D. M.	8	♀	ALL	3rd rel.	PR, WBC ↓
2	T. I.	6 6/12	♀	ALL	2nd rel.	?
3	B. M.	6 7/12	♂	ALL	2nd rel.	?
4	M. M.	8 4/12	♂	ALL	2nd rel.	CR
5	J. O.	3 6/12	♂	ALL	2nd rel.	?
6	N. A.	10	♀	ALL	2nd rel.	CR
7	B. C.	5 8/12	♀	ALL	3rd rel.	CR
8	N. N.	7 11/12	♂	ALL	3rd rel.	?
9	M. A	12	♂	ALL	1st rel.	?
10	M. I.	8 6/12	♂	ALL	1st rel.	?
11	K. C.	11 3/12	♀	ALL	4th rel.	?
12	K. K.	10 9/12	♀	ALL	5th rel.	PR
13	L. J.	7	♂	ALL	3rd rel.	WBC ↓
14	F. D.	2 3/12	♂	ALL	init. ph.	WBC ↓
15	S. B.	7 11/12	♀	ALL	init. ph.	WBC ↓
16	H. B.	12 5/12	♀	ALL	init. ph.	WBC ↓
17	S. N.	4	♀	AML	init. ph.	WBC ↓
18	W. H.	5 6/12	♂	AML	init. ph.	WBC ↓

PR = partial remission; CR = complete remission.
Reprinted with permission from (22).

Of the 13 terminal cases, complete remission was achieved in three, and a partial remission achieved in two other cases. Fever, occasionally accompanied by shivering, was frequently observed under MPC treatment in the first hour after injection. However, these symptoms never lasted more than a few hours, and no other side effects were observed. The recent toxicological studies on MPC (Chandra et al., unpublished results) have shown that this substance is non-mutagenic, non-teratogenic,and doses as high as 50 mg/kg body weight were well tolerated by rats.

On the basis of our experience with MPC in terminal cases, we were motivated to give MPC a clinical trial at the beginning of leukemia. A monotherapy with MPC, as devised for terminal cases is, however, not possible. We therefore decided to introduce MPC (0.5 mg/kg body weight) therapy at the beginning of treatment of cases which at the time of diagnosis had high leucocytosis. This initial treatment, a single injection of MPC, was then followed up by polychemotherapeutic protocol, adopted by the university hospitals of Berlin, Frankfurt and Münster (BFM-protocol, ref. 36). As shown in Table 2 (cases 13 to 18), 24 hrs after MPC injection, there was a significant reduction of leukemic cells in all these cases. The clinical studies have been extended to a total number of 24 patients with similar observations.

2. Studies with polyadenylic acid analogs

The fact that polyadenylic acid (poly A) is easily transcribed by the retroviral reverse transcriptase has led to the development of several poly A-analogs (reviewed in ref. 21 and 22). In the following pages we would like to describe our recent studies on the inhibition of various DNA polymerases with some poly(A)analogs. These analogs are (Fig. 5): poly (2-methylthioadenylic acid) or $(ms^2A)_n$; poly (2-ethylthioadenylic acid) or $(es^2A)_n$; and poly 2′-fluoro-deoxy-adenylic acid or $(dAfl)_n$.

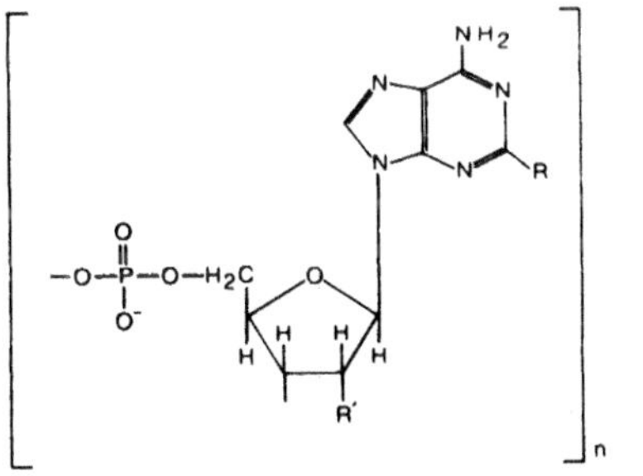

$R = -S-CH_3$, $R' = -OH$ poly(2-methylthioadenylic acid), $(ms^2A)_n$

$R = -S-CH_2CH_3$, $R' = -OH$ poly (2-ethylthioadenylic acid), $(es^2A)_n$

$R = -H$, $R' = -F$ poly (2-fluoro-2-deoxyadenylic acid), $(dAfl)_n$

FIGURE 5: Chemical Structures of Polyadenylic Acid Analogs.

The effect of these three poly-A analogs on $(rA)_n \cdot (dT)_{12}$-

primed DNA polymerase activity of RLV is shown in Fig.6. As follows from the results, $(ms^2A)_n$ shows a maximum inhibition of the enzyme activity followed by $(es^2A)_n$. On the contrary, $(dAfl)_n$ showed a concentration-dependent stimulation of the reverse-transcriptase reaction. However, in the presence of $(rC)_n \cdot (dG)_{12}$ as template-primer, the reverse transcriptase reaction was significantly inhibited by $(dAfl)_n$, as shown in Fig. 7 . This biphasic effect is understandable since $(dAfl)_n$ acts in the latter situation as an antitemplate; whereas, this analog does not compete with $(rA)_n$ as an antitemplate. Probably, $(dAfl)_n$ acts in this reaction like poly(A).

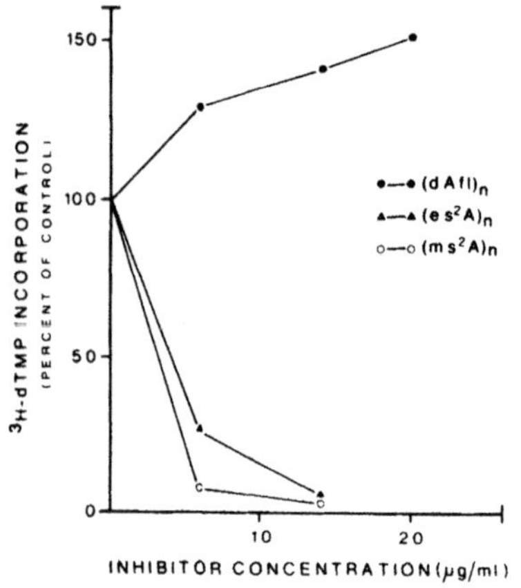

FIGURE 6: Effect of 2- and 2′-substituted poly(A) analogs on the $(rA)_n \cdot (dT)_{12}$-dependent activity of reverse transcriptase from RLV-infected mouse spleen.

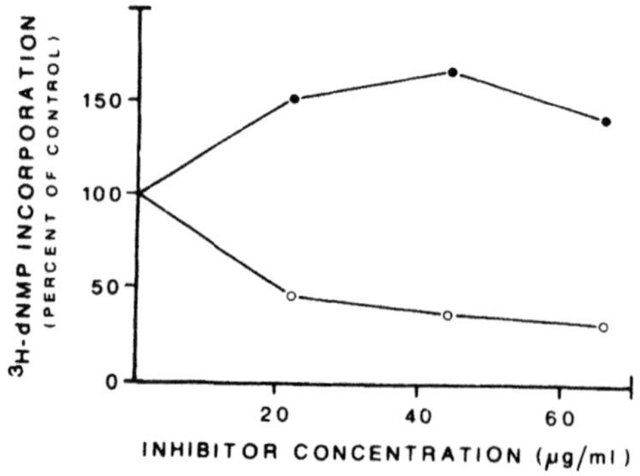

FIGURE 7: Effect of $(dAfl)_n$ on the $(rA)_n \cdot (dT)_{12}$-catalyzed activity of reverse transcriptase (RLV).

●——● $(rA)_n \cdot (dT)_{12}$ O——O $(rC)_n \cdot (dG)_{12}$

Reprinted with permission from (23).

A very similar spectrum of inhibition was observed in the DNA polymerase-γ reaction, catalyzed by $(rA)_n \cdot (dT)_{12}$. Both the 2-substituted derivatives inhibited the reaction strongly, whereby the $(ms^2A)_n$ analog was a better inhibitor. The 2′-substituted derivative, $(dAfl)_n$ showed a concentration dependent stimulation of the DNA polymerase-γ reaction.

The main conclusion from these experiments is that, $(es^2A)_n$ and $(ms^2A)_n$ inhibit the $(rA)_n \cdot (dT)_{12}$-catalyzed reactions, whereas, the $(dAfl)_n$ derivative stimulates the $(rA)_n \cdot (dT)_{12}$-dependent reaction. It was therefore of interest to look for their relative activity in the DNA polymerase-ß reaction, catalyzed by $(dA)_n \cdot (dT)_{12}$. We were able to show (23) that DNA polymerase-ß reaction is inhibited by all the three analogs. The fact that $(dAfl)_n$ inhibits the $(dA)_n \cdot (dT)_{12}$-catalyzed reaction indicates that $(dAfl)_n$ may have a secondary structure closer to $(rA)_n$, than to $(dA)_n$.

The effect of $(ms^2A)_n$ and $(es^2A)_n$ on the poly-dA catalyzed reaction of terminal transferase is shown in Fig. 8. Both the compounds inhibit Terminal transferase (TdT) activity, but in this case $(es^2A)_n$ is a better inhibitor. Lineweaver-Burk plot of the kinetic data of TdT reaction, in the presence of $(es^2A)_n$ showed that $(es^2A)_n$ inhibits the TdT reaction competitively.

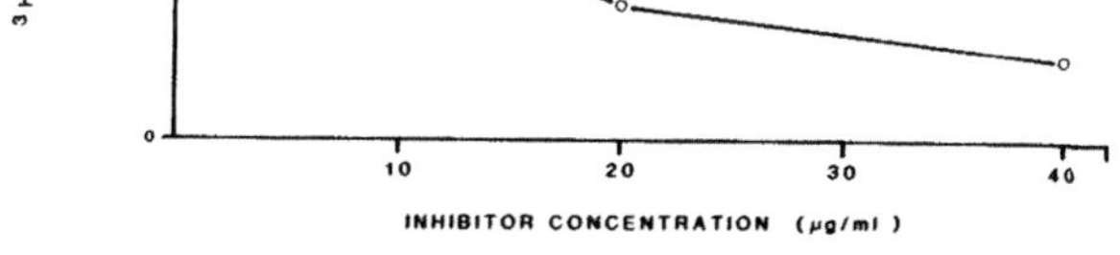

FIGURE 8 : Inhibition of $(dA)_n$-catalyzed terminal transferase reaction by 2-substituted poly(A)analogs. Reprinted with permission from (23).

The effect of $(dAfl)_n$ and other polynucleotides on the $(dA)_n$ catalyzed TdT reaction is shown in Fig. 9. Unexpectedly, we found a concentration- dependent stimulation of the TdT reaction when $(U)_n$ was added into the reaction mixture. This may be due to some conformational effect due to interaction between the two complimentary strands. To our knowledge, this type of effect has not been reported so far. Poly-rA and $(dAfl)_n$ which have no complimentarity to the starter, remain almost ineffective when added to the reaction mixture.

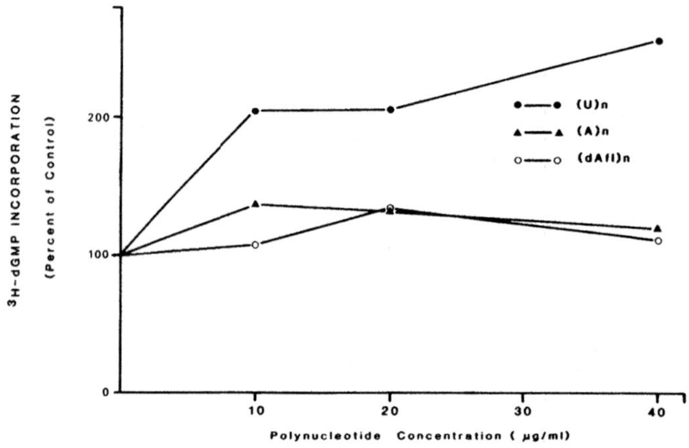

FIGURE 9: Effect of $(dAfl)_n$ and other polynucleotides on the $(dA)_n$-catalyzed reaction of terminal transferase. Reprinted with permission from (23).

To determine the selectivity of $(dAfl)_n$ action on polymerases, we have also studied its effect on DNA polymerase-α from calf thymus. As shown in Table 3, $(dAfl)_n$ derivative stimulates the reaction, whereas, $(es^2A)_n$ shows a slight inhibition of the reaction. The $(ms^2A)_n$ derivative did not show any significant effect at the inhibitor concentrations used. Thus, the only polymerase which is sensitive to $(dAfl)_n$ is the polymerase-ß. In this respect, it may serve as a good indicator for the detection of polymerase-ß activity.

CONCLUSION

In spite of a great variety of reverse transcriptase inhi-

Table 3

INFLUENCE OF 2- AND 2'-SUBSTITUTED POLYADENYLIC ACID ANALOGS ON THE DNA POLYMERASE-α ACTIVITY FROM CALF THYMUS

System	[^{3}H]dTMP incorporation into DNA (cpm/reaction mix)	% of control
Without enzyme	262	
Complete	9469	100 (9207)[a]
Inhibitor[b]		
$(ms^2A)_n$	7544	82
$(es^2A)_n$	6576	71
$(dAfl)_n$	13462	146

[a] Denotes the dTMP incorporation in complete system after the value of the blank probe (without enzyme) was subtracted. Values for inhibitor experiments were similarly corrected for blank.

[b] The inhibitor concentration was 20 μg/ml of the reaction mixture.

Reprinted with permission from (23).

bitors available today (21), an absolute specific inhibitor of reverse transcriptase is not yet available. The specificity is relative and often varies with the system studied. Thus, an inhibitor may have a Ki value of less than 10^{-5}M in the reverse transcriptase system, but a Ki of 10^{-3}M with other cellular DNA polymerases. As a result, it makes sense to evaluate the biological activity of such inhibitors under carefully controlled conditions. The question is: Is it possible to fractionate the antiviral and anticellular activities of such compounds in a particular dose range? We know from our own experience that the cytotoxicity and the antiviral activity of a compound can be differently influenced by structural modifications of the compound (21,22). The past experience has shown that compounds which bind to the enzyme are more specific as inhibitors, than the compounds which interact with the templates. In this sense, the development of polynucleotide analogs is a useful approach in designing specific inhibitors of retroviral reverse transcriptase

ACKNOWLEDGEMENTS

This work was supported by grants from Stiftung Volkswagenwerk (Grant No. 14 0305). We thank Dr. Erik de Clercq (Louvain) for the gift of poly(A) analogs.

REFERENCES

1. Temin, M.H. and Mizutani,S.: Nature 226, 1211, 1970.
2. Baltimore, D.: Nature 226, 1209, 1970.
3. Gallo, R.C., Yang,S.S. and Ting, R.C.: Nature 228,927, 1970
4. Gallo, R.C., Gallagher, R.E., Miller, N.R., Mondal, H., Saxinger W.C., Mayer, R.J., Smith, R.G. and Gillespie, D.H.: Cold Spring Harb. Symp.Quant.Biol. 39, 933, 1975.
5. Chandra,P. In: Proceedings of the XIIth International Cancer Congress, Buenos Aires, Vol.I, pp. 29, Pergamon Press,1978.
6. Chandra, P.: Advances in Opthalmology 38,234, 1979(Karger).
7. Chandra,P. In: Proceedings of the Federation of European Biochemical Societies, Dubrovnik, Yugoslavia (ed. P. Mildner), Vol. 61, p. 215. Pergamon Press, Oxford, 1979.
8. Chandra,P. and Steel, L.K.: Biochem.J. (Lond.) 167,513, 1977.
9. Chandra, P., Balikcioglu, S. and Mildner, B.: Cancer Lett. 5, 299, 1978.
10. Chandra,P., Steel, L.K. and Cavdar, A.O. In: Modern Trends in Human Leukemia (Edt. R. Neth et al.) vol. III, p. 497, 1979. Springer-Verlag, Berlin-Heidelberg-New York.
11. Chandra,P., Steel, L.K., Laube, H. and Kornhuber, B. In: Antiviral Mechanisms in the Control of Neoplasia (Ed. P. Chandra), p. 177, 1979, Plenum Press, New York.
12. Chandra, P., Steel., L.K., Laube, H., Balikcioglu, S., Ebener, U. and Welte,K. In: Proceedings of the 8th International symposium on the Biological Characterization of Human Tumors. Athens, vol. 5, p. 359, 1980, Excerpta Medica, Amsterdam.
13. Welte,K., Ebener, U. and Chandra,P.: Cancer Lett. 7,189,1979.
14. Chandra,P., Steel, L.K., Laube, H., Balikcioglu, S., Mildner, B., Ebener, U., Welte, K. and Vogel, A. In: Cold Spring Harb. Conference on Cell Proliferation (Viruses in Naturally Occuring Cancers) vol. 7, 775, 1980.
15. Chandra,P., Balikcioglu, S. and Mildner, B.: Cell. and Mol. Biol. 27, 239, 1981.
16. Chandra,P. : Editorial Revie, Survey of Immunologic Research, in press.
17. Ohno, T. and Spiegelman, S. : Proc.Natl.Acad.Sci.,U.S.A. 74, 2144, 1977.
18. Poiesz, B.J., Ruscetti, F.W., Gazdar, A.F., Bunn, P.A., Minna, J.D. and Gallo, R.C.: Proc.Natl.Acad.Sci.,U.S.A. 77,7415,1980.
19. Gallo, R.C. and Wong-Staal, F.: Blood (in press)
20. Chandra,P., Ebener,U., Steel, L.K., Laube,H., Gericke, D., Mildner,B., Bardos, T.J., Ho, Y.K. and Götz, A.: Ann. N.Y. Acad. Sci., 284, 444, 1977.
21. Chandra, P., Steel, L.K., Ebener, U., Woltersdorf, M., Laube, H., Kornhuber, B., Mildner, B. and Götz, A.: Pharmacol. and Ther. A 1, 231, 1977.
22. Chandra, P., Steel, L.K., Ebener, U., Woltersdorf, M., Laube, H., Kornhuber, B., Mildner, B. and Götz, A. In: Inhibitors of DNA and RNA polymerases (Edts. P.S. Sarin and R.C. Gallo), pp 47, Pergamon Press, Oxford, 1980.
23. Chandra,P., Demirhan,I. and Ebener,U. In: Terminal Transferase in Immunobiology and Leukemia (Edts. U.Bertazzoni and F.J. Bollum) p.87, 1982, Plenum Press, New York.

24. Smith, R.G. and Gallo, R.C.: Life Sciences 15,1711,1974.
25. Chandra, P. and Bardos, T.J.:Res. Commun. Chem. Path. and Pharmacol. 4, 615, 1972.
26. Chandra, P., Ebener, U. and Götz, A.: FEBS-Lett. 53,10,1975.
27. Chandra, P. In: Topics in Current Chemistry 52, 99, 1974, Springer-Verlag, Berlin-Heidelberg-New York.
28. Chandra,P., Ebener,U., Bardos, T.J., Gericke, D., Kornhuber, B. and Götz, A. In: Modulation of host immune resistance in the prevention or treatment of cancer (Edt. M.A. Chirigos), Fogarty International Center Proceedings (USA) vol.28, 169, 1977.
29. Chandra, P. In: Antimetabolites in Biochemistry, Biology and Medicine (Edts. J. Skoda and P. Langen) p. 249, 1979, Pergamon Press, Oxford.
30. Bertazzoni, U. and Bollum, F.J.: Terminal Transferase in Immunobiology and Leukemia, Plenum Publ. Corporation, New York, 1982.
31. Ebener, U. :Doctoral Thesis (Frankfurt University) 1977.
32. Kotler, M. and Becker, Y.: FEBS-Lett. 22, 222, 1972.
33. Chandra, P., Kornhuber,B., Gericke,D., Götz,A. and Ebener, U.: Z. Krebsforchg. u. Klin. Onkol. 83, 239, 1975.
34. Chandra, P., Kornhuber,B. and Ebener,U. In: Modern Trends in Human Leukemia (Edts. R. Neth et al.) vol. III, p.145, 1979, Springer-Verlag, Berlin-Heidelberg-New York.
35. Kornhuber,B. and Chandra,P. In: Antiviral Mechanisms in the Control of Neoplasia (Edt. P. Chandra), p.577, 1979, Plenum Press, New York.
36. Rhiem, H., Gardner,H., Jessenberger,K. and Tariverdian, G.: Proc.Amer.Assoc.Cancer Res. 15, 58, 1974.

13
THE ANTIPOXVIRUS ACTIVITY OF ISATIN-β-THIOSEMICARBAZONE (IBT)

EHUD KATZ

The Chanock Center for Virology, The Hebrew University-Hadassah Medical School, Jerusalem, Israel

SUMMARY

Two mutants of vaccinia virus, one resistant and the other dependent on isatin-β-thiosemicarbazone (IBT), were isolated and characterized. Complementation took place when these mutants and the wild-type strain were co-cultivated. Genetic recombination was carried out between the IBT-resistant and a temperature-sensitive mutant of vaccinia virus. IBT affects vaccinia virus growth at a step during the maturation of the virus particles; cleavage of high molecular weight precursors, giving rise to viral polypeptides, is inhibited under these conditions. Biochemical and electron microscopy observations suggested that IBT blocks the maturation of the wild-type strain of vaccinia virus later than rifampin. The growth of the IBT-dependent mutant is interrrupted in the absence of IBT at a similar stage of maturation. All seventeen compounds studied until now, that are related in their structure to IBT, which inhibited the growth of vaccinia virus and to which the IBT-resistant mutant showed resistance in its growth, were able to support the growth of the IBT-dependent mutant.

All these findings suggest that the wild-type and the IBT-dependent strains differ in the genetic locus involved with IBT, in such a way that maturation of the wild-type strain is inhibited on the one hand while growth of the IBT-dependent mutant is supported on the other hand.

INTRODUCTION

Methyl-isatin-β-thiosemicarbazone (Me-IBT, Marboran) is an antiviral drug which is effective essentially only against poxviruses. This drug has been used as a prophylactic agent against smallpox in India, Africa and South America (1). The best criterion for determining that a compound is specific against a particular virus and is potentially an antiviral drug, is the appearance of virus mutants that are resistant to the compound (2). The genetic information residing within the nucleic acid of the resistant mutant, is responsible for finding an alternate route to circumvent the wild-type (w.t.) developmental stage blocked by the drug. Such mutants are excellent tools for investigating the mechanism of action of drugs, since they will be inhibited at drug concentrations that affect the metabolism of the host cells. Furthermore, when a virus strain is resistant to one compound and not to another, it is clear then that the two compounds do not share a common mode of action.

Virus strains, resistant to antiviral drugs that may appear during treatment of infected persons or animals require the use of a combination of antiviral drugs with different modes of action to prevent or significantly reduce the risk involved in the appearance of such mutants.

Isolation and characterization of IBT mutants

To study the mechanism of action of IBT in the inhibition of the growth of poxviruses, two mutants were isolated. One is an IBT-resistant mutant (IBT^R), able to grow at IBT concentrations that inhibit the growth of the w.t. strain, and the other is an IBT-dependent mutant (IBT^D), needing IBT for its optimal growth (3). These two mutants were isolated after growing the w.t. strain of vaccinia virus in cultures of chick fibroblasts in the presence of 14 μM of IBT and 5 μg/ml of 5-iodo-2'-deoxyuridine (IUdR), the latter being used as a mutagenic agent. Both mutants were plaque-purified and their resistance or dependence on IBT were shown in HeLa, BSC-1 and chick embryo cells. The resistant mutant was able to grow in the presence of a wide range of IBT concentrations, up to concentrations that were toxic for the host cells (>40 μM). The IBT^D mutant reached its highest yield at a concentration of 3 to 4 μM of IBT. When this virus was grown in the absence of IBT, 2% revertants appeared, which showed either IBT resistance or IBT sensitivity (3).

No significant differences in virion morphology among the w.t. and the two mutants could be detected by electron microscopy of purified virus preparations, negatively stained with phosphotungstic acid. The mutants IBT^D and IBT^R were immunologically similar to the w.t. strain and their infectivity could be neutralized by human-antivaccinia immunoglobulin (3). The mutants did not differ from the w.t. strain in their sedimentation in sucrose gradients and only minor differences, most of them quantitative, were observed when the structural polypeptides of the viruses were analyzed by sodium-dodecyl-sulphate (SDS) polyacrylamide gel electrophoresis (4). Another drug-resistant mutant of vaccinia virus which we isolated in the presence of γ-thiochromanone-4-thiosemicarbazone, a compound related in its structure to IBT, was found to be resistant to IBT as well (5).

Complementation and recombination

IBT inhibits the growth of the w.t. strain of vaccinia virus but not the IBT^D and IBT^R mutants. Co-infection of w.t.-infected cells with either the IBT^R or IBT^D mutants in the presence of IBT revealed that each mutant was able to support the growth of the w.t. strain under these unfavorable growth conditions. Furthermore, co-infection of IBT^D-infected cells in the absence of IBT, with the w.t. strain or the IBT^R mutant, showed that these two strains could support the growth of the IBT^D mutant in the absence of IBT (4).

In order to determine whether the characteristic of resistance to IBT can be genetically transferred from one strain of poxvirus to another, a recombination experiment was performed. The two virus strains used in this study were the IBT^R mutant and a temperature-sensitive (ts) mutant of vaccinia virus (kindly donated by Dr. A. Kirn, Strasburg, France). They were grown in HeLa cells at 26°C and the progeny virus was plated in cultures containing IBT. Single plaques were picked out and enriched at the permissive temperature. They were tested for their plaque efficiency at both 30°C and 39°C (the latter being the non-permissive temperature for the mutant). Out of 22 plaque isolates which were tested, 12 exhibited the behavior of ts-IBT^R, indicating that they were derived from a recombination event that occurred between the IBT^R and the ts mutants (6).

Effect of IBT on the growth of vaccinia virus

In the growth cycle of poxviruses "early" and "late" periods are defined. In the "early" period, several structural polypeptides and enzymes, some of which are involved in the synthesis of the DNA of the virus, are formed. This period ends when initiation of the synthesis of viral DNA takes place. During the "late" period, synthesis of most of the viral structural polypeptides is carried out.

Previous studies have shown that IBT does not affect "early" events which occur during the growth cycle of the virus: uncoating is normal, "early" enzymes are synthesized (7), a number of "early" structural viral proteins are made (8) and viral DNA replicates (7, 9). The rate of transcription of mRNA is not affected and the formed mRNA binds to ribosomes to form polyribosomes (7). However, polyribosomes which are formed by late mRNA are rapidly broken down (7). On the basis of these findings, Woodson and Joklik (7) suggested that the deficiency in "late" structural viral proteins is the target for the inhibition of poxviruses by IBT.

The availability of the mutants and more developed techniques for analysis of mRNA molecules and polypeptides, enabled us to further explore the mechanism of inhibition of poxviruses by IBT. First, the synthesis of "early" polypeptides of vaccinia virus in the presence of IBT was analyzed by SDS-polyacrylamide gel electrophoresis. No interruption in the normal synthesis of "early" vaccinia viral polypeptides was found in the presence of IBT (10). The rate of viral DNA synthesis in the cytoplasm of infected cells was found to be only somewhat reduced. We studied the synthesis of the viral mRNA during the "late" period of vaccinia virus infection. The results previously observed by Woodson and Joklik (7) showing that a portion of the virus-specific RNA appears to be degraded, were now confirmed. However, the amount of viral RNA that was capped, properly methylated and polyadenylated was reduced only by approximately 50%, as compared to untreated-infected cultures. RNA from IBT-treated cells stimulated cell-free protein synthesis to one-half the level obtained with RNA from control cells. Polyacrylamide gel electrophoresis further demonstrated that this RNA translated "late" viral proteins in a reticulocyte cell-free system (11).

The synthesis of viral polypeptides was inhibited by 93% at six hours postinfection in infected spinner HeLa cells in the presence of IBT. However, at the beginning of the "late" period (between two to four hours after infection) almost no inhibition in the rate of protein synthesis could be detected. These results indicate that the inhibitory effect of IBT is manifested only after synthesis of "late" viral proteins has already commenced (11). In a separate series of experiments, using IBT-treated monolayer HeLa cells instead of suspension cultures, we noted minimal, if any, enhanced mRNA degradation or reduction in methylation, under conditions when polysome disaggregation and a 50% inhibition in protein synthesis occurred (E. Katz and B. Moss, unpublished results).

The viral DNA starts to be coated with proteins and acquires resistance to deoxyribonuclease, approximately five hours after infection (12). In the presence of IBT most of the viral DNA remains sensitive to this enzyme; only following removal of IBT, does the DNA become resistant to deoxyribonuclease (10).

Two major vaccinia virus structural polypeptides (4a and 4b) were shown by pulse-chase experiments to be formed from higher molecular-weight precursors (P4a and P4b)(13). These cleavages, which are associated with the maturation of the virus particles (14), are greatly inhibited in the presence of IBT. When DNA-protein complexes, which are formed in the presence of IBT, were analyzed by SDS-polyacrylamide gel electrophoresis, it was found that both 4a and 4b were almost entirely missing (15). It is possible that the failure of these two polypeptides normally associated with the core of the virus, to assemble with the viral DNA, is the reason for the sensitivity of the DNA to deoxyribonuclease.

The effect of IBT on the intracellular development of the w.t. strain of the virus was studied by electron microscopy. Infected cells in the absence of IBT contained mostly condensed forms with low electron-dense centers, that appeared to be mature infectious virus (16). However, in cells exposed to IBT, there was a great number of large, almost spherical immature forms, most of which appeared to have complete membranes (4). This finding was in agreement with that previously reported by Easterbrook (17). When cells, infected with the IBT-resistant mutant in the presence of IBT, were examined, most of the virus particles seen were mature (4).

Does the block caused by IBT during vaccinia virus growth occur later than that by rifampin?

In order to examine whether the block occurring during the growth of vaccinia virus in the presence of IBT takes place earlier or later than that caused by rifampin, another inhibitor of poxvirus growth (18, 19), infected cells were treated first with one drug and then with the other. The morphology of the viral structures accumulated under these conditions was followed by electron microscopy, since the viral structures formed under normal infection and those accumulated in the presence of either one of these inhibitors, are very clearly defined (4, 9, 19, 20). The more common viral strucutres found in each culture of the experiment are shown in Fig. 1 and the quantitative results are summarized in Table 1.

Table 1. Structures of vaccinia virus in infected HeLa cells treated with IBT and rifampin

Culture	Time of treatment (hr pi)			Time of fixation	Types of viral structures			
	0	7.5	8	(hr pi)	1→3	4→MV	MLD	1→3+MLD/4→MV
a	+Rif			8	0	0	102	∞
b	+Rif	+IBT	-Rif	12	104	34	0	3.05
c	+Rif		-Rif	12	149	66	0	2.25
d	+IBT			8	110	16	0	6.88
e	+IBT	+Rif	-IBT	12	63	50	11	1.48
f	+IBT		-IBT	12	103	41	0	2.51
g				12	60	47	0	1.28

HeLa cells were infected with vaccinia virus at a multiplicity of 30 pfu/cell. Following adsorption for 30 min at 37°C, the cells were washed and M199 medium containing rifampin (100 μg/ml) or IBT (14 μM) was added. To cultures b and e, IBT and rifampin, respectively, were added at 7.5 hr pi. At 8 hr pi rifampin was washed from cultures b and c and IBT from e and f. The cells were fixed and processed for electron microscopy, essentially as previously described (4). Rif - rifampin; MLD - membrane limited domain; MV - mature virion. 1→4 sequential stages during the development of the virus, as defined by deHarven and Yohn (20).

The numbers presented in the righthand column of Table 1 are comparative ratios that indicate the state of maturation of the viral structures in the cells. When the ratio is low, mature structures are more common than immature ones and when the ratio is high, mature forms are rare. In the control cultures (g), the ratio is relatively low (1.28). The highest ratios were obtained under conditions of continuous inhibition, either by rifampin (a = ∞) or IBT (d = 6.88). However, when the cultures

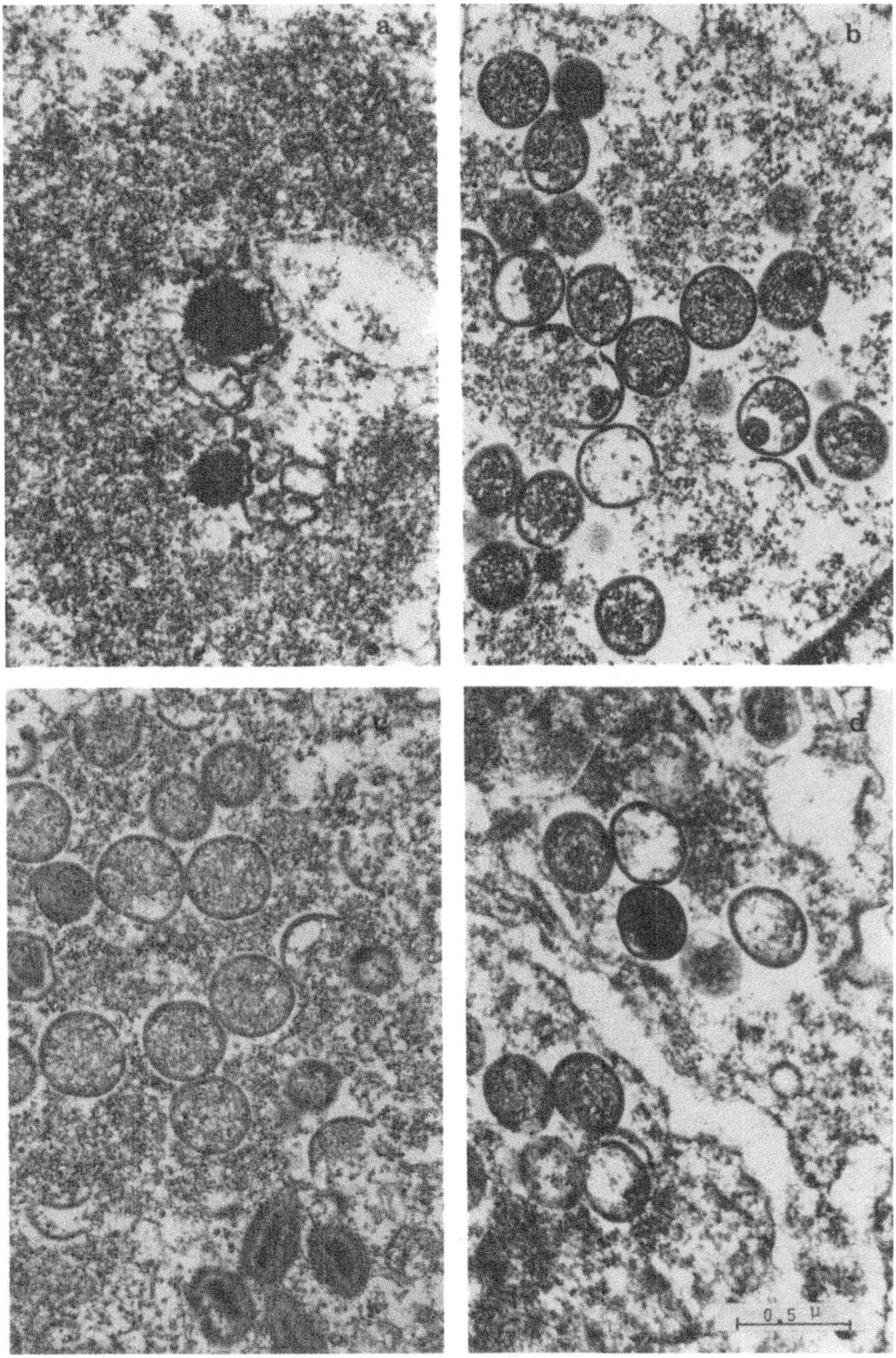
a
b
c
d
0.5 μ

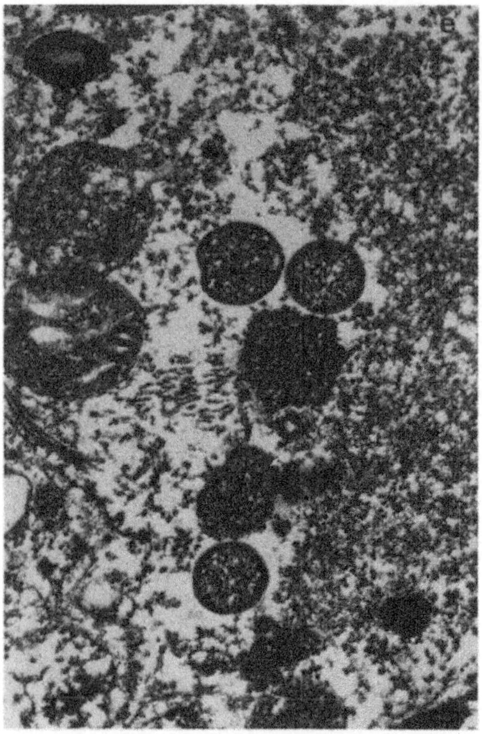

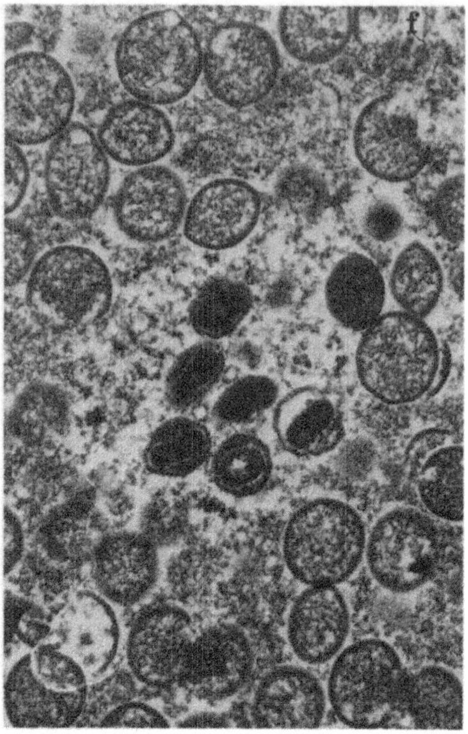

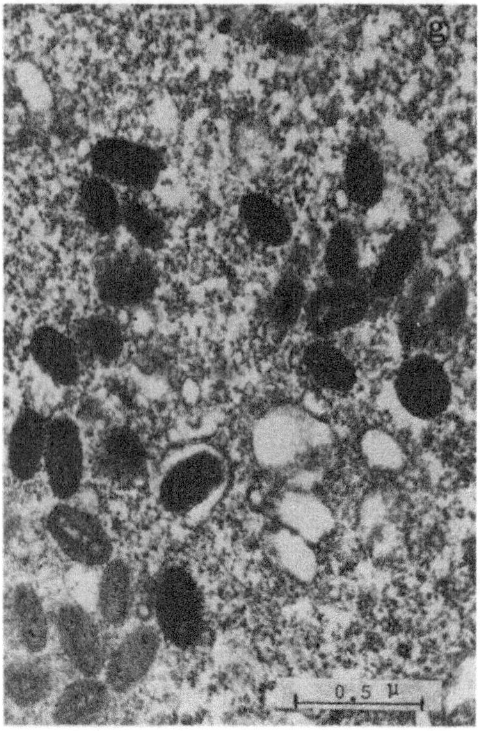

FIGURE 1. Electron microscopy of vaccinia virus infecting HeLa cells under different growth conditions, described in Table 1.

were treated with either rifampin or IBT and then washed (c and f), the ratios were significantly lower (2.25 and 2.51 respectively), indicating the reversibility of the antiviral effect caused by these two compounds. When cultures were treated first with rifampin and then with IBT (b), the ratio of the inhibition was relatively high (3.05); however, when cultures were treated first with IBT and then with rifampin (e) the ratio was significantly lower (1.48) and only slightly higher than that of untreated cultures (1.28). It is suggested that inhibition caused by IBT during the growth of vaccinia virus occurs later than with rifampin, since when rifampin is added after IBT, the inhibition is very limited.

<u>Stage at which IBT is required by IBT^D mutant</u>

Studies on the synthesis of viral macromolecules during the growth of the IBT^D mutant in the absence of IBT showed that "early" polypeptides were synthesized normally. The synthesis of the viral DNA was not inhibited in the absence of IBT, but this DNA did not acquire resistance to deoxyribonuclease. Although synthesis of "late" viral polypeptides took place, cleavage of P4a to 4a and of P4b to 4b was inhibited under these conditions. Cleavage and formation of deoxyribonuclease-resistant viral particles resumed only after addition of IBT to IBT^D-infected cells (22). These findings suggest that inhibition of the IBT^D mutant in the absence of IBT occurs at a stage very similar to that affected by IBT during the growth of the w.t. strain (10). Support for this conclusion comes from electron microscopy studies. The immature virus particles which appear in the IBT^D-infected cells in the absence of IBT, are very similar to those of the w.t. strain formed in the presence of IBT (4).

<u>Active site of IBT molecule necessary for inhibition of w.t. strain and for growth of IBT^D mutant</u>

To define the site of the IBT molecule that determines its anti-poxvirus activity, compounds related in their structure to IBT were examined. It was possible to divide these compounds into three groups: a) compounds which behave similarly to IBT, namely, inhibit the growth of the w.t. strain, do not affect the IBT^R mutant and support the growth of the IBT^D mutant, b) compounds that do not inhibit the growth of the w.t. strain and do not support the growth of the IBT^D virus, and c) compounds that inhibit the growth of all three virus strains.

The first group of compounds, all of which behave similarly to IBT, includes 14 compounds which were studied by us earlier (23, 24) and three additional compounds: 3,5-dichloro-4-hydroxybenzaldehyde-thiosemicarbazone, 3,5-dichloro-2-methoxybenzaldehyde-thiosemicarbazone and 3,5-dichloro-2-hydroxybenzaldehyde-thiosemicarbazone. All these 17 compounds are similar in that they consist of a side-chain covalently linked to an aromatic ring. The side chain in these compounds is a thiosemicarbazone ($-CNNHCSNH_2$), while the structure of the aromatic ring differs from one compound to another. This difference determines the activity of the compound in inhibiting the growth of the w.t. strain or in supporting the growth of the IBT^D mutant. The second group of compounds, which do not affect the growth of any of the three virus strains, includes: thiourea, thiosemicarbazide, semicarbazide, formyl-thiosemicarbazide, isatin-4'-methyl-3-thiosemicarbazide and 1-methylisatin-3-semicarbazone (23). Among these compounds one can find a compound which consists just of the side-chain of IBT (thiosemicarbazide), a compound which is a shortened side-chain (thiourea) and a compound which is a modified thiosemicarbazide, containing oxygen instead of sulphur (semicarbazide). This group of compounds also includes formyl-thiosemicarbazide, which has a formyl group instead of isatin and compounds which although containing an aromatic ring, have chemical modifications in the thiosemicarbazone side chain. The two compounds that were found to inhibit the growth of all three virus strains are 5-hydroxy-picolinadehyde-thiosemicarbazone and 1-formyl-isoquinoline-thiosemicarbazone. It was found that these two compounds interfere with the synthesis of the viral DNA (23). Since this synthesis is an earlier step in the virus growth cycle than the step affected by IBT, all three virus strains were inhibited.

ACKNOWLEDGEMENTS

The helpful assistance of S. Kitov, in performance of the electron microscopy studies, and N. Tasman in preparation of this manuscript, is greatly appreciated.

REFERENCES

1. Bauer, D.J. In: Modern Trends in Medical Virology, Butterworths, London, 1967, pp.49.
2. Herrmann, E.C. and Herrmann, J.A. Ann. N.Y. Acad. Sci. 284:632-637, 1977.
3. Katz, E., Winer, B., Margalith, E. and Goldblum, N. J. Gen. Virol. 19:161-164, 1973.
4. Katz, E., Margalith, E., Winer, B. and Lazar, A. J. Gen. Virol. 21:469-475, 1973.
5. Katz, E., Margalith, E. and Winer, B. Antimicrob. Agent. Chemother. 7:85-90, 1975.
6. Katz, E., Margalith, E. and Winer, B. J. Antimicrob. Chemother. 4:159-162, 1978.
7. Woodson, B. and Joklik, W.K. Proc. Natl. Acad. Sci. U.S.A. 54: 946-953, 1965.
8. Appleyard, G., Hume, V.B.M. and Westwood, J.C.N. Ann. N.Y. Acad. Aci. 130:92-104, 1965.
9. Easterbrook, K.B. Virology 17:245-251, 1961.
10. Katz, E., Margalith, E., Winer, B. and Goldblum, N. Antimicrob. Agents Chemother. 4:44-48, 1973.
11. Cooper, J.A., Moss, B. and Katz, E. Virology 96:381-392, 1979.
12. Joklik, W.K. and Becker, Y. J. Mol. Biol. 10:452-474, 1964.
13. Katz, E. and Moss, B. Proc. Natl. Acad. Sci. U.S.A. 66:677-684, 1970.
14. Katz, E. and Moss, B. J. Virol. 6:717-726, 1970.
15. Katz, E., Margalith, E. and Winer, B. J. Gen. Virol. 40:695-699, 1978.
16. Dales, S. and Siminovitch, L. J. Biophys. Biochem. Cytol. 10:475-503, 1961.
17. Easterbrook, K.B. Virology 17:245-251, 1962.
18. Heller, E., Argaman, M., Levy, H. and Goldblum, N. Nature 222:273-274, 1969.
19. Subak-Sharpe, J.H., Timbury, M.C. and Williams, J.F. Nature 222:341-345, 1969.
20. deHarven, E. and Yohn, D.S. Cancer Res. 26:995-1008, 1966.
21. Moss, B., Rosenblum, E.N. and Katz, E. Nature 224:1280-1284, 1969.
22. Katz, E., Margalith, E. and Winer, B. J. Gen. Virol. 21:477-484, 1973.

23. Katz, E., Margalith, E. and Winer, B. J. Gen. Virol. 25:239-244, 1974.
24. Katz, E., Margalith, E. and Winer, B. J. Antimicrob. Chemother. 2: 255-260, 1976.

14

HALOGENATED RIBOFURANOSYLBENZIMIDAZOLES

Igor Tamm, Pravinkumar B. Sehgal, Robert A. Lamb and Allan R. Goldberg

The Rockefeller University New York, New York 10021

INTRODUCTION

Halogenated ribofuranosylbenzimidazoles are the only known reversible and selective inhibitors of transcription of cellular or viral RNA by RNA polymerase II. The reversible inhibition of mRNA synthesis in eukaryotic cells is biologically the most important consequence of the novel action of these compounds. In different systems, 5,6-dichloro-1-ß-D-ribofuranosylbenzimidazole (DRB) (Fig. 1) blocks initiation or causes premature termination of transcripts synthesized by RNA polymerase II. Although DRB inhibits essentially completely the synthesis of most mRNAs, the synthesis of certain viral and cellular mRNAs is resistant. Indeed, 20-30% of the total cellular heterogeneous nuclear RNA (hnRNA) synthesis is resistant to inhibition by DRB even at very high concentrations. DRB itself, and not a phosphorylated derivative of DRB, inhibits RNA synthesis in eukaryotic cells. The ribofuranose moiety in the ß configuration is essential for selective activity, which requires a free 3' hydroxyl group. Activity of 1-ß-D-ribofuranosylbenzimidazoles increases with multiple substitution of halogen atoms in the benzenoid ring. The most active derivatives inhibit RNA synthesis in the 1-2 µM concentration range.

Historical background

The biological activity of DRB was discovered in experiments on chemical inhibition of influenza virus multiplication (1). DRB (50-75 µM) markedly inhibited the replication of influenza A and B viruses and also the proliferation of chick embryo cells in culture. The studies of several classes of benzimidazole derivatives were initiated because it was thought that derivatives could be identified that would affect regulatory mechanisms governing nucleic acid biosynthesis and not act as antagonists of nucleotide metabolism (2, reviewed in 3). Benzimidazole derivatives were selected for study because it had been established that vitamin B_{12} contained a benzimidazole moiety (5,6-

Y. Becker (ed.), Antiviral Drugs and Interferon: The Molecular Basis of Their Activity.

dimethyl-1-α-D-ribofuranosylbenzimidazole) and played a role in nucleic acid synthesis. Structure-activity studies with benzimidazole derivatives and influenza virus established early the critical role of the β-linked ribofuranosyl moiety at position N_1 in the imidazole ring and of multiple halogen substituents in the benzenoid nucleus of the bicyclic benzimidazole molecule (reviewed in 4, 5). No evidence was found of an antagonism between vitamin B_{12} and DRB.

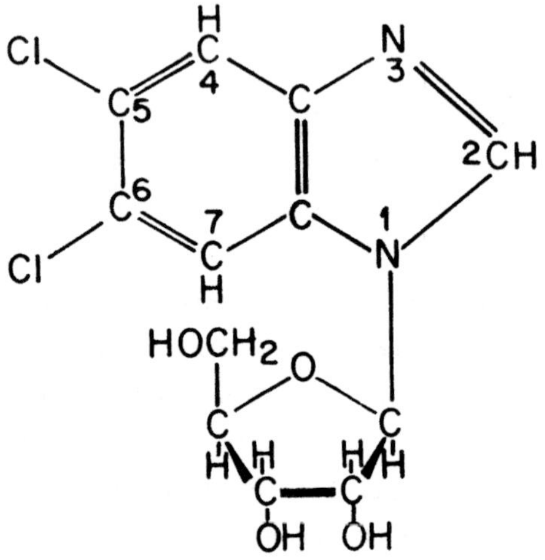

DRB
(5,6-Dichloro-1-β-D-ribofuranosylbenzimidazole)

Fig. 1. The structure of 5,6-dichloro-1-β-D-ribofuranosylbenzimidazole (DRB).

The early experiments on the inhibition by benzimidazole derivatives of the multiplication of influenza virus and of the synthesis of the soluble complement-fixing antigen (i.e. viral nucleocapsid) were done in an <u>in vitro</u> system which employed portions of the chick chorioallantoic membrane as the host tissue, and permitted precise and reproducible assays of the biological activities of the derivatives (6). The latent period of influenza virus in this system is 4 hr. This is followed by a rapid and exponential rise in nucleocapsid and virus, which lasts 4 hr and after which multiplication continues for 16 more hr at a reduced exponential rate. It was demonstrated that the DRB-sensitive process is limited in time to the first 4 hr of the viral growth cycle, as illustrated in Fig. 2.

The initial findings with DRB led to the hypothesis that host cell RNA may play an important role in the replication of some viruses (1). It has subsequently been demonstrated that cellular RNA polymerase II is involved in influenza virus replication (7) and that host RNAs, newly synthesized by RNA polymerase II, donate their 5' caps and 10-13 nucleotides to the influenza virus mRNAs (8).

Extension of the early studies of the antiviral activity of DRB to other viruses revealed that adenovirus 4 is as sensitive to DRB as influenza A and B viruses (9), whereas poliovirus 2 is less sensitive (10). The basis for these findings can now also be understood: adenoviruses depend for their transcription on the cellular RNA polymerase II, whereas picornaviruses do not (11-14).

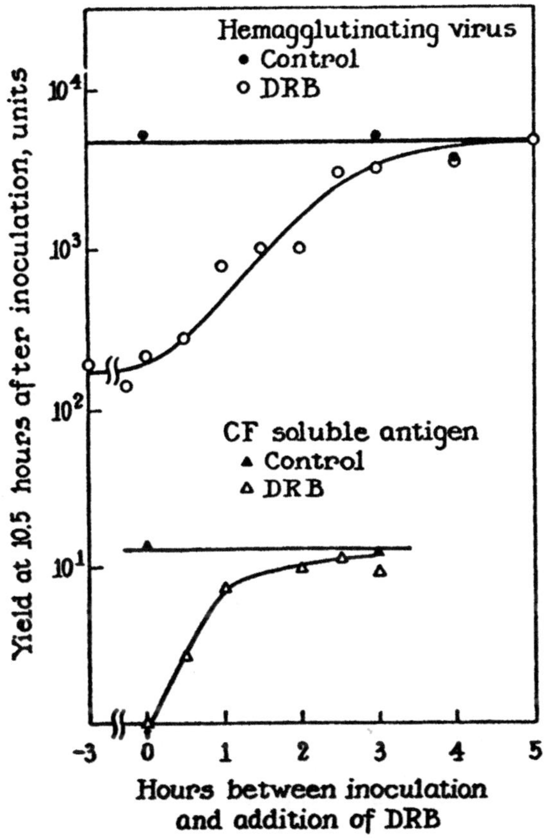

Fig. 2. Inhibition of production of influenza virus particles (hemagglutinating virus) and nucleocapsids (complement-fixing soluble antigen) in the chick chorioallantoic membrane *in vitro* relative to the time of addition of DRB (55 μM). From Tamm and Tyrrell (6). (Reprinted with permission).

Table 1. Sensitivity of viruses to growth inhibition by halogenated ribofuranosylbenzimidazoles

Virus, strain	Compound	Sensitivity to inhibition
Class I		
Influenza A_0, PR8	DRB	marked
Influenza A_1,FM1	DRB	marked
Influenza A_0, Mel	DRB	marked
Fowl plague, Herts	TRB	marked
Influenza B, Lee	DRB	marked
Influenza B, Lee	TRB	marked
Influenza B, MB	DRB	marked
Influenza B, 1760	DRB	marked
Oncorna, RSV	DRB	marked
Papova, SV40, polyoma	DRB	marked
Adeno 2	DRB	marked
Adeno 4	DRB	marked
Pox, vaccinia	DRB & TRB	marked
Class II		
Parainfluenza 1, Sendai	TRB	none[2]
Parainfluenza 1, Sendai	TRB	slight[2]
Newcastle disease	TRB	slight
Toga A, Semliki Forest	TRB	slight
Reo 3, Dearing	DBRB[3]	moderate
Polio 2, MEFI	DRB	moderate
Polio 2, MEF1	TRB	slight
Encephalomyocarditis	TRB	slight

[1]Sensitivity relative to the marked sensitivity of cellular hnRNA synthesis to inhibition by DRB or TRB (trichloro derivative).

[2]Different reports.

[3]5,6-Dibromo-1-β-D-ribofuranosylbenzimidazole.

Antiviral spectrum

Table 1 summarizes information on the sensitivity of viruses to inhibition by halogenated ribofuranosylbenzimidazoles. Class I comprises viruses whose replication is as sensitive to inhibition as is the transcription of the bulk of cellular hnRNA, and includes orthomyxo-, retro-, papova-, adeno-, and poxviruses. Class II comprises viruses whose replication is less sensitive, and includes paramyxo-, toga-, reo-, and picornaviruses; the sensitivity of these viruses covers the range from none to moderate.

Although it appears that all viruses in class I are dependent for their multiplication on cellular RNA polymerase II, the mechanism of this dependence varies and does not necessarily involve the synthesis of viral mRNA. Vaccinia virus multiplication is blocked by DRB at the virion assembly stage (17). The precise nature of the nuclear function required for the assembly of vaccinia virions is not known. The varying DRB sensitivity of viruses in class II may mainly reflect differences among these viruses in their dependence on a variety of host cell functions that become secondarily affected in DRB-treated cells.

Because halogenated ribofuranosylbenzimidazoles selectively inhibit the function of cellular RNA polymerase II, we shall first summarize results of studies which bear on the action of DRB and related derivatives on cellular RNA synthesis. Next we will describe the action of these compounds on viral RNA synthesis and certain other processes in viral replication, and finally touch on the use of halogenated ribofuranosylbenzimidazoles as tools in studies of the control and regulation of gene expression.

This chapter occupies its assigned position in the volume on the basis of the demonstrated selective action of DRB and related derivatives on the function of cellular RNA polymerase II. Halogenated ribofuranosylbenzimidazoles are highly active inhibitors of the multiplication of a wide range of DNA and some RNA viruses because cellular RNA polymerase II function plays a crucial role in the multiplication of these viruses, and not because DRB and related derivatives recognize virus-specific targets.

INHIBITION OF CELLULAR RNA SYNTHESIS

DRB inhibits cellular RNA synthesis (9, 18) without interfering with cellular energy metabolism (1) or causing a rapid and marked inhibition of cellular protein synthesis (9). DRB has been shown to inhibit RNA synthesis in a wide variety of cells, including human, monkey, murine, avian, insect, amphibian, and echinoderm. Extensive work on DRB action has been done in human, mouse, chicken, and chironomid cells, in all of which marked inhibition is obtained at 50 μM concentration and near maximal or maximal inhibition requires

>65 μM DRB. In mammalian cells some inhibition of RNA synthesis can be detected at 5 μM concentration.

Selective inhibition of cellular RNA synthesis catalyzed by RNA polymerase II

The demonstration in starfish oocytes and chironomid salivary gland cells that DRB inhibits nucleoplasmic (chromosomal) RNA labeling more than nucleolar RNA labeling (19, 20) led to biochemical experiments which established the selective effect of DRB on hnRNA synthesis in the chironomid system (22, 23, 24)

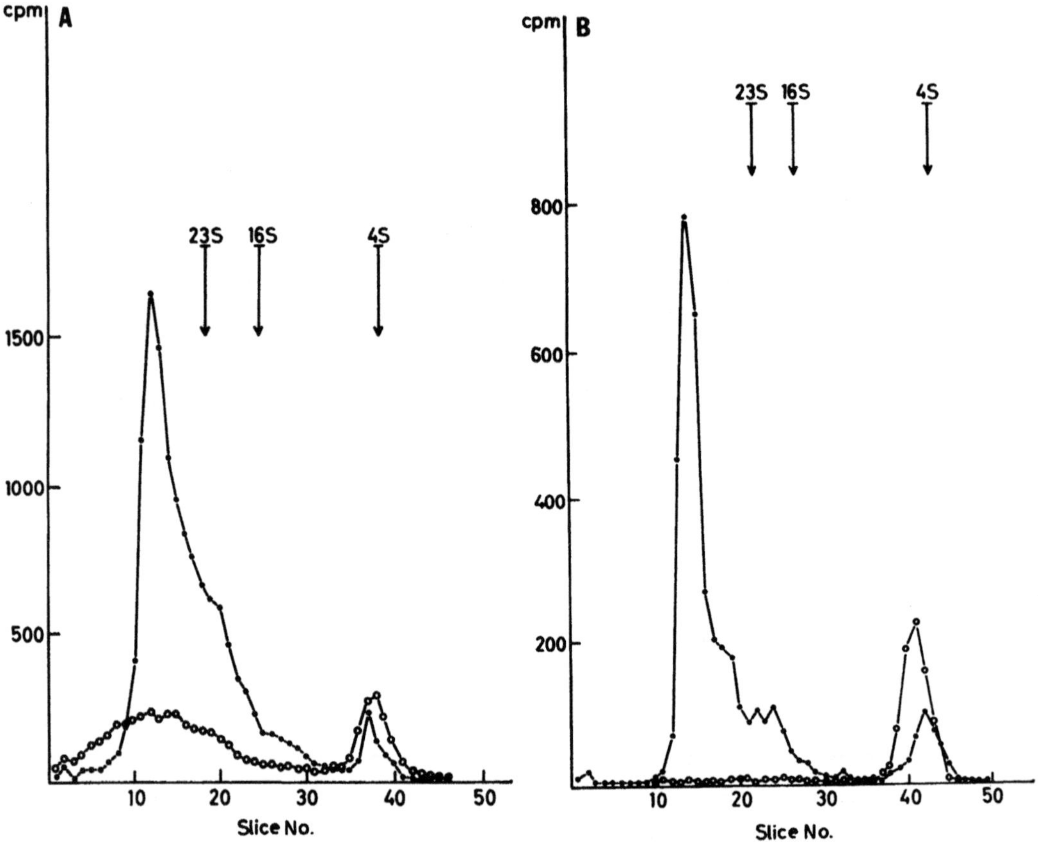

Fig. 3. Selective inhibition of hnRNA synthesis in salivary gland cells of larval Chironomus tentans by DRB (63 μM). (A) Control; (B) DRB. Two glands each were incubated at 18°C for 60 min in Cannon's modified insect medium without or with DRB, and then labeled with 100 μCi each of [^{3}H]uridine and [^{3}H]cytidine in a volume of 25 μl of fresh control or DRB-containing medium. Chromosomes and nuclei were isolated from 25 cells per group by dissection. RNA was extracted by pronase-SDS treatment and analyzed in 2% agarose gels. Chromosomes o—o; nucleoli ●—●. From Egyházi et al. (22). (Reprinted with permission).

and in avian (25), and mammalian cells (26). Fig. 3A and B illustrate the selective inhibition of hnRNA synthesis by DRB in the isolated salivary glands of Chironomus tentans (22). After treatment and labeling with tritiated

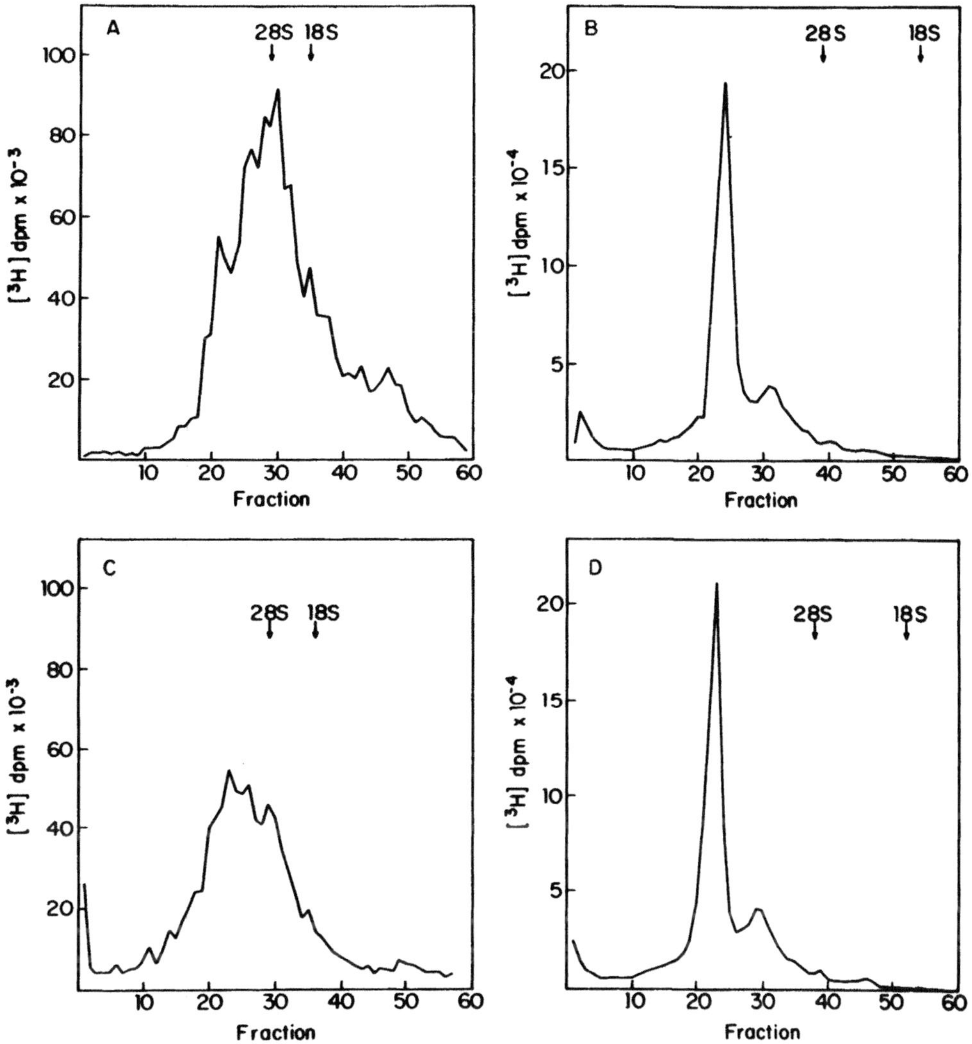

Fig. 4. Selective inhibition of hnRNA synthesis in mouse L-929 cells by DRB (12 μM). L-929 cells in suspension culture in serum-free modified Eagle's minimum essential medium were incubated with or without DRB (12 μM) for 15 min and then labeled with [^{3}H]uridine (15 μCi/ml) for 15 min in the continued presence or absence of DRB (12 μM). The nuclei were then isolated and separated into nucleoplasmic and nucleolar fractions. RNA from nucleoplasmic fractions was analyzed in 1% polyacrylamide + 1% agarose gels, whereas that from nucleolar fractions was analyzed in 2% polyacrylamide + 1% agarose gels. (A) Nucleoplasmic, control; (B) nucleolar, control; (C) nucleoplasmic, DRB; (D) nucleolar, DRB. From Tamm et al. (26). (Reprinted with permission).

cytidine and uridine, using 4 glands per variable, the glands were fixed and nucleoli, chromosomes, and nuclear sap were isolated by micromanipulation. Fig. 3A shows that, in untreated control cells, the chromosomes contain newly labeled, heterodisperse high molecular weight RNA as well as low molecular weight RNA, whereas the nucleoli contain a large amount of 38S RNA and also low molecular weight RNA. After DRB (63 μM) treatment (Fig. 3B), the chromosomal extract contains label only in the low molecular weight region, whereas the nucleolar fraction still contains both 38S and low molecular weight RNA.

Fig. 4 shows that 12 μM DRB selectively inhibits hnRNA synthesis in mouse L-929 cells, without a significant effect on preribosomal RNA synthesis (26). At higher concentrations, some inhibition of preribosomal RNA synthesis becomes evident (26), but at all concentrations (incl. 150 μM), the inhibition of hnRNA

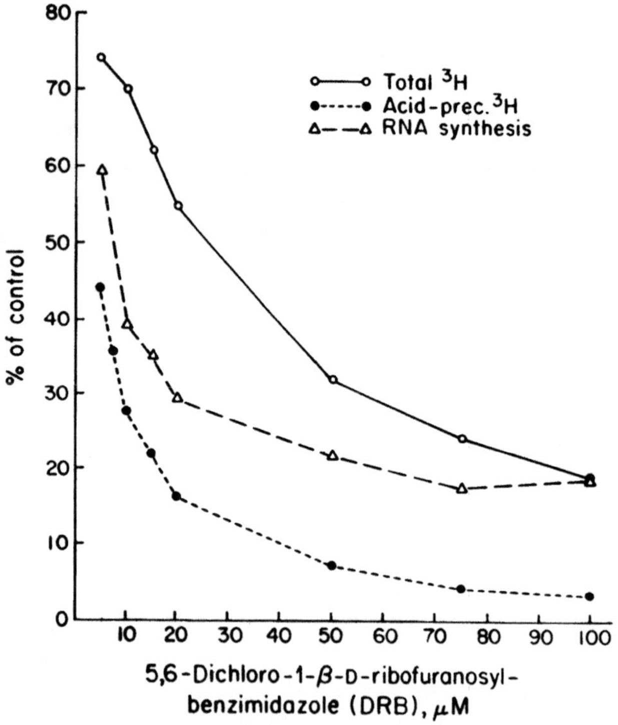

Fig. 5. Relationship between concentration of DRB and inhibition of uridine uptake and hnRNA synthesis in human HeLa cells. Cells in monolayer cultures in reinforced Eagle's medium containing 5% serum were treated for 15 min with DRB at varying concentrations and pulse-labeled with [^{3}H]uridine (2.5 μCi/ml) for 15 min in the continued presence of DRB. Actinomycin D (0.04 μg/ml) was present for 25 min before and also during the labeling to suppress preribosomal RNA synthesis. o——o total cell-associated radioactivity; ●---● acid-precipitable cell-associated radioactivity; Δ---Δ RNA synthesis. From Tamm (27).(Reprinted with permission).

synthesis is greater than that of preribosomal RNA. Furthermore, as was mentioned above, not all of the synthesis of the broadly distributed hnRNA is sensitive to inhibition by DRB in mammalian cells. Fig. 5 demonstrates for human HeLa-S3 cells that 20-30% of hnRNA synthesis is refractory to inhibition even at 100 μM DRB (27). It is evident that a part of hnRNA synthesis is highly sensitive to inhibition by DRB and shows a steep dose dependence in the 5-15 μM range. As the concentration of DRB is increased, a resistant fraction becomes apparent. The DRB-sensitive hnRNA is much more sensitive to inhibition by DRB than is preribosomal RNA. In chironomid cells, too, hnRNA synthesis is more sensitive to DRB than preribosomal RNA synthesis (22). In these cells no DRB-resistant fraction of hnRNA has been reported.

Functionally, the most important aspect of DRB action concerns its effect on mRNA appearance in the cytoplasm (28). Fig. 6 shows that 75 μM DRB inhibits the labeling of cytoplasmic poly(A^+) mRNA by ~95%, although in agreement with results in Fig. 5, it inhibits the labeling of nuclear hnRNA only ~70%. Evidence has been obtained that DRB also markedly inhibits the synthesis of poly(A^-) mRNA (28).

These results are consistent with the view that the DRB-sensitive fraction of hnRNA gives rise to the bulk of mRNAs. They leave open the possibility that the DRB-resistant fraction may give rise to some mRNAs (29, 30).If most of the DRB-resistant RNA is not related to cytoplasmic mRNA, it might be poor in unique mRNA coding sequences, or DRB-resistant RNA may contain an abundance of coding sequences, but yet have some characteristic which prevents its processing and transport to the cytoplasm. There is some evidence that DRB-resistant hnRNA is relatively poor in message-containing sequences (31). It has recently been reported that between half and two-thirds of the long (>700 bases) capped hnRNA molecules do not contribute a capped polysomal derivative to the cytoplasm (32), and that many long capped hnRNA molecules are of a different sequence category from those molecules that are successfully processed into mRNA (33). It is of interest that the only DRB-resistant mRNA in adenovirus (that for protein IX) is successfully transported to the cytoplasm and translated in the presence of DRB (34).

No significant differences have been observed between DRB-resistant and bulk hnRNA. The DRB-resistant fraction can be demonstrated by labeling cells with [^{3}H]uridine, [^{3}H]adenosine, ^{32}P, or L-[methyl-^{3}H]methionine. DRB-resistant RNA is initiated with both ATP and GTP in the presence of the drug (35). The size of the fraction of hnRNA that is resistant to DRB is independent

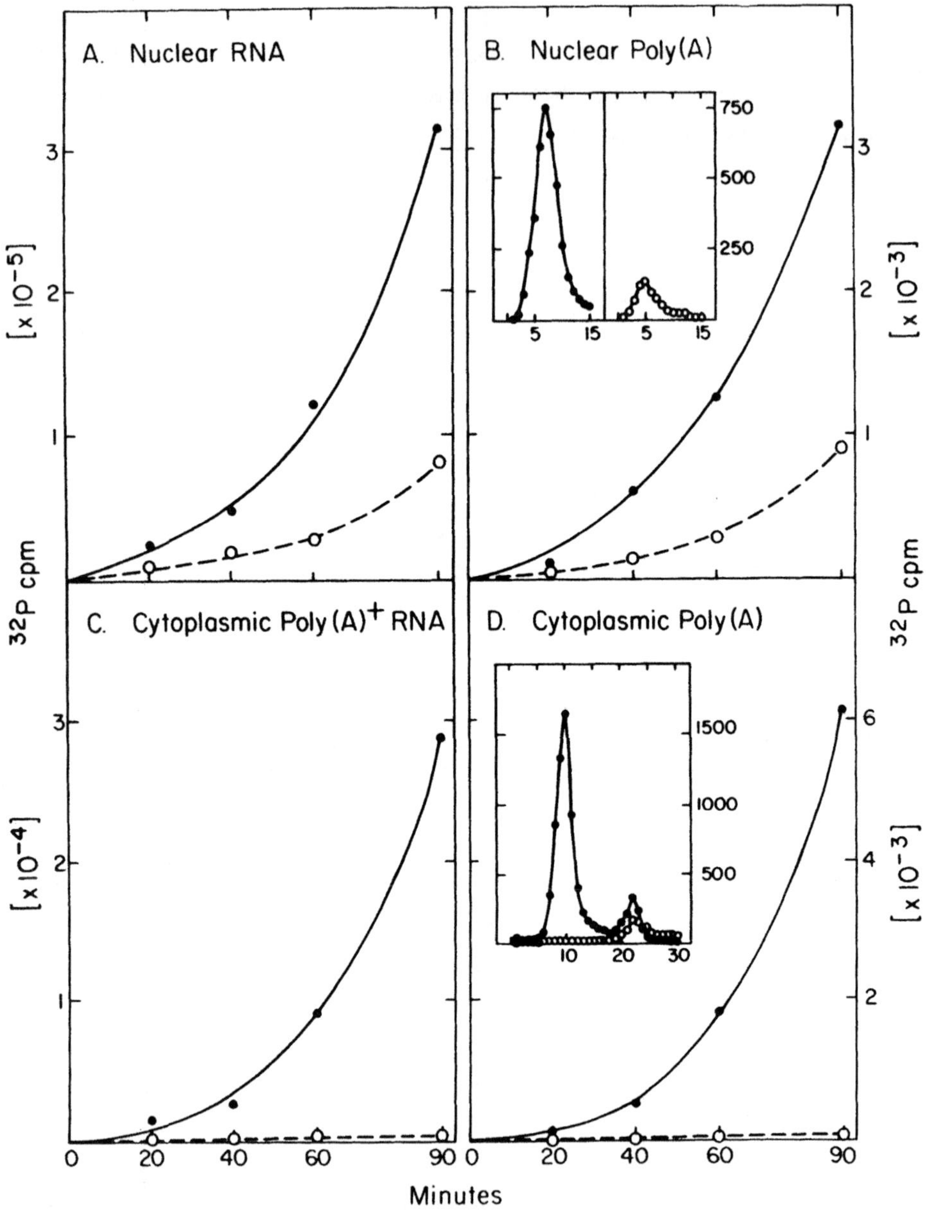

Fig. 6. Inhibition of hnRNA and mRNA synthesis in HeLa-S3 cells by DRB (75 μM). Cultures of 2 x 10^8 HeLa cells were treated with actinomycin D (0.04 μg/ml) for 25 min in PO_4-free growth medium and were then exposed to DRB (75 μM) for 15 min, at the end of which time they were labeled with $^{32}PO_4$ (5 mCi) in 100 ml vol in the continued presence of the drugs. Aliquots containing 5 x 10^7 cells were removed 20, 40, 60 and 90 min later, the cells were fractionated, and the RNA was extracted. (A) Nuclear RNA as determined by solubilization of ^{32}P radioactivity by T1 and pancreatic RNAases in the absence of NaCl. (B) Nuclear poly(A) as determined by ^{32}P radioactivity in a T1 digest of RNA which bound to

and eluted from a poly(U)-Sepharose column by a 10-90% v/v formamide wash. The 90 min samples were analyzed further on a 15% polyacrylamide gel (inset). (C) Cytoplasmic poly(A)-containing RNA was determined by ^{32}P radioactivity bound to and eluted from a poly(U)-Sepharose column by a 10-90% v/v formamide wash in the post-mitochondrial cytoplasmic supernate. (D) Cytoplasmic poly(A) as determined by ^{32}P radioactivity in T1 and pancreatic RNase digests of total cytoplasmic RNA migrating in the position of poly(A)$_{\sim 115}$ in 15% polyacrylamide gels. The gel profiles of the 90 min samples are shown in the inset. (●) control culture without DRB; (o) cells treated with DRB. From Sehgal et al. (28). (Reprinted with permission).

of the duration of DRB treatment (beyond 15 min) and of the duration of labeling which suggests that metabolic turnover of DRB-resistant RNA is similar to that of bulk hnRNA. No significant differences have been observed in the broad size distribution profiles of bulk hnRNA and DRB-resistant hnRNA molecules >1000 nucleotides long (26, 28, 30, 36). The two populations are also similar with respect to the presence of 3' terminal poly(A) and length of poly(A) chains (28); presence of 5' terminal cap 1 structures (28, 37, 38); oligo(U) content (39); and presence of internal methyl groups (40). It is possible that DRB treatment decreases labeling of m^6A in hnRNA with [methyl-^3H]methionine more than labeling of hnRNA with [^3H]uridine (37).

The selectivity of the action of DRB is further emphasized by the findings that it has no significant effects on transfer RNA, 5S RNA (HeLa), and 8S RNA (chironomid) synthesis by RNA polymerase III (25, 41, 42, 43), and that it only slightly inhibits the synthesis of mitochondrial poly(A)$^+$ RNA (28, 43). In keeping with the findings that preribosomal RNA synthesis by RNA polymerase I is inhibited in DRB-treated cells, although less so than hnRNA synthesis (see above), the synthesis of 5.8S rRNA, also a product of polymerase I, is sensitive to inhibition (43). DRB does inhibit the synthesis of small homodisperse nuclear RNAs, such as U_1 and U_2 (also designated D and C) by RNA polymerase II to approximately the same extent as hnRNA (37, 44). Hellung-Larsen et al. (44) have shown that the inhibition of D, C, and A (U_1, U_2, and U_3) RNA is greater than that of ribosomal RNA, which in turn is more inhibited than tRNA and 5S RNA.

Taken together, these results indicate that the mechanism of DRB action does not involve alterations in ribonucleotide metabolism. Harlow and Molloy (43) have demonstrated that, in HeLa cells, DRB does not affect the phosphorylaton of any ribonucleotides to triphosphates, the cellular conversion of [^3H]uridine to UTP or the size of the UTP and ATP pools.

Kinetics and reversibility of action

Fig. 7 shows the time course of inhibition of >10S hnRNA synthesis in HeLa cells in suspension cultures treated with 75 µM DRB for varying periods and labeled with [^{3}H]uridine for 30-35 sec in the continued presence of the drug (28). Some inhibition is evident already within 1 min, following which the rate of hnRNA synthesis decreases progressively. Maximal inhibition is reached between 5 and 15 min. The level of residual synthesis (30-35%) signifies the continued transcription of a subfraction of DRB-resistant hnRNA (see above).

Table 2 illustrates the prompt reversibility of the inhibition of RNA synthesis by DRB (60 µM) in monolayer cultures of L-929 cells (26). After 1-hr treatment of cells with DRB, measurement of RNA synthesis by labeling with [^{3}H]uridine for 15 min in the continued presence of DRB shows marked inhibition, which is completely reversed when the labeling is done in the absence of DRB. Similar prompt reversibility has been observed after 4-hr treatment with DRB. After 16 or 24 hr treatment of human diploid fibroblasts (FS-4 cells), full recovery of RNA synthesis requires a few hr (45). Reversibility of the

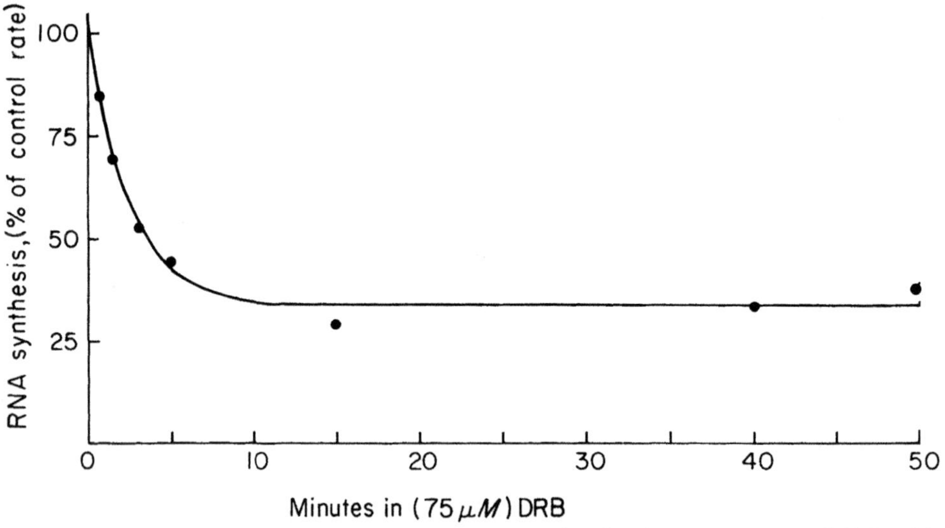

Fig. 7. Decline in the rate of hnRNA synthesis during exposure of HeLa-S3 cells to DRB (75 µM). Cultures of 1 x 10^8 cells were treated with actinomycin D (0.04 µg/ml) for 25 min in growth medium and were then exposed to DRB (75 µM). At different times thereafter, the cultures were pulsed with [^{3}H]uridine (3-5 mCi) in a volume of 50 ml for 30-45 sec in the continued presence of the drugs. The hnRNA was extracted and analyzed on sucrose gradients. The ^{3}H cpm incorporated in the presence of DRB have been corrected for the inhibition of precursor transport and are expressed as fractions of the incorporation obtained in control cultures that were not treated with DRB but were exposed to actinomycin D for the same length of time as the DRB-treated experimental cultures. The figure presents the pooled data from several separate experiments. From Sehgal et al. (28). (Reprinted with permission).

inhibition of RNA synthesis by DRB has also been demonstrated in chironomid salivary gland cells (20). Pulse-labeling of cells with RNA precursors immediately after washing the salivary gland explants in DRB-free medium has revealed incorporation into short nascent RNA molecules (24).

RNA chain initiation and elongation

Egyházi (24, 46, 47) demonstrated in chironomid salivary gland cells that early during inhibition of >10S hnRNA synthesis by DRB, the relative amount of label appearing in longer molecules increased progressively as if DRB were blocking hnRNA chain initiation or terminating hnRNA chains close to sites of initiation. Fig. 8A,B shows the results of chase type experiments in which cells were labeled for 45 min with tritiated cytidine and uridine and after which labeling was continued for 45 (A) or 90 (B) min in the presence of 63 μM DRB (46). Presence of DRB for 45 min resulted in a reduction of labeling of smaller sized hnRNA (<75S) while labeling of larger hnRNA molecules continued. After 90-min DRB treatment, labeling of the larger hnRNA species was also reduced. Incorporation in the 4-8S range and in the 16-30S range appeared to be resistant to inhibition by DRB.

Table 2. Reversibility of the inhibitory effect of 60 μm DRB on uridine incorporation in L cells in monolayer

Medium		[^{3}H]uridine incorporation[1]			
Before pulse	During 15-min pulse	cpm	μg protein	cpm/μg protein	% of control
Control	Control	2,295	89	25.8	100
DRB, 60 min	DRB	882	89	9.9	38
DRB, 60 min	Control	2,249	72	31.2	121

[1]Plastic petri dishes (60 mm) were seeded with 2.5 x 10^5 L cells per dish in MEM with 5% fetal calf serum. Approximately 22 hr later, the medium was removed and the incomplete monolayers of L cells were incubated with or without 30 μM DRB in MEM without serum. The medium was then replaced with either control or DRB-containing medium, and [^{3}H]uridine (sp. act. 29.65 Ci/mmol) was added to a concentration of 2.5 μCi/ml for pulse-labeling. [^{3}H]uridine incorporated into acid-precipitable material was determined and expressed in cpm/mg protein. From Tamm et al. (26).(Reprinted with permission).

Early experiments in HeLa cells showed that RNA transcripts that had already reached the length of about 1,000 nucleotides before DRB treatment of cells was begun continued to elongate, and presumably grew to full size (48). These studies in chironomid and mammalian cells established that DRB does not act as a general hnRNA chain terminator.

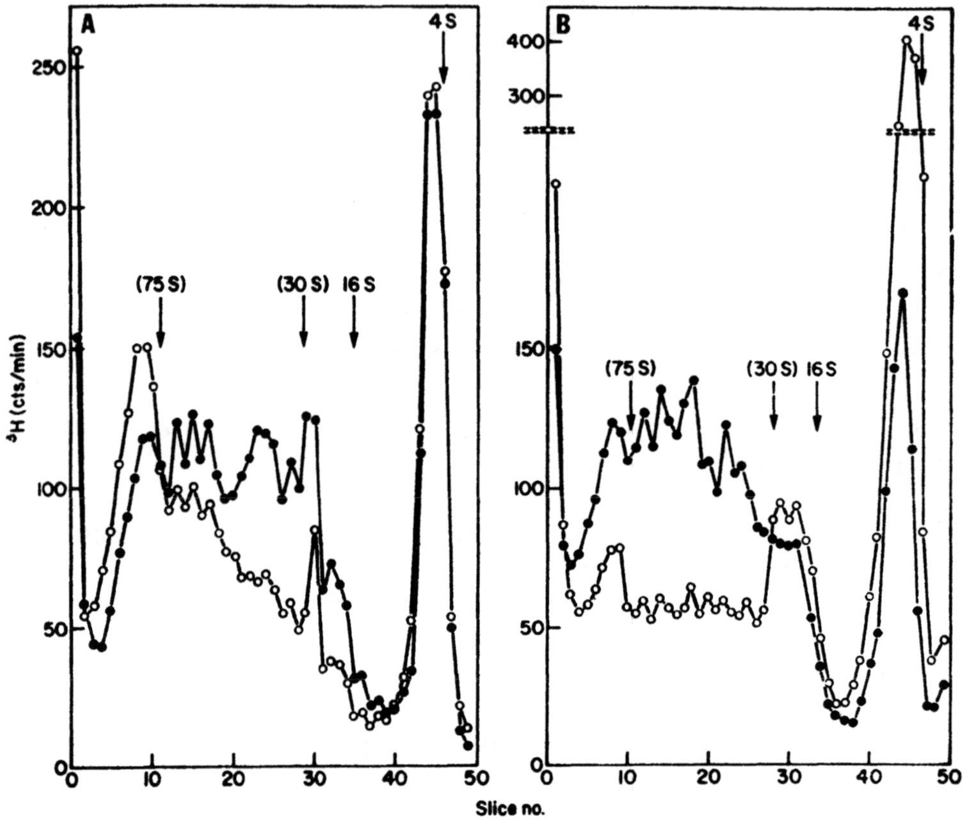

Fig. 8. Sequence of inhibition of hnRNA synthesis in the chromosomes of the salivary glands of larval Chironomus tentans after exposure to DRB (63 μM). In each of two experiments glands were obtained from 4 different animals and two groups set up, with glands in one group representing sisters of glands in the other. Each group of glands was incubated in 50 μl of medium containing 100 μCi each of [^{3}H]cytidine and [^{3}H]uridine at 18°C for 45 min. The control glands (●—●) were then collected whereas experimental glands (o—o) received DRB (63 μM) and were incubated for an additional period of 45 min (A) or 90 min (B). All glands were then fixed and from each group 6 sets of chromosomes I to III were isolated and the labeled RNA in the chromosomes was analyzed in 1% agarose gels. From Egyházi (46). (Reprinted with permission).

In chironomids, no evidence of accumulation of prematurely terminated hnRNA transcripts was found in further studies in which RNA was labeled with [α-^{32}P]ATP which is taken up by the salivary gland cells, and analyzed on 10% polyacrylamide gels to resolve low molecular weight RNA chains in the 30-1,000-

nucleotide range (49). It has therefore been suggested that inhibition of hnRNA transcription by DRB in chironomid cells takes place at the level of chain initiation (49).

In mammalian cells, the use of a combined in vitro-in vivo procedure permitted the first detection in DRB-treated as well as in control cells of an abundant population of RNA chains 100-300 nucleotides long, synthesized by RNA polymerase II (36, 50). Whole HeLa-S3 cells in suspension cultures were incubated with or without DRB, and then the nuclei were isolated and the nascent chains of RNA labeled with [^{3}H]UTP in vitro. In the absence of initiation of new chains by RNA polymerase II in the isolated nuclei incubated in vitro, end-labeling of the growing chains, followed by size analysis, gives a profile which is indicative of the abundancies of different sized chains immediately prior to the isolation of the nuclei from DRB-treated or control cells. This approach is successful because chains that are blocked or delayed in their growth in vivo can apparently elongate in vitro in the isolated nuclei, regardless of whether DRB was or was not present in vivo. In nucleoprotein complexes isolated from DRB-treated adenovirus 2-infected cells, nascent chains started in vivo are also successfully elongated in vitro (51) even though the in vivo treatment with DRB results in a blockade of transcription at promoter proximal sites (52-54). Presence of DRB in the in vitro reaction mixture containing isolated nuclei has no effect on the elongation of RNA chains by RNA polymerase II (36, 55)

As illustrated in Fig. 9, initial experiments of this kind revealed the existence, in apparent molar excess, of a population of short RNA chains in addition to the broadly distributed classical population of hnRNA chains (36). Treatment of HeLa-S3 cells with DRB for 10 or 40 min did not decrease the population of abundant short chains, although it decreased the number of hnRNA molecules >1,000 nucleotides long in a time-dependent manner (36). The profiles in Fig. 9 were obtained after labeling of the isolated nuclei in vitro for 15 min and analyzing the RNA on sucrose gradients. To obtain estimates of the size distribution of the abundant population of short chains as they existed in vivo prior to isolation of the nuclei, end-labeling in vitro can be carried out for varying periods, the RNA samples analyzed by polyacrylamide gel electrophoresis, and the in vitro chain elongation curves extrapolated to zero time. Such experiments have given length estimates of 100-300 nucleotides for the abundant nascent molecules whose growth is apparently delayed or prematurely terminated in vivo. By analogy with findings in the adenovirus system (52-54) these results indicate that DRB inhibits the synthesis of hnRNA by causing premature termination of nascent chains at short distances downstream from initiation

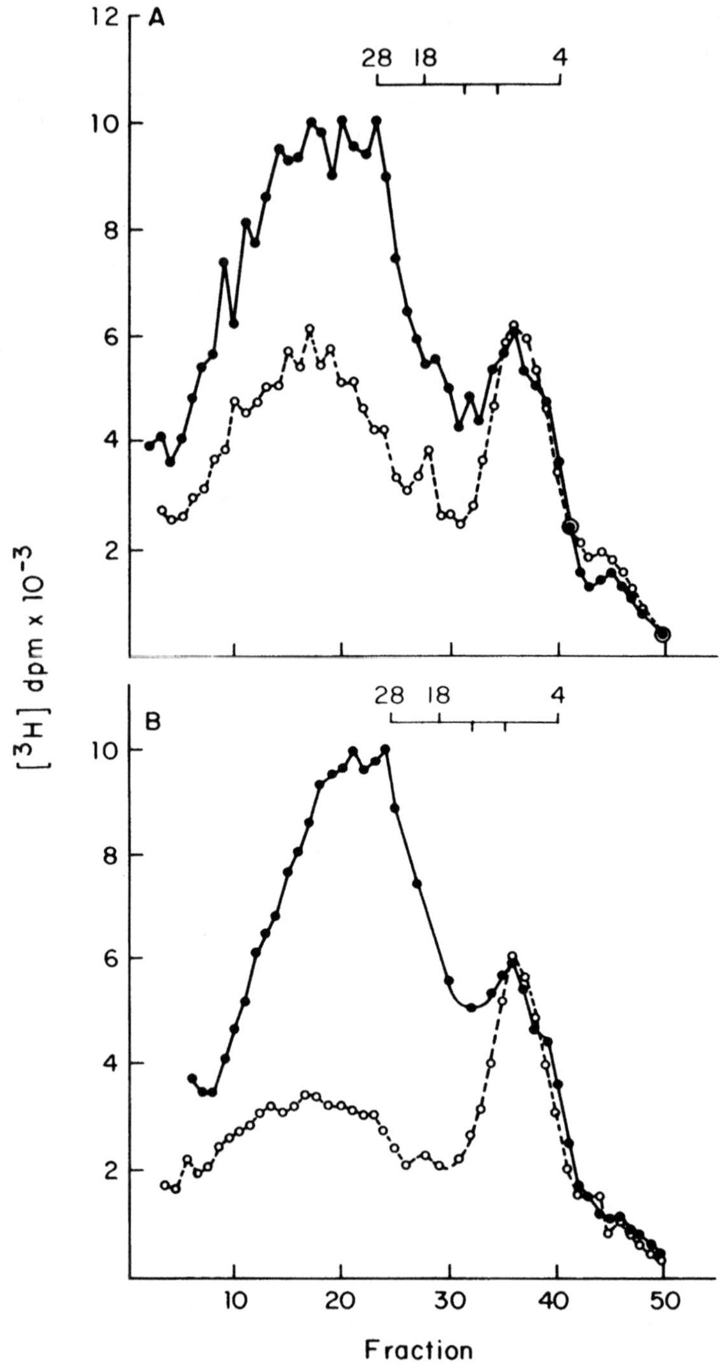
A
B
28 18 4
28 18 4
12
10
8
6
4
2
[³H] dpm x 10⁻³
10 20 30 40 50
Fraction

Fig. 9. Effects of pretreatment of HeLa cells with DRB on the synthesis c nucleoplasmic RNA species in isolated nuclei. Cultures were treated wit actinomycin D (0.04 μg/ml) for 25 min and exposed to DRB (60 μM) for 10 c 40 min at 37°C. Nuclei were isolated by swelling the cells in hypotonic buffer rupturing with a Dounce homogenizer, and centrifugation. The isolated nucle were incubated for 15 min at 26°C in a reaction mix for RNA synthesis containin [^{3}H]UTP. The nucleoplasmic and nucleolar fractions were separated by zona centrifugation. The RNA was extracted from the nucleoplasmic fractions t phenol-chloroform and analyzed on 1.1% polyacrylamide/1% agarose gels (A) ●—●, Control; o--o, 60 μM DRB, 10 min (in vivo). (B) ●—●, Control; o--c 60 μM DRB, 40 min (in vivo). The two marks below each bar are interpolate positions for RNA of 250,000 daltons (740 nucleotides) and 110,000 dalton (330 nucleotides). From Tamm (36). (Reprinted with permission).

sites (36, 50). Furthermore, as there appears to be a molar excess of shor chains even in HeLa cells not treated with DRB, the results are compatible wit the hypothesis that premature termination occurs with significant frequency i control cells and that the effect of DRB is to increase the frequency of suc premature termination. Although the experiment in Fig. 9 did not provid evidence of accumulation of short chains in DRB-treated cells in excess of th amounts found in control cells (36), subsequent studies have revealed that suc accumulation does occur (37, 50). This has also been observed in the adenoviru system (54).

In nuclei of lysed Ehrlich ascites cells, RNA polymerase II activity show two optima: one at 0.025 M and the other at 0.3 M $(NH_4)_2SO_4$ concentration (55) Pretreatment of cells for 15 min with 80 μM DRB eliminates all of th polymerase II activity in the low salt, but not in the high salt assay. Afte 30-min pretreatment of cells with 48 μM DRB, almost all of the polymerase I activity with the low salt optimum is lost, whereas 50% of the activity with th high salt optimum is retained. These results suggest that RNA polymerase II ma exist in cells in two functional states, and that DRB selectively affects RN polymerase II which is in the state in which it displays a low salt optimum i vitro (55). The different states of the enzyme may relate to differences i template structure.

Dreyer and Hausen (55) have investigated the question whether "free" RN polymerase II molecules accumulate in DRB-treated cells. No polymerase capabl of initiating transcription on added poly[d(A-T)] was detected in lysates o either control or DRB-treated cells. This template is effectively used b purified RNA polymerase II.

Measurement of the abundance of different sized RNA molecules containing 5 cap I ($m^7GpppN_1mpN_2p$) oligonucleotide structures characteristic of hnRNA ha provided strong support for the hypothesis that DRB increases the frequency of premature termination of nascent hnRNA chains (37). 5' cap I structures are added rapidly after initiation of transcription of new hnRNA chains. Fig. 10 demonstrates the effect of 40-min pretreatment of HeLa S3 cells with 75 μM DRB on the labeling (for 30 min) of nuclear RNA with L-[methyl-^{3}H]methionine in the continued presence of DRB. The overall labeling of short RNA chains was somewhat reduced after DRB treatment, however, the labeling of chains >2,000 nucleotides long was markedly reduced. L-[methyl-^{3}H]methionine labels not only the cap I structures characteristic of hnRNA, but also cap II structures of the type $m^{2,2,7}$GpppAmpUmpNp, which form 5' termini in certain homodisperse small nuclear RNA species, such as U_1, U_2, and U_3, also known as D, C, and A (56, 57). Analysis for the two types of cap structures in the abundant short RNA chains showed that DRB caused an approximately 2-fold increase in the labeling of m^7G-capped short chains, but at the same time also a 3-fold decrease in $m^{2,2,7}$-G-capped short chains (37).

The bar graph in Fig. 10 summarizes the distribution of cap I structures in hnRNA molecules of various sizes. It is evident that short RNA chains (100-500 nucleotides), bearing cap I structures characteristic of hnRNA, accumulate in DRB-treated cells in quantities much greater than those in untreated cells. Concurrently, labeling of cap I structures in long hnRNA is reduced on the average by 80% in DRB-treated cells. This reciprocal relationship between the apparent increase in short chains and the decrease in long chains is consistent with the hypothesis that DRB inhibits hnRNA synthesis by increasing the frequency of premature termination of newly initiated chains. It appears that analysis of the abundance of m^7G-capped short chains could provide a basis for quantitative estimation of the overall frequency of premature termination of short nascent hnRNA chains. The fact that premature chain termination occurs does not exclude the possibility that DRB may also have some effect on hnRNA chain initiation in mammalian cells.

Tweeten and Molloy (58) have reported that DRB causes premature termination of transcription within the β-globin gene. Using restriction enzyme fragments of the cloned β-globin gene, no full length transcripts of this gene were detected in nuclei of DRB-treated (125 μM) cells. Nuclei of control cells contained such transcripts. Furthermore, in vitro labeling of RNA in isolated nuclei and hybridization to restriction enzyme fragments from both the 5' and the 3' regions of the gene revealed the following: RNA in nuclei from DRB-treated cells hybridized only to the promoter-proximal gene fragment, whereas RNA in nuclei from control cells hybridized to gene fragments from both the 5' and 3' regions.

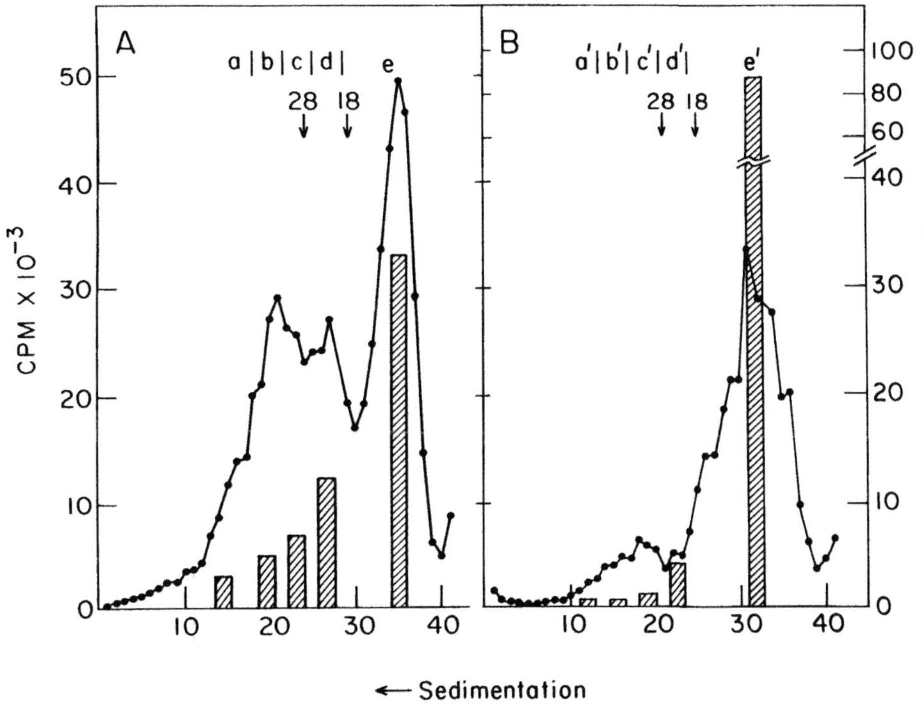

Fig. 10. Accumulation of short m^7G-capped RNA chains and reduction in long m^7G-capped hnRNA chains in HeLa-S3 cells treated with DRB (75 µM). Sucrose-Me_2SO gradient profiles of nuclear RNA labeled with L-[methyl-^3H]methionine (30 min) in control and DRB-treated cells and the abundance of cap I structures in RNA of various sizes. (A) Control; (B) DRB (75 µM, present for 40 min before labeling and also during labeling). For analysis of cap structures, gradients were divided as indicated: a, b, c, and d refer to portions of the gradient in the >18S region; e refers to the <18S portion. The bar graphs represent m^7G-containing caps. The cpm x 10^{-3} values on the left and right ordinates refer both to the cpm in sucrose-Me_2SO gradient fractions and to the total cpm in m^7G-containing caps in pools of fractions indicated by the graphs. From Tamm et al. (37). (Reprinted with permission).

The finding that DRB inhibits by ~70% the synthesis of short RNA capped with cap II structures of the type $m^{2,2,7}GpppN_1mpN_2mp$ suggests that small homodisperse nuclear RNA species such as U_1 and U_2 (alternatively known as D and C) are cotranscribed with hnRNA. This is consistent with the earlier observations that D and C, and also A are synthesized by RNA polymerase II (59, 60) and that U_1 and U_2 are derived from large transcription units up to 5,000 nucleotides long (61).

Ribosome formation and stability

As discussed above, the synthesis of preribosomal RNA is sensitive to inhibition by DRB although considerably less so than the synthesis of hnRNA (22, 25, 26). In *Chironomus tentans*, DRB (65 µM) does not affect the processing of

38S preribosomal RNA into 30S and 23S RNA (22, 62), or of the 30S intermediate into 28S ribosomal RNA (62). Processing of the 23S intermediate into 18S RNA in the nucleolus is however inhibited. Ringborg and Rydlander (62) have proposed that this inhibition may reflect a requirement for chromosomal RNA synthesis in the processing of nucleolar 23S RNA to ribosomal 18S RNA.

In chick embryo fibroblasts treated with DRB at high concentration (157 μM), ribosome production is altered in a complex manner (25, 63). Although the rate of appearance of 45S preribosomal RNA is reduced in DRB-treated cells, over the course of 6 hr the pool of 45S RNA builds up to a value about twofold higher than that seen in control cells (25). The reduced rate of appearance of 18S RNA parallels that of 45S RNA. The outstanding finding is the extensive degradation of the 32S and 28S RNA-containing preribosomes. After 5-7 hr labeling in the presence of DRB, the (32S + 28S)/18S ratio is 0.20-0.25, whereas in controls this ratio is nearly 1.0. Most of the degradation probably occurs before or at the 32S preribosome stage, and may be due to the formation of defective 45S RNA preribosomes rather than to interference with the action of the preribosome processing enzyme. Granick (25) has suggested that the defect in the 45S RNA preribosomes produced in the presence of DRB resides in the protein content of the particle rather than in its 45S RNA, which appeared normal with respect to size and methyl content. It is well known that protein-deficient preribosomes are subject to degradation during processing steps (64, 65). As a consequence of almost complete inhibition of mRNA synthesis in DRB-treated cells, certain ribosomal RNA species may become limiting. If so, after cleavage of the 45S RNA preribosome by the RNA processing enzyme, protein-deficient regions of the precursor 32S RNA preribosome may be exposed to random attack (25).

Although DRB treatment affects ribosome formation both in chironomid and avian cells, different ribosomal subunits are involved, i.e. the 18S subunit in the chironomid, and the 28S subunit in the avian cells.

Structural changes in the cell nucleus

DRB (100-200 μM) as well as adenosine, toyocamycin, and bromotubericidin cause within 6 hr striking changes in the nucleoli of chicken embryo fibroblasts (63). The round nucleoli lose material and unravel to become beaded strands, 20 μm in length. This change is reversible in that by 6 hr after change to drug-free medium, the normal nucleolar spheres have reformed. Chick embryo liver cells respond in a similar manner, but HeLa and L cells show no "nucleolar necklace" formation (63), although fragmentation of nucleoli occurs in 25% of the cells.

In contrast, actinomycin D, miracil D (leucanthone), and cordycepin, which are drugs that inhibit rRNA synthesis completely, cause a contraction and a collapse of chicken embryo fibroblast nucleoli into dense spheres, and also cause the contraction of DRB-induced long beaded necklaces into one or several dense spheres.

Inhibitors of protein synthesis such as acetoxycycloheximide, which slow 45S RNA synthesis but do not change the level of the 32S rRNA pool, do not cause nucleolar necklace formation, and alter the structure of nucleoli in only a minor way in 20% of chick embryo fibroblasts (63).

Combined biochemical and electron microscopic evidence suggests that depletion of the 32S RNA pool in DRB-treated chick embryo fibroblasts is an important element in nuclear necklace formation (25, 63). It is also apparent that continued 45S rRNA synthesis, even if at a reduced rate, is necessary to permit the development and maintenance of the beaded strand structure of the nucleolar necklaces, presumably because maintenance of the matrices of nascent rRNA chains that are being transcribed from the ribosomal DNA is required for beaded strand formation.

Treatment of rat liver cells for 2-6 hr with 63 μM DRB causes fragmentation of the nucleolus with segregation of its fibrillar and granular components and accumulation of many newly formed nucleolar perichromatin granules in the nucleolar area (66, 67, 68).

The interchromatin area of most nuclei contains interchromatin granules which have a diameter of 200-250 Å and are organized into clusters within which they are connected by thin fibrils (68). The clusters are irregularly distributed. The chemical constitution and the functional role of the granules are not known. However experimental alteration of RNA metabolism alters the distribution of the granules. DRB causes clumping of interchromatin granules and the simultaneous disappearance of most fibrillar RNP structures in the extranucleolar area in rat liver cells in culture (66). Electron microscopic autoradiography of [^{3}H]uridine labeled cells has shown that DRB causes accumulation in interchromatin granule clusters of RNA molecules synthesized elsewhere in the nucleus.

MECHANISMS OF INHIBITION OF VIRAL REPLICATION

Available evidence indicates that DRB blocks the replication of the class I viruses listed in Table 1 by inhibiting RNA synthesis mediated by cellular RNA polymerase II. In several instances, i.e. adeno, SV40, polyoma and the oncornaviruses, DRB appears to block viral RNA synthesis by inhibiting the transcription by RNA polymerase II of viral DNA templates. In other instances, i.e. influenza and vaccinia viruses, DRB, by blocking RNA polymerase II transcription of cellular DNA templates, interferes with cellular functions necessary for the replication of these viruses. Detailed studies of the mechanism by which DRB blocks adenovirus 2 (Ad2) transcription from the large late transcription unit (25 kb from map units 16.4 to ~99) in infected cells have provided evidence that this compound blocks transcription a short distance from the promoter leading to the accumulation of promoter-proximal, prematurely-terminated RNA chains 100-900 nucleotides long (52, 54). These studies have subsequently been extended to several Ad2 early transcription units (53), as well as to the SV40 (69, 70) and polyoma late transcription units (71). In each instance evidence has been obtained that DRB blocks RNA polymerase II-mediated transcription by enhancing attenuation of RNA synthesis close to the viral promoters (i.e. within 100-900 nucleotides). Several of these specific viral attenuation sites also function in the absence of DRB (54). Thus, the inference that DRB is a probe for attenuation of RNA transcription in mammalian cells has received strong support from studies of viral RNA transcription in living cells. However, it has recently been demonstrated that in a cell-free system for RNA synthesis (72), DRB blocks the initiation of new chains of RNA transcribed from exogenous truncated templates, such as those in a plasmid containing adenovirus promoters (73, 74).

Inhibition of viral RNA synthesis catalyzed by cellular RNA polymerase II

Adenovirus. DRB effectively blocks the replication of human adenovirus types 2 and 4 (75, 76). Treatment of Ad2-infected human KB cells 16-18 hr post infection ('late') with DRB (150 μM) has no apparent effect on Ad-2 specific DNA synthesis (76). Thus, DRB does not directly interfere with Ad2 DNA replication. However treatment of Ad2-infected KB cells with DRB beginning within the first 4 hr of infection inhibits "early" Ad2-specific nuclear RNA and mRNA accumulation and results in a marked inhibition of Ad2 replication. These studies indicate that the primary effect of DRB is an inhibition of Ad2-specific RNA synthesis.

The large late Ad2 transcription unit extends rightwards for 25 kb from map units 16.4 to ~99 (1 map unit = 350 bp; 100 units = 35000 bp = Ad2 genome) and constitutes the major transcription unit functioning late in infection. Treatment of Ad2-infected HeLa cells with DRB (75 μM) late in infection leads to a reduction of Ad2-specific nuclear transcription as monitored by 2 min pulse-labeling of cells with [^{3}H]uridine. DRB eventually blocks ~90% of Ad2-specific transcription. However, DRB blocks Ad2-specific mRNA labeling by >95% (52).

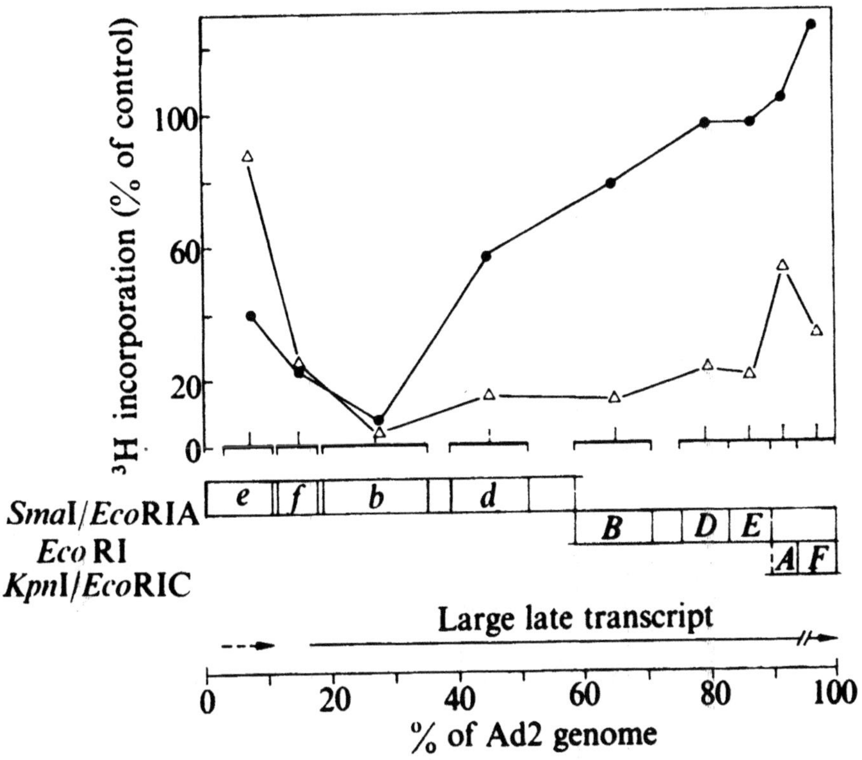

Fig. 11. Hybridization across the adenovirus 2 genome of [^{3}H]uridine pulse labeled nuclear RNA late in infection after brief treatment with DRB. Infecte cells were treated with DRB (75 μM) for 7 (●) or 40 (Δ) min 16 hr afte infection, and pulse-labeled with [^{3}H]uridine (0.4 mCi/ml) for 2 min. Nuclea RNA was extracted, fragmented by treatment with alkali to a length of ∿200-30 nucleotides, and hybridized to restriction fragments of adenovirus 2 DNA. Th total radioactivity hybridized was corrected for inhibition of [^{3}H]uridin transport by DRB. From Fraser et al. (52). (Reprinted with permission).

Fig. 11 describes some observations which have had a major effect on ou thinking about the mode of action of DRB (52). A gradient of decreasing inhibi tion was observed across the transcription unit when Ad2-infected HeLa cell were treated with DRB (75 μM) for 7 or 40 min beginning 16 hr after infectio ('late' in infection), pulse-labeled with [^{3}H]uridine for 2 min, and the nuclea RNA extracted and hybridized with excess amounts of filter-bound Ad2 DN fragments that span the large late transcription unit. The key finding was tha labeling of RNA derived from the DNA fragment containing the large late Ad promoter (at 16.4; SmaI f) was not inhibited as strongly as that of RNA derive from the adjacent downstream DNA fragment (SmaI b). Sucrose gradien sedimentation analyses (Fig. 12) and, subsequently, polyacrylamide-urea gel

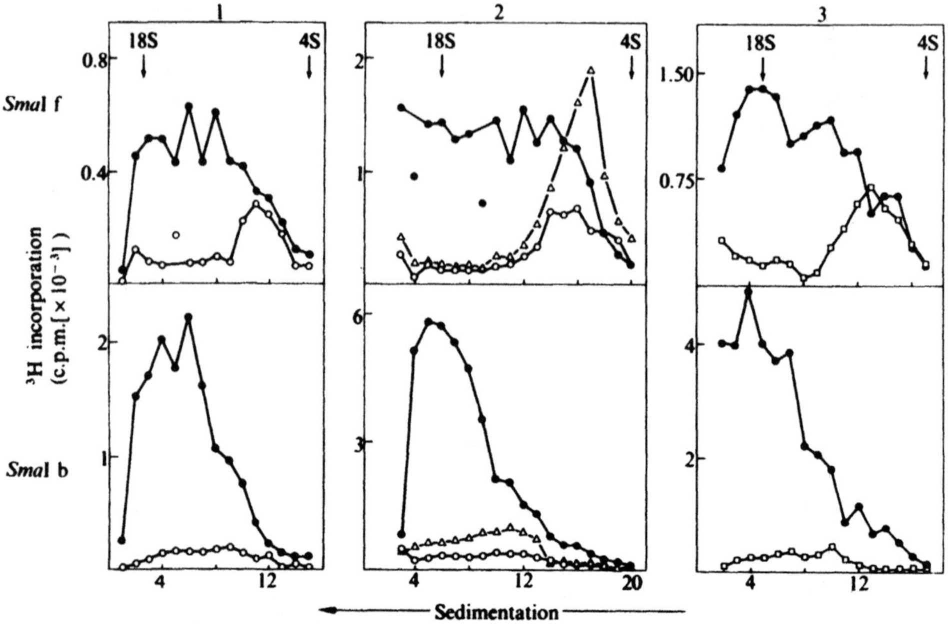

Fig. 12. Size of DRB-resistant SmaIf-specific late Ad2 RNA. Ad2-infected cells were exposed to DRB (75 μM) for 7 (o), 40 (Δ) or 60 (□) min late in infection, and pulse-labeled with ^{3}H-uridine for 2 min. An untreated sample of infected cells served as control (•). The nuclear RNA was extracted and sedimented through a 15-30% sucrose gradient in 0.05 NETS in a SW40 rotor, such that 18S RNA sedimented close to the bottom of the gradient. RNA from each fraction ≥ 4S was broken with alkali and hybridized to duplicate 3 mm square nitrocellulose filters containing 1 μg equivalent of Ad2 DNA restriction fragments Sma1f (11.6-18.2) and Sma1b (18.8-36.7) for 40 hr at 68°C in 2x TESS. The RNA in experiments 2 and 3 was also hybridized to SmaIe (3.0-11.1). [^{3}H] incorporation in the presence of DRB was corrected for the reduction in the transport of [^{3}H]uridine into cells (a factor of 1.6-1.9). Data presented in the three panels of this figure represent 3 different experiments. From Fraser et al. (52). (Reprinted with permission).

electrophoretic analyses showed that the DRB-resistant SmaI f-specific RNA was short (100-900 nucleotides) and consisted of multiple discrete species (54). There was a marked increase in the accumulation of discrete SmaI f-specific short RNA in the presence of DRB (Fig. 1 54). Furthermore, RNA fingerprint analyses showed that the DRB-resistant short RNA chains were capped but poly(A)-free and corresponded in sequence to the first ~350 nucleotides of the major late Ad2 transcription unit (54). Thus, these RNA molecules were chemically indistinguishable from the 5' ends of primary transcripts that are processed into mRNA.

In several instances discrete RNA species corresponding in length to those observed in the presence of DRB were also observed in Ad2-infected HeLa cells not treated with DRB albeit in lesser amounts. In addition, it was observed that the molarity of nascent, rightward-RNA chains in infected cells not treated with DRB was approximately 5-fold higher in the promoter-containing (SmaI f) DNA fragment than from regions further downstream (52, 54). These observations indicate (i) that significant premature RNA chain termination occurs at discrete sites in Ad2-infected HeLa cells and (ii) that DRB enhances this process, which results in an increased accumulation of promoter-proximal short RNA and a reduction in transcription from the rest of the transcription unit. Thus, in intact cells DRB appears to block hnRNA synthesis a short distance from the promoter sites.

Many of these observations have been independently confirmed (34, 51) and have been extended to a second late Ad2-transcription unit in the DNA fragment SmaI e (52, 54). Again, the DRB-resistant RNA has been found to be short and to consist of discrete species, and to accumulate in the presence of the compound (Fig. 1 of 54). However, not all late Ad2 mRNA synthesis is sensitive to inhibition by DRB. The mRNA for polypeptide IX, which maps on the rightward-reading strand between 9.7 and 10.9, appears to be DRB-resistant even though it is transcribed by RNA polymerase II and is capped (34). This observation suggests that not all Ad2 transcription units may be subject to DRB-mediated attenuation - an inference that is reminiscent of the detection of DRB-resistant cellular hnRNA (26, 28) and that of specific DRB-resistant cellular mRNA (78).

Promoter-proximal DRB-resistant Ad2-specific RNA derived from several Ad2 early transcription units has also been detected (53). Thus several of the early Ad2 transcription units appear to be subject to DRB-enhanced attenuation of transcription. Indeed, the detection of DRB-resistant short Ad2-specific early RNA and the mapping of this RNA onto the Ad2 genome helped to confirm the location of several of the early Ad2 promoter sites which had been deduced by using ultraviolet irradiation-inactivation and short RNA pulse-labeling procedures (53).

DRB strongly (>95%) inhibits the synthesis of Ad2-specific mRNA in Ad2-transformed rat cells (line 8617) (N. W. Fraser and P. B. Sehgal, unpublished). In preliminary experiments, pulse-labeled DRB-resistant virus-specific short nuclear RNA has been detected in these cells also.

To summarize, studies on the effects of DRB on RNA synthesis in the Ad2-infected HeLa and KB cells have provided strong evidence that DRB inhibits Ad2 virus replication by causing enhanced attenuation of transcription at multiple discrete sites close to the promoter regions. Furthermore, these attenuation sites also appear to function in the absence of DRB. Finally, the Ad2 experiments showed that nuclear RNA synthesis from the distal regions of individual defined transcription units could be inhibited by DRB (75 μM) by >95% and that at least one transcription unit (that for protein IX mRNA) was DRB-resistant. Thus, the detection of DRB-resistant long cellular hnRNA in drug-treated, uninfected HeLa cells probably does not result from "leakiness" of the inhibitory effect of DRB on hnRNA synthesis, but is likely to reflect an underlying difference in the transcription units for the DRB-sensitive and the resistant hnRNA.

Simian vacuolating virus 40 (SV40). DRB markedly inhibits the replication of SV40 in productively infected CV1 monkey cells (M.-T. Hsu et al., unpublished). Treatment of SV40-infected CV1 cells late in infection with DRB (75 μM) leads to an 85-90% inhibition of SV40 nuclear transcription as monitored by 2-min [^{3}H]uridine pulse labeling (N. W. Fraser and P.B. Sehgal, unpublished).

The effect of DRB in the SV40-BSC1 monkey cell system late in infection has been explored in some detail (69, 70, 79). In this system, DRB-resistant RNA (pulse-labeled in vivo with [^{3}H]uridine) hybridizes mainly to the DNA fragment (0.67 to 0.76) which contains the late SV40 promoter. Thus, it appears that DRB causes premature termination of RNA chains close to the late SV40 promoter in vivo. Treatment of infected cells with DRB (80 μM) for 45 min followed by in vitro labeling of SV40 RNA in nuclear transcription assays or in viral transcription complexes also results in the labeling of DRB-resistant viral RNA hybridizing primarily to the region of the late SV40 promoter (0.67-0.76). The in vitro transcription product appears to be primarily a promoter-proximal RNA of length 93-95 nucleotides. This short RNA is also observed, although to a lesser extent, in in vitro labeling assays with the transcription complex derived from infected cells not treated with DRB. These results have been interpreted as indicative of DRB-enhanced attenuation of SV40 RNA synthesis 93-95 nucleotides from the major late promoter at a site which appears to function physiologically and at which there is a stretch of U residues (70).

The in vivo DRB-imposed block in RNA synthesis in uninfected as well as in Ad2-infected HeLa cells has been found not to be operative in a number of in vitro transcription systems (50, 51). It is therefore surprising that the putative DRB-imposed in vivo block (at 93-95 nucleotides downstream) remains in force in the in vitro SV40 transcription system.

It has been shown recently that up to 70% of [^{3}H]uridine-pulse-labeled (3-4 min) SV40-specific RNA in CV1 cells late (45-48 hr) in infection is associated with a viral nucleoprotein complex that rapidly leaches out of isolated nuclei agitated under isotonic conditions (80, 81). Llopis and Stark (81) have suggested that this complex represents a structure which does not lead to productive mRNA synthesis and could be similar to the transcriptional complexes investigated by Aloni and his colleagues (69, 70, 79). It is of interest that transcription of the promoter-containing SV40 DNA fragment (0.67-0.76), as determined by [^{3}H]uridine labeling of SV40-specific RNA in vivo, is not completely or even largely resistant to the inhibitory effect of DRB (69). We cannot exclude the possibility that DRB blocks RNA chain initiation per se in the SV40-polyoma system in addition to enhancing attenuation a short distance from the promoter sites.

3. Polyoma virus. The effect of DRB on polyoma-specific late RNA synthesis has been investigated in infected mouse embryo fibroblast cultures (71). Exposure of mouse kidney cells to DRB (75 μM) during the late phase of infection by polyoma virus resulted in a ~90% inhibition of virus-specific RNA synthesis. Sedimentation analyses of DRB-resistant RNA revealed that labeling of short RNA (3-7S) was inhibited by DRB only 40-50% whereas that of high molecular weight viral RNA was markedly (>95%) inhibited. The short virus-specific RNA derived either from untreated cells or from those treated with DRB hybridized predominantly with the DNA fragment containing the major late polyoma promoter (0.58 to 0.72) and was complementary to the correct L (late) strand of virus DNA. These data are consistent with the view that DRB blocks late polyoma RNA synthesis close to the major late promoter. The available data do not, however, exclude an additional effect of DRB on polyoma-specific RNA chain initiation per se.

4. Rous sarcoma virus. DRB inhibits the replication of Rous sarcoma virus (RSV, Schmidt-Ruppin strain) in chick embryo fibroblast cultures in a reversible manner. 70 μM DRB, present throughout infection, inhibits RSV yield >99%, whereas 50 μM DRB inhibits RSV yield by ~90%. In these experiments the virus yield was measured over a 4-hr period from 30 to 34 hr from the beginning of infection. When treatment of cells with DRB (75 μM) is begun 12 hr after RSV infection, it still results in a >99% inhibition of virus yield. However, when DRB treatment is delayed for 18 hr after infection, a variable, although reduced amount of virus is produced. Although infected cultures treated with DRB (75 μM) between 12-24 hr after infection show a >99% inhibition of virus yield (measured at 24 hr), removal of DRB from these cultures at 24 hr leads to a virus yield (as measured between 40 and 44 hr after infection), which is approximately 25% of that in control cultures. Thus, the inhibitory effect of DRB appears to be largely reversible.

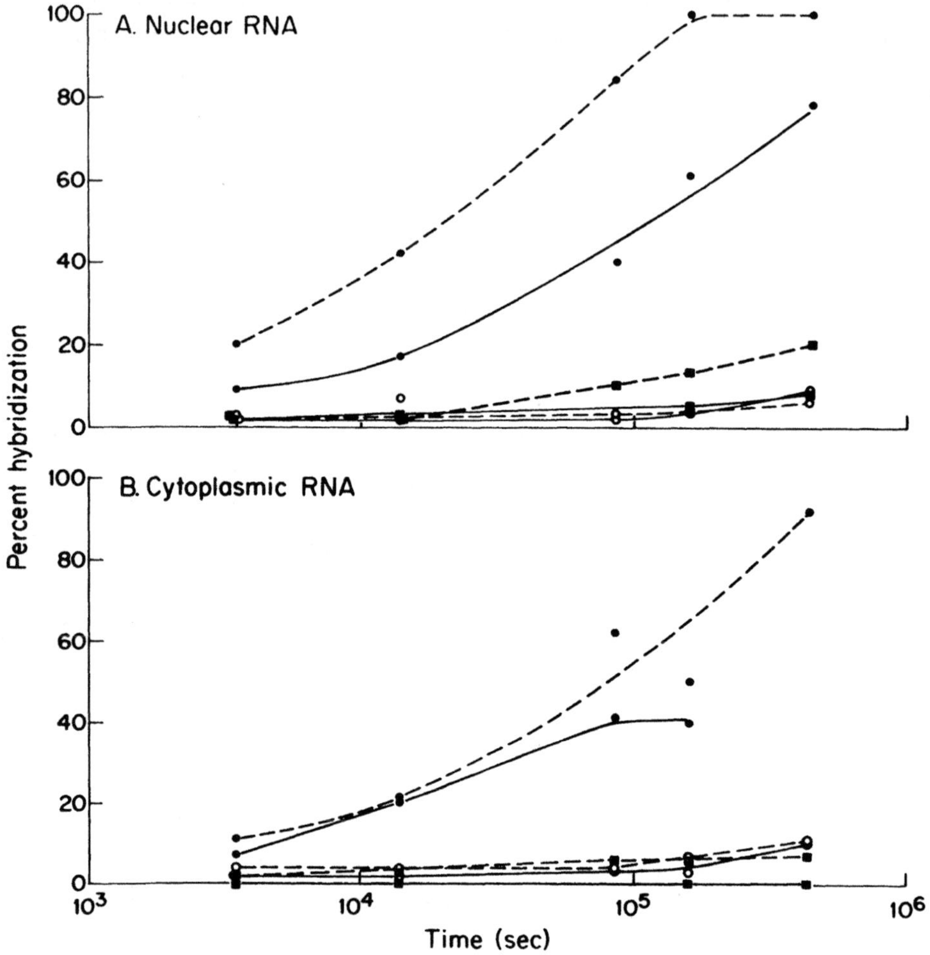

Fig. 13. Effect of DRB on accumulation of RSV-specific RNA in chick embryo fibroblasts. The steady-state levels of RSV-specific RNA were measured by hybridizing 5,000 cpm of ^{3}H-labeled RSV cDNA (representative of the whole genome) with nuclear (panel A) and cytoplasmic (panel B) poly(A)-containing (solid line) and poly(A)-free (broken line) RNA derived from uninfected (o), RSV-infected (●), and RSV-infected, DRB-treated (75 μM throughout infection,■) cells harvested 24 hr after the beginning of infection (82). 10 cultures of chick cells in 100 mm petri dishes were used in each of the three groups and each hybridization was carried out in duplicate.

DRB blocks RSV replication because it strongly inhibits RSV-specific RNA accumulation in infected cells, as shown in Fig. 13. In this experiment chick embryo fibroblast cultures were treated with DRB (75 µM) beginning immediately after adsorption of SR-RSV and the cells were harvested by trypsinization 24 hr later. Nuclear and cytoplasmic fractions were prepared, RNA was extracted from the two fractions and each sample separated into poly(A)-containing and poly(A)-free RNA preparations by poly(U)-sepharose chromatography. In order to estimate the amount of steady-state RSV-specific RNA, each RNA preparation was hybridized to a [^{3}H]-labeled cDNA probe corresponding to the entire RSV genome. The results summarized in Fig. 13 show that treatment of infected cells with DRB strongly inhibits the accumulation of RSV-specific RNA (poly(A^+) as well as poly(A^-)) in both the nuclear and the cytoplasmic compartments. These data suggest that DRB inhibits RSV replication by inhibiting virus-specific transcription mediated by cellular RNA polymerase II.

Inhibition of viral replication through inhibition of cellular RNA synthesis catalyzed by RNA polymerase II.

1. Influenza viruses. The antiviral effect of DRB and its congeners was discovered in experiments with influenza viruses replicating in chorioallantoic membrane pieces in culture (1). These early experiments indicated that exposure of chick cells to DRB early in infection markedly inhibits influenza virus multiplication (6). DRB did not directly inactivate the virus but rather affected cellular RNA synthesis (9, 18) which apparently resulted in a block in influenza virus replication. That influenza viruses need a host cell function to enable them to replicate has been verified using a variety of procedures (reviewed in 7). It has been demonstrated that a functional cellular RNA polymerase II is required for influenza virus replication (7) and that host RNAs donate their 5' caps and 10-13 adjacent nucleotides to the influenza virus mRNAs (8). Thus one possibility is that DRB blocks influenza virus replication by inhibiting the transcription of those cellular RNAs whose 5' ends are ordinarily donated to viral mRNA. It is clear that a great deal more work is necessary in order to pinpoint the specific host cell event through which DRB blocks influenza virus replication.

In influenza virus-infected HeLa cells labeled with [^{35}S]-methionine the following viral polypeptides can readily be detected: 3 P polypeptides (PB1, PA, PB2) associated with RNA transcriptase activity; the hemagglutinin (HA); the nucleocapsid protein (NP); the neuraminidase (NA), which in most gel electrophoresis systems comigrates with NP; the membrane protein (M_1); non-structural proteins NS_1 and NS_2. The non-structural protein M_2 cannot be

Fig. 14. Effect of DRB (100 μM) on the time-course of influenza virus protein synthesis. HeLa cells were treated with DRB (100 μM) for 30 min before infection with influenza virus, and the treatment was continued after infection. At the times indicated (hr) cultures were labeled with $[^{35}S]$methionine (20 μCi/ml) for 30 min in the continued presence of DRB and processed for polyacrylamide gel electrophorsis and autoradiography. Samples were run on a 17.5% polyacrylamide gel containing 4 M urea (83). U = uninfected cells labeled 3 hr after start of experiment; 1 = infected cells labeled 1 hr after infection; 3 = infected cells labeled 3 hr after infection. The part of the negative containing the virus-specified nonstructural protein NS_2 required a longer exposure time than the rest.

detected readily using [^{35}S]-methionine as a label.

DRB preferentially inhibits the synthesis of HA, NA, and M_1; the synthesis of NP, NS_1, and NS_2 is resistant. For example, some inhibition of HA, but not of NP, can be detected at 20 µM DRB concentration. Inhibition of HA and M_1 synthesis is moderate at 50 µM DRB concentration, marked at 75 µM, and very marked at 100 µM. In these experiments HeLa cells were treated with DRB at varying concentrations for 30 min, infected with the A/WSN/33 strain of influenza virus at an input multiplicity of 25 PFU/cell, and incubated for 3 hr in the continued presence of DRB. Cultures were then labeled with [^{35}S]methionine (20 µCi/ml) for 30 min and processed for polyacrylamide gel electrophoresis and autoradiography as previously described (83).

The inhibition of HA synthesis in DRB-treated (100 µM) cells is detectable 1 hr after infection as is illustrated in Fig. 14. By 3 hr, there is a very marked difference in the abundance of HA in the DRB-treated vs. the control cells. In contrast, the amounts of NP appear to be similar in control and DRB-treated influenza virus-infected cells.

To determine whether DRB had an effect on "primary transcription" of the viral RNA, cultures were treated with cycloheximide, which permits transcription and accumulation of viral mRNAs from the input virion RNA, but prevents their translation. Immediately after removal of the cycloheximide, some cultures were labeled with [^{35}S]-methionine to evaluate indirectly the abundancies of the mRNAs synthesized by the virion transcriptase. Other cultures were further incubated to allow amplification of the mRNAs after virion RNA synthesis and secondary transcription, and then labeled with [^{35}S]-methionine. Fig. 15 shows the effects of 100 µM DRB on the synthesis of influenza virus-specific polypeptides in chick embryo fibroblasts (CEF) which had or had not been treated with cycloheximide. The top panel shows a 17.5% polyacrylamide gel which resolves all influenza virus polypeptides except NA. The bottom panel shows an 8% polyacrylamide gel for the better resolution of PB1, PA, PB2, HA, NA and NP. Comparison of CEF infected control cultures with DRB-treated infected CEF cultures shows that the synthesis of HA, NA and M_1 is selectively inhibited by DRB, whereas the synthesis of PB1 is enhanced. After allowing primary transcript mRNAs to accumulate for 3 hr in the presence of cycloheximide, and then permitting protein synthesis, all the influenza virus polypeptides with the possible exception of NS_2, can be detected (84, 85). However, the synthesis of HA and NA under these conditions is less than that of the other polypeptides as discussed previously (85). Inclusion of DRB in this protocol markedly reduces the amounts of HA, NA and M_1 synthesized and also slightly reduces the synthesis

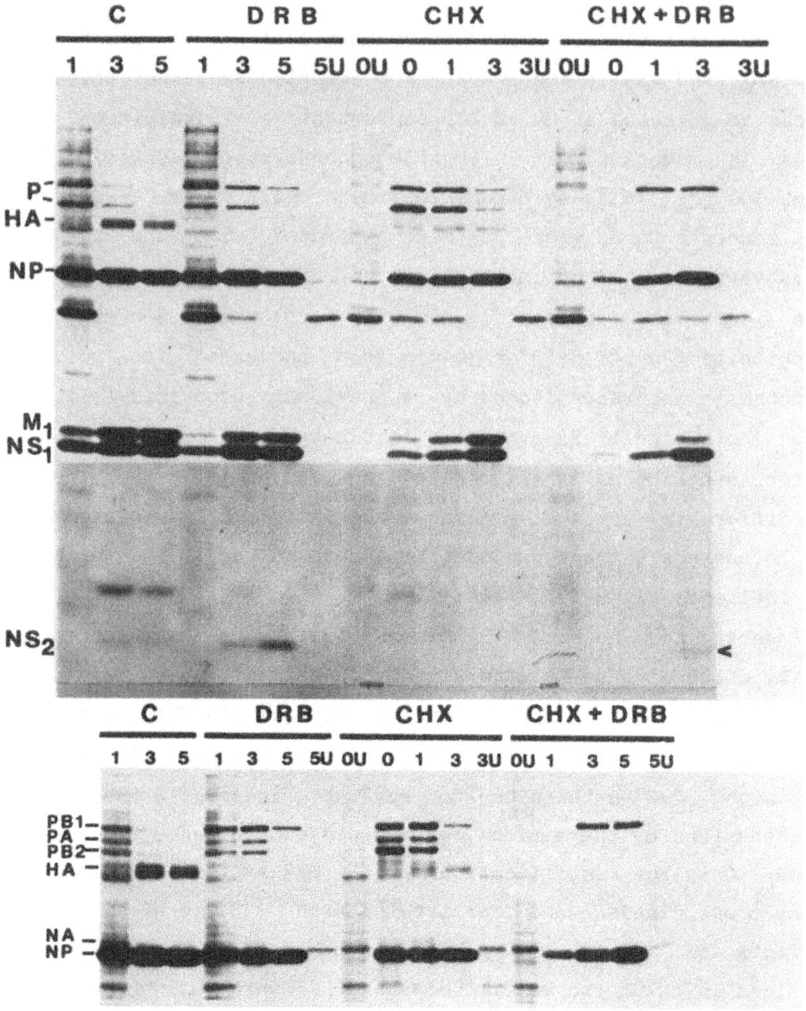

Fig. 15. The effect of DRB (100 μM) on the synthesis of polypeptides in influenza virus-infected chick embryo fibroblasts after temporary inhibition of protein synthesis by cycloheximide. C = control cultures, DRB = cultures treated with DRB (100 μM) for 30 min before and throughout infection. CHX = cultures treated with cycloheximide (100 μg/ml) 30 min before and during the first 3 hr of infection. CHX + DRB = cycloheximide and DRB treatment as above. Cycloheximide was removed by extensive washing with Eagle's medium with or without DRB (100 μM) and incubation continued in the same media. At the times indicated, cultures were labeled with [^{35}S] methionine (20 μCi/ml) for 30 min and then processed for polyacrylamide gel electrophoresis and autoradiography. U = uninfected cultures. Numbers = hr after infection or hr after cycloheximide removal when cells were labeled. Top panel: Samples subjected to electrophoresis on a 17.5% polyacrylamide gel containing 4 M urea (83). Bottom panel: Samples subjected to electrophoresis on an 8% polyacrylamide gel containing 4 M urea (84). The part of the negative containing NS_2 required a longer exposure time than the rest.

of PA and PB_2. However, the synthesis of PB_1 is enhanced (compare the points after cycloheximide removal with or without DRB). These data indicate that DRB interferes either with the synthesis or accumulation of specific mRNAs produced in the course of primary transcription.

In summary, DRB selectively inhibits the synthesis of influenza virus polypeptides HA, NA and M_1. It appears likely, based on all available information, that the effect is largely due to reduction in virus-specific mRNA synthesis, however other mechanisms cannot be excluded.

The effects of camptothecin on influenza virus replication (86) are similar to those observed with DRB. Biochemically, the action of camptothecin on RNA synthesis is in certain respects unlike that of DRB in that it preferentially inhibits ribosomal precursor RNA synthesis, and at the same time also inhibits DNA synthesis (87, 88, 89; also see above). These differences in the biochemical actions of camptothecin and DRB appear to be irrelevant with respect to their similar selective effects on influenza virus replication. What appears relevant is that the action of camptothecin on hnRNA synthesis is similar to that of DRB in that both compounds inhibit hnRNA synthesis markedly, but incompletely. Camptothecin, like DRB, has no effect or only a minor effect on the synthesis of 4S, 5S, or mitochondrial RNA (88, 90, 91). Neither compound has an immediate effect on the level of cellular protein synthesis (89). Inhibition of protein synthesis in treated cells appears to be secondary to the inhibition of mRNA synthesis.

One possible explanation may be that the selective effects of DRB and camptothecin are related to sequence specificities of the short host RNA segments synthesized by RNA polymerase II and donated to viral mRNAs as 5' terminal sequences. Thus, the synthesis of HA, N, and M proteins would be expected to be sensitive to DRB if the donor sequences for the mRNAs for these proteins were derived from sequences transcribed as part of the 70-80% of hnRNA that is DRB sensitive, whereas the synthesis of NP, NS_1, and NS_2 would be insensitive if the donor sequences were derived from the 20-30% hnRNA that is DRB resistant. Evidence has recently been obtained that the 5' terminal sequences of different influenza mRNAs are different (M. W. Shaw and R. A. Lamb, unpublished). It would be of interest to determine whether the DRB-resistant hnRNA contains 5'-terminal sequences for the NP, NS_1, and NS_2 mRNAs, but not for HA, NA, and M_1 mRNAs.

2. Vaccinia virus. DRB markedly inhibits the multiplication of vaccinia virus in the chick chorioallantoic membrane (92) and Ehrlich ascites tumor cells (17). The vaccinia-Ehrlich ascites cell system has been investigated in some

detail (17). DRB (90 μM), present throughout infection, does not inhibit cytoplasmic vaccinia DNA replication or cytoplasmic viral (early and late) RNA synthesis. The RNA synthesized in the presence of DRB appears to be of normal length. The labeling of early, intermediate and late viral proteins is largely unaffected by inclusion of DRB in the infected cultures. However, viral DNA obtained from DRB-treated, infected cells late in infection is mainly in a DNAse-sensitive form although that obtained from DRB-free, infected cells is largely DNAse-resistant. Thus the assembly of vaccinia virus is markedly inhibited in DRB-treated cells (17). It appears likely that DRB blocks the expression of a cellular function that is necessary for vaccinia virus maturation and assembly, and that in consequence, vaccinia virus replication is strongly inhibited.

EFFECTS AT THE CELL MEMBRANE, TRANSPORT INTO CELLS, AND PHOSPHORYLATION

DRB inhibits the uptake of [^{3}H]uridine (25, 26, 28, 45, 93) and [^{3}H]thymidine (76, 94) into avian and mammalian cells, but has no effect on [^{3}H]leucine uptake in short-term experiments (26, 28). The extent of inhibition of [^{3}H]uridine uptake depends on the cells used, and is considerably greater in HeLa than L cells. As would be expected, the difference in acid-extractable radioactivity between DRB-treated and control cells is most marked after short periods (1-30 min) of labeling. After several hr, equilibrium conditions are reached and significant differences are no longer observed.

DRB is not phosphorylated in mammalian cells (35, 43). Indeed, two kinds of results have recently been obtained which raise the possibility that DRB may not need to enter the cells or may be transported only in minute amounts (95). First, the inhibition by DRB of the formation of cytoplasmic poly(A)$^+$ RNA is as rapid and marked in mutant cells (S49-AE$_1$) deficient in the transport of purine and pyrimidine nucleosides as in the parent wild-type cells (S49). Second, neither the mutant nor the parent cells appear to accumulate [^{3}H]DRB to a significant extent, although the parent cells do accumulate [^{3}H]adenosine in a linear manner (Fig. 16). It should be noted, however, that in experiments with compounds which form weak and readily reversible bonds with their respective target sites, it is difficult to control for the possibility that during the exposure of cells to the labeled compound, such as [^{3}H]DRB, the compound enters cells, but is lost from cells during washing of the cells prior to determination of cell-associated radioactivity, even when the washing is done with a solution containing unlabeled compound.

Chironomid salivary gland cells take DRB up rapidly and with kinetics which suggest a facilitated diffusion mechanism (96). The great majority of DRB

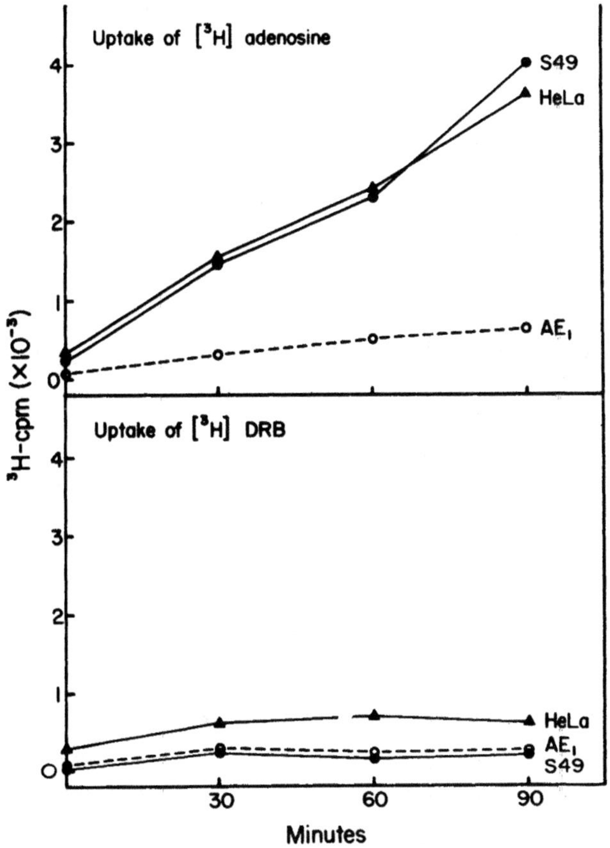

Fig. 16. Long-term uptake of $[^3H]$DRB and $[^3H]$adenosine by S49, AE_1, and HeLa cells. Five ml cultures at 1 x 10^6 cells/ml were incubated with 60 µM $[^3H]$adenosine (A) or 60 µM $[^3H]$DRB (B) at 37°C. At varying times, duplicate 0.8-ml samples were removed and washed three times with PBS containing 200 µM unlabeled DRB to prevent the loss of $[^3H]$DRB from cells, and the acid-soluble radioactivity was determined. Uptake by S49 (●), AE_1 (o), and HeLa cells (▲). The uptake at 1.5 hr of $[^3H]$adenosine and $[^3H]$DRB, respectively, for the S49 cells was 4,073 and 205 cpm; for the AE_1 cells, it was 697 and 228 cpm. From Tweeten and Molloy (95). (Reprinted with permission).

molecules transported into these cells become phosphorylated to 2'- and 5'-DRB monophosphate isomers, but not to the di- or the triphosphate of DRB (96, 97). The rate of phosphorylation is considerably slower than that of transport. It is unlikely that the DRB monophosphate isomers are the active compounds in the inhibition of hnRNA synthesis, because 1) Balbiani ring RNA synthesis is abolished within 5 min after addition of DRB to cells whereas DRB-monophosphate compounds appear in cells in measurable amounts only after 5 to 10 min, 2) the transcription block is rapidly reversed when DRB is removed from the medium, in spite of the fact that the monophosphate derivatives of DRB are trapped in the cells (96), and 3) adenosine and 3'-deoxyadenosine inhibit the intracellular phosphorylation of DRB to 2'-DRB-P and 5'-DRB-P without lowering the intracellular concentration of [^{3}H]DRB or significantly affecting the inhibition of hnRNA synthesis (98).

In summary, the available evidence indicates that DRB in its unphosphorylated form inhibits hnRNA synthesis and that it has effects at the cell surface.

STRUCTURE-ACTIVITY RELATIONSHIPS

The unique biological activity of DRB and several other halogenated 1-β-D-ribofuranosylbenzimidazoles was discovered in structure-activity relationship studies with a wide range of benzimidazole derivatives using two criteria for identification of interesting compounds: 1) high inhibitory activity on influenza virus multiplication and, 2) in terms of the ratio of inhibitory to toxic concentrations, relatively low non-specific cytotoxicity for cells of the host tissue, which was the isolated chorioallantoic membrane of the chicken embryo (1, 99, 100). Subsequent work from several laboratories, including our own, has shown that the results obtained in the influenza virus-chorioallantoic membrane system were accurately reflective of the action of halogenated 1-β-D-ribofuranosylbenzimidazoles as selective inhibitors of cellular hnRNA transcription. The early work established the key requirement for the 1-β-D-ribofuranosyl moiety by demonstrating that the corresponding 1-β-D-ribofuranosyl derivatives of halogenated benzimidazoles were only 1/10th-1/5th as active, although of similar cytotoxicity (1, 99, reviewed in 5). A number of the glycosyl derivatives and several glycityl derivatives were even less active. The halogenated benzimidazoles themselves were of low activity and selectivity. It was thus demonstrated that conversion of such compounds to corresponding 1-α-D-ribofuranosyl derivatives imparts selective activity. The activity of the halogenated 1-β-D-ribofuranosylbenzimidazoles increases with multiple substitution of halogen atoms in the benzenoid ring of the bicyclic benzimidazole

Table 3. Influenza virus inhibitory activity and relative cytotoxicity of benzimidazole derivatives*

	Molecular Weight	Concentration (μM) at which virus yield is reduced 75%	Relative activity		
			Virus Inhibition†	Cytotoxicity CAM‡	Cytotoxicity MKC§
Benzimidazole	118.13	3500	1	1	1
5-Chloro-	152.6	750	4.7	6	
5,6 Dichloro-	187.03	∿250	∿14		
1-β-D-Ribofuranosylbenzimidazole					
5-(or 6-)Chloro-	284.5	280	13	4.6	
5,6-Dichloro-	319.15	38	92	39	47
5-(or 6-)Bromo-4,6-(or 5,7-)dichloro-	398.06	1.8	1950	∿330	∿620
1-α-D-Ribofuranosylbenzimidazole					
5-(or 6-)Bromo-4,6-(or5,7-)dichloro-	398.06	14	250		

*Based on data referenced in Sehgal and Tamm (5).

†Unsubstituted benzimidazole as the reference compound was considered to have a virus inhibitory activity equivalent to 1.

‡CAM = chlorioallantoic membrane from embryonated chicken eggs incubated in vitro. Benzimidazole causes 2+ macroscopic damage at a concentration of 12,000 μM (10).

§MKC = primary cultures of rhesus monkey kidney cells. Benzimidazole causes 3+ microscopic damage at a concentration of 5,200 μM (10).

nucleus. In a number of instances, the bromo-substituted compounds are more active than the corresponding chloro derivatives. These structures activity relationships are illustrated in Table 3.

The most active derivative so far investigated is 5-(or 6)-bromo-4,6-(or 5,7)-dichloro-1-β-D-ribofuranosylbenzimidazole (100), which is 18-22 times more active, on a molar basis, than DRB in inhibiting RNA synthesis, influenza virus replication, and cell proliferation, as well as in causing interferon superinduction (101). Thus, the 1-β-D-ribofuranoside of mono-bromo-dichloro-benzimidazole shows marked activity in the 1-2 μM concentration range. Qualitatively, the actions of the different halogenated 1-β-D-ribofuranosides appear to be similar. DRB has been used in most investigations, because it is more soluble than triply halogenated derivatives. We have found that the hnRNA-inhibitory activity of the readily soluble DRB-5'-monophosphate, ammonium salt, is indistinguishable from that of DRB in dose-response experiments carried out as described in the legend for Fig. 5 (27).

Investigation of the structure-activity relationships have recently been extended to five additional analogs of DRB, all with modified carbohydrate moieties (102). The results indicate that a free 3'-OH in the ribose moiety is important for the selective activity of DRB on transcription of hnRNA in salivary glands of Chironomus tentans. Methylation of the 3'-OH largely abolishes activity. In contrast, the 2'-O-methyl derivative of DRB is similar to DRB in its activity.

DRB-RESISTANT CELL MUTANTS

DRB-resistant mutants of Chinese hamster ovary (CHO) cells (103, 104) and human diploid fibroblasts (103) have been isolated after treatment with the mutagen ethylmethane sulfonate. In addition, resistant mutants of CHO (103) and HeLa cells (I. Tamm, unpublished) have been isolated without mutagen treatment.

In CHO cells there is a linear relationship between the concentration of the mutagen ethylmethane sulfonate in the range of 20-300 μg/ml and the frequency of occurrence of DRB-resistant mutants (103). The mutations to DRB resistance are expressed codominantly in somatic cell hybrids of resistant and sensitive cell lines (103, 104). DRB-resistant mutant cells, which have largely lost sensitivity to the growth-inhibitory effects of DRB, are still sensitive to the inhibitory effect of the compound on nucleoside transport (103).

Funanage (104) showed that the RNA polymerase II activity of cell lysates (55) from DRB-treated (60 μM) parent wild type CHO cells was reduced by a maximum of 60% compared to the activity of cell lysates from control untreated wild type cells. This agrees with earlier results in which RNA polymerase II activity was measured in nuclei isolated from DRB-treated and control HeLa cells (36). In contrast, treatment with 300 μM DRB was required to reduce the *in vitro* RNA polymerase activity of the lysate of mutant cells by 60% (104), which indicates considerable resistance of the mutant cells to the hnRNA inhibitory activity of DRB. As the activity of the isolated RNA polymerase II was not determined, the results cannot be interpreted in terms of DRB action on the enzyme per se. Rather they reflect differences in the *in vivo* abundancies of hnRNA molecules undergoing elongation in DRB-treated mutant as opposed to wild-type cells.

The response of the DRB-resistant cells to the *in vivo* action of DRB is biphasic as revealed by *in vitro* RNA polymerase assays with cell lysates (104). About 50% of the DRB-sensitive RNA polymerase II activity (as defined in wild-type cells), showed a wild-type sensitivity to DRB, but the remainder was depressed only at high concentrations of DRB (200-300 μM), which confirmed that the drb^R mutation is expressed codominantly.

The quantitative relationships between survival of DRB-treated cells and the level of RNA polymerase II activity demonstrable in cell lysates are of interest. 90 μM DRB did not affect the survival of mutant cells in spite of the fact that the DRB-sensitive RNA polymerase II activity in lysates of such cells was reduced by 50% (104). Similarly, 12 μM DRB reduced the survival of parent wild type cells only by 8% although the polymerase II activity in lysates was reduced by about 45%. If the polymerase II activity observed in cell lysates does reflect the number of growing chains of hnRNA in DRB-treated and in control cells, then it would follow that some reduction in hnRNA transcription may be compatible with the survival of CHO cells, which agrees with findings in HeLa cells (see below; 27, 105). It appears that the initial inhibitory effect of DRB on hnRNA synthesis becomes partially reversed on continued incubation of cells in the presence of DRB (105).

As will be discussed below, it has recently been reported (73) that DRB directly inhibits RNA polymerase II-specific transcription when added to an extract of whole HeLa cells of the kind in which RNA polymerase II is capable of initiating transcription at promoter sites in DNA (72, 106). In such an *in vitro* system, DRB blocks initiation of new RNA chains (73, 74). A stable DRB-resistant mutant of HeLa cells has also been isolated after mutagenization of cells with N-methyl-N'-nitro-nitrosoguanidine (107). The transcription of RNA in the mutant cells is resistant to inhibition by DRB, as is transcription of ε-globin and the adenovirus 2 (major late promoter) genes in extracts prepared from the mutant cells. As reported previously (103), uridine uptake by DRB-resistant mutant cells is sensitive to inhibition by DRB (107). No in vivo results have as yet been reported establishing that DRB is capable of inhibiting initiation of RNA chains by RNA polymerase II in intact HeLa cells. However, the findings of Mittleman et al. (107) clearly indicate that the DRB-resistant HeLa cell line possesses an altered transcriptional machinery, and therefore may prove useful for the identification of the primary biochemical target for the action of DRB.

MODULATION OF DRB ACTION ON CELLS BY SERUM IN LONG-TERM EXPERIMENTS

Serum counteracts the deleterious effects of DRB (40 or 60 μM) on the cycling and survival of HeLa cells in a dose-dependent manner within a range of serum concentrations (2.5-30%) within which untreated control cells proliferate at the same maximal rate (105, 108). Time-lapse cinemicrographic analysis of cell populations incubated with 5 or 15% fetal calf serum has shown that in DRB-treated cultures observed over the course of 5 days, the frequency of mitoses is markedly and directly dependent on serum concentration, whereas the frequency of deaths is inversely dependent (108). The prolongation of the intermitotic interval in cells succeeding to cycle in the presence of DRB is greater in 5% than in 15% serum. Both nondialyzable and dialyzable serum components play a role in promoting the proliferation of DRB-treated cell populations (105).

After treatment of HeLa cells with 60 μM DRB for 24 hr or longer in either 5 or 15% serum, the sensitivity of bulk hnRNA synthesis to inhibition by DRB becomes reduced, as compared to cells treated for a short period (30 min-5 hr) (105, 108). The change is somewhat greater at the higher serum concentration. It has not yet been determined whether the synthesis of different species of hnRNA may be differentially affected by serum concentration in cells treated with DRB for a prolonged period. When treatment is discontinued by removal of DRB from the medium, hnRNA synthesis returns to the control level. Supernatant DRB-containing medium from cultures which had been incubated for 3-5 days is fully active in inhibiting RNA synthesis when added to a new set of cultures (101, 108). Thus, it appears that when cells are treated with DRB for a prolonged period, growth factors in serum can partially block the effects of DRB at one or more levels of regulation of cellular biosynthesis. The simplest hypothesis is that, as the result of the action of serum component(s), the access of DRB to its putative target sites becomes restricted. Alternatively, serum components could affect the expression of the action of DRB after it had bound to its receptors. That serum can affect the action of DRB at the plasma membrane is shown by the findings that 48-hr treatment of cells with 60 μM DRB causes a greater inhibition of the uptake of [^{3}H]uridine and [^{3}H]leucine in 5%

serum (92% and 57%, respectively) than in 15% serum (76% and 29%, respectively) (105).

It is of interest that the recovery of normal proliferative activity of cells after treatment with 60 µM DRB for 24 hr in 5% or 15% serum is independent of serum concentration in the medium (27). This is additional evidence that significant serum effects on DRB-treated cells are not of a generally supportive kind, but relate to the action of DRB. The time required for recovery to begin is directly related to the concentration of DRB used to treat cells and the duration of treatment, and inversely related to serum concentration during treatment (27). After treatment with 40 µM DRB for 24 hr in medium supplemented with 5% serum, cultures regain within 1 day a normal rate of increase in cell number. However, after treatment with 80 µM DRB for 40 hr, more than a week is required for recovery. Microscopic examination of the cultures during the "dormant" period has shown an increased proportion of large cells.

ACTION IN CELL-FREE SYSTEMS

Several attempts have been made to evaluate the effects of DRB and its phosphorylated derivatives in cell-free transcription systems derived from mammalian cells. DRB does not inhibit synthesis of RNA by purified RNA polymerase when poly[d(A-T)] or denatured calf thymus DNA is used as a template (reviewed in 15; 73). Furthermore, no evidence of a direct effect of DRB on RNA transcription has been obtained in nuclear systems derived from human HeLa (36, 50) or murine Ehrlich ascites cells (35, 55) in which new chains of RNA are not initiated by RNA polymerase II and nascent chains are not prematurely terminated.

Dreyer and Hausen (35) did observe an inhibitory effect of DRB triphosphate at high concentrations (millimolar) on RNA polymerase II and to a lesser extent on RNA polymerase I in both the nuclear and the purified polymerase systems. The effect of DRB triphosphate appeared to be a general inhibition of RNA synthesis and not a specific block at the initiation step. Furthermore, the high concentration of DRB-triphosphate required to achieve this inhibitory effect and the inability to detect any phosphorylated derivatives of DRB in treated cells led these investigators to conclude that their cell-free observations using DRB-triphosphate were not relevant to the mechanism by which DRB inhibits transcription in intact cells.

Winicov (109) has observed a small (25%) inhibitory effect of 100 µM DRB on transcription of hnRNA by RNA polymerase II in isolated nuclei from mouse L cells. Subsequently Winicov and Button (110) carried out _in vitro_ experimemts with 350 µM DRB in which γ-thio analogs of ATP and GTP were used as affinity probes to measure RNA synthesis initiated in isolated L cell nuclei. DRB at this very high concentration markedly inhibited the incorporation of [^{35}S]GTP[S] and [^{35}]ATP[S] into RNA of all size classes. 350 μM rather than 100 μM DRB was used, because the results were more reproducible at the higher concentration, however, because of the very high concentration of DRB used, it is difficult to evaluate the significance of the results.

The development of cell-free systems for RNA transcription in which faithful initiation occurs very efficiently (72, 106) has opened up new avenues of approach to the study of the mechanism of action of DRB. Zandomeni et al. (73) have used a whole cell extract prepared from HeLa cells (72) and investigated the effects of DRB on transcription by RNA polymerase II from the human ε-globin gene cut with Bam HI. In this system, RNA initiated at the cap site elongates to a run-off product 460 nucleotides long. When DRB was present in the reaction mix from the start of the in vitro reaction, dose-dependent inhibition of synthesis of the size-specific run-off product was observed in the 10-60 μM concentration range of DRB. Higher concentrations of inhibitor were required to obtain comparable levels of inhibition with DRB-5'-monophosphate and DRB-5'-triphosphate. DRB and its phosphorylated derivatives had no inhibitory effects on the transcription of the VA gene of adenovirus by RNA polymerase III. Thus with respect both to the dose-response relationship and the selectivity of action, the results of Zandomeni et al. (73) in the in vitro system, which is capable of correct initiation of new chains, are in close agreement with those previously obtained in vivo (see above). Both in vivo and in the in vitro system, 50% inhibition of transcription by RNA polymerase II is obtained at DRB concentrations <10 μM. However, in the in vitro reaction no evidence was obtained of discrete premature termination products larger than 100 nucleotides either in the presence or absence of DRB (73). As pointed out by Zandomeni et al. (73), RNAs smaller than 100 nucleotides would not have been detected because of the background inherent in this in vitro transcriptional system, nor would have premature termination products of varying size within the 100-460-nucleotide range been detected, if present.

Zandomeni et al. (74) have recently obtained direct evidence that in the transcriptionally active cell extracts DRB blocks initiation of transcription from exogenous truncated templates, such as those in a plasmid containing both adenovirus 2 major late promoter and the early region IV promoter. In these experiments, transcription complexes were permitted to form in the presence of only two initial ribonucleotide triphosphates which, for the promoters used, are ATP and CTP. Chain elongation was dependent on addition of the other two nucleoside triphosphates, one of which was radioactively labeled ([α-^{32}P]GTP). When DRB (60 μM) was present during the entire incubation period, including the initial incubation with ATP and CTP only, transcription was blocked. However, when the transcription initiation complex was first permitted to form in the absence of DRB, and the compound added subsequently along with UTP and [α-^{32}P]GTP, elongation proceeded despite the presence of DRB, with the production of runoff products of the same length as in DRB-free controls. Zandomeni et al. (74) made additional efforts to detect premature termination products in reactions conducted in the presence of DRB, but found no evidence for such products nor were any prematurely terminated chains detected in the absence of DRB.

Although the findings of Zandomeni et al. (73,74) and Mittleman et al. (107) are of great interest, they have been obtained in a system which does not display the full complexity of the transcription process by RNA polymerase II as it occurs *in vivo*. It should be noted that in the cell extract experiments with Ad2, naked plasmid DNA containing truncated adenovirus transcription units was used and that DRB was added to the reaction mixture in ethanol rather than in an aqueous solution (73,74,107). Control samples also contained 0.5% ethanol. In the *in vitro* system, all of the adenovirus transcription units which have been examined, including that which yields the mRNA for polypeptide IX, have been found highly sensitive to inhibition by DRB (R. Weinmann, personal communication), whereas the *in vivo* synthesis of the polypeptide IX mRNA is at least partially resistant (34).

All of the *in vivo* observations discussed in preceding sections, are consistent with the view that DRB blocks an early event in the synthesis of RNA by RNA polymerase II, and, furthermore, that in cells of higher organisms, 20-30% of total hnRNA synthesis is resistant to inhibition by DRB. Zandomeni et al. (73,74) have suggested that premature termination may occur independently of DRB and only in the DRB-resistant fraction of hnRNA. It has been further suggested that inhibition of hnRNA synthesis by DRB, brought about by blocking of initiation of DRB-sensitive chains, results in greater availability of labeled precursors for the labeling of prematurely terminating chains of DRB-resistant RNA, which are terminated independently of the presence of DRB. The apparent accumulation of short chains in DRB-treated cultures (37,50) thus would not reflect a DRB-induced increase in the frequency of premature termination of DRB-sensitive chains, but would merely be a reflection of increased labeling of DRB-resistant chains. This interpretation fails to account for the markedly enhanced accumulation of discrete, promoter-proximal SmaI f- and SmaI e-specific short nuclear RNA transcribed from two of the late Ad2 transcription units in infected HeLa cells treated with DRB and labeled with $^{32}PO_4$ for 4.5-5 hr (Fig. 1 in Fraser et al. (54)). In the same experiments, the labeling of the DRB-resistant Ad2 VA RNA was approximately equal to that in DRB-free cells. (Fig. 1 in Fraser et al. (54)).

Enhanced accumulation of several discrete species of promoter-proximal Ad2-specific short RNA chains has also been observed after treatment of cells with DRB in vivo and labeling of nascent chains for 45 sec in isolated transcription complexes in vitro (S. Chen-Kiang, unpublished). Furthermore, these abundant populations of short RNA chains are able to elongate to full length and be polyadenylated and processed in vitro. Thus, the in vivo DRB-blocked short RNA can be elongated to productive hnRNA chains at least in this cell-free system (a "precursor-product" relationship). It should be emphasized that the specific pattern of discrete, short late Ad2-specific DRB-blocked RNA observed by Fraser et al. (54) following a 4.5-5 hr $^{32}PO_4$ labeling of intact, infected cells is also observed when transcription complexes isolated from DRB-treated, infected cells are labeled in vitro (S. Chen-Kiang, unpublished). Thus, the specific pattern of stop or pause sites is highly reproducible.

It is clear that further experiments are needed to elucidate the complex transcriptional mechanisms that are affected by DRB.

DRB AND CONGENERS AS TOOLS IN THE ANALYSIS OF THE REGULATION OF MAMMALIAN GENE EXPRESSION

Two key properties of DRB make it an unusual research tool in cell biology. First, DRB (60-75 µM) rapidly and selectively inhibits >95% of cellular and viral mRNA synthesis with only a minimal effect on protein synthesis. Second, the inhibitory effect of DRB on RNA synthesis is rapidly reversible even after treatment of cells with DRB for several hours. These two features have made DRB a very useful research tool in experimental situations which require a selective and temporary block of mRNA synthesis. For example, Fig. 17 shows an experiment in which cells treated with interferon for 1 hr in the presence of DRB retain the potential to enter an antiviral state if DRB is removed within 3 hr after termination of the 1-hr interferon treatment (111). However maintenance of DRB in the cultures for 4-5 hr or longer leads to a decay of the signal generated by the interaction of interferon molecules with their cell surface receptors. Thus the generation of the antiviral state appears to remain programmed for approximately 3 hr after interaction of the cells with interferon. DRB has also been used extensively in gaining a better understanding of the regulation of human interferon production (reviewed in 16),

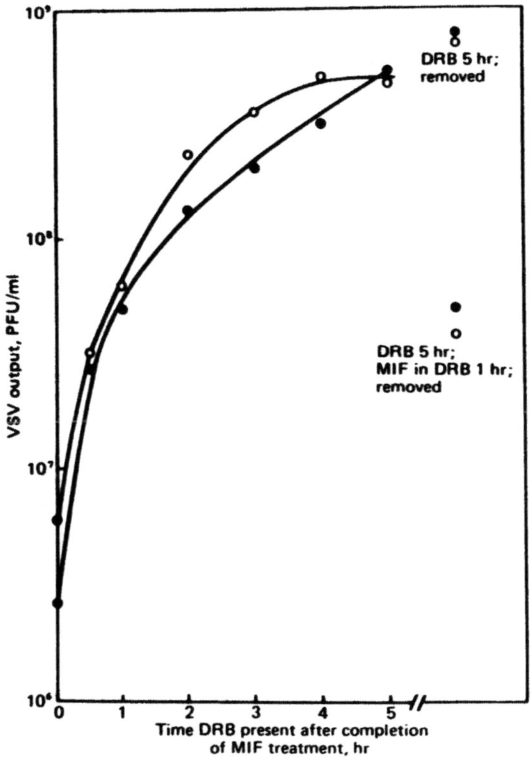

Fig. 17. Decay in resistance to virus with increasing duration of exposure to DRB after completion of a mouse interferon (MIF) pulse delivered to DRB-inhibited L-929 cells. Two separate experiments are shown (• and o). They were done in triplicate. MIF and anti-MIF antibody always contained 100 μM DRB made up in BME/2% FCS. Coverslip monolayers containing approximately 8 x 10^4 cells were successively treated as follows: 100 μM DRB for 1 hr; MIF for 1 hr, removed by five washes of 100 μM DRB; anti-MIF antibody for 15 min; and continued exposure to 100 μM DRB until the DRB was removed at different intervals and replaced (after wash) with BME/2% FCS. Cells were then incubated at 37°C for 18-20 hr to allow development of antiviral activity, inoculated with 10^3 PFU of VSV, and incubated again for 24 hr. Supernatant fluids from triplicate coverslips were assayed for VSV output. The output of coverslips treated with DRB for 5 hr but not exposed to an MIF pulse is labeled "DRB 5 hr; removed". It closely approximated the output of coverslips treated with neither DRB nor MIF (not shown). The output of coverslips treated only with an MIF pulse (not shown) approximated the plotted output of those coverslips treated with both DRB and MIF from which DRB was removed at time zero, the end of the MIF pulse. Also plotted is the output of coverslips exposed to DRB for 5 hr before being treated with MIF for 1 hr and anti-MIF antibody for 15 min. The DRB was then removed by washing, as the label indicates, and further incubation and assay for VSV output were carried out as described above. From Gordon and Stevenson (111). (Reprinted with permission).

of the requirement of early adenovirus mRNA transcription in viral DNA replication (76), of the G1 phase control mechanism over S phase DNA synthesis (112), of gene expression in response to platelet-derived growth factor (113), and of other processes which involve mRNA synthesis.

CONCLUDING REMARKS

Halogenated ribofuranosylbenzimidazoles, such as 5,6-dichloro-1-β-D-ribofuranosylbenzimidazole (DRB), are of continuing interest because they selectively inhibit the transcription of viral and cellular RNA by cellular RNA polymerase II.

There now exist several excellent systems for the further study of the action of these compounds on viral and cellular RNA synthesis. Furthermore, biochemical techniques are available whose full application should lead to the elucidation of the precise mechanism of action of DRB and related derivatives. Such a mechanism will have to account for the following outstanding findings, which, for simplicity, are enumerated in relation to DRB, the compound which so far has been studied most extensively: 1) The β-linked ribofuranose moiety with a free 3' hydroxyl is required for selective inhibitory activity on transcription by RNA polymerase II, however DRB is active in the unphosphorylated state; 2) The effect of DRB on polymerase II-catalyzed RNA synthesis is immediate and completely reversible upon termination of treatment; 3) In different experimental systems DRB blocks initiation of new RNA chains or their elongation past specific promoter-proximal attenuation sites; it does not act as a general chain terminator; 4) DRB prevents the synthesis of most, but not all mRNAs; 5) A substantial portion (20-30%) of the synthesis of m^7G-capped hnRNA in cells of higher organisms is completely resistant to DRB; a similar portion of the synthesis of small homodisperse $m^{2,2,7}G$-capped nuclear RNAs such as U_1 and U_2 (D and C) is also resistant; 6) DRB has some inhibitory effect on the synthesis of preribosomal RNA and 5.8S RNA by RNA polymerase I; 7) DRB action does not involve alterations in ribonucleotide metabolism; 8) In the course of long-term treatment of cells with DRB, cellular hnRNA synthesis becomes less inhibited by DRB, which is not due to inactivation of DRB in the medium; in such long-term experiments, serum counteracts the action of DRB, although in short-term experiments, inhibition of RNA synthesis is not affected by serum concentration.

It is apparent that the mechanism of action of DRB is tied in with fundamental questions concerned with RNA polymerase II function and its regulation in living cells.

ACKNOWLEDGEMENTS

We thank Doctors E. Egyházi, G. R. Molloy and I. Gordon for permission to use previously published figures, and Doctors R. Weinmann and S. Chen-Kiang for making available results prior to publication. We are indebted to Mrs. K. Pickering, Ms. A. Gifford, and Mrs. C. Doktor for the expert preparation of a printer-ready copy of the manuscript. This investigation was supported by USPHS Research Grants CA-18608, AI-16262, AI-05600 and CA-13362, and by Program Project Grant CA-18213. P.B.S. was also supported by an American Cancer Society Junior Faculty Research Award and an Irma T. Hirschl Award.

REFERENCES

1. Tamm, I., Folkers, K., Shunk, C. H. and Horsfall, F. L. Jr. J. Exp. Med. 99:227-250, 1954.
2. Tamm, I., Folkers, K. and Horsfall, F. L. Jr. Yale J. Biol. Med. 24:559-567, 1952.
3. Tamm, I. and Caliguiri, L. A. In: International Encyclopedia of Pharmacology and Therapeutics. Chemotherapy of Viral Diseases, vol. 1 (Ed. D. J. Bauer), Pergamon Press, Oxford, 1972, pp. 181-230.
4. Horsfall, F. L. Jr. and Tamm, I. Ann. Rev. Microbiol. 11:339-370, 1957.
5. Sehgal, P. B. and Tamm, I. Biochem. Pharmacol. 27:2475-2485, 1978.
6. Tamm, I. and Tyrrell, D. A. J. J. Exp. Med. 100:541-562, 1954.
7. Lamb, R. A. and Choppin, P. W. J. Virol. 23:816-819, 1977.
8. Herz, C., Stavnezer, E. and Krug, R. M. Cell 26:391-400, 1981.
9. Tamm, I., Nemes, M. M. and Osterhout, S. J. Exp. Med. 111:339-349, 1960.
10. Tamm, I. and Nemes, M. M. Virology 4:483-498, 1957.
11. Ledinko, N. Nature New Biol. 233:247-248, 1971.
12. Wallace, R. D. and Kates, J. J. Virol. 9:627-635, 1972.
13. Baltimore, D. and Franklin, R. M. Biochem. Biophys. Res. Commun. 9:388-392, 1962.
14. Baltimore, D. and Franklin, R. M. J. Biol. Chem. 238:3395-3400, 1963.
15. Tamm, I. and Sehgal, P. B. Adv. Virus Res. 22:187-258, 1978.
16. Sehgal, P. B. and Tamm, I. In: Antibiotics and Chemotherapy vol. 27 (Ed. F. E. Hahn), S. Karger, Basel, 1980, pp. 93-138.
17. Pothier, P., Dru, A. and Beaud, G. J. Gen. Virol. 55:87-94, 1981.
18. Tamm, I. Science 126:1235-1236, 1957.
19. Geuskens, M. Arch. Biol. 72:153-171, 1961.
20. Sirlin, J. L. and Jacob, J. Nature 204:545-547, 1964.
21. Sirlin, J. L., Jacob, J. and Birnstiel, M. L. Natl. Cancer Inst. Monogr. 23:255-270, 1966.
22. Egyházi, E., Daneholt, B., Edström, J.-E., Lambert, B. and Ringborg, U. J. Cell Biol. 47:516-520, 1970.
23. Egyházi, E. Cell 7:507-515, 1976.
24. Egyházi, E. Nature (London) 262:319-321, 1976.
25. Granick, D. J. Cell Biol. 65:389-417, 1975.
26. Tamm, I., Hand, R. and Caliguiri, L. A. J. Cell Biol. 69:229-240, 1976.
27. Tamm, I. J. Cellular Physiol., submitted, 1982.
28. Sehgal, P. B., Darnell, J. E., Jr. and Tamm, I. Cell 9:473-480, 1976.
29. Tamm, I. and Sehgal, P. B. In: Alfred Benzon Symposium 13: Specific Eukaryotic Genes (Eds. J. Engberg, H. Klenow and V. Leick), Munksgaard, Copenhagen, 1979, pp. 424-439.
30. Tamm, I. and Sehgal, P. B. In: Drug Effects on the Cell Nucleus (Eds. H. Busch, S. T. Crooke, and Y. Daskal), Academic Press, New York, 1979, pp. 251-274.

31. Herman, R. C. Biochemistry 18:916-920, 1979.
32. Salditt-Georgieff, M., Harpold, M. M., Wilson, M. C. and Darnell, J. E., Jr. Molec. Cell. Biol. 1:179-187, 1981.
33. Salditt-Georgieff, M. and Darnell, J. E., Jr. Molec. Cell. Biol. 2:701-707, 1982.
34. Vennström, B., Persson, H., Pettersson, U. and Philipson, L. Nucl. Acids Res. 7:1405-1418, 1979.
35. Dreyer, C. and Hausen, P. Nucl. Acids Res. 5:3325-3335, 1978.
36. Tamm, I. Proc. Natl. Acad. Sci. USA 74:5011-5015, 1977.
37. Tamm, I., Kikuchi, T., Darnell, J. E., Jr. and Salditt-Georgieff, M. Biochemistry 19:2743-2748, 1980.
38. Sehgal, P. B. Ph.D. Thesis, The Rockefeller University, 1977.
39. Harlow, P. and Molloy, G. R. Arch. Biochem. Biophys. 207:380-391, 1981.
40. Sommer, S., Lavi, U. and Darnell, J. E., Jr. J. Mol. Biol. 124:487-499, 1978.
41. Egyházi, E., Daneholt, B., Edström, J.-E., Lambert, B. and Ringborg, U. J. Cell Biol. 48:120-127, 1971.
42. Egyházi, E., D'Monte, B. and Edström, J.-E. J. Cell Biol. 53:523-531, 1972.
43. Harlow, P. and Molloy, G. R. Arch. Biochem. Biophys. 203:764-773, 1980.
44. Hellung-Larsen, P., Gram Jensen, E. and Frederiksen, S. Biochem. Biophys. Res. Commun. 99:1303-1310, 1981.
45. Sehgal, P. B. and Tamm, I. Proc. Natl. Acad. Sci.. USA 73:1621-1625, 1976.
46. Egyházi, E. J. Mol. Biol. 84:173-183, 1974.
47. Egyházi, E. Proc. Natl. Acad. Sci. USA 72:947-950, 1975.
48. Sehgal, P. B., Derman, E., Molloy, G. R., Tamm, I. and Darnell, J. E., Jr. Science 194:431-433, 1976.
49. Egyházi, E., Pigon, A. and Rydlander, L. Eur. J. Biochem. 122:445-451, 1982.
50. Tamm, I. and Kikuchi, T. Proc. Natl. Acad. Sci. USA 76:5750-5754, 1979.
51. Chen-Kiang, S., Wolgemuth, D. J., Hsu, M.-T. and Darnell, J. E., Jr. Cell 28:575-584, 1982.
52. Fraser, N. W., Sehgal, P. B. and Darnell, J. E., Jr. Nature (London) 272:590-593, 1978.
53. Sehgal, P. B., Fraser, N. W. and Darnell. J. E., Jr. Virology 94:185-191, 1979.
54. Fraser, N. W., Sehgal, P. B. and Darnell, J. E., Jr. Proc. Natl. Acad. Sci. USA 76:2571-2575, 1979.
55. Dreyer, C. and Hausen, P. Eur. J. Biochem. 86:241-253, 1978.
56. Ro-Choi, T. S. and Busch, H. In: The Cell Nucleus, vol. III, (Ed. H. Busch), Academic Press, New York, 1974, pp. 151-208.
57. Fernandez-Munoz, R., Lavi, U. and Darnell, J. E., Jr. Nucl. Acids Res. 4:3357-3369, 1977.
58. Tweeten, K. A. and Molloy, G. R. Nucl. Acids Res. 9:3307-3319, 1981.
59. Frederiksen, S., Hellung-Larsen, P. and Gram Jensen, E. FEBS Lett. 87:227-231, 1978.
60. Frederiksen, S. and Hellung-Larsen, P. In: Alfred Benzon Symposium 13: Specific Eukaryotic Genes (Eds. J. Engberg, H. Klenow and V. Leick), Munksgaard, Copenhagen, 1979, pp. 457-469.
61. Eliceiri, G. L. Nature (London) 279: 80-81, 1979.
62. Ringborg, U. and Rydlander, L. J. Cell Biol. 51:355-368, 1971.
63. Granick, D. J. Cell Biol. 65:389-417, 1975.
64. Craig, N. C. and Perry, R. P. J. Cell Biol. 45:554-564, 1970.
65. Pederson, T. and Kumar, H. J. Mol. Biol. 61:655-668, 1971.
66. Puvion, E., Bachellerie, J.-P. and Burglen, M.-J. J. Ultrastruct. Res. 69:1-12, 1979.

67. Puvion, E., Puvion-Dutilleul, F. and Leduc, E. H. J. Ultrastruct. Res. 76:181-190, 1981.
68. Puvion, E. Biol. Cell 43:147-150, 1982.
69. Laub, O., Jakobovits, E. B. and Aloni, Y. Proc. Natl. Acad. Sci. USA 77:3297-3301, 1980.
70. Hay, N., Skolnik-David, H. and Aloni, Y. Cell 29:183-193, 1982.
71. Montandon, P.-E. and Acheson, N. H. J. Gen. Virol. 59:367-376, 1982.
72. Manley, J. L., Fire, A., Cano, A., Sharp, P. A. and Gefter, M. L. Proc. Natl. Acad. Sci USA 77:3855-3859, 1980.
73. Zandomeni, R., Mittleman, B., Bunick, D., Ackerman, S. and Weinmann, R. Proc. Natl. Acad. Sci USA 79:3167-3170, 1982.
74. Zandomeni, R., Bunick, D., Ackerman, S., Mittleman, B. and Weinmann, R. J. Mol. Biol., submitted, 1982.
75. Tamm, I., Eggers, H. J., Bablanian, R., Wagner, A. F. and Folkers, K. Nature (London) 223:785-788, 1969.
76. Brötz, M., Doerfler, W. and Tamm, I. Virology 86:516-529, 1978.
77. Evans, R., Weber, J., Ziff, E. and Darnell, J. E., Jr. Nature (London) 278:367-370, 1979.
78. Sehgal, P. B., Sagar, A. D. and Braude, I. A. Science 214:803-805, 1981.
79. Skolnik-David, H., Hay, N. and Aloni, Y. Proc. Natl. Acad. Sci. USA 79:2743-2747, 1982.
80. Llopis, R. and Stark, G. R. J. Virol. 38:91-103, 1981.
81. Llopis, R. and Stark, G. R. J. Virol. 44:864-870, 1982.
82. Hayward, W. S. and Hanafusa, H. J. Virol. 11:157-167, 1973.
83. Lamb, R. A. and Choppin, P. W. Proc. Natl. Acad. Sci. USA 76:4908-4912, 1979.
84. Lamb, R. A. and Choppin, P. W. Virology 74:504-519, 1976.
85. Lamb, R. A. and Choppin, P. W. In: Negative Strand Viruses and the Host Cell (Eds. B. W. J. Mahy and R. D. Barry), Academic Press, London, 1978, pp. 227-237.
86. Minor, P. D. and Dimmock, N. J. Virology 78:393-406, 1977.
87. Bossman, H. B. Biochem. Biophys. Res. Commun. 41:1412-1420, 1970.
88. Wu, R. S., Kumar, A. and Warner, J. R. Proc. Natl. Acad. Sci. USA 68:3009-3014, 1971.
89. Horwitz, S. B., Chang, C.-K. and Grollman, A. P. Molec. Pharmacol. 7:632-644, 1971.
90. Kessel, D. Biochim. Biophys. Acta 246:225-232, 1971.
91. Abelson, H. T. and Penman, S. Nature New Biol. 237:144-146, 1972.
92. Tamm, I. and Overman, J. R. Virology 3:185-196, 1957.
93. Sehgal, P. B., Tamm, I. and Vilcek, J. Science 190:282-284, 1975.
94. Hand, R. and Tamm, I. Exp. Cell Res. 107:343-354, 1977.
95. Tweeten, K. A. and Molloy, G. R. Arch. Biochem. Biophys. 217:332-340, 1982.
96. Egyházi, E., Ossoinak, A., Holst, M., Rosendahl, K. and Tayip, U. J. Biol. Chem. 255:7807-7812, 1980.
97. Egyházi, E. and Shugar, D. FEBS Lett. 107:431-435, 1979.
98. Egyházi, E., Holst, M. and Tayip, U. Eur. J. Biochem. in press, 1983.
99. Tamm, I. J. Bacteriol. 72:42-53, 1956.
100. Tamm, I., Folkers, K. and Shunk, C. H. J. Bacteriol. 72:54-58, 1956.
101. Tamm, I. and Sehgal, P. B. J. Exp. Med. 145:339-349, 1977.
102. Egyházi, E., Ossoinak, A., Tayip, U., Kazimierczuk, Z. and Shugar, D. Biochim. Biophys. Acta 697:213-220, 1982.
103. Gupta, R. S. and Siminovitch, L. Somat. Cell Genet. 6:151-169, 1980.
104. Funanage, V. L. Molec. Cell. Biol. 2:467-477, 1982.
105. Tamm, I. Molec. Pharmacol., submitted, 1982.
106. Weil, P., Luse, D., Segall, J. and Roeder, R. Cell 18:469-484, 1979.

107. Mittleman, B., Zandomeni, R. and Weinmann, R. J. Mol. Biol., submitted, 1982.
108. Tamm, I., Kikuchi, T. and Murphy, J. S. Proc. Natl. Acad. Sci. USA 79:2569-2573, 1982.
109. Winicov, I. Biochemistry 18:1575-1581, 1979.
110. Winicov, I. and Button, J. D. Eur. J. Biochem. 124:239-244, 1982.
111. Gordon, I. and Stevenson, D. Proc. Natl. Acad. Sci. USA 77:452-456, 1980.
112. Chadwick, D. E., Ignotz, G. G., Ignotz, R. A. and Lieberman, I. J. Cell. Physiol. 104:61-72, 1980.
113. Pledger, W. J., Scher, C. D., Smith, J. C., Goon, D. S. and Stiles, C. D. In: Miami Winter Symposia vol. 18: Cellular Responses to Molecular Modulators (Eds. L. W. Mozes, J. Schultz, W. A. Scott and R. Werner), Academic Press, New York, 1981, pp. 61-78.

15
RIBAVIRIN

Roberts A. Smith, Department of Chemistry & Biochemistry, University of California at Los Angeles, and Peter Canonico, Department of Antiviral Studies, U.S. Army Medical Research Institute of Diseases.

Ribavirin (1-β-D-ribofuranosyl-1,2,4-triazole-3-carboxamide) is a broad spectrum viral chemotherapeutic agent, demonstrating antiviral activity against a large number of both DNA-and RNA-containing viruses. Studies demonstrating the broad array of virus replication inhibited by ribavirin in cell culture have been carefully summarized by Sidwell (1,2) and most of the <u>in vivo</u> evidence in animal models has been summarized by Allen (3). In addition, Canonico (4) has summarized current results since the earlier publications in a review focusing on several aspects of ribavirin. Possibly the most significant recent clinical results with ribavirin have been obtained in two different studies; one conducted by Knight et al (5) at Baylor College of Medicine, and the other under the direction of McCormick (6) of the U.S. Center for Disease Control. In the Knight study, ribavirin was administered by small particle aerosol in an outbreak of influenza virus A/England/333/80 (HINI) infections in college students. The 14 randomly selected patients retained an average estimated 1.15g of ribavirin in 23 h of treatment over three (3) days. As judged by the highly significant reduction in height and duration of fever, reduction in systemic illness, and disappearance of influenza virus from the respiratory secretions, ribavirin had a highly significant therapeutic effect over the 17 control patients.

In the McCormick study, patients suffering from a Lassa fever viremia of $>10^4$ pfu/ml of serum and having SGOT values 150 IU/dl were treated with iv ribavirin alone or in combination with immune convalescent serum. Convalescent serum alone had

Y. Becker (ed.), Antiviral Drugs and Interferon: The Molecular Basis of Their Activity.

little if any effect on the course of the disease, but in combination with ribavirin, all the patients recovered. Even with ribavirin alone, all patients recovered from the disease. Without treatment, patients were at high risk with only a 20% chance of survival. Thus ribavirin clearly has a therapeutic role in this life-threatening disease.

Ribavirin may appear to the casual observer more as a pyrimidine analogue than any other structure. But by single crystal X-ray diffraction analysis Prusiner and Sundaralingam (7) have clearly established its close resemblance in structure to guanosine (Figure 1). The geometry about the carboxamide function renders ribavirin similar to guanosine. The carbonyl oxygen at position 7 and the amide nitrogen at position 8 of ribavirin occupy stereochemically similar positions to the carbonyl oxygen 10 and the ring nitrogen 1 of guanosine. This

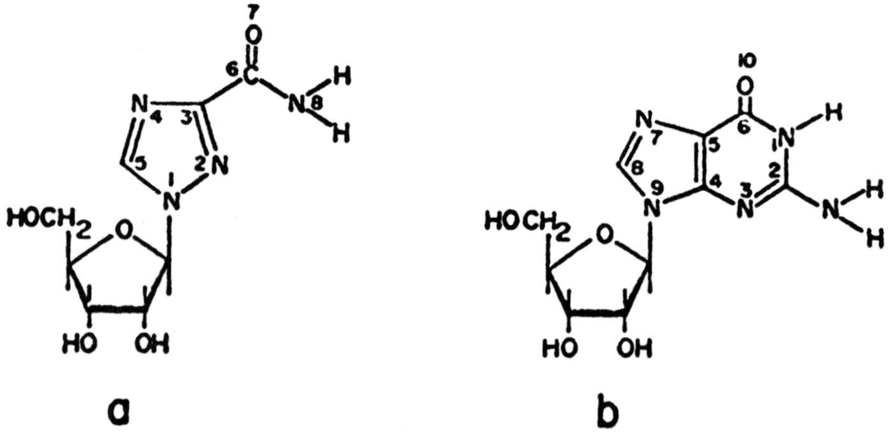

FIGURE 1. (a) Ribavirin; (b) Guanosine.

structural similarity has been a guiding principle in studies on the mode of action of ribavirin suggesting an involvement of guanosine nucleotides in some fashions.

A possible mechanism of action of ribavirin involving guanine nucleotides was further supported by the studies of Streeter et al (8), who showed clearly that the virustatic effect of ribavirin against measles virus grown in Vero cells, could be abolished by the addition of xanthosine or guanosine, or to a lesser extent inosine to the growth medium at 200 μg/ml. Adenosine, deoxyadenosine, aminoimidazolecarboxamide riboside, cytidine, uridine or thymidine added to the system had no effect. Browne (9) also demonstrated the complete reversal of the action of ribavirin by guanosine in the production of influenza A/Port Chalmers/1/73 H_3N_2 virus hemagglutinin in MDCK cells. Malinoski and Stollar (10) however, showed that xanthosine, but not guanosine, reversed the ribavirin inhibition of Sindbis virus replication in Aedes albopictus cells. These authors also state that guanosine alone inhibits virus replication in these cells and thus its effect on ribavirin is masked.

Despite the physical, chemical and cell culture evidence linking ribavirin and guanosine nucleotides, in vivo results obtained by Knight et al (11), employing an aerosolized form of ribavirin in the treatment of influenza A infection in mice could not be blocked by guanosine or guanosine-5'-monophosphate. Unfortunately, it is not known if the dosages of guanosine and GMP given to mice were sufficient to yield pharmacologically active levels of guanosine nucleotides in blood or tissues. Guanosine however, was capable of markedly reducing the antiviral effect of ribavirin against the same influenza virus grown in MDCK cells.

Ribavirin, until recently, has been remarkably difficult to quantitate. Its UV absorption spectrum has not led to sensitive measurements (max at 20 nm extinction coeff. = $1.17 \times 10^{-3} l\ mol^{-1} cm^{-1}$) of metabolites and unchanged compound. Most pharmacokinetic studies have depended on use of either ^{14}C-or ^{3}H-labelled material and have been reported in terms of radioactivity in tissues and fluids. The early studies in rats (12) suggested that ribavirin is both deamidated and deribosylated, thus both ribosylated and non-ribosylated triazole carboxylate as well as ribavirin, and the triazole carboxamide were seen in urine. However, employing 5-^{14}C-ribavirin, Ferrara et al (13) have shown that very little (ca.3%) and Catlin et al (14) showed that less than 2% of the ^{14}C administered to rats in ribavirin was expired as $^{14}CO_2$. Thus the triazole ring may be considered metabolically very stable. Ferrara et al (13) showed that 60% of injected ^{14}C-ribavirin was eliminated in the urine of monkeys in 72 h while 82% was eliminated in the urine of rats in 24 h. In comparison with rats, tissues of the rhesus monkey retained much more ribavirin and its metabolites. This was particularly evident in red blood cells where the monkeys showed a continual increase in ribavirin for the first 36h. following injection. In contrast, the drug content in red blood cells of rats decreased with time, demonstrating a disappearance curve similar to that seen in rat plasma.

In humans, a mean of 53% of the orally administered ribavirin was eliminated in 72 h (14) and like rhesus monkeys, humans tend to concentrate ribavirin in red blood cells. At 72 h , about 3% of the total dose administered is present in human red cells, the later phase of the plasma disappearance curve is 48 h or more (14). Unlike rats, where a number of metabolites show up in the urine (12), our studies employing a high performance liquid chromatographic technique (15) have revealed only 1,2,4 triazole-3-carboxamide and unchanged ribavirin in the urine of nine subjects treated with ^{14}C-ribavirin, thus unlike the rat, humans do not deaminate ribavirin or its free base. A new study under way (James Connor, personal communication) promises to reveal more information, especially about the very difficult to assess plasma levels of ribavirin in humans, since it will use a recently developed radioimmunoassay which considerably increases the sensitivity of measurement of the drug.

Ribavirin is phosphorylated in red blood cells and as shown by the comparative study of Zimmerman & Deeprose (16), is very rapidly converted to its 5' triphosphate. Their results (16) also show that in human red blood cells, ribavirin exists principally as the 5'triphosphate and the ratio of the tri , di-, and monophosphate of ribavirin is about the same as that of the adenosine nucleotides (about 25:5:1). Since human red blood cells are particularly weak with respect to phosphatase activity, it may be possible that the relatively long time required for complete elimination of ribavirin from the body is related to its sequestration in tissues in the phosphorylated form, with rapid release occurring only from those tissues endowed with reasonably active phosphatases.

Figure 2 outlines the potential paths for the formation of ribavirin-5'-monophosphate from ribavirin. Streeter et al (17) have shown that ribavirin is a substrate for purine nucleoside phosphorylase which readily equilibrates ribavirin, inorganic phosphate, ribose-1-phosphate and triazole carboxamide. Since this reaction is readily reversible, it may play a prominent role

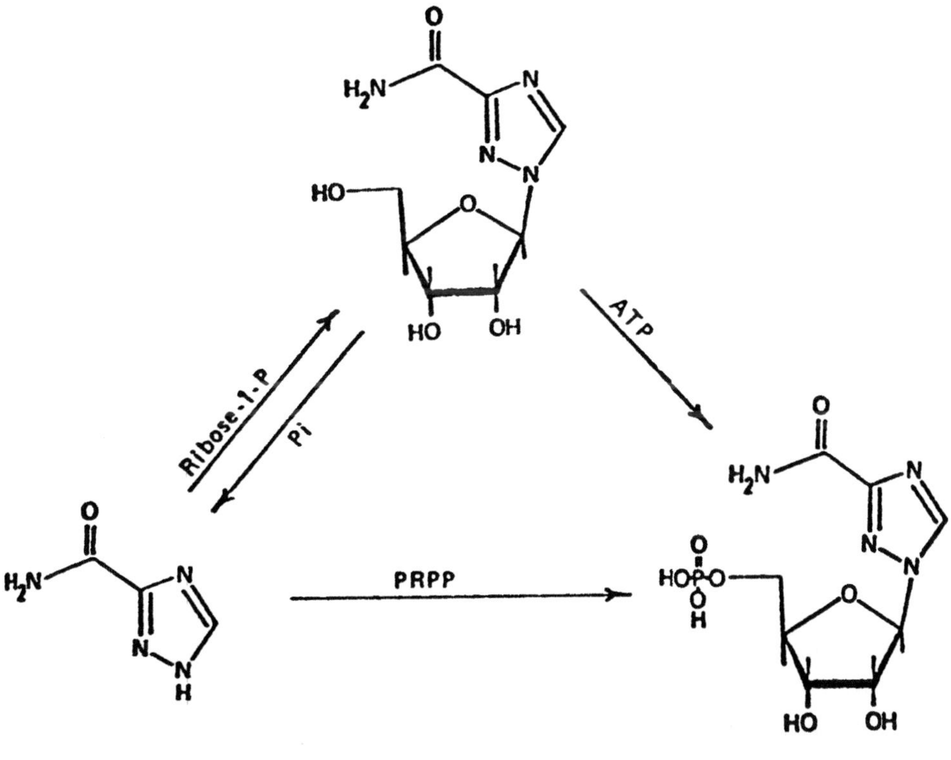

FIGURE 2. Potential routes leading to ribavirin-5'-monophosphate.

in "in vivo" deribosylation of ribavirin and it could account for the rapid formation of the free base in animals and man (14,15). This reaction also presumably accounts for the antiviral activity demonstrated by the free base (21), since evidence now shows the chief intracellular form of the drug is in its 5'-phosphorylated form.

Efforts to demonstrate the formation of ribavirin-5'-monophosphate from triazolecarboxamide and phosphoribosyl-pyrophosphate via the hypoxanthine guanine phosphoribosyl-pyrophosphate transferase have failed (17). Thus, 1,2,4-triazole-carboxamide is not a guanine analogue in all respects. These studies were limited to enzymes available in Ehrlich's ascites cells and in rat liver extracts. While this pathway may function in other systems, it is unlikely to represent a major route of transformation.

A partially purified kinase from rat liver of relatively narrow specificity has been shown by Streeter et al (18) to phosphorylate ribavirin to its 5' monophosphate. This fraction, when purified about 30-fold, was capable of phosphorylating ribavirin, deoxyadenosine and adenosine in an approximate ratio of 1:0.5:1. A wide variety of other nucleosides tested, including guanosine, deoxyguanosine, cytidine, uridine, deoxythymidine, deoxycytidine were either not active or only slightly phosphorylated by this preparation. Table I outlines a fascinating result obtained with this liver preparation, in which the ability to phosphorylate various analogues of ribavirin was examined. Any variation in the sugar moiety led to remarkably less phosphorylation and the 2' deoxyribose analogue of ribavirin, which at 1 mM was only poorly phosphorylated (about 10%) as well as ribavirin has been shown to be devoid of antiviral properties (20). Antiviral activity is not seen with the xylyl or arabinosyl derivative either. Variation in the heterocyclic structure likewise leads to a loss of antiviral activity and again the 1,4,5-triazole and the 1,2,5-triazole are relatively poorly phosphorylated (Table I). 1-β-D-ribofuranosyl-1,2,4-triazole-3-carboxamidine is not at all phosphorylated by the partially purified rat liver kinase, but it is active in many in vitro systems as an antiviral agent.

TABLE I. Phosphorylation of Ribavirin and Analogues by Rat Liver Kinase

Structural Variation in the Sugar		% of Ribavirin Phosphorylation	Structural Variation in the Triazole		% of Ribavirin Phosphorylation
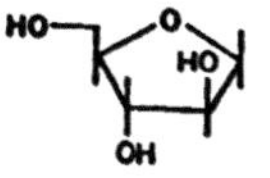	arabinosyl	8	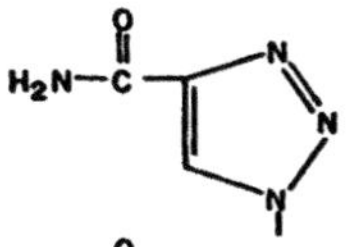	1,4,5-triazole	4
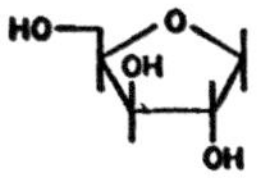	xylyl	1	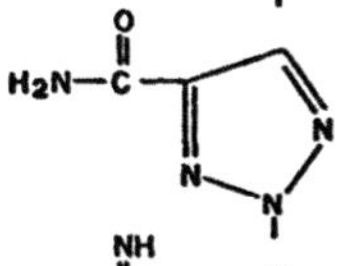	1,2,5-triazole	19
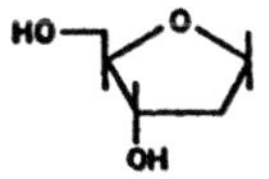	2-deoxyribosyl	11	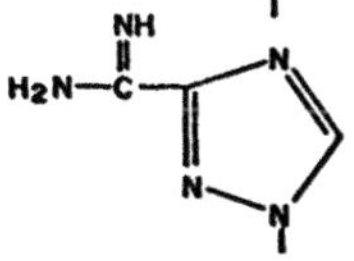	3-carboxamidino-1,2,4-triazole	0

All ribavirin analogues were tested at 1 mM and extent of their phosphorylation compared to that of ribavirin. α-^{32}P-ATP was employed and the nucleoside monophosphates separated by Dowex-I chromatography. Reaction conditions as outlined in Streeter et al. (1974).

Presumably, the carboxamidine is biologically hydrolyzed to ribavirin in these systems.

Although Streeter et al (18) could not precisely identify the rat liver kinase capable of converting ribavirin to its 5'-monophosphate, Willis et al (22) demonstrated that human lymphoblasts deficient in adenosine kinase but possessing deoxyadenosine kinase could not phosphorylate the drug. Fenwick (23) also concluded that it is adenosine kinase in Chinese hamster fibroblasts which converts ribavirin to its 5'-monophosphate.

Ribavirin-5'-monophosphate is a powerful competitive inhibitor of inosine monophosphate (IMP) dehydrogenase, and since this enzyme is involved in the final steps of the de novo synthesis of guanosine monophosphate (GMP) and is normally regulated "in vivo" by GMP, one would expect ribavirin to have a major effect on intracellular guanosine nucleotide pools. Streeter et al (8) measured the Ki for ribavirin-5'-monophosphate at 2.5×10^{-7} M while that for GMP is about 2.2×10^{-4} M. Thus ribavirin-5'-monophosphate is about 1,000 fold more potent than GMP in regulating this enzyme. In Table II some other derivatives and isomers of ribavirin monophosphate are listed and their relative inhibition

TABLE II. IMP deH_2 Inhibition by Analogues of Ribavirin-5-Phosphate.

Structural Variation		$I_{50}/[S]$	Structural Variation		$I_{50}/[S]$
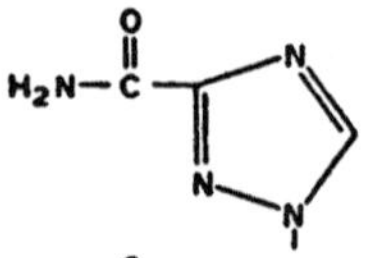	none	0.016	$-CH_2-P(=O)O_2^-$	methylphosphonate	0.60
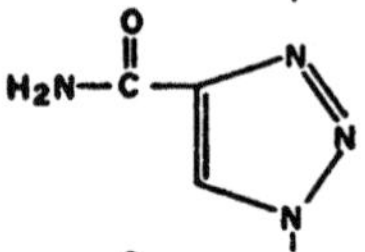	1,4,5-triazole	0.016	$-CH_2(CH_2)_2CO_2^-$	5'-butyrate	10+
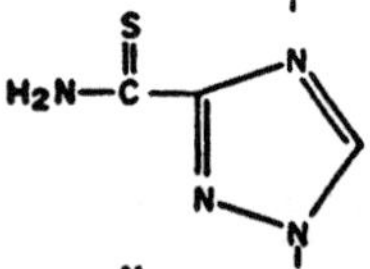	thiocarboxamide	10+	$-O-P(=O)(O^-)-NH_3^+$	5'-phosphoramidate	1.9
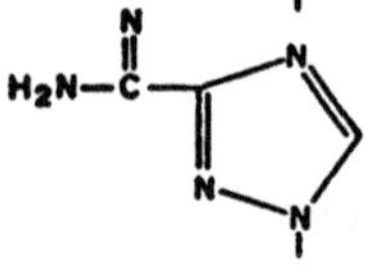	carboxamidino	1.0	$-S(=O)(O^-)-NH_2$	5'-sulfamate	1+

All compounds are compared with ribavirin-5-monophosphate using the ratio of inhibitor concentration causing 50% inhibition to the substrate (IMP) concentration (Streeter, 1975, unpublished).

of IMP dehydrogenase is shown. It is of distinct interest that 1-(β-D-ribofuranosyl) 1,4,5-triazole-3-carboxamide-5'-monophosphate, the nucleoside of which has no antiviral properties, is apparently as potent an IMP dehydrogenase inhibitor as is ribavirin-5'-monophosphate. It is entirely possible that the relatively low rate of phosphorylation of the 1,4,5-triazole nucleoside accounts for its lack of antiviral properties. The thiocarboxamide and carboxamidino derivatives of ribavirin monophosphate are about 60 and 600 times less powerful IMP dehydrogenase inhibitors, although both compounds are respectable viral chemotherapeutic agents (20). In examining modification at the 5' phosphoryl group, it is clear that any change from the phosphomonoester structure reduces the effectiveness of its inhibition of IMP dehydrogenase. Another nucleoside which is a potent inhibitor of IMP dehydrogenase (25), but which shows only very weak antiviral properties (1,24) is 2-β-ribofuranosyl-thiazole-4-carboxamide.

Clearly the work of Zimmerman & Deeprose (16), and the earlier work of Lowe et al (26), demonstrates a major effect on the ribonucleotide and the deoxyribonucleotide pools in L5178Y cells exposed to ribavirin in their growth medium. By far the largest effect is seen on the dTTP pool size (26) which increases 250% in these cells. Increases as great as 800% have been seen in dTTP levels in KB cells (Drach, J.C., personal communication). As expected, the level of guanosine nucleotides was considerably altered in the L5178Y cells (16), reducing as much as 50% the GTP pool and 70% the dGTP. However, while increasing the concentration of ribavirin from 100 μM in the growth medium resulted in a proportional increase of ribavirin 5'-triphosphate (from 103±29 p.moles/10^6 cells to 473 p.moles/10^6 cells) it only shifted the GTP pool size from 48% to 43% of control. The only other significant alteration (ie > 20% shift in pool size) was a 44% increase in the UTP pool size at 100 μM ribavirin. Such large changes in nucleotide pool sizes must have enormous effects on the regulation of metabolic events within the cell. Aside from the effect of ribavirin-5'-monophosphate on IMP dehydrogenase which can directly regulate the intracellular level of GMP, the mechanism by which other nucleotide levels are perturbed, remains presently unknown.

Thymidine kinase from a number of sources, including certain viral enzymes and all isozymes so far examined, has shown remarkably sensitive regulation by dTTP levels (27). Thus in the presence of ribavirin with increases in the intracellular levels of dTTP ranging up to 8-fold, it is not surprising that cellular uptake and phosphorylation of ^{3}H-thymidine in KB cells is powerfully inhibited by ribavirin (I_{50}=4.3μM). Thus as Drach et al (28) concluded, measurement of DNA synthesis exclusively by incorporation of ^{3}H-thymidine may cause serious misinterpretation of data, as perhaps was the case in the conclusion concerning the mode of action of ribavirin reached by a number of earlier investigators (1,29,30). Therefore, while investigators have reached the conclusion that ribavirin at 2 to 4 μM causes a 50%

inhibition of DNA synthesis (31), KB cells showed no inhibition of DNA synthesis as measured by flow cytometry, and by ^{32}Pi incorporation with up to 200 μM ribavirin in the growth medium(28). Since neither ribavirin nor its 5' mono-, di-, or triphosphate are inhibitors of thymidine kinase (Drach, J.C., personal communication; Cheng, Y.C., personal communication) it is probable that the elevated intracellular level of dTTP resulting from addition of ribavirin to the growth medium, because of the tight control this substance has on thymidine kinase, inhibits the phosphorylation of ^{3}H-thymidine. While the molecular events leading to elevated levels of dTTP are not understood, the consequential inhibition of thymidine phosphorylation has been clearly misinterpreted as inhibition of DNA synthesis.

To some extent, but to a lesser degree, the same problem exists with uptake and incorporation of ^{3}H-uridine into RNA. Canonico et al (32) reported that while ribavirin does not inhibit RNA synthesis it shows a marked concentration-dependent inhibition on the uptake of exogenous ^{3}H-uridine by BHK-21 cells. Thus at low levels of ribavirin (25 μg/ml) an apparent inhibition of ^{3}H-uridine incorporation into TCA soluble and insoluble material is seen and at a high level (800 μg/ml) reversal of the inhibition occurs. At 100 μg/ml of ribavirin in the growth medium, a 35% inhibition of ^{3}H-uridine incorporation into TCA-precipitable material in the BHK-21 cell system was seen (32). Goswami et al (33) also showed that ribavirin inhibited the uptake of ^{3}H-uridine in mouse L cells, but with ^{32}Pi they saw a stimulation of both viral specific and cellular poly(A) RNA synthesis (Table III).

TABLE III. Effect of ribavirin on synthesis and methylation of poly(A)-containing RNA in vaccinia virus infected mouse L-cells

Ribavirin μg/ml	RNA fraction	CPMx10^{-2} [^{3}H]	[^{32}P]	[^{3}H]/[^{32}P]	methylation inhibition %
0	Poly(A)+	2,090	3,805	0.55	-
100	Poly(A)+	3,271	9,256	0.35	36
0	Virus	255	268	0.95	-
100	Virus	690	1,074	0.64	36
0	Poly(A)+	1,968	7,214	0.27	-
500	Poly(A)+	2,530	15,560	0.16	41
0	Virus	230	73	0.31	-
500	Virus	448	2,431	0.18	42

Cells were pretreated with ribavirin for 2 to 4.5hr. then infected in a medium devoid of methionine and containing one tenth the normal phosphate level. After 30 min. cells were collected and labelled for 90 min. with [^{3}H]-methionine and [^{32}Pi]. Controls were treated the same except ribavirin was omitted.

mRNA of a large number of viruses and eukaryotic cells is processed after synthesis by elaboration of an unusual structure

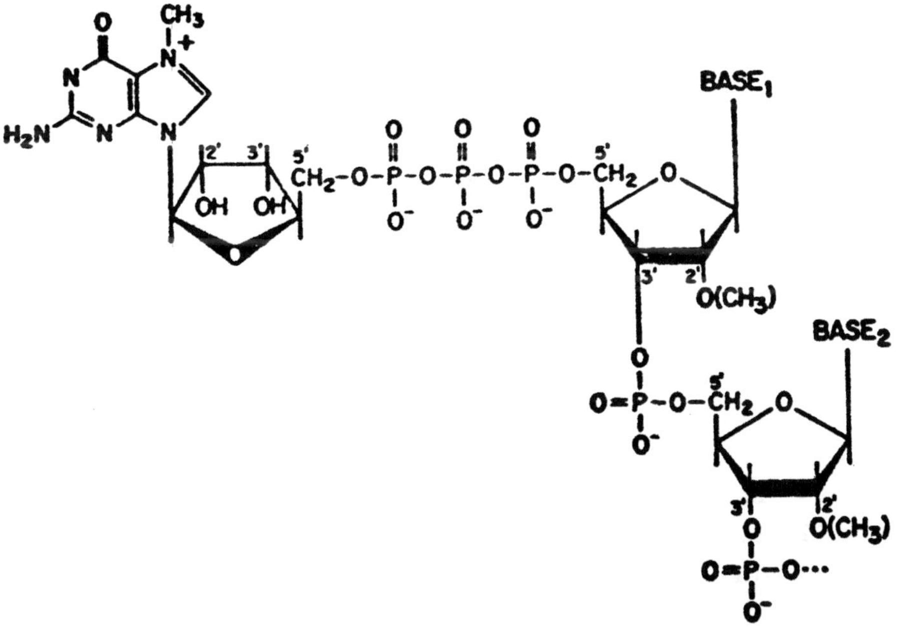

FIGURE 3. Structure of 5' "cap".

consisting of guanosine methylated on position 7 and attached through its 5' hydroxyl to a triphosphate bridge to the 5' hydroxyl of the penultimate nucleotide which is usually methylated on its 2' hydroxyl group (Figure 3).

The series of reactions involved in elaborating the "cap" are outlined in (Figure 4) and are largely the result of very careful work of Shatkin and his associates (34) and Moss and his colleagues (35). They involve first a transfer of GMP from GTP to the 5' end of existing mRNA, followed secondly by methylation of the 7 position of the terminal guanosine utilizing S-adenosylmethionine and thirdly, by methylation of the penultimate nucleotide. Moss et al (35) showed clearly that

$$\text{pppG + ppN...} \xrightarrow{\text{Guanylyltransferase}} \text{G(5')ppp(5')N...}$$

$$\text{G(5')ppp(5')N... + Ado Met} \xrightarrow{\text{Methyltransferase}} \text{m}^7\text{G(5')ppp(5')N... + Ado Hcy}$$

$$\text{m}^7\text{G(5')ppp(5')N... + Ado Met} \xrightarrow{\text{Methyltransferase}} \text{m}^7\text{G(5')ppp(5')N}^{\text{m}}\text{... + Ado Hcy}$$

FIGURE 4. Enzymatic synthesis of "cap".

vaccinia virus contains all the necessary capping enzymes in the virion and also that its mRNA contains both $m^7G(5')pppG^m$ and $m^7G(5')pppA^m$. Goswami et al (36) examined the effect of ribavirin on both transcription and the 5' terminal modification of vaccinia mRNA. Neither ribavirin nor any of its 5' phosphorylated derivatives, inhibited the vaccinia virus RNA polymerase (36). However, the 5'-triphosphate of ribavirin was a potent inhibitor of the in vitro 5'-guanylylation of synthesized uncapped mRNA. As shown in Figure 5 the inhibition is competitive with GTP and the Ki for RTP is 32 μM, while the Km for GTP is 22 μM. Thus, the Ki/Km ratio of 1.45 indicates a powerful inhibitory effect of RTP on the guanylylation reaction.

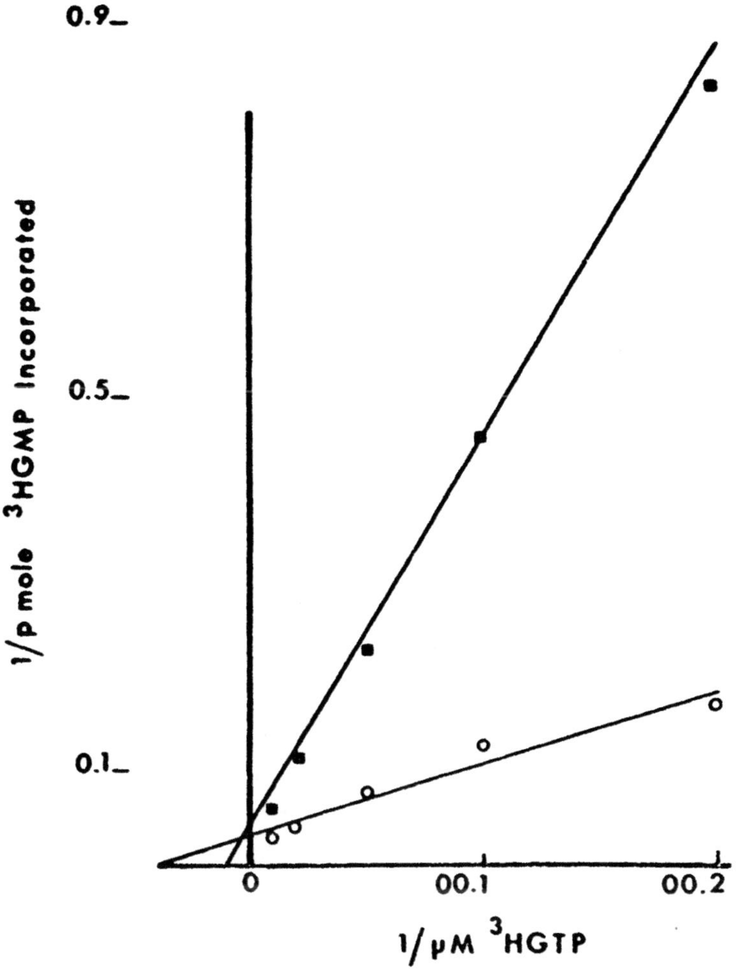

FIGURE 5. Lineweaver-Burk plots showing competitive inhibition of viral guanylyltransferase by ribavirin 5'-triphosphate. The incubation mixture in a total volume of 100 μl contained: 50 mM Tris-HCl (pH 7.8), 1 mM DTT, 2.5 mM $MgCl_2$, saturating amounts (30 μg) of RNA substrate, 5 μg of enzyme protein, 5-100 μM [3H]-GTP (11.2 Ci/mMole) and 100 μM Ribavirin 5'-triphosphate, where indicated. The reactions were incubated at 37°C for 30 minutes and stopped by the addition of 1 ml ice cold 10% TCA. After 15 minutes in ice, the precipitates were collected on glass fiber filter discs washed with 5% TCA and ethanol, dried and counted. The lines were fitted to the points by the method of least squares using a Smith-Corona Marchant 1016 computer with an IOTA-1 programmer. O—O, no ribavirin 5'-triphosphate; ■—■, 100 μM ribavirin 5'-triphosphate.

Goswami et al (33) have also shown with purified vaccinia virus capping enzymes, that ribavirin-5'-triphosphate is an effective inhibitor of the viral mRNA (guanine-7-)methyltransferase (70% inhibition is obtained with 25 μM RTP). Ribavirin triphosphate does not inhibit the vaccinia viral mRNA (nucleoside-2'-) methyltransferase, unlike S-adenosylhomocysteine analogues (37) which appear to be non-specific in their suppression of methylation. Whether the inhibition of methylation of the 7 position of guanine in the 5' terminal of the "cap" is accounted for by actual incorporation of ribavirin into that position instead of guanine, is not now known. In any case, it is clear that ribavirin would be an unlikely substrate for methylation once incorporated into the "cap" (R.K. Robins, personal communication).

In another study employing Venezuelan equine encephalitis (VEE) virus grown in BHK-21 cells, Canonico et al (32) have also shown a major effect of ribavirin on the formation of 5' cap on the viral mRNA. While ribavirin, added to the culture medium of BHK-21 cells infected with VEE virus inhibits replication by more than 99%, it is possible to isolate mRNA from such cells and in fact treatment of such cells with up to 300 μg/ml of ribavirin did not diminish the quantity of mRNA produced. Employing ^{3}H-guanine, Canonico et al (32) demonstrated that the amount of 7mGppp "cap" isolable from RNase digests of viral mRNA decreased proportionally with the amount of ribavirin added to the growth medium. At 300 μg ribavirin/ml, nearly 10-fold less "cap" was seen.

Most importantly, the capacity of mRNA from virus-infected ribavirin-treated BHK-21 cells to support protein synthesis in an in vitro rabbit reticulocyte cell-free translation system was remarkably reduced (Figure 6), thus with mRNA from infected cells treated with 100 μg/ml of ribavirin, the incorporation of ^{3}H-leucine into TCA precipitable material was reduced almost 75%.

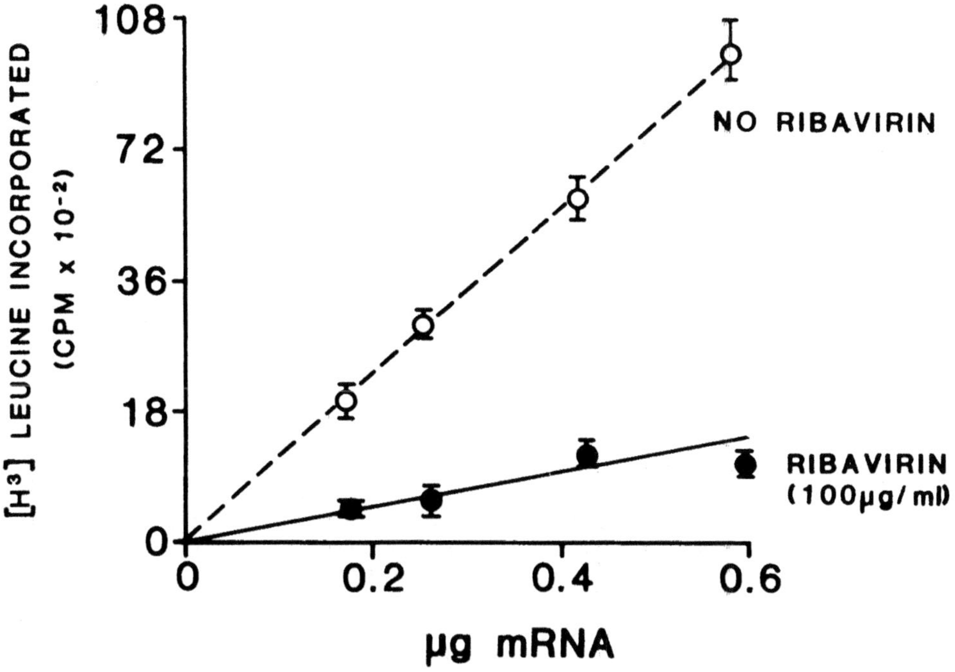

FIGURE 6. In vitro translation of mRNA isolated from ribavirin-treated and untreated VEE-infected cells.

Thus we conclude that a major mode of action of ribavirin as a viral chemotherapeutic agent lies in its ability to alter nucleotide pools and the packaging of mRNA. While this process may not be totally viral specific, a certain selectivity can result, since viral infected cells generally produce much more mRNA than infected cells.

REFERENCES

1. Sidwell RW, Huffman JH, Khare GP, Allen LB, Witkowski JT and Robins RK. Science, 177, 705-706 (1972).
2. Sidwell, RW, in Ribavirin, A Broad Spectrum Antiviral Agent, ed. by Smith RA and Kirkpatrick W. Academic Press (1980) pp. 23-42.

3. Allen LB, in Ribavirin, A Broad Spectrum Antiviral Agent, ed. by Smith RA and Kirkpatrick W. Academic Press (1980) pp. 43-58.
4. Canonico PG, in Antibiotics VI, ed. by Hahn FE, Springer-Verlag (1982) in press.
5. Knight V, Wilson SZ, Quarles JM, Greggs SE, McClung HW, Waters BK, Cameron RW, Zerwas JM and Couch RB. Lancet, (1981) pp. 945 to 949.
6. McCormick J, personal communication.
7. Prusiner P and Sundaralingam M. Nature New Biol. 244, 116 (1973).
8. Streeter DG, Witowski JT, Khare GP, Sidwell RW, Bauer RJ, Robins RK and Simon LN. Proc. Nat'l. Acad. of Science 70, 1174 (1973).
9. Browne MJ. Antimicrob. Agents & Chemother. 15, 747 (1978).
10. Malinoski F and Stollar V. Virology 110, 281 (1980).
11. Knight V, Wilson SZ, Alling DW, Moore RV and Longoria RM. Antimicrob. agents and Chemother. 20, 477 (1981)
12. Miller JP, Kigwana LJ, Streeter DG, Robins RK, Simon LN and Roboz J. N.Y. Acad. Sci. 284, 211 (1977).
13. Ferrara EA, Oishi JS, Wannemacher RW, Jr. and Stephen EL. Antimicrob. Agents and Chemotherap. 19, 1042 (1981).
14. Catlin DH, Smith RA, Samuels AI in Ribavirin, a Broad Spectrum Antiviral Agent, ed. by Smith RA and Kirkpatrick W. Academic Press (1980) pp. 83-98.
15. Smith RA and Yazdany C. Unpublished experiments 1981.
16. Zimmerman TP and Deeprose RD. Biochem. Pharmacol, 27, 709 (1978).
17. Streeter DG, Miller JP, Robins RK and Simon LN. N.Y. Acad. Sci. 284, 201 (1977).
18. Streeter DG, Simon LN, Robins RK and Miller JP. Biochem. 13, 4543 (1974).
19. Smith RA, in Ribavirin, A Broad Spectrum Antiviral Agent, ed. by Smith RA and Kirkpatrick W. Academic Press (1980) pp. 99 to 118.
20. Harris S and Robins RK in Ribavirin, A Broad Spectrum Antiviral Agent, ed. by Smith RA and Kirkpatrick W. Academic Press (1980) pp. 1 to 22.
21. Sidwell RW, Simon LN, Witkowski JT and Robins RK, in Progress in Chemotherapy, vol. II, ed. by Daikos GK. Athens, (1974) pp. 889-903.
22. Willis RC, Carson DA and Seegmiller JE. Proc. Nat'l. Acad. Sci. 138, 249 (1978).
23. Fenwick RG, Jr. (1979) XIth Int. Congress of Biochem. Toronto p.710.
24. Witkowski JT, Robins RK, Sidwell RW, Simon LN. J.Med. Chem. 15, 1150 (1972).
25. Streeter DG and Miller JP. Biochem. and Biophys. Res. Comm. 103, 1409 (1981).
26. Lowe JK, Brox B and Henderson JF. Cancer Res. 37, 736 (1977).
27. Kit S, Leung WC, Trakula D, Torgensen G and Dubbs DR. Int. J. Cancer, 15, 598 (1974).
28. Drach JC, Thomas MA, Barnett JW, Smith SH, Shipman C, Jr. Science, 212, 549 (1981).

29. Swallow DL. Ann. N.Y. Acad. Sci. 284, 289 (1977).
30. Drach JC and Shipman C, Jr. Ann. N.Y. Acad. Sci. 284, 369 (1977).
31. Larsson A, Stenberg K and Oberg B. Antimicrob. Agents and Chemotherapy, 13, 154 (1978).
32. Canonico PG, Little JS, Jahrling PB and Stephen EL. Proceedings of 11th ICC & 19th ICAC, Current Chemotherapy & Infectious Disease, pp. 1370-1372 (1980).
33. Goswami BB, Sharma O, Borek E and Smith RA. Current Chemotherapy and Immunotherapy: Proc. 12th International Congress of Chemotherapy. 1982 (in press).
34. Shatkin, AJ. Cell, 9, 645 (1976).
35. Moss B, Martin SA, Ensinger M, Boone RF and Wei C. Prog. in Nucleic Acid Res. Mol. Biol. 19, 63-81 (1976).
36. Goswami BB, Borek E, Sharma O, Fujitaki J and Smith RA. Biochem. and Biophys. Res. Comm. 89, 830 (1979).
37. Pugh CSG and Berchardt RT. Biochemistry, 21, 1535 (1982).

16
MOLECULAR BASIS OF GUANIDINE ACTION

M. SCHROM and L. A. CALIGUIRI
Department of Microbiology and Immunology, Albany Medical College of Union University, Albany, N.Y. 12208

INTRODUCTION

Guanidine and guanidinium compounds such as arginine, citrulline, ornithine, and creatine are natural constituents of animal cells and serum. It was discovered independently by Rightsel et al. (1), Crowther and Melnick (2), and Ueda et al. (3) that the replication of polioviruses in tissue culture was inhibited by guanidine at millimolar (mM) concentrations. Guanidine is a basic compound which exists at physiological pH as a resonance hybrid with a net positive charge. It has long been used at molar concentrations as a protein denaturant and dissociating agent, and it has therefore been inferred that its specific antiviral activity is the result of an induced conformational change in a viral product which is extremely sensitive to the denaturant.

Spectrum of Activity

As shown in Table 1, polioviruses are the most sensitive to the compound, being inhibited by concentrations of 0.1-2 mM (4). Most other enteroviruses are also sensitive to guanidine, although some of the echoviruses and coxsackieviruses are resistant (5). The rhinoviruses show a wide variation in susceptibility to guanidine (6), as does foot and mouth disease virus (FMDV) (7); however, these observations are complicated by the observations that several compounds, including amino acids commonly found in differing concentrations in different tissue culture media, act as guanidine antagonists, a phenomenon to be dealt with later (8).

The only other groups of viruses which have been shown to have any sensitivity to guanidine are two togaviruses, Sindbis and Semliki Forest (9), and two plant viruses, tobacco necrosis virus (10) and tobacco mosaic virus (11), although these viruses all required concentrations of 10-50 mM guanidine to demonstrate significant inhibition by this drug. These viruses share with the picornaviruses the property of positive-strand genomes, the significance of which is unknown, although the negative-strand viruses which have been examined, such as influenza and parainfluenza, are resistant, as are the double-stranded reoviruses (6). Among the DNA viruses which have been examined, all are resistant, including poxviruses, polyoma virus, adenoviruses, and herpesviruses (6).

As the majority of studies on the mechanism of action of guanidine have been done with poliovirus, this review will focus mainly on these studies; therefore, a brief description of the important features of poliovirus replication is in order.

Table I. Spectrum of Antiviral Activity of Guanidine

Sensitive to Guanidine	Resistant to Guanidine
Poliovirus 1	Echovirus 22
Poliovirus 2	Echovirus 23
Poliovirus 3	Echovirus 28
Echovirus 1	Rhinovirus B632
Echovirus 5	Apthovirus A[1]
Coxsackievirus A7	Apthovirus C[1]
Coxsackievirus A11	Apthovirus O[1]
Coxsackievirus A13	Reovirus
Coxsackievirus A16	Influenza A2
Coxsackievirus A18	Parainfluenza 1
Coxsackievirus B3	Parainfluenza SV5
Coxsackievirus B5	Parainfluenza 3
Rhinovirus HGP	Measles
Apthovirus SAT 1	Adenovirus Types 1-7
Apthovirus SAT 2	Togaviruses[2]
Apthovirus SAT 3	Tobacco Necrosis[3]
Apthovirus Asia 1	Tobacco Moxaic[3]

1 Sensitive at high concentrations (6 mM).

2 Sindbis and Semliki Forest viruses sensitive at high concentrations (10 mM).

3 Sensitive at 10-50 mM.

References: 4-7, 9-11

Strategy of Picornavirus Replication

The initial interaction of poliovirus with susceptible cells involves the specific attachment of the virus to lipoprotein receptors (12,13,14), probably by interaction with VP1. There follows a conformational change which results in loss of VP4 and exposure of VP3 on the surface of the particle (15,16,17). The particle also becomes lipophilic which may enhance its interaction with the cell membrane (17). Although still controversial, most evidence indicates that the virus is taken into the cell by pinocytosis (18), after which it liberates the viral RNA into the cytoplasm where it functions both as mRNA for the synthesis of viral proteins and template for replication of the viral RNA. Like most cellular mRNA's, poliovirus RNA contains a 3' poly A tract (19), but poliovirus RNA differs from cellular mRNA in that it does not contain an inverted cap structure on the 5' end (20). The virion RNA contains a genome-linked protein with a phosphodiester bond linking the terminal U to a tyrosine residue on the 2,200 molecular weight protein, VPg (20). When isolated from polyribosomes, the viral RNA is freed of the VPg and contains a free pUp (21), this cleavage presumably carried out by a cellular enzyme (22). The lack of the cap structure is significant to poliovirus in that the virus rapidly inhibits cellular protein synthesis, the mechanism of which appears to be by inactivation of a cellular factor, the 34,000 dalton cap binding protein, required for translation of capped mRNAs (22a). Interestingly, the cardioviruses which also contain VPg do not inactivate the cap-binding protein and appear to inhibit cellular protein synthesis by competition with cellular mRNA for available ribosomes (23).

The most unique feature of picornaviral protein synthesis is that there is only a single initiation site on the viral mRNA (24). The genome is translated into a primary polyprotein whose size, 247,000 daltons, roughly corresponds to the entire coding capacity of the genome (25). The primary product, cleaved as a nascent peptide, is not seen except in the presence of amino acid analogs or inhibitors of proteolytic enzymes (26,27). The three primary cleavage products are found in infected cells and correspond to the precursors of the capsid proteins, viral proteases, and viral polymerase (28,29). Although the cleavage seems to be at least partially mediated by cellular proteases (27), there is evidence for intramolecular cleavage of the protease precursor (30), suggesting that the viral protein may be entirely capable of self cleavage.

The polymerase region contains the sequences for VPg and this suggests that cleavage of this protein may be involved in the process of initiation of RNA synthesis (31). It has long been known that a viral RNA-dependent RNA polymerase could be isolated from infected cells (32) as a complex of cellular membranes, viral RNA, and viral and cellular proteins (33-36). Early preparations were template-independent, capable merely of completing chains initiated in vivo (37). Purified preparations consisting of NCVP 4b and NCVP 7c are template-dependent, but require primers such as oligo U (38,39). This suggests that VPg acts as a primer in vivo and that its cleavage from the NH terminal end of NCVP 7c may constitute the initiation event.

There are three species of viral RNA found in infected cells. A 35S single-stranded RNA corresponds to the viral genome. A partially double-stranded form with

4-7 nascent single strands called the replicative intermediate (RI) is most likely the precursor of the ssRNA (40,41). It sediments heterogeneously from 18-60S. Also found in infected cells is a completely double-stranded 18S replicative form (RF) whose role is unclear and may represent a dead end by-product of virus replication (42). All of the structures contain VPg except that portion of the 35S RNA which is found engaged in protein synthesis (43,44).

The replication of poliovirus occurs in association with cytoplasmic membranes (34). At the time that the bulk of viral RNA and protein synthesis become detectable biochemically, electron microscopic observation reveals marked proliferation of smooth cytoplasmic membranes in the perinuclear region of infected cells (45). These structures are associated with viral RNA synthesis and can be separated by sucrose gradient centrifugation from the rough membranes which are associated with viral protein synthesis (34).

The observed membrane proliferation is associated with a marked incorporation of $^{32}PO_4$ and ^{3}H-choline, indicating that the membrane lipids are newly synthesized (46). The chemical content of the newly synthesized membranes differs as the phospholipid: protein ratio nearly doubles during the course of an infection (47). The lipid content differs substantially, there being a reduced amount of sphingomyelin in the newly synthesized membranes (48). There is an increase in both lipid and protein content which occurs, despite the fact that cellular protein synthesis is greatly inhibited, suggesting that the proteins of the newly synthesized membranes are either entirely viral or pre-existing cellular proteins (48).

The association of viral replication with membranes appears to be critical for proper action of the viral RNA replication complex, as lysis of the membranes of crude replication complexes renders the bulk of the RNA product double stranded (49). Tightly associated to the RNA replication complex are the polypeptides which are precursors of the structural proteins in the virion (36). The exact process of assembly of virions remains unclear, but appears to involve the continued association of the primary cleavage product which is the structural protein precursor with smooth membranes and its cleavage to the 5S complex of VPO, VP1, VP3 (50,51). This complex appears to assemble into a 14S pentamer, which has been shown to assemble to a 74S empty capsid _in vitro_ (52). This 74S empty capsid may not represent the naturally occurring precursor of mature virions, as different 74S structures have been obtained when the assembly reaction is performed in the presence and absence of cytoplasmic extracts of infected cells (53). Using inhibitors of viral replication, the 74S procapsids have been chased into mature virions (54), but it is not clear that this is the natural route of assembly. Mature 150S virions have not been formed _in vitro_ from empty capsids and free RNA.

Polovirus Effects on Host Cells

Picornavirus infection results in severe cytopathic effects in infected cells (55,56). Included are extensive morphological changes such as cell rounding, cytoplasmic and nuclear alterations, and chromosomal aberrations. These are thought to be caused by newly synthesized viral proteins as inhibitors of protein synthesis prevent such changes (57), although it has also been suggested that double-stranded RNA can cause cell death (58). There is evidence, however, that double-stranded RNA may not be the ultimate cause of the poliovirus-induced cytopathic effects. Synthetic double-stranded RNAs can also produce cytopathic effects non-specifically and the quantity of double-stranded RNA necessary to produce cytopathic effects in vitro is not synthesized in vivo until considerably after cytopathic effects have begun.

Among the biochemical alterations seen in infected cells, and apart from the stimulation of membrane biosynthesis described above, are inhibitions of cellular DNA, RNA, and protein synthesis (57,59,60,61). The latter is by far the more severe, and it has been suggested that the inhibitions of DNA and RNA are consequences of the inhibition of protein synthesis. It has also been suggested that double-stranded RNA may be involved in the inhibition of protein synthesis (62) but, as is the case for double-stranded RNA-induced morphological lesions, levels of such RNA which are effective in vitro are not reached in vivo until long after the inhibition is complete. It is clear that the ultimate effect with poliovirus-infected cells is the inactivation of the 34,000 dalton cellular cap binding protein required for initiation of translation to capped mRNAs (22a). Curiously, the cardioviruses inhibit protein synthesis in infected cells as well or better than the enteroviruses, but do so without inactivating the cap binding protein (23). It has therefore been suggested that these viruses inhibit cellular protein synthesis by virtue of a higher binding affinity for ribosomes than the cellular mRNAs.

INHIBITION OF VIRUS REPLICATION BY GUANIDINE

The viruses which are sensitive to guanidine do not appear to be directly inactivated by the drug, as prolonged incubation in the presence of guanidine is without effect (2,4). One of the most useful tools in the study of the mechanism of action of guanidine has been virus variants which are either resistant to, or dependent on the presence of the drug (63,64). From studies of these variants, as well as the sensitive virus, it is clear that the inhibitory effect on sensitive virus, as well as the required presence for dependent virus, does not occur for the first hour of infection; that is, during absorption, penetration, or uncoating (65). The effects of the drug are manifest starting at 1-2 h post-infection or mid-way through the latent period, and continue through most of the replicative cycle (66).

Most available evidence indicates that the primary effect of guanidine is the inhibition of viral RNA synthesis. During the exponential phase of viral growth, the addition of guanidine results in a rapid shutdown of viral RNA synthesis (67,68). Protein synthesis is inhibited much more slowly, most likely as a consequence of the inhibition of RNA synthesis. Empty capsids, devoid of RNA, accumulate in the cytoplasm of infected cells and upon removal of the drug, RNA synthesis resumes and the empty capsids are chased into mature virions (54). Another indication that the stage of viral replication affected is RNA synthesis is the observation that the development of UV resistance by infective centers, which requires RNA synthesis, is also inhibited by guanidine (69). In addition, the onset of RNA synthesis, which can be extrapolated back to about 1 h after infection, corresponds temporally to the time of development of guanidine sensitivity and the requirement for the drug by dependent virus.

Guanidine inhibits other virus-induced processes such as choline incorporation (70), virion morphogenesis (54), and inhibition of cellular protein synthesis under certain conditions (57). It is possible, however, to reconcile these effects with a primary lesion in viral RNA synthesis (49). Choline incorporation into membranes is tightly related to viral RNA synthesis by the fact that the poliovirus RNA replication complex is membrane bound and dissociation of the membrane component from replication complexes in vitro results in aberrant RNA synthesis. In addition, choline itself is an antagonist of guanidine action (66), a phenomenon to be dealt with below.

If virion morphogenesis takes place via association of capsid subunits or procapsids with viral RNA, then a primary effect on viral RNA synthesis would predict a secondary effect on virion morphogenesis and an accumulation of capsid subunits or empty capsids, both of which have been demonstrated experimentally (54).

An effect of guanidine on inhibition of cellular protein synthesis is more difficult to explain. Experimentally, at multiplicities of infection of less than 10 PFU/cell, such an effect can be demonstrated (71). At higher multiplicities there is inhibition of protein synthesis (57,71), but no viral protein synthesis can be detected. Since at least part of the evidence for the hypothesis that viral protein synthesis is responsible for the inhibition of cellular protein synthesis comes from studies with guanidine, one must be careful to avoid circular reasoning. However, if one assumes that the input viral genome is translated in the presence of guanidine, but only a limited number of times (so as to preclude easy detection of the product even at relatively high multiplicities), then it is possible that the small quantities of viral protein synthesized may be capable of inhibiting cellular protein synthesis. The explanation for the apparent lack of viral protein synthesis is that the bulk of such products are translated from progeny genomes, the synthesis of which would be prevented if guanidine inhibits viral RNA synthesis. At low multiplicities of infection, the number of viral genomes present would be insufficient to produce enough viral proteins to result in an inhibition of cellular protein synthesis.

It has also been suggested that the inhibition of cellular protein synthesis in picornavirus-infected cells is the result of membrane perturbations resulting in an influx of sodium (72). Since guanidine inhibits this observed influx and since choline, which has been used as a substitute for sodium in transport studies, blocks guanidine action, it was suggested that the effect of guanidine was an effect on cation-dependent phenomena (73). However, this is difficult to reconcile with the observation that guanidine prevents the influx of sodium, yet cellular protein synthesis can still become inhibited in the presence of guanidine. This suggests that the inhibition of cellular protein synthesis is not dependent on sodium influx.

Thus, the available evidence suggests that the major effect of guanidine is to inhibit viral RNA synthesis. A number of studies have suggested, by indirect means, that the initiation of new nucleotide chains is the point at which RNA synthesis is blocked (67,74), as studies in vitro with RNA polymerase preparations, which fail to initiate the synthesis of new RNA molecules, are insensitive to guanidine (32). In those preparations in which initiation is presumed to occur, the synthesis of complete 35S molecules is most sensitive to the drug (75). Interestingly, one observation which is unexplained by this hypothesis is that in guanidine-treated, infected cells, the RNA polymerase complexes are larger, which suggests a defect in the release of newly synthesized chains (76).

Since initiation of RNA synthesis is apparently the site of action of guanidine, considerable attention has been focused on the identification of the RNA polymerase and the identification of the sensitive moiety. But, by far, the more difficult observation to reconcile with the site of action of guanidine being RNA synthesis is the genetic evidence that in guanidine sensitivity, locus resides in the capsid protein "cistron" (77), rather than the polymerase region. The alterations in guanidine sensitivity in drug-resistant and -dependent variants have been correlated with changes in the biochemical properties of the capsid polypeptides, although interestingly not with a single protein (78). This relationship between capsid proteins and guanidine sensitivity has also been suggested by several other observations. Viral polypeptides are synthesized on rough membranes and the beginning stages of proposed virion assembly, the formation of procapsids, appears to occur there. Viral RNA synthesis occurs on smooth membranes. The intracellular locations of the viral subunits (34) suggests the need for transport of the viral protein and RNA synthesizing structures, and it has been suggested that the cytoskeleton plays a role in such transport (79). One of the effects of guanidine is to inhibit the association of viral RNA and procapsid (80), which may suggest that viral RNA synthesis requires the presence of viral capsid protein.

The primary sequence of poliovirus type 1 has recently been elucidated (25) and an interpretation of the sequence, the deduced genetic map, and the known features of poliovirus replication may allow some clarification of this paradox.

The nucleotide sequences which specify VPg are located in the polymerase "cistron" adjacent to the amino terminus of the polypeptide which has been purified as the template and primer-dependent RNA polymerase (31). Since during their synthesis, at least, all RNA molecules appear to contain VPg (43,44), it is logical to assume that VPg is necessary for initiation of RNA synthesis. We would suggest that it may be that VPg, which has never been isolated free in the cell may, in fact, never exist unattached to either RNA or its precursor protein. The initiation step in RNA synthesis may be cleavage of VPg, which remains enzyme bound, and transfer of the terminal uridine residue by the polymerase and subsequent elongation by the cleaved polymerase product. While this does require that one polymerase precursor molecule be expended for each RNA molecule generated, it could explain the failure of the cleaved product to initiate in vitro, and the observation that only crude cytoplasmic extracts (which presumably have polymerase precursor and protease) have been observed to initiate RNA synthesis in vitro (75).

Such a hypothesis might also explain the relationship of the capsid proteins to RNA synthesis. Since the polymerase region of the genome is located at the 3' end of the mRNA, in order to translate VPg and the polymerase molecule the capsid proteins must have been translated first. In addition, since the poly-protein has been suggested to be capable of intramolecular cleavage, its secondary structure may be influenced by the capsid protein region which may, in turn, influence the polymerase region. If the capsid proteins are altered by their interaction with guanidine, the effect on the protease and polymerase may be due to alterations in the secondary structure of the nascent polyprotein. This is suggested by the observation that apthovirus variants resistant to guanidine have altered nonstructural P34 (81), which is analogous to NCVP X of poliovirus, the protease. It is possible that alterations in the proteins upstream of the polymerase region may affect the secondary structure if their interaction with guanidine results in a sufficient conformational change.

This hypothesis may also explain what had been an additional inconsistency in the suggestion that the capsid proteins were involved in guanidine sensitivity. The defective interfering particles of poliovirus have equal sensitivity of their RNA synthesis to guanidine, despite their loss of varying portions of the capsid protein region of the genome (82). Most such particles produce an unstable precursor of the capsid proteins (82), and the fact that it is unstable may preclude any effect on the secondary structure of the protease and polymerase region, which may be necessary to confer resistance. A guanidine-resistant defective interfering particle has recently been described and it is of interest that it induces the synthesis of two stable fragments of the capsid region (83). The secondary structure of these fragments may protect the remaining portion of the polyprotein product from the effects of the drug.

It should be emphasized that it is only speculation that the effects of guanidine are related to it's ability to induce conformational changes in proteins. It is clear that this capacity exists at molar concentrations, as these changes are readily observed. However, one would have to conclude that at millimolar concentrations, some very specific viral proteins are subject to conformational changes, while the vast majority of cellular (and other viral) proteins are not susceptible.

There is some experimental support for this contention. Guanidine reduces the optimum temperature for the growth of apthovirus by 3-5°C and at suboptimal temperatures enhances the growth of some virus strains by as much as 10-fold (84). In the same study, it was indicated that D_2O would reverse the effects of guanidine, suggesting the importance of protein conformation.

Guanidine Antagonists

In an attempt to deduce the possible mode of action of guanidine, compounds were screened for virus inhibitory activity as it relates to structure. Several other compounds including arginine, ornithine, citrulline, and carbazide were shown to have virus inhibitory activities (85). These compounds seem to share the same active site as guanidine, as guanidine-resistant mutants are also less sensitive to these compounds. However, during the course of these studies, several compounds were identified which could reverse the action of guanidine. Such compounds include choline, methionine, valine, leucine, threonine, ethanolamine, and dimethylpropanolamine (85). These compounds seem to antagonize guanidine with drug-sensitive virus (66) and fail to replace guanidine as a necessary metabolite for drug-dependent virus (86). In fact, they blocked the action of guanidine with drug-dependent virus. Since no clear picture of the mechanism of guanidine action has emerged, it is difficult to speculate on the mode of action of compounds which antagonize it; however, a few pertinent observations can be made. Choline and dimethylethanolamine act additively with the aforementioned amino acids. This would suggest that whatever the site of action of guanidine, at least two sites exist which can block it (66).

It appears unlikely that the guanidine antagonizing agents act by a specific interaction with the drug via an irreversible modification or inactivation of the guanidine molecule (86,87). Although it might be noted, in passing, that the majority of substituted guanidines which have been examined have lower activity than the parent compound (74). The more plausible explanations would suggest that these antagonists interact loosely with the guanidine molecule and thus reduce the interaction of the drug with the viral target. Alternatively, assuming that the virus inhibitory activity of guanidine is due to an induced conformational change in the viral proteins, that compounds which reverse the action of guanidine might reverse the conformational changes. However, this would suggest that the guanidine antagonists ought to induce conformational changes on their own, a phenomenon which does not appear to be a common property.

One of the most important by-products of the studies on antiguanidines has been an explanation for some of the variation between different viruses in their sensitivities to guanidine. At least some of these differences may be explainable by differences in the media used by different investigators. Varying amounts of arginine, which mimics guanidine, and methionine, valine, leucine, and threonine, which block guanidine, may dramatically affect the amount of guanidine necessary to inhibit the growth of a particular virus (8,88).

CONCLUSION

In the 20 years since the phenomenon of guanidine inhibition of poliovirus replication was first described, our understanding of the replication of the picornaviruses has increased dramatically. Yet the mechanism of action of guanidine remains an enigma.

The existing evidence suggests that the known denaturant properties of guanidine at molar concentrations might also be active on a limited number of very specific substrates at millimolar concentrations. It is clear that the major effect of guanidine appears to be the inhibition of viral RNA synthesis, yet the genetic and biochemical evidence suggests that its major effect is on the capsid proteins. In order to accommodate this, it is necessary to conclude that as the nascent polyprotein is synthesized, proper secondary structure of the capsid protein region on the NH end is necessary for proper function of the polymerase region on the COOH end, and that guanidine specifically alters this secondary structure. This is a reasonable possibility, especially if the polyprotein molecule is capable of intramolecular cleavage and if the cleavage of VPg from the polymerase molecule is, in fact, the initiation event in viral RNA synthesis. Verification of this hypothesis must await clarification of these steps in the replication process and the demonstration that guanidine is, indeed, capable of inducing conformational changes at millimolar concentrations.

ACKNOWLEDGEMENT

The authors are deeply grateful to Kathleen Cavanagh for her patience and expert typing.

REFERENCES

1. Rightsel, W.A., Dice, J.R., McAlpine, R.J., Timm, E.A., McLean, I.W. Jr., Dixon, G.J., and Schabel, F.M. Jr. Science 134:558-559, 1961.
2. Crowther, D. and Melnick, J.L. Virology 15:65-74, 1961.
3. Veda, T., Toyoshima, S., Tsuti, T., Seto, Y., and Nomoto, J. Keio J. Med. 10:257-265, 1961.
4. Loddo, B., Ferrari, W., Brotzu, G., and Spanedda, A. Nature 193:97-98, 1962.
5. Tamm, I. and Eggers, H.J. Virology 18:439-447, 1962.
6. Caliguiri, L.A. and Tamm, I. In: International Encyclopedia of Pharmacology and Therapeutics (Ed. D.J. Bauer), Pergamon Press, Oxford and N.Y., 1972, pp. 181-230.
7. Pringle, C.R. Nature 204:1012-1012, 1964.
8. Dinter, Z. and Bengtsson, S. Virology 24:254-261, 1964.
9. Friedman, R.M. J. Virol. 6:628-636, 1970.
10. Varma, J.P. Virology 36:305-308, 1968.
11. Dawson, W.O. Intervirology 6:83-89, 1975.
12. McLaren, L.C., Holland, J.J., and Syverton, J.T. J. Exp. Med. 109:475-486, 1959.
13. Holland, J.J. and McLaren, L.C. J. Exp. Med. 109:187, 1959.
14. Holland, J.J. and McLaren, L.C. J. Exp. Med. 114:161-171, 1961.
15. Crowell, R.L. and Philipson, L. J. Virol. 8:509-515, 1971.
16. Lonberg-Holm, K. and Korant, B.D. J. Virol. 9:29-40, 1972.
17. Rueckert, R.R. In: Comprehensive Virology (Eds. H. Fraenkel-Conrat and R.R. Wagner), Vol. 6, Plenum Press, N.Y., 1976, pp. 131-213.
18. Mandel, B. Symp. Quant. Biol. 27:123-131, 1962.
19. Spector, D.H. and Baltimore, D. Proc. Nat'l. Acad. Sci. USA 71:2983-2987, 1974.
20. Lee, Y.F., Nomoto, A., Detjen, B.M., and Wimmer, E. Proc. Nat'l. Acad. Sci. 74:59-63, 1977.
21. Hewlett, M.J., Rose, J.K., and Baltimore, D. Proc. Nat'l. Acad. Sci. USA 73:327-330, 1976.
22. Ambros, V., Petterson, R.F., and Baltimore, D. Cell 15:1439-1446, 1978.
22a. Hansen, J., Etchison, D., Hershey, J.W.B., and Ehrenfeld, E. J. Virol. 42:200-207, 1982.
23. Jen, G., Detjen, B.M., and Thach, R.E. J. Virol. 35:150-156, 1980.
24. Kozak, M. Cell 15:1109-1123, 1978.
25. Kitamura, N., Semler, B.L., Rothberg, P.G., Larsen, G.R., Adler, C.J., Dorner, A.J., Emini, E.A., Hanecak, R., Lee, J.J., van der Werf, S., Anderson, C.W., and Wimmer, E. Nature 291:547-553, 1981.
26. Jacobson, M.F., Asso, J., and Baltimore, D. J. Mol. Biol. 49:657-669, 1970.
27. Korant, B. J. Virol. 10:751-759, 1972.
28. Summers, D.R. and Maizel, J.V. Jr. Proc. Nat'l. Acad. Sci. 68:2852-2856, 1971.
29. Taber, R., Rekosh, D., and Baltimore, D. J. Virol. 8:395-401, 1971.
30. Palmenberg, A.C. and Rueckert, R.R. J. Virol. 41:244-249, 1982.
31. Pallansch, M.A., Kew, O.M., Palmenberg, A.C., Golini, F., Wimmer, E., and Rueckert, R.R. J. Virol. 35:414-419, 1980.
32. Baltimore, D., Eggers, H.J., Franklin, R.M., and Tamm, I. Proc. Nat'l. Acad. Sci. 49:843-849, 1963.
33. Caliguiri, L.A. and Tamm, I. Science 166:885-886, 1969.
34. Caliguiri, L.A. and Tamm, I. Virology 42:100-111, 1970.

35. Caliguiri, L.A. and Tamm, I. Virology 42:112-122, 1970.
36. Caliguiri, L.A. and Mosser, A.G. Virology 46:375-386, 1971.
37. Girard, M. J. Virol. 3:376-384, 1969.
38. Flanegan, H.B. and Baltimore, D. Proc. Nat'l. Acad. Sci. USA 74:3677-3680, 1977.
39. Flanegan, J. and Van dyke, T. J. Virol. 32:155-166, 1980.
40. Baltimore, D. and Girard, M. Proc. Nat'l. Acad. Sci. 56:741-748, 1966.
41. Baltimore, D. J. Mol. Biol. 32:359-368, 1968.
42. Bishop, M.J. and Koch, G. J. Biol. Chem. 242:1736-1743, 1967.
43. Nomoto, A., Detjen, B., Pozzatti, R., and Wimmer, E. Nature 268:208-213, 1977.
44. Nomoto, A., Kitamura, N., Golini, F., and Wimmer, E. Proc. Nat'l. Acad. Sci. USA 74:5345-5349, 1977.
45. Dales, S., Eggers, H.J., Tamm, I., and Palade, G.E. Virology 26:379-389, 1965.
46. Penman, S. Virology 25:148-152, 1965.
47. Mosser, A.G., Caliguiri, L.A., and Tamm, I. Virology 47:39-47, 1972.
48. Mosser, A.G., Caliguiri, L.A., Scheid, A.S., and Tamm, I. Virology 47:30-38, 1972.
49. Plagemann, P.G.W. and Swim, H.E. J. Mol. Biol. 35:13-35, 1968.
50. Jacobson, M.F. and Baltimore, D. Proc. Nat'l. Acad. Sci. 61:77-84, 1968.
51. Phillips, B.A., Summers, D.F., and Maizel, J.V. Jr. Virology 35:216-226, 1968.
52. Phillips, B.A. Virology 44:307-316, 1971.
53. Phillips, B.A. and Wiemert, S. Virology 88:92-104, 1978.
54. Jacobson, M.F. and Baltimore, D. J. Mol. Biol. 33:369-378, 1968.
55. Bablanian, R. Progr. Med. Virol. 19:40-83, 1975.
56. Schrom, M. and Bablanian, R. Arch. Virol. 70:173-187, 1981.
57. Bablanian, R., Eggers, H.J., and Tamm, I. Virology 26:100-113, 1965.
58. Cordell-Stewart, B. and Taylor, M.W. Proc. Nat'l. Acad. Sci. 68:1326-1330.
59. Ensminger, W.D. and Tamm, I. Virology 40:152-165, 1970.
60. Hand, R. and Tamm, I. Virology 47:331-337, 1972.
61. McCormick, W. and Penman, S. Virology 31:135-141, 1967.
62. Cordell-Stewart, B. and Taylor, M.W. J. Virol. 11:232-237, 1973.
63. Loddo, B., Ferrari, W., Spanedda, A., and Brotzu, G. Experentia 18:518-519, 1962.
64. Nakano, M., Iwami, S., and Tagava, I. Virology 21:264-266, 1963.
65. Eggers, H.J., Reich, E., and Tamm, I. Proc. Nat'l. Acad. Sci. USA 50:183-190, 1963.
66. Mosser, A.G., Caliguiri, L.A., and Tamm, I. Virology 45:653-663, 1971.
67. Caliguiri, L.A. and Tamm, I. Virology 35:408-417, 1968.
68. Noble, J. and Levintow, L. Virology 40:634-642, 1970.
69. Eggers, H.J., Ikegami, N., and Tamm, I. Virology 25:475-478, 1965.
70. Penman, S. and Summers, D. Virology 27:614-620, 1965.
71. Holland, J.J. J. Mol. Biol. 8:574-581, 1964.
72. Carrasco, L. and Smith, A.E. Nature 264:807-809, 1976.
73. Nair, C.N., Stowers, J.W., and Singfield, B. J. Virol. 31:184-189, 1979.
74. Caliguiri, L.A. and Tamm, I. In: Selective Inhibitors of Viral Function (Ed. W. Carter) Chemical Rubber Co., Cleveland, Ohio, 1973, pp. 257-293.
75. Tershak, D.R. J. Virol. 41:313-318, 1982.
76. Huang, A.S. and Baltimore, D. J. Mol. Biol. 47:275-291, 1970.
77. Cooper, P.D., Wentworth, B.B., and McCahon, D. Virology 40:486-493, 1970.
78. Korant, B.D. Virology 81:25-36, 1977.
79. Lenk, R. and Penman, S. Cell 16:289-301, 1979.
80. Yin, F.H. Virology 82:299-307, 1977.
81. Saunders, K. and King, A.M.Q. J. Virol. 42:389-394, 1982.
82. Cole, C.N. and Baltimore, D. J. Mol. Biol. 76:325-343, 1973.
83. Tershak, D.R. J. Virol. 41:615-625, 1982.

84. Nick, H. and Ahl, R. Arch. Virol. 57:71-83, 1976.
85. Lwoff, A. Biochem. J. 96:289-301, 1965.
86. Philipson, L. Bengtsson, S., and Dinter, Z. Virology 29:317-329, 1966.
87. Lwoff, A. and Lwoff, M. Compt. Rend. Acac. Sci. 260:4116-4118, 1965.
88. Lwoff, A. and Lwoff, M. Compt. Rend. Acad. Sci. 259:949-952, 1964.

17

MECHANISM OF ACTION OF 2-(α-HYDROXYBENZYL)-BENZIMIDAZOLE (HBB)

HANS J. EGGERS
Institut für Virologie der Universität zu Köln, Fürst-Pückler-Str. 56, D-5000 Köln 41, Germany

SUMMARY

2-(α-Hydroxybenzyl)-benzimidazole (HBB) selectively inhibits many members of the picornavirus group. It does not affect members of other virus families. The compound leaves unimpaired vital functions of the cell including cellular RNA and protein synthesis as well as cell multiplication. The antiviral activity of HBB can be demonstrated in cell culture and in virus-infected animals.

HBB does not interfere with early virus-cell interactions. It specificly inhibits viral RNA synthesis, whereas inhibition of viral polypeptide synthesis appears to be a secondary effect due to inhibition of viral RNA synthesis.

New evidence is presented to support the hypothesis that inhibition of viral RNA synthesis is mediated by inhibition of initiation of RNA synthesis at the viral RNA polymerase.

INTRODUCTION

Since 1952 the antiviral activity of benzimidazoles has been studied extensively (1). A major stimulus during this work has been the central finding that virus-inhibiting activity and cell toxicity of chemical derivatives of benzimidazoles may vary independently (2). The introduction of 2-(α-hydroxybenzyl)-benzimidazole (HBB) (Fig. 1) as an antiviral agent marked a most important step of the investigations on selective inhibition of viral replication (3,4). It was found that HBB inhibits multiplication of poliovirus 2 in monkey kidney cell culture at concentrations which caused no microscopic changes in the cells. On the other hand, influenza virus multiplication remained unaffected (4).

Y. Becker (ed.), Antiviral Drugs and Interferon: The Molecular Basis of Their Activity.

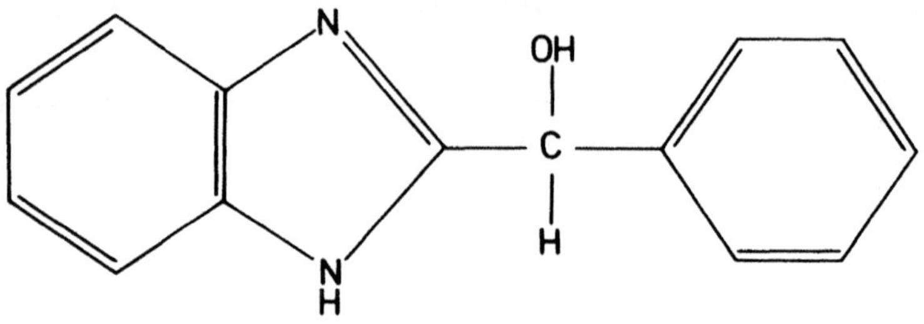

FIGURE 1. 2-(α-Hydroxybenzyl)-benzimidazole (HBB).

During the last two decades a wealth of information has been accumulated on the antiviral action of HBB and some of its chemical derivatives. It is now well established that, contrary to common belief some 20 years ago, virus multiplication can be selectively inhibited by low molecular weight compounds. In the present discussion I shall focus on investigations on the mechanism of action of HBB.

ANTIVIRAL ACTIVITY OF HBB IN CELL CULTURE

Virus-inhibitory spectrum and effects of HBB on uninfected cells

The antiviral spectrum of HBB exhibits a remarkable pattern: at concentrations of HBB nontoxic to cells, viz. 220 µM, the multiplication of many members of the genus Enterovirus is being inhibited, whereas members of other virus families are HBB-insusceptible (5,6). Details concerning the antiviral spectrum of HBB can be found in previous reviews (7,8). Suffice it to point out here that the pattern of HBB susceptibility or insusceptibility even within the picornavirus family follows almost exclusively taxonomic characteristics (8,9).

As to be expected for any inhibitor, the susceptibility of various picornaviruses to HBB varies considerably (5,6). However, these differences appear to be dependent on the virus. So far, no example of varying susceptibility of picornaviruses due to the host cell is known.

As indicated above, concentrations of HBB highly inhibitory

to susceptible enteroviruses (220 µM) did not affect the virus yields of insusceptible viruses. This finding strongly suggested that essential metabolic pathways of cells might not be affected by HBB. This conclusion has been confirmed in further investigations on the effects of HBB on metabolic reactions of uninfected primary monkey kidney cells. Concentrations of up to 493 µM HBB caused no effects on the following metabolic activities (up to 3 hr): cumulative O_2 uptake, glucose utilization, lactic acid production, ^{14}C-adenosine uptake into RNA and ^{14}C-L-alanine uptake into proteins (5). At 219 µM HBB, the ^{3}H-uridine uptake into RNA was also unaffected during the duration of the experiment (6.5 hr) (10). In addition, extensive morphological observations of HBB-treated cells (219 µM) demonstrated no significant differences from untreated cultures during the duration of the experiment (5 to 7 days). Minimal morphological changes were seen at 493 µM concentrations, but even at 876 µM concentrations morphological changes were only moderate (5,11, 12,13). Cell multiplication - a broad and sensitive indication of cellular metabolism - was also found to be unaffected in the presence of 219 µM HBB. This has been demonstrated for primary monkey kidney cells as well as HeLa cells (5,10).

All of these data substantiate the thesis that HBB selectively inhibits members of the picornavirus family.

Effects of HBB on the viral replication cycle

The time course of the HBB-sensitive process on the viral multiplication cycle has been determined by removing or adding the compound at various times after virus infection. Detailed studies with echovirus 12 and coxsackievirus A9 replicating in monkey kidney cell cultures under single cycle conditions have revealed that no HBB-inhibitable process takes place during the first 2 1/2 hr of the latent period. The HBB-inhibitable period begins during the last 1 1/2 hr of the latent phase and extends until the latest rapid increase phase of the virus (10,14,15). Once the HBB-inhibitable period has begun, addition of the compound has an immediate inhibitory effect on the production of infectious virus: latest by 15 min, infectious virus production has come to a complete stop. These observations made in mass culture hold true also for individual cells: HBB stops ongoing virus multiplication (16).

The described kinetics of the antiviral action of HBB revealed clearly that early steps of virus-cell interaction might not be affected by the compound. In direct experiments this has been shown to be the case. Virus adsorption, penetration, and uncoating remained unaffected by the presence of HBB (10, 17,18).

Since picornaviral RNA and protein synthesis are continuing in the untreated, infected cells throughout the rapid increase phase of virus production (19), the kinetics of the HBB-inhibitable process suggested an effect on picornavirus macromolecular synthesis. This turned out to be the case: synthesis of infectious viral RNA was demonstrated to be inhibited by HBB (10). Inhibition of synthesis of viral RNA by HBB has also been demonstrated through measurements of the rates of incorporation of ^{3}H-uridine into virus-specific RNA in actinomycin-treated cells (20). This finding excludes the theoretical possibility that HBB might in some way inhibit only the acquisition of infectivity of viral RNA which has already been synthesized.

It should be pointed out here that HBB could never be shown to have any direct inactivating effect on the infectivity of HBB-susceptible virions or their RNA (5,10).

As indicated above, HBB exhibited an immediate effect on ongoing synthesis of infectious virus (see also Fig. 3). Similar observations were made concerning synthesis of viral RNA. Within 30 min after addition of 200 μM HBB to cells infected with echovirus 12 during various times of the rapid increase phase of virus, no viral RNA synthesis was detectable in ^{3}H-uridine pulses (Fig. 2).

Stop of viral RNA synthesis eventually stops synthesis of viral polypeptides. It was possible, however, to demonstrate a differential effect of HBB on viral protein and virion synthesis by use of the hemagglutinating echovirus 12. It could be shown that the increase in virus hemagglutinin for a period of ca. 1 to 1 1/2 hr after addition of HBB was exclusively due to formation of empty capsids, whereas synthesis of the RNA-

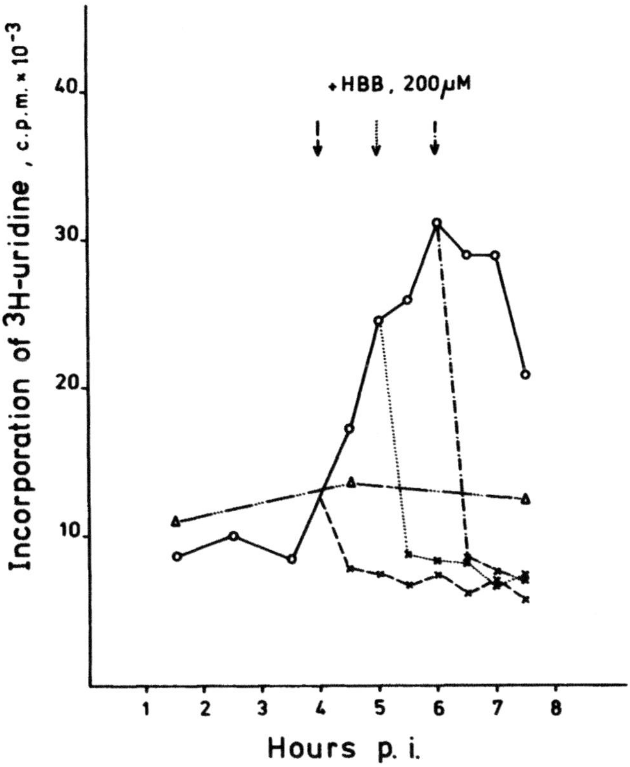

FIGURE 2. Rapid stop of echovirus 12 RNA synthesis in actinomycin-treated GMK cells by 200 µM D-HBB·HCl at various times during the exponential increase period of the virus. Curves are indicated by: circles (infected); triangles (uninfected); crosses indicate addition of HBB at the following times: dashed curve (4 hr); dotted curve (5 hr); dashed-dotted curve (6 hr). B. Rosenwirth and H.J. Eggers, unpublished work 1976.

containing infectious virus had come to the above-mentioned rapid stop (Fig. 3) (10,15). In addition it could be shown that the empty capsids produced after addition of HBB were comprised of polypeptides which had been synthesized after addition of the compound (15). This experiment clearly proves that HBB does not directly inhibit synthesis of viral capsid proteins.

The evidence so far presented suggests that at least some aspects of the virus-inhibitory activity of HBB can be explained by its inhibiting effects on viral RNA synthesis. The fact that viral RNA synthesis was strongly inhibited by HBB, whereas

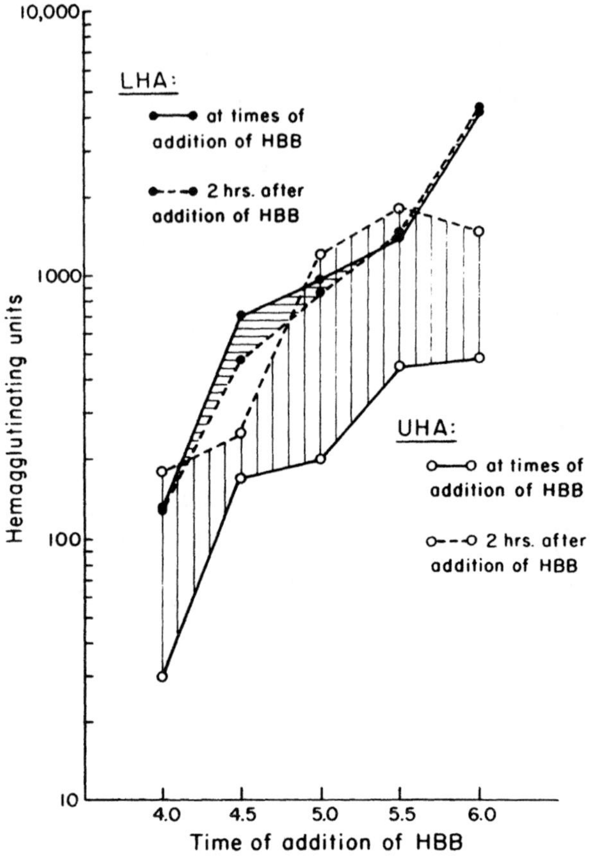

FIGURE 3. Changes in UHA (upper hemagglutinin: empty capsids) and LHA (lower hemagglutinin: complete infectious virus) in a 2 hr period of incubation with 219 μM HBB, added at various times in the viral replication cycle. Curves are indicated by: full circles (LHA); open circles (UHA); full curves (at time of addition of HBB); broken curves (2 hr after addition of HBB). According to Halperen et al. (15).

cellular RNA synthesis remained unimpaired by the compound - the first demonstration of specific chemical inhibition of the synthesis of viral nucleic acid - led, besides other considerations, to a search for a virus-induced RNA polymerase. Such a virus-induced RNA polymerase activity was, in fact, detected in picornavirus-infected cells (21,22,23). This finding prompted a study on the effects of HBB on the appearance in enterovirus-infected cells of the virus RNA polymerase activity. Addition of HBB to poliovirus-infected cells during the early rapid increase phase of virus multiplication caused a decrease in viral polymerase activity to almost background level, as demonstrated in an assay in cell-free extracts. On the other hand, the compound appeared to have no effect on the activity of the viral enzyme preparation when added to the cell-free RNA polymerase assay (22).

At the time, these findings have been interpreted as follows: HBB inhibits the synthesis of a virus-directed RNA polymerase, the enzyme responsible for the replication of viral RNA. With inhibition of synthesis of the RNA polymerase the replication of viral RNA also comes to a stop which, in turn, finally leads to inhibition of synthesis of viral capsid polypeptides.

This interpretation, however, in the light of our present knowledge, raises a number of difficulties. e.g., it would imply a differential effect on viral polypeptide synthesis, since capsid protein synthesis continues for at least 1 hr after addition of HBB to infected cells. After addition of the compound to enterovirus-infected cells, a central finding is the rapid halt not only of multiplication of infectious virus, but also of viral RNA synthesis (see Figs. 2 and 3). This observation makes it unlikely that inhibition of viral RNA synthesis is mediated directly through inhibition of synthesis of viral RNA polymerase. Granted, enterovirus RNA polymerase activity appears unstable during the viral replication cycle (23). However, the rapid and complete stop of viral RNA synthesis after addition of HBB cannot be explained by this mechanism, viz., inhibition of synthesis of viral RNA polymerase with subsequent disappearance of enzyme activity by metabolic turnover. Rather, a direct effect of HBB on viral RNA synthesis appears much more plausible.

This interpretation appears to be in conflict with the observation that the in vitro enzyme activity is apparently unimpaired by HBB (22). However, the following considerations must be kept in mind. The preparations originally used were very crude, reacting probably under nonoptimal conditions.

The maximum reaction time was 15 min. It has been suggested, therefore, that incorporation of precursors into RNA in this system may, in a large part, be due to a completion of nascent RNA chains rather than initiation of new chains (24,25). As to the antiviral action of guanidine, this thesis appears to be supported by the detailed investigations of Tershak (26, 27).

We have carried out similar studies with HBB on the effects of this compound on echovirus 12 polymerase in vitro. Echovirus 12 polymerase was prepared essentially by the proce-

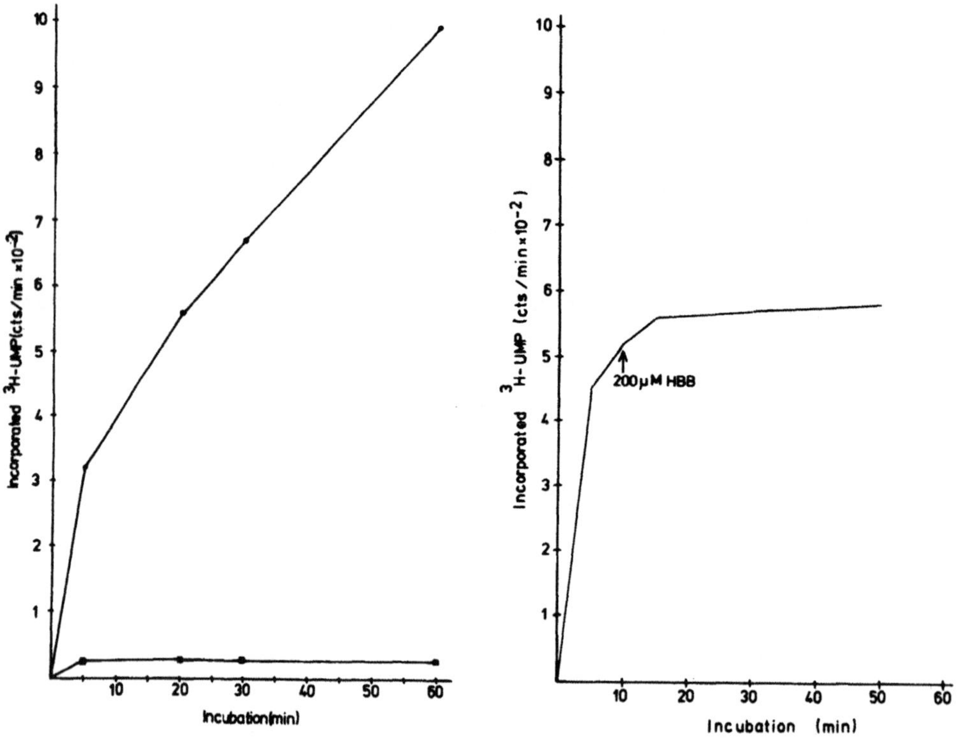

FIGURE 4. a) Echovirus 12-induced RNA polymerase activity in cell-free extracts. Full circles: preparation from infected cells. Squares: preparation from mock-infected cells. Methods according to Tershak (26), modified. The reaction mixtures contained 150 µM ATP, CTP, and GTP, and 2 µM UTP. Phosphocreatine was used up to 60 mM and creatine phosphokinase up to 1000 µg/ml.

b) Effect of D-HBB·HCl added 10 min after start of the reaction. R. Neumann and H.J. Eggers, unpublished work 1982.

dure of Tershak (26), and incorporation of ^{3}H-UMP into acid-insoluble polynucleotide was measured in the absence and presence of 200 µM HBB (Fig. 4). Incorporation in the absence of inhibitor (Fig. 4a) appeared steep linear for about 5 to 10 min, then incorporation decreased, though synthesis continued for at least 60 min, the longest period of time tested. Addition of HBB to the system after 10 min incubation stopped incorporation of radioactivity almost immediately (Fig. 4b). A similar observation - data not shown - was made when HBB had been added after 30 min incubation.

When HBB was present in the reaction mixture beginning with the start of incubation (Fig. 5), the initial steep incorporation of ^{3}H-UMP remained uninhibited, but further incorporation did not occur, in contrast to the control mixture without HBB.

In a further experiment (data not shown), HBB was added to the RNA polymerase preparation in different concentrations (50 µM, 75 µM, 100 µM, 200 µM). Again, the initial incorporation remained unaffected, but further incorporation was inhibited by 75 µM HBB and higher concentrations, whereas 50 µM HBB was without apparent effect. This dose-effect relationship fairly well parallels that observed in virus-infected cells. The products synthesized in the echovirus 12 RNA polymerase preparation have not yet been sufficiently characterized. Relevant investigations are in progress in our laboratory.

In analogy to the results obtained by Tershak in the poliovirus system with guanidine (26,27) and considering the presently known data on the antiviral action of HBB we propose the following interpretation of the experimental findings. The initiation step of RNA synthesis appears to be the most probable site of the antiviral action of HBB. The initial steep incorporation curve in the cell-free RNA polymerase system perhaps reflects completion of RNA chains already initiated; therefore, this part of UMP incorporation is insusceptible to HBB. The more flat part of the incorporation curve might signify the complete process of RNA replication including initiation of RNA synthesis. Accordingly, this part of RNA synthesis is susceptible to the action of HBB.

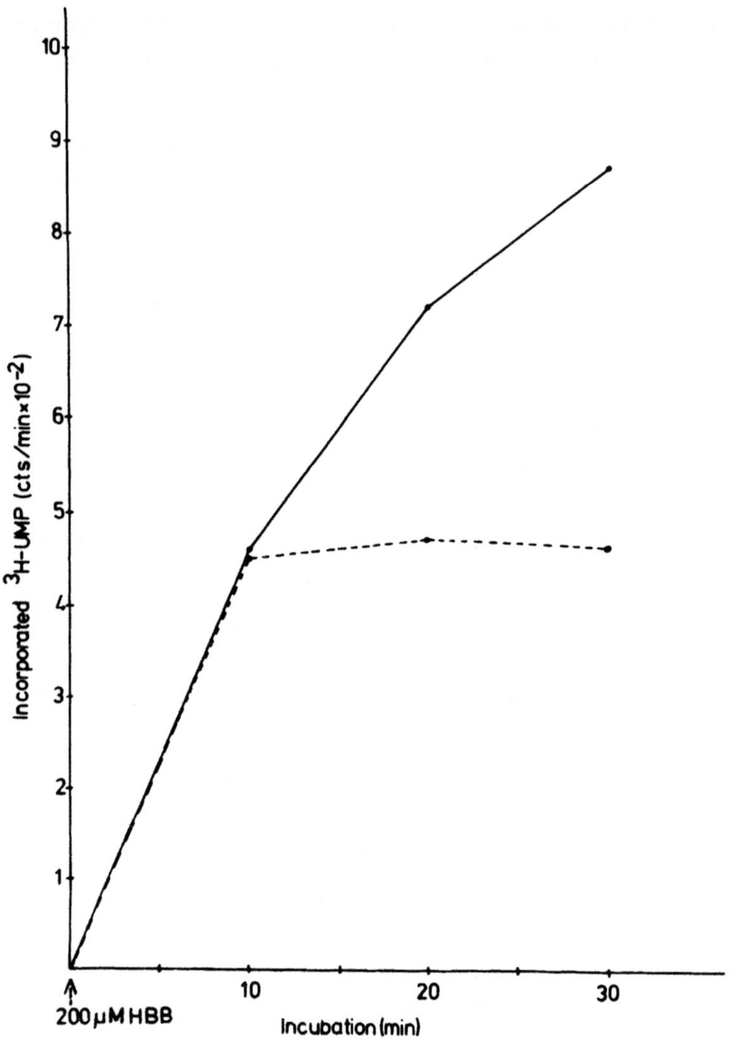

FIGURE 5. Echovirus 12-induced RNA polymerase activity in cell-free extracts, D-HBB·HCl being present beginning with the start of the reaction. Methods see Fig. 4. R. Neumann and H.J. Eggers, unpublished work 1982

In conclusion, all experimental data available so far are compatible with the hypothesis that HBB inhibits initiation of viral RNA synthesis. In molecular terms, however, the precise site of action of HBB is not yet known. The complexity of the replication process of picornaviruses makes a precise analysis rather difficult (27,28,29,30,31). We consider a study of conformational modifications of viral proteins by various effectors a fruitful approach (32).

HBB resistance and dependence.

Passage of drug-sensitive virus in the presence of HBB readily yields HBB-resistant virus (5). Clones of varying degrees of HBB resistance are being obtained. Sensitivity and resistance to HBB is a genetic property of the virion (33).

In addition to HBB-resistant mutants, HBB-dependent mutants of various enteroviruses have been isolated (34,35,36). This represented the first demonstration of drug dependence in viruses. Like HBB resistance also HBB dependence is genetically determined (36). HBB is not required for early interactions of dependent virus with the cells but for RNA synthesis of dependent virus (36). During the exponential increase period of dependent virus, the continued presence of compound is required (14). Present data available suggest that the drug-sensitive and drug-dependent processes may be biochemically analogous, but further experiments on this point are required.

HBB-dependent mutants give rise among their progeny to HBB-independent virus particles with considerable frequency. In a special system, apparently free of selective pressure, it was determined that the revertants from HBB dependence comprised mutants of a wide spectrum of HBB sensitivity and resistence (37). An estimate of the mutation frequency from HBB dependence to HBB independence revealed a - at the time - surprisingly high mutation frequency of about 10^{-4} mutations per replication (37), an order of magnitude, now well accepted for picornaviruses (38).

Rescue phenomena have been demonstrated with guanidine-sensitive, guanidine-resistant, and guanidine-dependent polioviruses (39,40,41,42,43). The results obtained with these systems strongly suggest that the assisting virus supplies a function for the rescued one, the viral RNA polymerase being a likely carrier of such a function.

Analogous experiments have been done with an HBB system. It has been demonstrated that HBB-dependent coxsackievirus A9 can be rescued in monkey kidney cells by HBB-sensitive echovirus 7 in the absence of the compound. Conversely, however, no significant rescue of HBB-sensitive echovirus 7 by HBB-dependent coxsackievirus A9 could be demonstrated in cultures treated with HBB (43).

As indicated above, a direct effect of HBB on viral RNA synthesis, not mediated by the effects on synthesis of viral RNA polymerase, appears plausible. Thus, rescue of an HBB-sensitive virus in the presence of HBB may not be possible. Experiments to clarify this problem are under way in our laboratory.

Effects of HBB on picornavirus-infected cells

HBB significantly inhibits the development of picornavirus-induced morphological changes in monkey kidney cells, but the ultimate degeneration of the infected cells cannot be prevented, even in the absence of detectable virus replication (44). These virus-induced morphological changes in the presence of HBB are quite different from those observed in infected, untreated controls. The most plausible hypothesis to explain the ultimate degeneration of infected, HBB-treated cells is derived from the experimental finding, that HBB does not prevent the early virus-induced shutoff of cellular protein and RNA synthesis (see also Fig. 6). The shutoff in the presence of 100 μM HBB - a concentration inhibiting completely echovirus 12 replication - occurs with kinetics indistinguishable from that of infected, untreated cells. It is obvious that under such conditions the cell cannot survive any length of time, though cytopathic effects are not visible before 24 hr postinfection. It should be stressed, however, that the cytopathic effects in untreated, infected cells, occurring within 6 hr after infection, are morphologically completely different (47,48,49,50).

All of these findings are fully compatible with the kinetics of the antiviral action of HBB described above: HBB leaves unaffected early virus-cell interactions including virus uncoating and virus-induced cellular shutoff. It interferes with viral RNA synthesis thereby ultimately inhibiting viral protein (capsid) synthesis which may be a most important factor causing conventional picornavirus-induced morphological alterations (48).

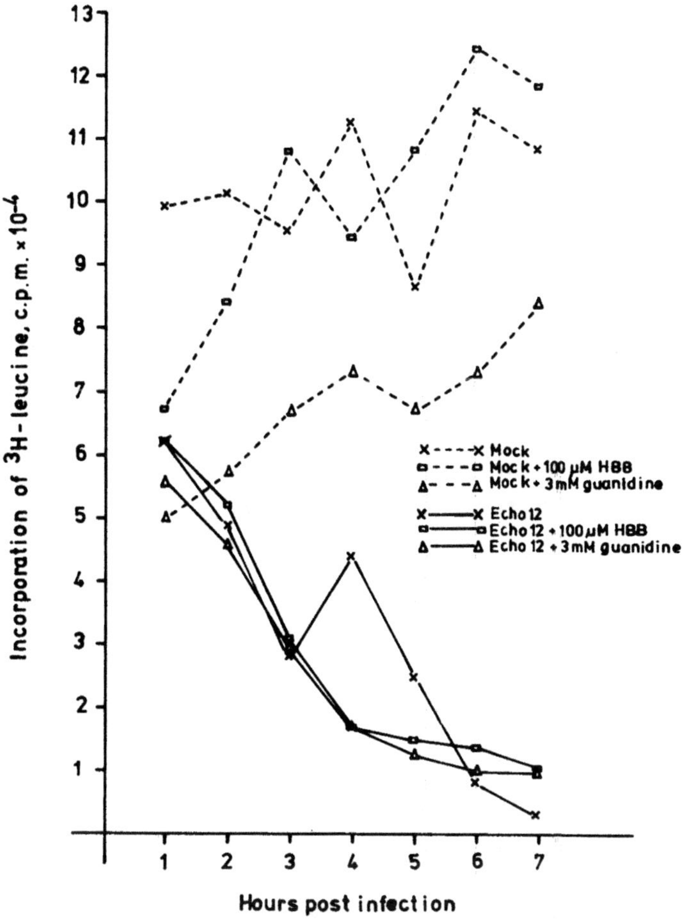

FIGURE 6. Host cell protein synthesis shutoff in GMK cells infected with echovirus 12. Lack of effect on shutoff by D-HBB·HCl or guanidine·HCl. Curves are indicated by: broken curves (control); full curves (echovirus 12); crosses (no treatment); squares (plus 100 µM HBB); triangles (plus 3 mM guanidine). B. Rosenwirth and H.J. Eggers, unpublished work 1978. Methods according to Rosenwirth and Eggers (45,46).

CONCLUSIONS

2-(α-Hydroxybenzyl)-benzimidazole (HBB) selectively inhibits many members of the picornavirus group. HBB does not interfere with early virus-cell interactions including virus adsorption, penetration, uncoating and virus-induced shutoff of cellular protein and RNA synthesis. The compound inhibits viral RNA synthesis, probably by interfering with initiation of RNA synthesis at the virus-induced RNA polymerase. Inhibition of viral protein synthesis, in all probability, is a secondary effect due to inhibition of viral RNA synthesis by HBB.

The virus selectivity of the compound is evident from the fact that - at virus-inhibitory concentrations - it leaves vital functions of the cell unaffected including cell multiplication. The virus selectivity of HBB is also manifest in that the compound does not affect members of families other than that of the picornavirus family.

Though it has been thought for a long time that HBB might not be protective in enterovirus-infected animals, possibly due to the emergence of HBB-resistant or HBB-dependent mutants, recent investigations on prophylactic and therapeutic effects in infected animals have been encouraging (8,51,52). It could be shown that the less dramatic antiviral effects of HBB (and its derivatives) in animals - as compared to those in cell culture - are due to metabolic inactivation of the compounds rather than due to the emergence of drug-resistant virus mutants (8). Therapeutic trials with benzimidazoles in virus-infected human volunteers are ongoing (53).

REFERENCES

1. Tamm, I., Folkers, K. and Horsfall, F.L., Jr. Yale J. Biol. Med. 24:559-567, 1952.
2. Tamm, I. Bull. N.Y. Acad. Med. 31:537-540, 1955.
3. Hollinshead, A.C. and Smith, P.K. J. Pharmacol. Exp. Ther. 123:54-62, 1958.
4. Tamm, I. and Nemes, M.M. J. Clin. Invest. 38:1047, 1959.
5. Eggers, H.J. and Tamm, I. J. Exp. Med. 113:657-682, 1961.
6. Eggers, H.J. and Tamm, I. Virology 13:545-546, 1961.
7. Tamm, I. and Caliguiri, L.A. In: The International Encyclopedia of Pharmacology and Therapeutics (Ed. D.J. Bauer), Vol. 1. Pergamon, Oxford, New York, 1972, pp. 115-179.
8. Eggers, H.J. In: Chemotherapy of Viral Infections (Eds. P.E. Came and L.A. Caliguiri), Handb. Exp. Pharm., Vol. 61. Springer, Berlin, Heidelberg, New York, 1982, pp. 377-417.
9. Siegl, G. and Eggers, H.J. J. Gen. Virol. 61:111-114, 1982.
10. Eggers, H.J. and Tamm, I. Virology 18:426-438, 1962.
11. Tamm, I., Bablanian, R., Nemes, M.M., Shunk, C.H., Robinson, F.M. and Folkers, K. J. Exp. Med. 113:625-656, 1961.

12. Tamm, I., Eggers, H.J., Bablanian, R., Wagner, A.F. and Folkers, K. Nature 223:785-788, 1969.
13. Tamm, I. and Eggers, H.J. Virology 18:439-447, 1962.
14. Eggers, H.J., Reich, E. and Tamm, I. Proc. Natl. Acad. Sci. USA 50:183-190, 1963.
15. Halperen, S., Eggers, H.J. and Tamm, I. Virology 24:36-46, 1964.
16. Caliguiri, L.A., Eggers, H.J., Ikegami, N. and Tamm, I. Virology 27:551-558, 1965.
17. Eggers, H.J. and Waidner, E. Nature 227:952-953, 1970.
18. Eggers, H.J. Virology 78:241-252, 1977.
19. Luria, S.E., Darnell, J.E., Jr., Baltimore, D. and Campbell, A. General Virology. 3rd ed. John Wiley and Sons, New York, Santa Barbara, Chichester, Brisbane, Toronto, 1978.
20. Eggers, H.J. and Tamm, I. Nature 197:1327-1328, 1963.
21. Baltimore, D. and Franklin, R.M. Biochem. Biophys. Res. Commun. 9:388-392, 1962.
22. Baltimore, D., Eggers, H.J., Franklin, R.M. and Tamm, I. Proc. Natl. Acad. Sci. USA 49:843-849, 1963.
23. Eggers, H.J., Baltimore, D. and Tamm, I. Virology 21:281-282, 1963.
24. Caliguiri, L.A. and Tamm, I. Virology 35:408-417, 1968.
25. Caliguiri, L.A. and Tamm, I. Virology 36:223-231, 1968.
26. Tershak, D.R. J. Virol. 41:313-318, 1982.
27. Tershak, D.R., Yin, F.H. and Korant, B.D. In: Chemotherapy of Viral Infections (Eds. P.E. Came and L.A. Caliguiri), Handb. Exp. Pharm., Vol. 61. Springer, Berlin, Heidelberg, New York, 1982, pp. 343-375.
28. Rueckert, R.R., Mathews, T.J., Kew, O.M., Pallansch, M., McLean, C. and Omilianowski, D. In: Molecular Biology of Picornaviruses (Ed. R. Perez-Bercoff). Plenum, New York, 1979, pp. 113-125.
29. Cooper, P.D., Wentworth, B.B. and McCahon, D. Virology 40:486-493, 1970.
30. Korant, B.D. Virology 81:17-28, 1977.
31. Adrian, Th., Rosenwirth, B. and Eggers, H.J. Virology 99:329-339, 1979.
32. Eggers, H.J. and Tamm, I. Annu. Rev. Pharmacol. 6:231-250, 1966.
33. Tamm, I. and Eggers, H.J. In: 2nd International Symposium of Chemotherapy (Eds. H.P. Kuemmerle, P. Preziosi and P. Rentchnick), Naples 1961, Part II. Karger, Basel, New York, 1963, pp. 88-118.
34. Eggers, H.J. Cold Spring Harbor Symp. Quant. Biol. 27:309, 1962.
35. Eggers, H.J. and Tamm, I. VIIIth Int. Congr. Microbiol. p. 85, 1962.
36. Eggers, H.J. and Tamm, I. Virology 20:62-74, 1963.
37. Eggers, H.J. and Tamm, I. Science 148:97-98, 1965.
38. Nottay, B.K., Kew, O.M., Hatch, M.H., Heyward, J.T. and Obijeski, J.F. Virology 108:405-423, 1981.
39. Cords, C.E. and Holland, J.J. Proc. Natl. Acad. Sci. USA 51:1080-1082, 1964.
40. Holland, J.J. and Cords, C.E. Proc. Natl. Acad. Sci. USA 51:1082-1085, 1964.

41. Wecker, E. and Lederhilger, G. Proc. Natl. Acad. Sci. USA 52:246-251, 1964.
42. Agol, V.I. and Shirman, G.A. Biochem. Biophys. Res. Commun. 17:28-33, 1964.
43. Ikegami, N., Eggers, H.J. and Tamm, I. Proc. Natl. Acad. Sci. USA 52:1419-1426, 1964.
44. Bablanian, R., Eggers, H.J. and Tamm, I. J. Bacteriol. 91:1289-1294, 1966.
45. Rosenwirth, B. and Eggers, H.J. Nature 267:370-371, 1977.
46. Rosenwirth, B. and Eggers, H.J. Eur. J. Biochem. 92:61-67, 1978.
47. Bablanian, R., Eggers, H.J. and Tamm, I. Virology 26:100-113, 1965.
48. Bablanian, R., Eggers, H.J. and Tamm, I. Virology 26:114-121, 1965.
49. Dales, S., Eggers, H.J., Tamm, I. and Palade, G.E. Virology 26:379-389, 1965.
50. Bienz, K., Egger, D., Rasser, Y. and Bossart, W. Virology 100:390-399, 1980.
51. Eggers, H.J. J. Exp. Med. 143:1367-1381, 1976.
52. Herrmann, E.C., Jr., Herrmann, J.A. and DeLong, D.C. Antiviral Research 1:301-314, 1981.
53. Phillpotts, R.J., DeLong, D.C., Wallace, J., Jones, R.W., Reed, S.E. and Tyrrell, D.A.J. Lancet I:1342-1344, 1981.

18

MULTIPLE ACTIONS OF AMANTADINE AGAINST INFLUENZA VIRUSES

ALAN J. HAY and MARY C. ZAMBON
Division of Virology, National Institute for Medical Research, Mill Hill, London, NW7 1AA.

SUMMARY

The data presented in this paper distinguish two actions of amantadine on the replication of fowl plague virus in tissue culture which occur at different concentrations. Low concentrations (approximately 5 μM) appear to inhibit the assembly of fowl plague virus by interacting with the virus haemagglutinin. In contrast, concentrations of 0.5 mM or greater, which have no effect on virus assembly, inhibit an early stage in infection which may involve low pH-dependent fusion between the virus envelope and the membrane of secondary lysosomes (1, 2). Whereas most influenza viruses display the latter sensitivity to amantadine they differ greatly in their susceptibility to low concentrations of the drug. The replication of certain human influenza viruses, e.g. A/Singapore/1/57(H_2N_2), is also affected by low drug concentrations at an early stage prior to primary transcription. This action which is dependent on properties of the matrix protein appears different from that of high concentrations of amantadine.

INTRODUCTION

Amantadine hydrochloride is effective as a prophylactic agent against influenza A virus infections of animals and man (3) and is licensed for clinical use. Studies of its mechanism of action in tissue culture have indicated that the drug does not affect binding of virus to cells but interferes with an early stage during replication prior to primary transcription, probably uncoating of the virus particle (1, 2, 4-8).

Y. Becker (ed.), Antiviral Drugs and Interferon: The Molecular Basis of Their Activity.

Influenza A viruses differ in the degree to which their replication is affected by the drug and amantadine-resistant variants can readily be isolated from sensitive virus strains (9-12). Susceptibility to the action of the drug can be transferred genetically (11-14) and analyses of recombinants formed between certain amantadine-sensitive and amantadine-resistant human viruses have indicated that their sensitivity to the drug is determined principally by the gene coding for the matrix protein (12, 15). Genetic analyses of recombinants of the avian influenza virus, fowl plague virus (FPV), on the other hand, gave a more complicated picture suggesting that susceptibility to amantadine may be influenced by other viral genes (16).

The data presented in this paper show that amantadine can inhibit a late stage in the replication of FPV by interfering with the formation of virus particles and that the genetic basis of this action is associated with the haemagglutinin gene.

RESULTS

The effects of amantadine on the replication of four different viruses have been compared. As previously observed (11, 16) the replication of A/Singapore/1/57 (H_2N_2) and fowl plague virus, Rostock strain (H_7N_1), FPV (R), are particularly sensitive to the drug. The plaque titre of FPV (R) was reduced by 10^3-10^4 by inclusion of amantadine (1 μg/ml) in the agar overlay. This concentration was optimal and plaques reappeared at higher concentrations,as observed by Scholtissek and Faulkner (16), and were approximately one-third in number and roughly half the size of control at 100 μg/ml amantadine. Plaque-purified A/Singapore contained approximately 5% of particles which produced plaques, one-quarter the size of control, in the presence of 1 μg/ml amantadine. The other two viruses, an amantadine-resistant variant of A/BEL/42 (H_1N_1) isolated by Appleyard (11) and fowl plague virus, Weybridge strain (H_7N_7), FPV(W), were chosen because of their resistance to 1 μg/ml amantadine,and have been used to produce recombinants

Table 1. Effect of amantadine on virus production

Amantadine treatment	Virus yield (% of control)			
	FPV(R)	A/Sing	FPV(W)	A/BEL
30 min before infection				
1 μg/ml	9(9)	9(9)	50	51
10 μg/ml	9	3	71	28
100 μg/ml	17	<2(<2)	<2(90)	<2(4)
60 min after infection				
1 μg/ml	8(8)	66	64	93
10 μg/ml	22	71	100	100
100 μg/ml	60	87	100	80

The experimental details are given in the legend to Fig. 1. The yield of ^{35}S virus from amantadine-treated cultures is expressed as a proportion of that produced by control cultures. Figures in parenthesis give the virus yield 11 hr following removal of the drug either 1 hr after infection from pre-treated cultures or 3 hr after infection when amantadine was added at 1 hr pi.

with the sensitive viruses. The plaque titres of these viruses were unaffected by 1 μg/ml amantadine although the plaque size was sometimes reduced by up to two-fold.

Inhibition of virus production and virus protein synthesis

Single replicative cycles of FPV(R) and A/Singapore following high multiplicity infection of either chick cells or chorioallantoic membranes were inhibited by low (1 μg/ml) as well as by high concentrations of amantadine whereas the replication of FPV(W) and A/BEL were relatively insensitive to 1 μg/ml of the drug. Comparable results were obtained by measuring either the haemagglutinin titre of culture medium or ^{35}S-labelled virus produced in 11 hr by infected cells labelled at 4 hr after infection with ^{35}S methionine. The yields of both FPV(R) and A/Singapore were reduced 10-fold following infection in the presence of 1 μg/ml amantadine whereas FPV(W) and BEL were affected by no more than two-fold (Table 1). Addition of amantadine (1 μg/ml) 30 min after infection, while inhibiting the yield of FPV(R) to a similar extent, had little effect on the production of A/Singapore.

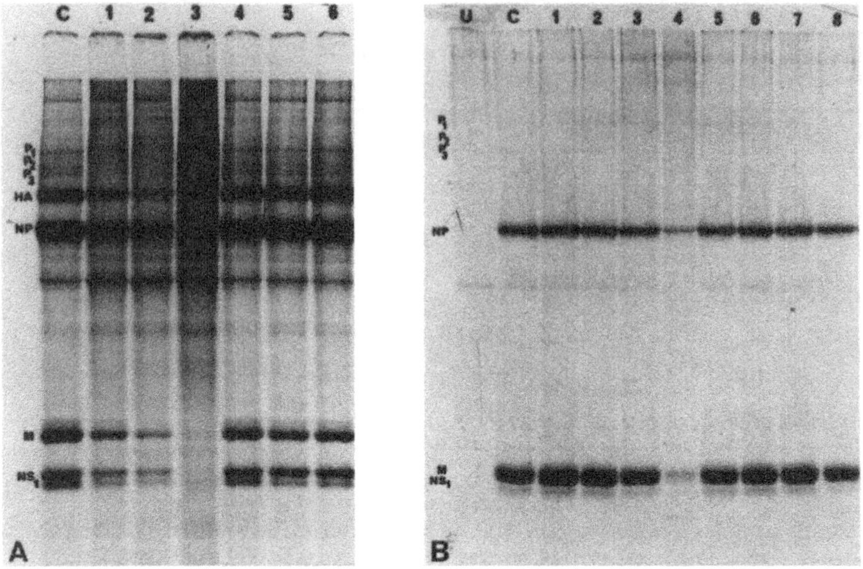

FIGURE 1. Effect of amantadine on polypeptide synthesis in cells infected with A/Singapore/1/57 (A) or FPV(R) (B). Chick cell cultures were infected at a multiplicity of approximately 50 pfu/cell and incubated in medium containing ^{35}S-methionine (10 μCi/ml) from 4 hr after infection. Amantadine at the appropriate concentration was added either 30 min prior to infection and included in the virus inoculum or at 60 min after infection. Culture medium was harvested at 11 hr pi and virus sedimented in a 15-40% sucrose gradient (18). ^{35}S-labelled polypeptides from cells were analysed by electrophoresis at 10 volts/cm for 20 hr in a 16% polyacrylamide gel followed by autoradiography (6), and are designated as previously described. A. Cells infected with A/Singapore/1/57 (C) and treated before (1-3) or after (4-6) infection with amantadine at 1 μg/ml (1, 4) 10 μg/ml (2,5) or 100 μg/ml (3,6). B. Cells infected with FPV(R) (C) and treated before (1-4) or after (5-8) infection with amantadine at 1 μg/ml (1,5), 10 μg/ml (2,6), 100 μg/ml (3,7), 500 μg/ml (4,8). U, uninfected cells.

Virus protein synthesis in A/Singapore-infected cells was also reduced, approximately three-fold, only after pretreatment with the drug; however, these concentrations of the drug had no significant effect on protein synthesis in FPV(R)-infected cells (Figure 1, Table 2). It is apparent, therefore, that whereas the replication of A/Singapore is affected early

Table 2. Effect of amantadine on virus protein synthesis

Amantadine treatment	Virus yield (% of control)			
	FPV(R)	A/Sing	FPV(W)	A/BEL
30 min before infection				
1 μg/ml	100	31(35)	100	83
10 μg/ml	100	20	90	74
100 μg/ml	80	<5(<5)	<5(80)	<5(40)
500 μg/ml	15(78)	<5	-	-
60 min after infection				
1 μg/ml	90	94	100	100
10 μg/ml	100	83	80	100
100 μg/ml	89	90	68	88
500 μg/ml	80	86	-	-

The experimental details are given in the legend to Fig. 1. The data were computed from microdensitometer scans of autoradiograms and are expressed as a percentage of radioactivity associated with the major viral polypeptides HA, NP, M and NS_1 from control cells. Figures in parenthesis give the results for cells from which amantadine was removed either 1 hr after infection of pretreated cultures or 3 hr after infection when the drug was added at 1 hr pi.

in infection the replication of FPV(R) is inhibited at some later stage.

High concentrations of amantadine inhibited the replication of all four viruses only when added prior to infection. Both virus yield and virus protein synthesis in cells infected with A/Singapore, FPV(W) or A/BEL were drastically curtailed by 100 μg/ml amantadine. The replication of FPV(R) was affected to a similar extent only by a higher concentration of approximately 500 μg/ml.

The inhibitory effects of high and low concentrations of amantadine on the replication of FPV(R) are thus clearly distinguishable - the optimum concentration for the latter action was approximately 1 μg/ml. The two actions also differ in reversibility. Whereas the inhibitions of FPV(R) (6) and FPV(W) infections by high concentrations were readily reversed by removal of the drug at later times, the action of 1 μg/ml on FPV(R) replication was unaffected (Tables 1 and 2). The inhibitory effects of both 100 μg/ml and 1 μg/ml amantadine on the replication of A/Singapore were also unaffected by removal of the drug while the effects of 100 μg/ml on A/BEL infection were partially reversible. These results suggest that the inhibition of virus infections by high concentration of amantadine differ in reversibility. However, the apparent irreversibility of the effect of a high concentration in the case of A/Singapore may be due to the reversible action being superseded by an irreversible inhibition by lower concentration after removal of the drug.

Effect of amantadine on viral RNA synthesis

Since it was apparent that infection *per se* by FPV(R) was not inhibited by 1 μg/ml amantadine, the effect of this concentration of the drug on a number of replicative processes was investigated. In agreement with the lack of any inhibitory effect on viral protein synthesis, the synthesis of all viral RNAs, mRNA (primary and secondary synthesis), unpolyadenylated cRNAs and vRNAs, was unaffected by such treatment (Figure 2).

Primary transcription in A/Singapore-infected cells was reduced only if the drug (1 μg/ml or 100 μg/ml) was added before or during the infection period and was unaffected if added once transcription had begun (Figure 3), as observed for FPV-infected cells treated with high concentrations of the drug (6). Thus contrary to some suggestions (17) the drug does not appear to affect directly viral RNA synthesis.

Fractionation of cellular components by differential centrifugation did not indicate any changes in the localization of viral components in FPV-infected cells as a result of drug treatment. The properties of viral ribonucleoproteins, including their density in metrizamide gradients and protein and RNA compositions, were also unaltered. Cleavage of the haemagglutinin polypeptide also occurred normally in drug-treated cells although whether it is inserted correctly into the plasma membrane has not been investigated.

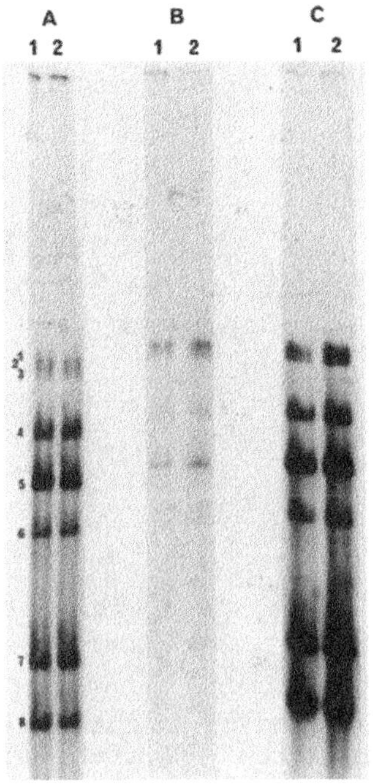

FIGURE 2

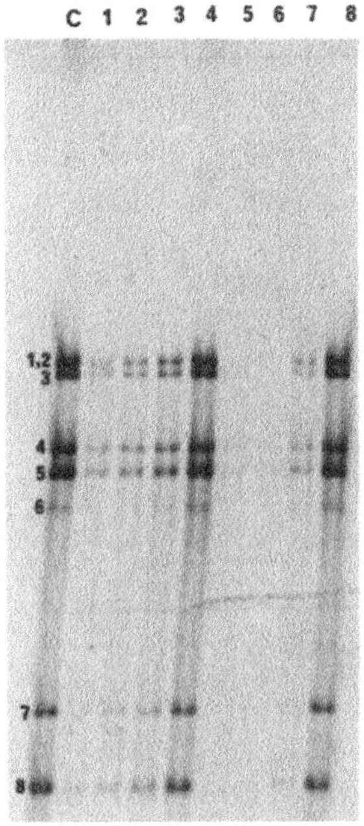

FIGURE 3

FIGURE 2. Effect of amantadine (1 μg/ml) on RNA synthesis in FPV-infected cells. Infected chick cells were incubated in medium containing ^{3}H-uridine (100 μCi/ml) between 2 and 3 hr pi and RNA was extracted and analysed on 4% polyacrylamide gels as previously described (19, 20). Primary transcription was measured in cells pretreated for ½ hr and incubated throughout infection in the presence of cycloheximide (100 μg/ml). A. mRNA synthesis; B. primary transcription; C. vRNA synthesis. 1. control cells; 2. cells incubated with amantadine (1 μg/ml).

FIGURE 3. Effect of amantadine on primary transcription in A/Singapore-infected cells. Cycloheximide (100 μg/ml) was added to chick cells ½ hr before infection with A/Singapore/1/57 (50 pfu/cell) at 22° for 20 min. Amantadine, 1 μg/ml (1-4) or 100 μg/ml (5-8), was added ½ hr before infection (1,5), at the start (2,6) or end of the infection period (3,7) or ½ hr pi (4,8). Cells were incubated with ^{3}H-uridine between 1½ and 2½ hr pi and mRNAs analysed as previously described (19). C, control culture.

Effect of amantadine (1 μg/ml) on the incorporation of proteins into FPV(R) virions

The incorporation into virus particles of viral proteins synthesized during the productive phase of infection (18),

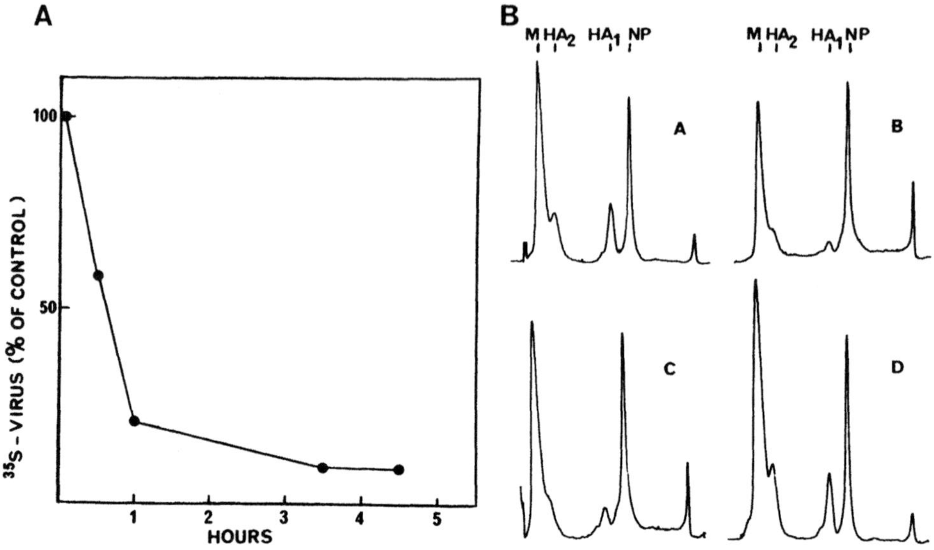

FIGURE 4. A. Effect on ^{35}S virus production of amantadine added at different times prior to a pulse with ^{35}S methionine. Amantadine was added to FPV-infected cells at 0, 1, 3½, 4 and 4½ hr pi. Cells were labelled with ^{35}S methionine (10 μCi/ml) between 4 and 4½ hr and washed and incubated in medium containing 1 mM methionine. The ^{35}S virus yield at 11 hr is expressed as a percentage of that from untreated cells and is plotted against time of addition of drug prior to the end of the pulse. B. Polypeptide composition of ^{35}S virus released from cells treated with amantadine at different times relative to the pulse. Microdensitometer scans are of autoradiographs obtained following different periods of exposure.
A, control; amantadine added 1 hr (B), 3½ hr (C) and 4½ hr (D) pi.

between 4 and 8 hr, appeared to be inhibited by amantadine (1 μg/ml) only if the drug was added prior to their synthesis. The effect of adding amantadine at different times prior to a 30 min pulse with ^{35}S-methionine on the subsequent production of ^{35}S virus is shown in Figure 4A. Approximately 60 min were required for the drug to exert 80% of its inhibitory effect and this time lag was not reduced by short treatments with higher concentrations of 10 or 100 μg/ml which were in fact less inhibitory.

Residual virus released from cells incubated throughout infection in the presence of amantadine (1 μg/ml) contained

Table 3. Effect of cyclooctylamine and related compounds on the yield of FPV

Compound	Virus yield (% of control)	
	1 μg/ml	10 μg/ml
amantadine	7%	8%
octylamine	100%	72%
cyclooctylamine	5%	5%
cycloheptylamine	30%	15%
cyclohexylamine	70%	30%
cyclopentylamine	100%	100%
cyclooctanol	100%	100%
1,2-diaminocyclohexane	100%	100%

a much lower proportion of HA_1 and HA_2, approximately 1/5 of that in virus from untreated cells, whereas the relative proportions of NP and M polypeptides were unaffected (Figure 4B). The effect was less apparent when the drug was added between ½ and 1 hr before the pulse with ^{35}S-methionine and undetectable when added after the pulse. In view of the small amounts of ^{35}S-virus released from cells incubated throughout infection in the presence of amantadine, information on the proportions of the minor polypeptide components P_1, P_2, P_3 and neuraminidase is not available. It appears, therefore, that of the proteins analysed the incorporation of the haemagglutinin into virus is selectively affected by amantadine. This observation together with the lack of effect noted when amantadine is added after the pulse-label suggest that the assembly of virus particles is inhibited rather than simply their release from cells.

Inhibitory action of cyclooctylamine and related compounds

Cyclooctylamine is similar to amantadine in its effects upon influenza virus infections and the differential sensitivities of viruses to the two drugs were observed to be covariant (14) The inhibitory actions of a number of related compounds were compared to determine which structural attributes are important in blocking virus formation in FPV(R)-infected cells.

Table 4. Genome compositions of amantadine-resistant recombinants between FPV(R) and FPV(W)

Recombinant	Gene Number 1	2	3	4	5	6	7	8
*1	W	W	W	W	W	W	R	W
2	W	W	W	W	W	R	R	W
3	W	W	W	W	W	R	W	R
4	R	W	W	W	R	W	W	W
5	W	W	W	W	R	R	W	R
6	W	R	W	W	W	R	R	W
7	R	W	W	W	W	R	W	R
8	W	R	W	W	R	W	R	W
9	R	R	W	W	W	R	R	W
*10	W	W	R	W	R	R	R	R
*11	R	R	R	W	W	R	W	R
*12	R	R	R	W	R	R	R	W
13	W	R	R	W	R	R	R	R
14	R	R	R	W	R	R	R	R

Chick cells were coinfected at a multiplicity of 10 with FPV(R) and FPV(W) at relative infectivities of between 1 and 25. Medium taken at 16 hr was titrated in the presence or absence of amantadine (1 µg/ml). Plaques were purified and the genome composition determined as described by Hay et al. (15). R and W denote genes (1-8) of FPV(R) and FPV(W), respectively. *Viruses isolated in the absence of drug and subsequently shown to be amantadine-resistant.

The optimum concentration of cyclooctylamine and its inhibitory effect were similar to those of amantadine. Homologues with a smaller ring size were progressively less effective (Table 3). Whereas higher concentrations (10 µg/ml) of cycloheptylamine and cyclohexylamine were more effective than 1 µg/ml, cyclopentylamine was ineffective at both concentrations. The straight chain analogue octylamine as well as related compounds lacking an amino group, eg cyclooctanol or cyclooctane, were ineffective at comparable concentrations and the presence of an extra amino group on diaminocyclohexane did not enhance its activity over that of cyclohexylamine.

Genetic analyses of recombinants

The genome compositions of several amantadine (1 μg/ml)-resistant recombinants of A/Singapore isolated following co-infection with either A/BEL or A/PR/8/34 confirmed that the matrix protein gene was the only gene required to confer resistance to 1 μg/ml amantadine. Analyses of amantadine-resistant recombinants formed between FPV(R) and FPV(W) showed that the only gene of FPV(W) common to all resistant viruses was gene 4 which codes for the virus haemagglutinin (Table 4). Scholtissek and Faulkner (16) reached a similar conclusion based on drug-susceptibility of recombinants measured in a single replicative cycle assay which was, however, not supported by drug susceptibility measured by plaque titration. One limitation of the present study was that all recombinants analysed contained the HA gene of FPV(W). It is evident, therefore, that the different actions of amantadine against FPV(R) and A/Singapore have different genetic bases.

DISCUSSION

The results of these investigations have shown that amantadine may exert different actions on the replication of influenza viruses depending on drug concentration and virus strain. Concentrations of amantadine greater than 100 μg/ml (0.5 mM) inhibit an early stage in the infection of cells by FPV(R) whereas lower concentrations with an optimum of approximately 1 μg/ml (5 μM) appear to affect the assembly of virus particles. No other aspects of virus replication were observed to be impaired in amantadine (1 μg/ml)-treated FPV-infected cells and addition of the drug prior to synthesis of proteins was required in order to affect their incorporation into virus. Genetic analyses conducted in the current investigations and those of Scholtissek and Faulkner (16) indicate that the latter action is principally dependent on properties of the viral haemagglutinin. Virus particles produced during infection in the presence of the drug contained a reduced amount of haemagglutinin suggesting a selective action against this viral component. Proteolytic cleavage of the haemagglutinin is not affected by presence of the drug; however, insufficient information is available to indicate whether the drug affects the insertion of the HA into the plasma membrane or its subsequent function.

There are no genetic data relating to the action of high concentrations of amantadine and it appears that all influenza viruses as well as other enveloped viruses such as VSV and SFV, may be sensitive to this action although the optimum concentrations differ. A number of recent investigations (1, 2, 21-23) have indicated that these concentrations of amantadine, approximately 0.5 mM, may affect a stage in virus infection involving fusion between the virus envelope and the membrane of endocytic vesicles. It is suggested that this process, which for influenza viruses has an optimum pH of between 5.0 and 5.5 depending on the virus strain, is inhibited by lysosomotropic agents due to an increase in pH within these vesicles. Amantadine, 0.5 mM, causes an increase in lysosomal pH from approximately 4.8 to 5.4 (24). The higher concentration of drug required to inhibit FPV as compared with A/Singapore may reflect different pH optima of cell fusion induced by these viruses, 5.5 for FPV and 5.1 for A/Japan/57, a close relative of A/Singapore (23).

It appears unlikely that lower concentrations of amantadine, approximately 1 μg/ml, which also inhibit an early stage in infection of cells by A/Singapore, act in this way. Genetic analyses (12, 15) have shown that this action is principally dependent on properties of the viral matrix protein whereas the viral haemagglutinin is the principal mediator of *in vitro* virus-induced fusion (25-28). An explanation for the action put forward by Bukrinskaya *et al*. (8, 29) is that rimantadine, a derivative of amantadine, inhibits a second stage in the uncoating of virus particles, involving the removal of the matrix protein from the ribonucleoprotein.

Other differences between the actions of low and high concentrations of cyclic amines in inhibiting early stages in virus infection are in the reversibility of the effects and in the structural requirements for activity. Whereas inhibition by high concentrations appears potentially reversible upon removal of the drug, the actions of low concentrations were observed to be irreversible whether against infection by A/Singapore or assembly of FPV. Both actions by low concentrations of cyclic alkyl amines require an eight-membered ring for optimum activity. In contrast, a wide range of amines, including e.g. methylamine and ammonium chloride, at concentrations of 1 mM or greater exhibit antiviral activity (30, 31). The structural requirements for the actions of low concentrations of cyclic alkyl amines imply a similarity in the sites of action e.g. in association with membranes. Information on the mechanisms of action is as yet very limited. Although amantadine can affect the phase-transition characteristics of lipid bilayers (32) it seems unlikely that such a general property would correlate either with the differing effects of the drug or with the selective nature of its action against different viruses.

Which antiviral effects are more relevant to the clinical use of amantadine and related compounds is difficult to assess without more information on local drug concentrations. Doses of the drug normally given to patients have been reported to give blood levels of 0.3-0.6 μg/ml while tissue levels in animals may exceed this concentration (33-35). The variability in response of infection by different viruses would, however, correlate more readily with effects elicited by low drug concentrations.

ACKNOWLEDGMENTS

We thank Wiktor Chicklowski and Paul Cassell for excellent assistance. Mary Zambon is the recipient of a research studentship from the Medical Research Council.

REFERENCES

1. Matlin, K.S., Reggio, H., Helenius, A. and Simons, K. J. Cell Biol. 91:605-615, 1981.
2. Yoshimura, A., Kuroda, K., Kawasaki, K., Yamashina, S., Maeda, T. and Ohnishi, S. J. Virol. 43:284-293, 1982.

3. Oxford, J.S. and Galbraith, A. Pharmacol. Ther. 11: 181-262, 1980.
4. Kato, N. and Eggers, H.J. Virology 37:632-641, 1969.
5. Dourmashkin, R.R. and Tyrrell, D.A.J. J. Gen. Virol. 24:129-141, 1974.
6. Skehel, J.J., Hay, A.J. and Armstrong, J.A. J. Gen. Virol. 38:97-110, 1978.
7. Koff, W.C. and Knight, V. J. Virol. 31:261-263, 1979.
8. Bukrinskaya, A.G., Vorkunova, N.K., Kornilayeva, G.V., Narmanbetova, R.A. and Vorkunova, G.K. J. Gen. Virol. 60:49-59, 1982.
9. Cochran, K.W., Maasab, H.F., Tsunoda, A. and Berlin, B.S. Ann. N.Y. Acad. Sci. 130:432-439, 1965.
10. Oxford, J.S., Logan, I.S. and Potter, C.W. Ann. N.Y. Acad. Sci. 173:300-313, 1970.
11. Appleyard, G. J. Gen. Virol. 36:249-255, 1977.
12. Lubeck, M.D., Schulman, J.L. and Palese, P. J. Virol. 28:710-716, 1978.
13. Tuckova, E., Vonka, V., Zavadova, A. and Kutinova, L. J. Biol. Stand. 1:341-346, 1973.
14. Appleyard, G. and Maber, H.B. J. Antimicrobial Chemotherapy 1(Suppl.):49-53, 1975.
15. Hay, A.J., Kennedy, N.C.T., Skehel, J.J. and Appleyard, G. J. Gen. Virol. 42:189-191, 1979.
16. Scholtissek, C. and Faulkner, G. J. Gen. Virol. 44: 807-815, 1979.
17. Indulen, M.K. and Kalninya, V.A. In:Developments in Antiviral Therapy (Eds. L.H. Collier and J. Oxford), Academic Press, 1980, pp. 107-118.
18. Hay, A.J. Virology 60:398-418, 1974.
19. Hay, A.J., Lomniczi, B., Bellamy, A.R. and Skehel, J.J. Virology 83:337-345, 1977.
20. Smith, G.L. and Hay, A.J. Virology 118:96-108, 1982.
21. Helenius, A., Kartenbeck, J., Simons, K. and Fries, E. J. Cell Biol. 84:404-420, 1980.
22. Helenius, A., Marsh, M. and White, J. J. Gen. Virol. 58:47-61, 1982.
23. White, J., Matlin, K. and Helenius, A. J. Cell Biol. 89:674-679, 1981.
24. Ohkuma, S. and Poole, B. Proc. Natl. Acad. Sci. USA 75:3327-3331, 1978.
25. Skehel, J.J., Bayley, P.M., Brown, E.B., Martin, S.R., Waterfield, M.D., White, J.M., Wilson, I.A. and Wiley, D.C. Proc. Natl. Acad. Sci. USA 79:968-972, 1982.
26. White, J., Kartenbeck, J. and Helenius, A. EMBO Journal 1:217-222, 1982.
27. Huang, R.T.C., Rott, R. and Klenk, H.-D. Virology 110: 243-247, 1981.
28. Maeda, T., Kawasaki, K. and Ohnishi, S. Proc. Natl. Acad. Sci. USA 78:4133-4137, 1981.
29. Bukrinskaya, A.G., Vorkunova, N.K. and Pushkarskaya, N.L. J. Gen. Virol. 60:61-66, 1982.
30. Jensen, E.M. and Liu, O.C. Proc. Soc. Exp. Biol. Med. 112:456-458, 1963.
31. Fletcher, R.D., Hirschfield, J.E. and Forbes, M. Nature 207:664-668, 1965.

32. Jain, M.K., Yen-Hin Wu, N., Morgan, T.K., Biggs, M.S. and Murray, R.K. Chem. Phys. Lipids 17:71-78, 1976.
33. Couch, R.B. and Jackson, G.G. J. Inf. Dis. 134:516-527, 1976.
34. Wood, T.R. Ann. N.Y. Acad. Sci. 130:419-431, 1965.
35. Bleidner, W.E., Harmon, J.B., Hewes, W.E., Lynes, T.E. and Hermann, E.C. J. Pharmacol. Exp. Ther. 150:484-490, 1965.

19

POLYNUCLEOTIDE INDUCERS OF INTERFERON: STUDIES WITH POLY(G)•POLY(C)

P. F. TORRENCE AND E. DE CLERCQ

SUMMARY

The premises for an active polynucleotide inducer of interferon are a double-helical structure, a sufficiently high molecular weight and thermal stability, and a specific conformation depending on the nature of the ribose-phosphate backbone and of the heterocyclic bases in the interior of the double helix. Following these premises, the double-stranded poly(G)• poly(C) complex should normally be an effective inducer of interferon. Yet, it was found that poly(G)•poly(C) was active as an interferon inducer only when annealled and assayed under the appropriate conditions. Its activity was critically influenced by the choice of the cells and assay method used, i.e. length of exposure to the cells, to measure its interferon inducing potency.

1. INTRODUCTION

A rather staggering diversity of substance of both man-made and natural origin are capable of inducing interferons (1). Included are substances such as endotoxins, polycarboxylates, low molecular weight compounds, T or B lymphocyte-stimulating agents and double-stranded RNA's. This diversity has virtually precluded attempts to relate interferon-inducing potential to any specific molecular structural feature. Deduction of structure-activity relationship has also been difficult within any given class inducers.

Double-stranded RNA's are of considerable interest in that they remain the most potent, least toxic and most widely active of all the different classes of interferon inducers. They may eventually find clinical application as interferon inducers when used alone or in combination with exogenous interferon. Examination of the relationship between polynucleotide structure and interferon-inducing potency therefore has been pursued as a means for obtaining a clinically useful interferon inducer with minimal toxicity and maximal activity (2,3).

A final, perhaps more esoteric reason for investigating the relationship of nucleic acid structure to interferon induction was that until recently, interferon induction was the only known biological response elicited by dsRNA. Only antibodies raised against double-stranded polynucleotides were available as models to evaluate the factors influencing the interaction of dsRNA with proteins (4-6). In the past several years, however, dsRNA's have been found to elicit a range of biological or biochemical responses including inhibition of protein synthesis, activation of protein P_1 kinase and activation of 2-5A (pppA2'p5'A2'p5'A) synthetase (7-10).

1.1. Characteristics of polynucleotide interferon inducers

Within the last decade, a substantial amount of information has been accumulated regarding the relationship of nucleic acid structure to interferon inducing ability. Progress also has been made in the attempt to relate interferon induction by polynucleotides to the physiochemical and conformational parameters of the nucleic acid. Most of these studies have been performed with synthetic dsRNA's such as poly(I)•poly(C) and poly(A)•poly(U).

1.1.1 Polynucleotide interferon inducers are double stranded and the melting temperature (Tm) of the double helix is a major factor in determining the magnitude of the interferon response (2). As a reflection of this dependence upon Tm, nucleic acid interferon inducers are relatively resistant to degradation by nuclease. For instance, substitution of bromine for hydrogen at pyrimidine C5 or substitution of sulfur for oxygen at pyrimidine C-2 of poly(I)•poly(C) gives duplexes with higher Tm's than poly(I)•poly(C) (11,12), and the resulting duplexes are also more resistant to nucleases (i.e. pancreatic ribonuclease A). For example, they resist digestion by the nucleases present in human serum, as reflected by the maintenance of their full interferon inducing potency whereas, under the same conditions, poly(I)•poly(C) itself loses all activity (13).

1.1.2. The efficiency of a nucleic acid as an inducer depends upon the nucleic acid's molecular weight, and, when the inducer is composed of separate polymer strands, induction ability also depends on the molecular weight of the individual strands (2). For poly(I)•poly(C) there is a critical range of sedimentation constants (2-5S) in which there is a nearly linear decrease of interferon inducing ability with decreasing molecular size of either poly(I) or poly(C) (14). Within this range, the interferon inducing activity of poly(I)•poly(C) is more dependent on maintaining a high molecular size of poly(I) than of poly(C).

1.1.3 Interferon induction of polynucleotides depends upon the nature of the ribose-phosphate backbone. Extensive studies with 2'-modified dsRNA's have demonstrated that the interferon-inducing activity of the polynucleotide depends upon the nature of the 2'-substituent, whether this substituent is introduced into poly(A)•poly(U) or poly(I)•poly(C), and upon which strand (purine or pyrimidine) of the duplex where the substituent has been introduced. Replacement of the 2'-hydroxyl by azido gives, in the poly(A)•poly(U) series, poly(dAz)•poly(U), poly(A)•poly(dUz) and poly(dAz)•poly(dUz), all inactive as interferon inducers (15,16), whereas, in the poly(I)• poly(C) series, poly(I)•poly(dCz) is inactive and poly(dIz)•poly(C) is active as

Table 1. Interferon-inducing activity of 2'-fluorinated poly(I)•poly(C) analogues[a]

Polynucleotide (μg/ml)	IFN titer ($\log_{10}$ units/ml) Polynucleotide concentration			
	5	0.5	0.5	0.005
Poly(I)•poly(C)	3.5	2.6	2.0	1.2
Poly(dIfl)•poly(C)	4.4	4.1	3.9	3.3
Poly(I)•poly(dCfl)	< 3.5	2.8	1.8	1.5
Poly(dIfl)•poly(dCfl)	< 1.0	< 1.0		
Poly(dI)•poly(C)	1.2	1.2		
Poly(I)•poly(dC)	< 1.0	< 1.0		
Poly(dI)•poly(dC)	< 1.0	• 1.0		
Poly(dIfl)•poly(dC)	1.2	1.2		
Poly(dI)•poly(dCfl)	1.0	1.0		

[a]As determined in primary kidney cells superinduced with cycloheximide and actinomycin D. Data taken from ref. 18.

an interferon inducer (16). Other 2'-substituted polynucleotides active as inducers include poly(dIcl)•poly(C), poly(dIfl)•poly(C) and poly(I)•poly(dCfl) (17,18). Poly(dIfl)•poly(C) is even more active as an interferon inducer than Poly(I)•poly(C), when assayed in primary rabbit kidney cells (Table 1). The high interferon inducing activity of poly(dIz)•poly(C), poly(dIcl)•poly(C) and poly(dIfl)•poly(C) suggests that the receptor site for interferon induction does not recognize the 2'-hydroxyls as such, but rather the steric configuration conferred upon the nucleic acid by the presence of these 2'-hydroxyls. Apparently, the conformational aspects involved in receptor recognition are not significantly affected by substitution of F, Cl or N_3 for OH in the purine strand of poly(I)•poly(C). The above data also suggest that the conformation of dsRNA's like poly(I)•poly(C) is determined predominantly by the pyrimidine strand since similar substitution, i.e. of Cl or N_3 for OH, in the pyrimidine strand lead to a complete loss of interferon inducing capacity.

1.1.4. The interferon-inducing activity of a nucleic acid is dependent on the nature of the heterocyclic bases in the interior of the double helix. A particular example is the substitution of CH for N-7 in the purine ring of either poly(A)•poly(U) or poly(I)•poly(C). When such a modification is introduced in poly(A)•poly(U) series, leading to duplexes based on poly(7-deazaadenylic acid), all the resulting complexes are without interferon-inducing activity (19). To the contrary, introduction of the same modification in the poly(I)•poly(C) series gives poly(7-deazaionosinic acid)-based duplexes which are good to excellent interferon inducers (11,20).

Two physicochemical studies have attempted to provide a physical basis for these results. Theoretical calculations on the conformational properties of adenosine, 7-deazaadenosine, inosine, 7-deazainosine and 2'-deoxyadenosine suggested that the energetic accessibility of some glycoside bond angles may determine the interferon-inducing potential of nucleic acids (21). Circular dichroism studies revealed that base tilt changes may play a major role in determining the nucleic acid conformation that is required for interferon induction (22).

1.1.5. Various modifications of poly(I)•poly(C) appear compatible with interferon-inducing activity. These modifications include, as outlined above, substitution of bromine for hydrogen at C-5 of cytosine (11,20), sulfur for oxygen at C-2 of cytosine (12), CH for N-7 of hypoxanthine (19,20), and fluorine chlorine or azido for hydroxyl at C'-2 of inosine (16-18). Two other modifications (Fig. 1) which do not abolish the interferon-inducing ability of poly(I)•poly(C) include the substitution of thiophosphate for phosphate in either the purine or pyrimidine strand, or both (23), and the substitution of amino for hydrogen at C-2 of hypoxanthine (leading to the conversion of poly(I)•poly(C) to poly(G)•poly(C)), as will be discussed below. Similarly, base-mismatching, as in poly(I_x,U)•poly(C), poly(I)•poly(C_x,U), and poly(I) poly(C_x,G), does not compromise the interferon-inducing potential of the dsRNA, provided x $\geq$ 10 (24-26). In fact, mismatched poly(I)•poly(C) analogues may eventually achieve a better therapeutic index than poly(I)•poly(C), since they would not be as toxic as poly(I)•poly(C) because of their shorter half-life in biological fluids.

1.2 Nature of the interferon-inducer receptor

A number of studies with polynucleotide inducers of interferon have provided further insight into the mode of action of such polynucleotides at the cellular level. In competition experiments (19, 27-29), single-stranded polynucleotides, while themselves inactive as interferon inducers, did not affect the interferon response to active inducers; on the other hand, certain inactive double- and triple-helical complexes were found able to prevent or reduce interferon production in response to active inducers. One interpretation of these data is that single-stranded polynucleotides cannot bind to the putative interferon inducer receptor, while certain inactive complexes may bind to the receptor without being able to trigger the interferon response.

Carrying this reasoning one step further, one may hypothesize that the interferon induction process occurs in a biphasic manner involving first recognition of a rather large segment of the polynucleotide to allow for proper binding to the receptor, followed by recognition of a much smaller region of the nucleic acid which would then serve as the actual trigger of the induction process (30). Binding of the large segment would be subject to conformational constraints such as those explained in sections 1.1.3 and 1.1.4. The final triggering unit would be composed of no more than one double-helical turn. The latter conclusion stems from a series of studies with partially 2'-O-methylated poly(I)•poly(C) (30), mismatched poly(I_x,U)•poly(C) (24) and spin—labelled poly(I)•poly(C) (31) analogues. Where the actual interferon inducer would be located (at the outer cell membrane or within the cell), is still a matter of conjecture.

2. MATERIALS AND METHODS

2.1 Sources and properties of nucleic acids

The poly(I)•poly(C) used as reference material in the various interferon induction assays was from PL Biochemicals (Milwaukee, Wisconsin). The poly(G) was purchased from Sigma Chemical Co. (St. Louis, Missouri) and had an $s_{20,w}$ = 13.6 S; alternatively poly(G) was obtained from PL Biochemicals in which case the $s_{20,w}$ was 9.0 S. Poly(C) was purchased from PL Biochemicals and its $s_{20,w}$ was 3.08, 8.2 or 13.2 S.

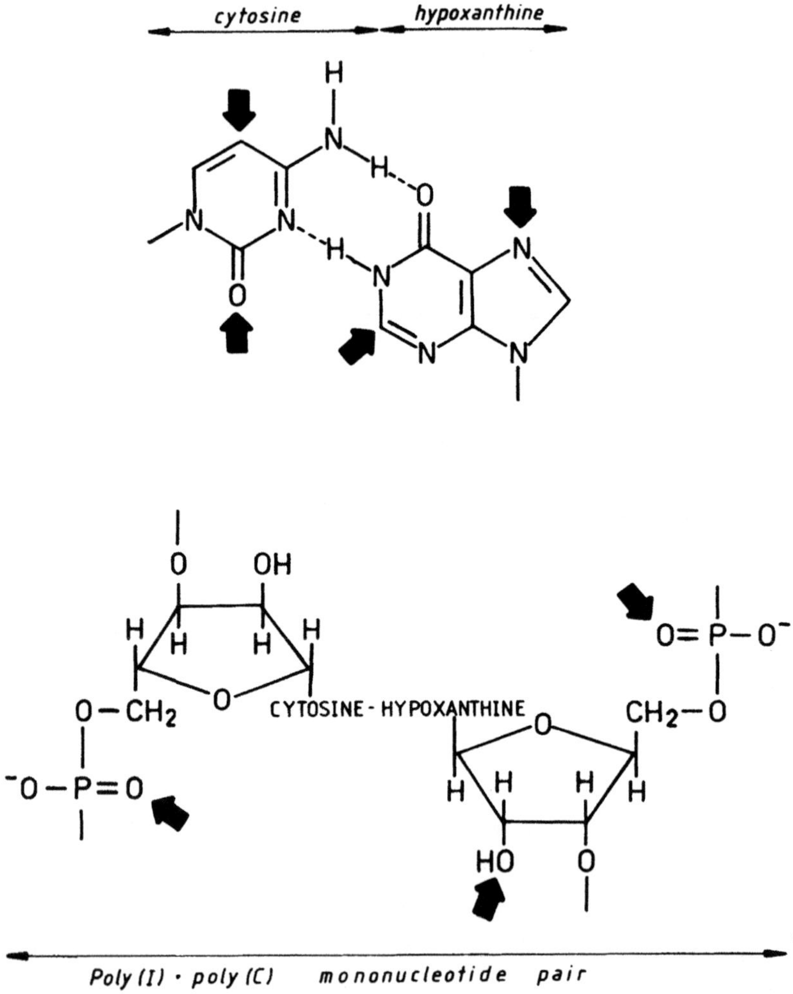

FIGURE 1. Substituents of poly(I)•poly(C) which have proven compatible with significant interferon inducing activity: Br at C-5 of cytosine, S at C-2 of cytosine, CH instead of N-7 of hypoxanthine, NH_2 at C-2 of hypoxanthine, F, Cl or N_3 at C'-2 of inosine and S at the phosphates of either poly(I) or poly(C).

2.2 Preparation of poly(G)·poly(C)

In a typical experiment, poly(G) ($s_{20,w}$ = 13.6 S, 475 μl of 21 mole/ml, 10 μmoles) was mixed at room temperature with poly(C) ($s_{20,w}$ = 13.2 S, 1.49 ml of 6.7 μmoles/ml, 10 μmoles). This mixture was then diluted to 5.0 ml with water and sufficient urea (2.4 g) was added to give a final urea molarity of 8 M and a final polynucleotide phosphate molarity of $4 \cdot 10^{-3}$ Mp. The mixture was then placed in a boiling water bath and heated (2-3 min) until the temperature of the solution reached > 95°. After one additional minute, potassium chloride (4 M) and HEPES (pH 7.5, 1 M) were added to give a final concentration of 0.01 M KCl, 0.01 M HEPES. The solution was removed from the water bath and allowed to cool slowly (∿ 30 min) to room temperature. The entire mixture was then transferred to a dialysis bag and dialyzed at 4° against three-two liter changes of water over a period of 48 to 72 hours. After dialysis was completed, the concentration of poly(G)·poly(C) solution was determined spectrophotometrically.

2.3 Determination of interferon-inducing activity

Interferon production was measured in primary rabbit kidney (PRK) cells, human fibroblasts, murine L-929 cells, rabbits and mice according to previously established procedures (32). The standard procedure for measuring interferon production in PRK cells was as follows. Confluent PRK cells were exposed for 1 h to the polynucleotide in serum-free Eagle's minimal essential medium (MEM) and subsequently treated ("superinduced") with cycloheximide (2 μg/ml) for 3 h and actinomycin D (3 μg/ml) for 30 min. Following this superinduction period, the cells were washed extensively with MEM and further incubated for 20 h with MEM containing 3% heated fetal calf serum. The cell culture fluid was then harvested for interferon titration. As an alternative measure, PRK cells could be exposed for 24 h to the polynucleotide before the superinduction period was initiated (see section 3). When the polynucleotides were exposed to the cells (either PRK, human fibroblast or murine L-929) in the presence of DEAE-dextran (100 μg/ml), the incubation period did not last for longer than 1 h and was immediately followed by the interferon production period (20 h); in this case the cells were not superinduced with cycloheximide and actinomycin D (see section 3).

Rabbit IFN titers were determined by inhibition of the cytopathic effect (CPE) of vesicular stomatitis virus (VSV) in PRK cells. Mouse IFN titers were determined by reduction of VSV plaque formation in murine L-929 cells, and human IFN titrations were carried out in human fibroblast (strain VGS) cultures by using a VSV CPE inhibition technique similar to that used for the rabbit IFN titrations. The human, rabbit and mouse IFN titers are expressed in terms of the U.S. National Institutes of Health (NIH) human fibroblast reference interferon G023-902-527, rabbit reference interferon G019-902-528 and mouse reference interferon G002-904-511 standards, respectively.

3. RESULTS AND DISCUSSION

Of all the base-modified nucleic acids studied as potential interferon inducers, the most elusive has been the poly(G)•poly(C) double helix. This can be considered as an analog of poly(I)•poly(C) in which the hydrogen at C-2 of the purine ring has been replaced by an amino group (Fig. 1). Of course, due to the presence of an additional hydrogen bond formed between the cytosine C-2 oxygen and the guanine amino group, the Tm of poly(G)•poly(C) is extremely high; for instance in 0.01 M phosphate buffer, poly(G)•poly(C) shows no Tm below 90° whereas poly(I)•poly(C) has a Tm of 58° (33).

The interferon-inducing ability of poly(G)•poly(C) has been the subject of considerable disagreement in the literature; for instance, Colby and Chamberlin (34), Aksenov *et al*. (35), Timkovsky *et al*. (36), Vilner *et al*. (37) and Novokhatsky *et al*. (38) reported poly(G)•poly(C) to be an active interferon inducer, but Field *et al*. (39), Field (40), Matsuda *et al*. (41), De Clercq *et al*. (42) and De Clercq and Torrence (43) found no interferon-inducing activity in their poly(G)•poly(C) preparations. The difficulties associated with studies of poly(G)•poly(C) may be closely related to the extraordinarily stable secondary structure of the homopolymer poly(G) (44,45). Aksenov *et al*. (35) attributed the failure of other authors to detect induction with poly(G)•poly(C) to an improper purification of the monopolymer preparations from protein and salt impurities. The authors (35-38) have claimed that to produce an active poly(G)•poly(C), special consideration should also be devoted to the molecular weight of the homopolymers and the annealling conditions used for complex formation. However, quite a number of attempts to prepare an active poly(G)•poly(C) duplex in our laboratories following the above recommendations or using the annealing procedures of Pochon and Michelson (46) and Englander et al. (47) invariably failed (Torrence and De Clercq: unpublished data). We have speculated that, since Colby and Chamberlin (34) prepared an active poly(G)•poly(C) duplex* by transcribing poly(G) from a poly(C) template, such direct transcription of native poly(G) on an existing poly(C) strand would be the only means to circumvent the problem of self-annealling of poly(G) and to obtain a proper poly(G)•poly(C) duplex (see also 48).

Recent observations from our laboratories have demonstrated the complexit of the poly(G)•poly(C) issue, but, as the same time, have provided a final resolution of the question: can poly(G)•poly(C) induce interferon?

3.1 Significant interferon-inducing ability of a poly(G)•poly(C) duplex depends upon the molecular weights of the component homopolymers.

Only poly(G)•poly(C) prepared from the highest available molecular weight homopolymers had maximum interferon inducing activity (Table 2). Poly(G)•poly(C) prepared from high molecular weight poly(G) and medium molecular weight poly(C) was of intermediate activity, whereas poly(G)•poly(C) prepared from poly(C) of low molecular weight or from poly(G) of intermediate molecular weight was devoid of detectable interferon inducing activity.

Table 2. Dependence of interferon inducing acitvity of poly(G)•poly(C) on the size of the individual homopolymers

Sedimentation constants ($s_{20,w}$)[a]		Concentration[b]	IFN titer[c,d]
poly(G)	poly(C)		($\log_{10}$ units/ml
13.6	13.2	10^{-4}	3.5
13.6	8.2	10^{-4}	2.0
13.6	3.08	10^{-4}	< 1.3
9.0	13.2	10^{-4}	< 1.3

[a] In Svedbergs.
[b] Molarity in phosphate residues (Mp).
[c] As determined in primary rabbit kidney cells superinduced with cycloheximide and actinomycin D. Polymers were exposed to the cells for 24 hours before th superinduction procedure was initiated.
[d] Under the same conditions, poly(I)•poly(C) induced 2.0 $\log_{10}$ IFN units/ml whe applied to the cells at 10^{-4} Mp for 24 hours.

*In this study (34), only virus resistance, and not interferon production per se, was measured.

Other authors (38) using a different annealling procedure and measuring Sindbis virus cytopathogenicity in chick embryo cells have also found that the activity of the poly(G)•poly(C) complex depended upon the molecular weight of the component homopolymers. These results are in general agreement with the data of other polynucleotide complexes such as poly(I)•poly(C) (as reviewed in 2 and 14) or poly(A)•poly(U) (49). From the data presented in Table 2 it is also clear that the interferon-inducing activity of the poly(G)•poly(C) complexes was more sensitive to a reduction of the molecular weight of poly(G) than of poly(C) (37).

3.2 The induction of interferon by the poly(G)•poly(C) duplex is dramatically dependent upon the manner in which the component homopolymers are annealled.

This is illustrated in Table 3 where it is demonstrated that the complex, if any, formed at room temperature in 8 M urea was devoid of interferon-inducing activity whereas that annealled in 8 M urea at a temperature > 90° possessed considerable inteferon inducing capacity. The annealling method employed here is a modification of the method of Englander *et al*. (47). Previous work led to the recommendation that the poly(G) and poly(C) mixture should be boiled in order to produce maximal antiviral activity (38). However, according to our own experience and that of Englander *et al*. (47) such treatment does not give a clear 1:1 stoichiometric complex as judged by mixing curves constructed by the method of continuous variation. On the other hand, the original urea technique introduced by Englander *et al*. (47) and its modification employed here does lead to a well-defined 1:1 complex stoichiometry.

3.3 The interferon-inducing ability of poly(G)•poly(C) depends upon the conditions of biological assay.

If poly(G)•poly(C) was prepared according to the above protocol and then exposed to primary rabbit kidney cells for 1 hour before superinduction was begun (the usual procedure for interferon induction by polynucleotide interferon inducers), then it was completely devoid of interferon inducing activity (Table 4). If, however, the complex was exposed to the cells for 24 hours before the superinduction procedure was initiated, then poly(G)•poly(C) stimulated the production of interferon (Table 4) in quantities comparable to that induced by poly(I)•poly(C) under the same conditions. Small but measurable interferon titers were obtained even in the absence of the superinduction procedures (data not shown).

High levels of interferon were obtained if poly(G)•poly(C) was exposed to the cells in the presence of DEAE-dextran of $CaCl_2$, but not poly(l-lysine) or poly(l-ornithine). In this case, no superinduction was required and the time of exposure to the cells could be reduced from 24 hours to 1 hour (Table 4). Priming of the cells with rabbit interferon did not increase the amounts of interferon induced by poly(G)•poly(C) (data not shown). If, after the 24 hour exposure period, the cell culture medium was harvested and exposed to fresh primary rabbit kidney cells for 1 hr, no interferon was produced; in addition,

Table 3. Interferon induction by poly(G)•poly(C) depends upon the method of complex formation

Procedure[a,b]	IFN titer of resulting complex (log_{10} units/ml)[c]
1. Poly(G) and poly(C) were mixed in 8 M urea, heated to 95° for 2 min; 4 M KCL and 1 M Hepes (pH 7.5) were added to a final concentration of 0.01 M each; the solution was cooled to ambient temperature over a period of 30 min and then dialyzed at 4° against 3 changes of water.	3.0
2. Same as above, but sample not heated.	< 1.3

[a]Poly (G) and poly(C) had $s_{20,w}$ values of 13.6 and 13.2 S, respectively. [b]Final concentration of poly(G)•poly(C) : 10^{-4} Mp.
[c]As determined in primary rabbit kidney cells exposed to the complex for 1 h in the presence of DEAE-dextran (100 μg/ml); cells not superinduced with cycloheximide and actinomycin D.

if poly(G)•poly(C) was incubated in MEM (without cells) for 24 hours and then exposed to the cells for 1 h, no interferon production was apparent (data not shown).

The interferon inducing ability of poly(G)•poly(C) was also critically influenced by the choice of the cell system (Table 5). Maximum interferon production was noted in primary rabbit kidney cells exposed to poly(G)•poly(C) for 1 h in the presence of DEAE-dextran. A similar procedure was also effective in triggering an interferon response in murine L-929 cells, but human skin fibroblasts (strain VGS), or embryo-skin muscle fibroblasts (strain E_6SM) or fibroblastoid cells [strain MG-63, derived from an osteosarcoma (50)] failed to yield any detectable interferon under these conditions (Table 5). However, the latter cell line showed a marked interferon response to poly(G)•poly(C), when exposed to the complex for 24 h and then superinduced (data not shown).

Table 4. Interferon induction by poly(G)•poly(C) depends upon the assay conditions

Procedure	IFN titer ($\log_{10}$ units/ml)[a]	
	poly(G)•poly(C)[b]	poly(I)•poly(C)[b]
1. Complex exposed to the cells for 1 hour	< 0.8	4.7
2. Complex exposed to the cells for 24 hours	2.3	2.0
3. Complex exposed to the cells for 1 h in the presence of DEAE-dextran (100 μg/ml).	3.0	3.5
4. Complex exposed to the cells for 1 h in the presence of $CaCl_2$ (5 mM)	3.5	4.5

[a]As determined in primary rabbit kidney cells, superinduced with cycloheximide and actinomycin D (procedures 1 and 2), or not superinduced (procedures 3 and 4).
[b]Final concentration : 10^{-5} Mp.

The interferon inducing activity of poly(G)•poly(C) was not confined to *in vitro* situations. From Fig. 2, it is readily apparent that poly(G)•poly(C) showed *in vivo* activity in both mice and rabbits. In mice, poly(G)•poly(C) was as effective as interferon inducer as was poly(I)•poly(C). In rabbits, it produced lower titers of interferon than did poly(I)•poly(C), but the serum interferon response was more protracted for poly(G)•poly(C) than for poly(I)•poly(C): *i. e.*, the interferon titers produced in response to poly(G)•poly(C) did not diminish over the first 24 hours following poly(G)•poly(C) injection, whereas the interferon titers in response to poly(I)•poly(C) peaked at 3-6 hours and declined rapidly thereafter.

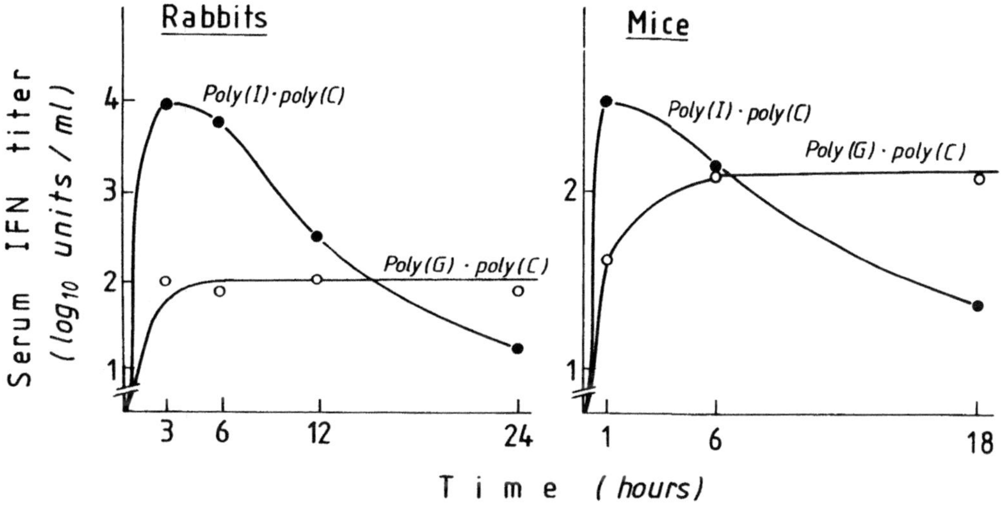

FIGURE 2. Interferon-inducing activity of poly(G)•poly(C) in rabbits and mice. Rabbits : Young albino white rabbits (weighing ∿ 1 kg) were injected intravenously with 1 ml Dulbecco's PBS containing either poly(G)•poly(C) or poly(I)•poly(C) at 10^{-4} Mp. Blood samples were removed at the indicated times. Mice : Female NMRI mice (weighing 20-25 g) were injected intravenously with 0.2 ml of Dulbecco's PBS containing either poly(G)•poly(C) or poly(I)•poly(C) at 10^{-5} Mp. Blood was withdrawn at the indicated times (3 mice/group). Serum interferon titers were determined as described in Materials and Methods.

4. CONCLUSIONS

The results described here resolve in a definitive manner the controversy on the interferon inducing ability of the double-stranded poly(G)•poly(C) complex. Thus, significant interferon-inducing activity of poly(G)•poly(C) depends upon a sufficiently high molecular weight of both homopolymers, the method by which the two complementary strands are annealled, and the manner by which the resulting complex is assayed for biological activity in cell culture. As to this latter criterion, our studies have established that, for the interferon-inducing potency of poly(G)•poly(C) to be expressed, it must be exposed to the cells for 24 hours, a much longer exposure time than is required for other active inducers such as poly(I)•poly(C). Alternatively, either DEAE-dextran or $CaCl_2$, when administered together with the poly(G)•poly(C), can eliminate the necessity of the 24 hour exposure. These results suggest, but do not establish, that poly(G)•poly(C) is poorly (or slowly) taken up by cells. Changes in the poly(G)•poly(C) structure do not appear to account for the peculiar behavior of poly(G)•poly(C) since the medium from cells exposed to poly(G)•poly(C) for 24 h produced no interferon when exposed to fresh cells for 1 hour.

Table 5. Interferon induction by poly(G)•poly(C) depends upon the choice of the cell system.

Cells[a]	IFN titer ($\log_{10}$ units/ml)	
	Poly(G)•poly(C)[b]	Poly(I)•poly(C)[b]
Primary rabbit kidney cells	3.2	4.1
Murine L-929 cells	1.6	2.8
Human fibroblasts		
VGS strain	< 0.5	2.6
E_6SM strain	< 0.7	2.0
MG-63 strain	< 1.0	2.3

[a]Exposed to the complex for 1 h in the presence of DEAE-dextran (100 μg/ml); cells not superinduced with cycloheximide and actinomycin D.
[b]Final concentration : 10^{-4} Mp.

Poly(G)•poly(C) is endowed with a number of properties which make it an atypical polynucleotide interferon inducer. First, it is active on primary rabbit kidney cells, usually exquisitely sensitive to polynucleotide interferon inducers, but only after prolonged exposure. Alternatively, it is active when added to the cells in the presence of DEAE-dextran or $CaCl_2$. Secondly, it shows a different degree of activity depending on the choice of the cell line. For instance, poly(G)•poly(C) stimulates the production of interferon in primary rabbit kidney cells, mouse L cells and in human MG-63 cells, mostly less effectively than does poly(I)•poly(C) under similar conditions. However, in human fibroblast lines such as E_6SM and VGS which respond well to interferon induction by poly(I)•poly(C), poly(G)•poly(C) does not induce detectable interferon. Thirdly, poly(G)•poly(C) is a good interferon inducer in vivo, in both mice and rabbits. Interferon induction by poly(G)•poly(C) differs from induction by poly(I)•poly(C) in the above animal species in that poly(I)•poly(C) gives interferon titers peaking shortly after injection of inducer, whereas the interferon response to poly(G)•poly(C) does not reveal such a peak but shows a rather protracted interferon production course.

The rather unusual behavior of poly(G)•poly(C) as an interferon inducer suggests it may be valuable to characterize further this nucleic acid with regard to the species (α, β, γ) and subspecies of interferon induced, the hyporeactivity following interferon induction, and the in vivo toxicity of poly(G)•poly(C) and its in vivo antiviral and antitumor potentials.

REFERENCES

1. Torrence, P. F. and De Clercq, E. Methods Enzymol. 78: 291-299, 1981.
2. Torrence, P. F. and De Clercq, E. Pharmacol. Ther. A2: 1-88, 1977.
3. Torrence, P. F. In: Hormone Receptors (Ed. L. D. Kohn), J. Wiley and Sons, Ltd., London 1982. pp 335-355.
4. Stollar, B. D. In: The Antigens (Ed. M. Sela), Academic Press, New York, 1973, vol. 1, pp. 1-85.
5. Stollar, B. D. Crit. Rev. Biochem. 3: 45-69, 1975.
6. Lacour, F., Nahon-Merlin, E. and Michelson, M. Curr. Top. Microbiol. 62: 1-39, 1973.
7. Revel, M. In: Interferon 1 (Ed. I. Gresser), Academic Press, New York, 1979, pp 101-163.
8. Baglioni, C. Cell 17: 255-264, 1979.
9. Hovanessian, A. G. Differentiation 15: 139-151, 1979.
10. Torrence, P. F. Mol. Aspects Med., 5, 129-171, 1982.
11. Torrence, P. F., De Clercq, E., Waters, J. A. and Witkop, B. Biochemistry 13: 4400-4408, 1974.
12. Reuss, K., Scheit, K.-H. and Saiko, O. Nucleic Acids Res. 3, 2861-2875, 1976.
13. De Clercq, E. Eur. J. Biochem. 93: 165-172, 1979.

14. De Clercq, E. Top. Curr. Chem. 52: 173-208, 1974.
15. Torrence, P. F., Waters, J. A., Buckler, C. E. and Witkop, B. Biochem. Biophys. Res. Commun. 52: 809-809, 1973.
16. De Clercq, E., Torrence, P. F., Stollar, B. D., Hobbs, J., Fukui, T., Kakiuchi, N. and Ikehara, M. Eur. J. Biochem. 88: 341-349, 1978.
17. De Clercq, E., Stollar, B. D. Hobbs, J., Fukui, T., Kakiuchi, N. and Ikehara, M. Eur. J. Biochem. 107: 279-288, 1980.
18. Kakiuchi, N., Marck, C., Rousseau, N., Leng, M., De Clercq, E. and Guschlbauer, W. J. Biol. Chem. 257: 1924-1928, 1982.
19. De Clercq, E., Torrence, P. F. and Witkop, B. Proc. Natl. Acad. Sci. USA 71: 182-186, 1974.
20. De Clercq, E., Edy, V. G., Torrence, P. F., Waters, J. A. and Witkop, B. Molec. Pharmacol. 12: 1045-1051, 1976.
21. Miles, D. L., Miles, D. W. and Eyring, H. Proc. Natl. Acad. Sci. USA 76: 1018-1021, 1979.
22. Bobst, A. M., Torrence, P. F., Kouidou, S. and Witkop, B. Proc. Natl. Acad. Sci. USA 73: 3788-3792, 1976.
23. Black, D. R., Eckstein, F., De Clercq, E. and Merigan, T. C. Antimicrob. Agents Chemother. 3: 198-206, 1973.
24. De Clercq, E., Huang, G.-F., Bhooshan, B., Ledley, G. and Torrence, P. F. Nucleic Acids Res. 7: 2033-2014, 1979.
25. Ts'o, P.O.P., Alderfer, J. L., Levy, J., Marshall, L. W., O'Malley, J., Horoszewicz, J. S. and Carter, W. A. Molec. Pharmacol. 12: 299-312, 1976.
26. Carter, W. A., O'Malley, J., Beeson, M., Cunningham, P., Kelvin, A., Vere-Hodge, A., Alderfer, J. L. and Ts'o, P. O.P. Molec. Pharmacol. 12: 440-453, 1976.
27. De Clercq, E., Torrence, P. F., Witkop, B. and De Somer, P. In: Effects of Interferon on Cells, Viruses and the Immune System (Ed.
28. A. Geraldes), Academic Press, New York, 1975, pp 215-236.
28. O'Malley, J. A., Ho, Y. K., Chakrabarti, P., Di Berardino, L., Chandra, P., Orinda, D. A. O., Byrd, D. M., Bardos, T. J. and Carter, W. J. Molec. Pharmacol. 11: 61-69, 1975.
29. Johnston, M. D., Atherton, K. T., Hutchinson, D. W. and Burke, D. C. Biochim. Biophys. Acta 435: 69-75, 1976.
30. Greene, J. J., Alderfer, J. L., Tazawa, I., Tazawa, S., Ts'o, P. O. P., O'Malley, J. A. and Carter, W. A. Biochemistry 17: 4214-4220, 1978.
31. Bobst, A. M., Langemeier, P. W., Torrence, P. F. and De Clercq, E. Biochemistry 20: 4798-4803, 1981.
32. De Clercq, E. Methods Enzymol. 78: 227-236, 1981.
33. Janik, B. In: Physiochemical Characteristics of Oligonucleotides and Polynucleotides. IFI/Plenum, New York, 1971.
34. Colby, C. and Chamberlin, M. J. Proc. Natl. Acad. Sci. USA 63: 160-167, 1969.
35. Aksenov, O. A., Timkowsky, A. L., Ageeva, O. N., Kogan, E. M., Bresler, S. E., Smorodinstev, A. A. and Tikhomirova-Sidorova, N. S. Vop. Virus 18: 345-349, 1973.
36. Timkovsky, A. L., Aksenov, O. A., Bresler, S. E., Kogan, E. M., Smorodintsev, A. A. and Tikhomirova-Sidorova, N. S. Vop. Virus 18: 345-349, 1973.
37. Vilner, L. M., Brodskaya, L. M., Kogan, E. M., Timkovsky, A. L., Tikhomirova-Sidorova, N. S. and Bresler, S. E. Vop. Virus 19: 45-49, 1974.
38. Novokhatsky, A. S., Ershov, F. I., Timkovsky, A. L., Bresler, S. E.,

Kogan, E. M. and Tikhomirova-Sidorova, N. S. Acta Virol. 19: 121-129, 1975.

39. Field, A. K., Tytell, A. A., Lampson, G. P. and Hilleman, M. R. Proc. Natl. Acad. Sci. USA 58: 1004-1009.
40. Field, A. K. In: Selective Inhibitors of Viral Functions (Ed. W. A. Carter), CRC Press, Cleveland, Ohio, 1973, pp 149-176.
41. Matsuda, S., Kida, M., Shirafuji, H., Yoneda, M. and Yaoi, H. Arch Ges. Virusforsch. 34: 105-118, 1971.
42. De Clercq, E., Eckstein, F. and Merigan, T. C. Ann. N. Y. Acad. Sci. 173: 444-461, 1970.
43. De Clercq, E. and Torrence, P. F. J. Gen. Virol. 37: 619-623, 977.
44. Fresco, J. R. and Massoulié, J. J. Amer. Chem. Soc. 85: 1352-1353, 1963.
45. Guschlbauer, W. In: The Jerusalem Symposium on Quantum Chemistry and Biochemistry IV. The Purines. Theory and Experiment. (Eds. E. D. Bergman and B. Pullman), Academic Press, New York, 1972, pp 247-309.
46. Pochon, F. and Michelson, A. M. Proc. Natl. Acad. Sci. USA 53: 1425-1430, 1965.
47. Englander, J. J., Kallenbach, N. R. and Englander, S. W. J. Mol. Biol. 63: 153-169, 1972.
48. Gomatos, P. J. and Kuechenthal, I. J. Virol. 23: 80-90, 1977.
49. Janik, B., De Clercq, E. and Sommer, R. G. Biochim. Biophys. Acta 517: 269-273, 1978.
50. Billiau, A., Edy, V. G., Heremans, H., Van Damme, J., Desmijter, J., Georgiades, J. A. and De Somer, P. Antimicrob. Agents Chemother. 12: 11-15, 1977.

20
INTERFERON AND INTERFERON-LIKE FACTORS IN PLANTS

ILAN SELA
Virus Laboratory, Faculty of Agriculture, Rehovot, Israel

INTRODUCTION AND BACKGROUND

In accordance with the general policy of this series, this communication summarizes the study of the antiviral factor (AVF) and related subjects as developed in my laboratory. It is not a review article describing the broader subject of resistance, localization and related protective phenomena. Hence, references are made only to material directly associated with the described subjects. Also, since ongoing research is also described, data of students' theses or articles in preparation for publication will also be referred to. For review articles giving a broader spectrum of related subjects and outlining the evolving concepts of the field of resistance against plant virus diseases, the reader is referred to Loebenstein (1) and to Sela (2; 3). Since the study of AVF is in its midst, the tentative models and speculations brought up are clearly so defined.

The story of AVF began with an unrelated project of purifying prune ringspot virus. Fulton (4) described the adsorption of this prune virus to hydrated calcium phosphate (HCP). We tried the procedure with a number of plant viruses and found that all, with the exception of potato virus X, adsorbed to HCP and could be centrifuged down with the gel, leaving a non-infectious supernatant fluid (5). Only then did we realize that this simple technique made it possible to compare saps from virus-infected and non-infected plants without introducing the factor of infectivity. When, in preliminary studies, non-infectious saps, derived from virus-infected plants, inhibited subsequent virus infection (5), the presence of an antiviral factor was assumed and its study initiated.

THE MEASUREMENT OF ANTIVIRAL ACTIVITY IN PLANTS

In some instances plants react to virus infection by producing local lesions (6). The number of inflicted lesions is proportional to the virus concentration in the inoculum. It is however almost impossible to achieve a linear relationship between lesion number and infectivity since, unlike the situation in animal and bacterial virology, clonal ("identical twin") cell cultures are not available and plant protoplasts do not react with local lesions. Hence, the information obtained from a local lesion test is not quantitative. Nevertheless, quantitative comparisons can be made on half-leaves, testing the infectivity of a certain inoculum relative to another (7).

Antiviral activity can be biologically measured in a number of ways: (i) The antiviral fraction is mixed with a known virus inoculum and applied to half-leaves of a locally reacting plant, while a control-treated inoculum is applied to the opposite halves of the same leaves. (ii) Systemically reacting plant leaves are inoculated with the virus. Leaf-discs are punched therefrom and homogenized. The infectivity of these homogenates is tested in half-leaves of a locally reacting plant. This method allows the application of the tested material at various times before and after inoculation, and the follow-up of growth curves of the virus rather than mere spot tests (for instance ref 8). A nice and promising variation of this method is the replacement of leaf-discs by protoplasts (for example, Loebenstein _et al._, ref 9). Although protoplasts do not comprise a clone of cells, they are far less heterogeneous than the various cells within a leaf disc and they allow a synchronous infection of practically all the cultured cells. The determination of infectivity level is, however, still dependent upon application to a locally reacting plant, and is, therefore, not quantitative.

For better quantification the virus content in leaf-discs or protoplasts must be directly measured. A step forward towards this goal is the employment of labelled specific antibodies. In recent studies of AVF and interferon in plants (10, 11, 12) we somewhat modified the enzyme-linked immunosorbent assay (ELISA) of Clark and Adams (13). Basically, specific antibodies against the virus are conjugated to an enzyme. The presence of the virus is thus detected through the enzymatic activity. Recently, automated measurements of ELISA plates became available, enabling kinetic reading of an ELISA reaction. Virus content is thus determined at a linear range, in exactly the same way that enzyme units are determined, and results can be expressed in actual nanograms of virus.

SEPARATION OF VARIOUS PLANT FRACTIONS AND THE PURIFICATION OF AVF

It was only for reasons of convenience that TMV-infected Nicotiana glutinosa leaves were chosen as a source for AVF purification. As described later, N. glutinosa carried a dominant N-gene which is responsible for the localization of TMV infection. The association of AVF with the N-gene was established only much later (14), thus, the accidental choice of this particular plat-virus combination led to the successful purification of AVF.

HCP-treated plant sap was designated "crude AVF" since it was the first TMV-devoid fraction that could be tested for antiviral activity (5). However, this single step, which removes 90-95% of the plant sap proteins leaving the AVF unadsorbed, amounts to about 10-20 fold purification. Crude AVF can be stored as a lyophilized powder, in a desiccator, at room temperature, for a few months.

As it turned out, some other activities, beside that of AVF, were retained in the HCP-treated plant sap and could be separated by chromatography on DEAE-cellulose. Hence, "crude AVF" became the starting material for purification of a number of active substances (15). A schematic representation of the various plant fractions is shown in Fig. 1.

AVF itself, upon further chromatography on DEAE-cellulose, is eluted as a broad zone between 0.48 and 0.60 M NaCl in 0.01 M sodium phosphate buffer, pH 7.6. This is quite a unique position for a protein to be eluted from such a column (16).

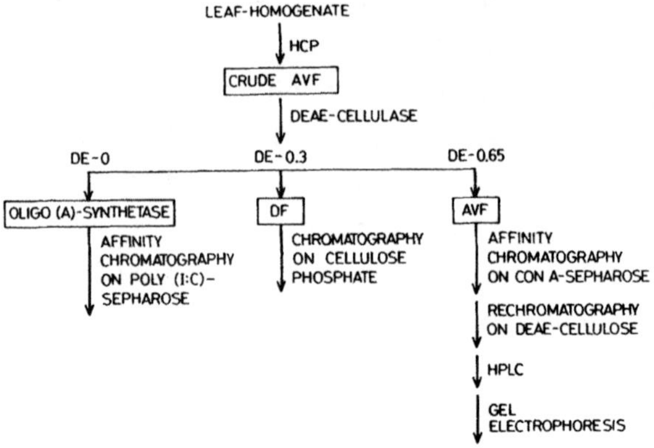

FIGURE 1. A schematic illustration of the preparation of various plant fractions employed throughout this study (from Perspectives in Virology XI:129-139, 1981). (Reprinted with permission).

In some cases an AVF-like material was identified as one which exhibits antiviral activity and chromatographs on DEAE-cellulose like AVF (17). For routine work AVF is eluted from this column, in a step-wise manner, between 0.45-0.60 M NaCl. When the other plant fractions are also required, step-wise elution is carried out, as shown in Fig. 1.

For further purification, AVF is adsorbed onto a column of concanavalin A bound to Sepharose. Being a glycoprotein (see below), AVF binds to the lectin and is eluted therefrom with α-methyl mannoside and α-methyl glucoside (18; 19).

AVF can be purified to a higher degree of homogeneity by high-performance liquid chromatography (HPLC). When the procedure for the purification of human leukocyte interferon (20; 21) is followed, a number of fractions, exhibiting antiviral activity, are identified. The most active of these reduces TMV multiplicity to "zero", i.e., to a level below the sensitivity of the ELISA test (22). When this fraction is electrophoresed on SDS-polyacrylamide gels and stained with coomassie blue, a single band of about 21-22,000 daltons appears. The material extracted from the corresponding position is highly active as an antiviral agent (Fig. 2).

A few comments are due at this point. We have isolated and described one substance with antiviral activity which is stimulated in a plant following TMV infection. At a number of

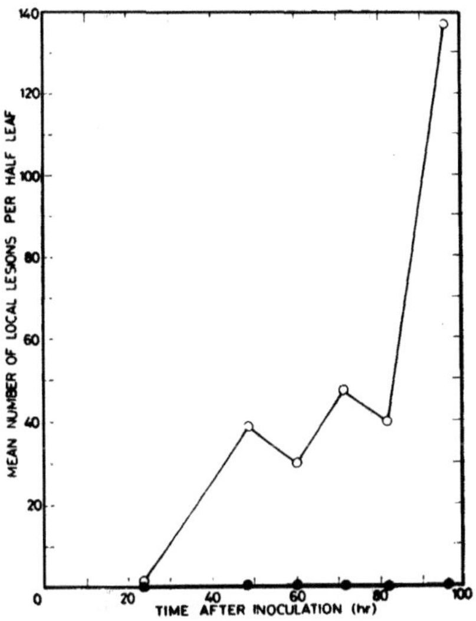

FIGURE 2. Growth curve of TMV in tobacco "Samsun" leaf-discs with (●) and without (o) AVF (100 pg protein/ml), eluted from a polyacrylamide gel after it had been separated by HPLC.

points along the process of purification we had to make selections. For instance, for logistic reasons we selected the material which does not adsorb to HCP, or when a number of HPLC fractions exhibited antiviral activity, we again selected only one. Hence, it would not be surprising if other, similar factors, are found. Furthermore, in almost every step (except for the final selection of the HPLC fraction) AVF is observed as a broad zone rather than a peak (8; 14; 18), indicating an internal heterogeneity. We do not know whether this heterogeneity results from the existence of a number of closely-related, but distinct, AVF subspecies, as in the case of human leukocyte interferon, or if it is due to various degrees of glycosylation, phosphorylation or other derivatives of the same protein core.

The HPLC fraction, further electrophoresed and eluted from a gel, can be considered homogeneous. Hence its activity can be estimated on a molar basis. The end point of AVF activity is about 10^{-14} to 10^{-17} M and the estimated number of cells in a standard leaf-disc test is 10^{6}-10^{7}. Thus, a few molecules of AVF (or even less) per cell are sufficient to induce an antiviral state in the plant tissue.

THE CHEMICAL NATURE OF AVF

AVF is eluted from DEAE-cellulose at a relatively high salt concentration and, upon extraction with phenol, separates into the aqueous phase (23), indicating its hydrophylic nature and negative charge.

N. glutinosa leaves were labelled with radioactive phosphorus. The labelled DEAE-eluted AVF was electrophoresed on SDS-polyacrylamide gels, resulting in a radioactive zone specific to fractions from infected plants and almost absent from non-infected plants (8). The antiviral activity co-electrophoresed with the radioactivity (14). AVF resists low pH conditions. It can be kept at pH 2.0 (18) or chromatographed at 1 M acetic acid (22) without losing its antiviral activity. It also remains active after being extracted from SDS-polyacrylamide gels (18).

AVF binds to concanavalin A. This is a specific binding of mannose and/or glucose to the lectin, not due to hydrophobic interactions as reported for several materials, including interferons (24; 25). This was concluded because α-methyl mannoside and α-methyl glycoside could compete with AVF and elute it off the concanavalin A, while ethylene glycol had no effect when applied either before or after the sugar derivatives (19).

Concanavalin A-eluted AVF had to be re-chromatographed on DEAE-cellulose in order to remove the lectin, some of which was "leaked" from the column. Since at this stage only nanogram amounts of protein remained in the active fraction, a small "leakage" of the lectin means 100-1000 times more concanavalin A than AVF. This AVF preparation, hydrolyzed and analyzed by gas-liquid chromatography, contained α-glucose, α-mannose, galactose, xylose and arabinose.

Concanavalin A-eluted AVF electrophoreses on polyacrylamide gels as a broad "smear" which can be stained by both coomassie blue, for protein, and the Schiff reagent, for carbohydrates (18).

A major band, sensitive to the cooperative action of α-glucosidase and alkaline phosphatase was, however, observed. This band migrated as a 22,000 dalton protein, nicely agreeing with the molecular of the above described HPLC fraction. AVF is also sensitive to pronase (23). Evidently AVF is a glycoprotein, possibly a phosphorylated one.

Recently, Orchansky treated AVF with a mixture of 12 different glycosidases attached to CNRr-activated Sepharose, in the same manner described for human fibroblast interferon (10). The glycosidase-treated AVF was dialyzed and tested for antiviral activity, while the liquid outside the dialysis bag was lyophilized and checked for sugar-residues by gas liquid chromatography, in order to ascertain that deglycosylation had indeed takn place. After a relatively short incubation (3 hr) with the glycosidases, AVF lost its antiviral activity, but regained it after further incubation, for 24 hr. This was true for both DEAE- and concanavalin A-eluted AVF. Surprisingly, mock DEAE-eluted samples, prepared from non-infected plants and lacking antiviral activity, gained such activity upon deglycosylation. Concanavalin A-eluted mock fractions, however, remained inactive even upon deglycosylation. The only solid conclusion that can be drawn from these results is that the core of deglycosylated AVF is still active. It is tempting, however, to assume a model into which the rest of the data will also fit: The protein core of AVF is by itself, active, while the carbohydrate moiety was developed simply to enable its secretion or to confer specific cell receptor recognition on AVF. Damaging the distal end of the carbohydrate chains abolishes recognition, but further degradation results in non-specific activity of the protein core. The relevance of this model may be tested only when sufficient amounts of pure AVF become available.

It is believed that a precursor of AVF (pre-AVF) is present in plants of the genus Nicotiana and is processed to become active AVF upon infection with TMV or other stimuli (described in detail in a latter chapter). If this pre-AVF is composed of the same protein core as AVF, only differently glycosylated, its deglycosylation will produce the same active AVF core. This may account for the antiviral activity of deglycosylated mock-AVF preparations, but does not necessarily indicate that this is the actual AVF processing pathway in the tissue. Pre-AVF must then be quite similar, in structure, to AVF, since it purifies with it up to the point of being co-eluted from DEAE-cellulose. However, AVF separates from this postulated pre-AVF at the stage of affinity chromatography. Mock concanavalin A-eluted preparations could not be activated by deglycosylation.

AVF can be induced in N. glutinosa by poly-ionisinic:polycytidylic acid (poly I:C; ref. 12). Even though unglycosylated AVF is active, tunicamycin, which inhibits protein glycosylation, prevents its induction by poly I:C in N. glutinosa. This may indicate either that other glycoproteins participate in the stimulation of AVF, the inhibition of which causes the cessation of AVF stimulation, or that the poly I:C induced AVF differs from the TMV-induced one.

AVF STIMULATION IN INFECTED PLANTS

TMV localization and the N-gene

Virus infection in plants can be of a systemic nature, where the virus spreads all over the plant, or localized, where, at a certain stage, virus multiplication gradually slows down until it ceases completely and the virus remains confined to a limited area around its site of entry. Localization usually results in the formation of local lesions, surrounded by a ring of tissue, which is resistant to subsequent viral infection (26; 27). The non-infected tissue, beyond these protected rings, is also partially resistant (28).

The fate of TMV infection in plants of the genus Nicotiana, whether it will be local or systemic, depends on a single host gene. Plants carrying the dominant allele (N), localize TMV infection, while plants carrying the recessive allele (n) allow systemic infection (29). The various genetic aspects of the N-gene and localization have been recently discussed in detail elsewhere (2).

The plant *N. glutinosa* carries the N-gene and localizes TMV infection. All native tobacco (*N. tabacum*) varieties carry the n allele and react systemically to TMV infection. Plant breeders have, however, succeeded in introducing the N-gene, from *N. glutinosa*, to tobacco. A number of pairs of tobacco cultivars, which differ from each other in carrying the N versus the n allele, are therefore available. One such pair is the systemically reacting tobacco variant "Samsun" and its N-gene-carrying cultivar, "Samsun NN", which localizes TMV infection.

It is a well established fact that in order to make a plant resistant to viral infection, a number of resistance genes, from different sources, must be introduced. This is due to the rapid rate of virus mutation, which quickly produces a virus strain which can escape single gene-derived resistance. It is surprising, therefore, that although TMV localization, in *Nicotiana*, depends on a single gene, none of the hundreds of naturally occurring, or artificially induced, TMV mutants has escaped localization (30). This usually means that any such mutation is lethal, implying that the viral event which affects the N-gene-directed localizing mechanism is so essential to TMV multiplication, that a virus mutation which causes such a variation that does not allow an escape from the regular N-gene-directed processes, cannot survive. Our tentative suggestion is that the replicative form of TMV-RNA, activates this protective machinery.

The association of the N-gene with AVF stimulation

When ^{32}P-labelled DEAE-eluted AVF, from TMV-infected *N. glutinosa* leaves, is electrophoresed on polyacrylamide gels, a typical pattern of radioactivity and antiviral activity is obtained. This is reduced to almost background levels if derived from a non-infected plant or from TMV-infected tobacco "Samsum". When the AVF fraction from TMV-infected tobacco "Samsun NN", which carries the N-gene, is electrophoresed, the *N. glutinosa*-typical pattern is reproduced (Fig. 3; ref. 8). Recently, Loebenstein and Gera (31) demonstrated the producttion of an AVF-like substance, designated "inhibitor of virus

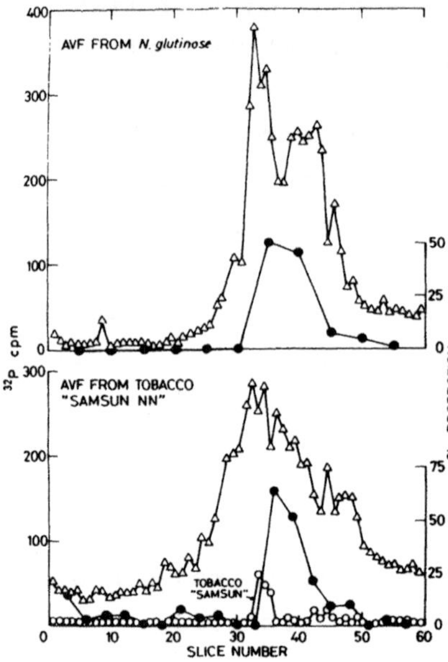

FIGURE 3. Patterns of radioactivity (Δ) and antiviral activity (●) following electrophoresis of DEAE-eluted AVF on SDS-polyacrylamide gels. Radioactivity from a similar fraction derived from tobacco "Samsun" is represented by (o) (from Antignus et al., J. Gen. Virol. 35:107-116, 1977). (Reprinted with permission).

replication" (IVR), in tobacco protoplasts. Again, IVR was produced only in TMV-infected protoplasts from tobacco "Samsun NN" and not from its homologous n-gene-carrying cultivar.

The stimulation of AVF and IVR is clearly associated with the presence of the N-gene in the cells (for the role of this gene in AVF stimulation see next section). Only preliminary studies have so far been carried out to determine whether the effect of the N versus the n gene is a "yes or no" situation or merely a quantitative one. Ratner (17) has recently determined as AVF-like the antiviral substance extractable from the "green islands" of tobacco "Samsun" (32; 33), systemically infected with TMV. Moreover, such a substance, at lower levels, was extracted from the "yellow islands" of the same leaves as well - both activities were compared to non-infected plants. AVF-like material could also be extracted from N. glutinosa systemically infected with cucumber (CMV) or alfalfa (AMV) mosaic virus. In all of these systemic infections, the antiviral substance appeared in relatively small amounts and late in infection. In the TMV-localizing Datura stramonium, however, a very strong AVF activity was detected within 48 hr of inoculation. In Chempodium amaranticolor, which localizes almost every viral infection, no antiviral activity which separates on DEAE-cellulose like AVF could be found. It seems that, at least in solanaceous plants, AVF is stimulated by viral infection in systemically reacting plants too but to a lesser degree and later in infection, and that the N-gene effect is of a quantitative nature only contributing to a better efficiency of AVF stimulation.

A good time correlation was found between the appearance of AVF and the beginning of TMV localization in N-gene-carrying plants. At the initial stages, TMV replication, in both systemic and localizing plants, proceeds parallelly, until its cessation in the N-gene-carrying cultivars (Fig. 4). In the latter plants TMV multiplication is halted at about 20-40 hr from inoculation, which is also the time span required for AVF stimulation (Fig. 5).

The mechanism of AVF stimulation

The appearance of AVF, as detected by both phosphate labelling and antiviral activity, depends on both the presence of the N-gene and TMV infection. The idea that the N-gene determines an inactive AVF precursor, which, following infection with TMV, is activated via phosphorylation, was therefore pursued. In a series of in vitro tests it was indeed found that antiviral activity and a new phosphorylated protein appeared (34) following the incubation of a crude fraction of mock AVF, from non-infected N. glutinosa , with a fraction of soluble proteins, from TMV-infected leaves of this plant, (both devoid of AVF activity) under conditions suitable for protein phosphorylation. The appearance of both the antiviral activity and the new phosphorylated protein, which could be separated from AVF, was totally dependent upon the presence of ATP, cyclic AMP (cAMP) and cyclic GMP (cGMP) in the reaction mixture. Hence, although in vitro AVF synthesis required ATP

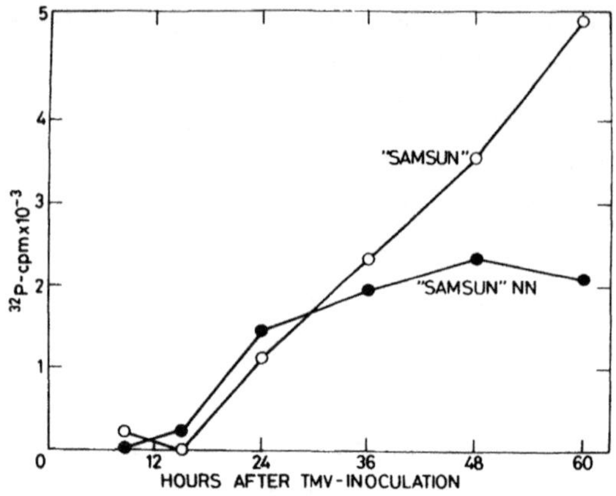

FIGURE 4. Accumulation of TMV-RNA, extracted from actinomycin D-treated leaves at various times after TMV inoculation and electrophoresed on polyacrylamide gels. Radioactive counts of gel-slices with TMV-RNA are shown.

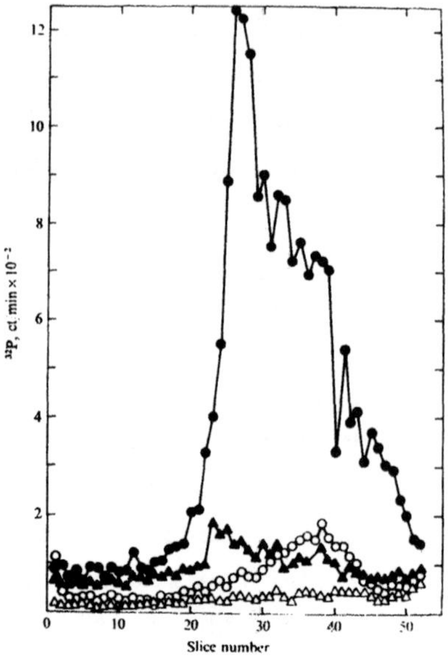

FIGURE 5. Polyacrylamide gel patterns of ^{32}P-labelled DEAE-eluted AVF. AVF was extracted from leaves 1 (Δ), 18 (o), 24 (▲) and 48 (●) hr after TMV inoculation (from Antignus et al., J. Gen. Virol. 35:107-116, 1977).

and cyclic nucleotides it was not directly phosphorylated, implying coupling to some other process of phosphorylation. Further studies indicated that the crude fraction, presumably containing pre-AVF, could be extracted from the n-gene-carrying tobacco "Samsun" as well. The fraction of soluble proteins, however, must have come from an N-gene-carrying plant. Surprisingly, it was found that in these *in vitro* experiments the need for TMV infectivity was bypassed, so both fractions could derive from non-infected plants. Thus, the external supplementation of ATP and cyclic nucleotides replaced TMV infection. In another series of *in vivo* tests the situation was further clarified. First it was found that *N. glutinosa* leaves or callus-cultures could, sometimes, produce AVF, following the introduction of the dibutyryl derivatives of cAMP and cGMP. AVF stimulation, however, was much stronger and more consistent when the tissue was treated with double-stranded RNA (dsRNA) in the form of poly I:C (12). Poly I:C has previously been reported to cause resistance to viral infection in N-gene-carrying plants (35). It is of interest to find out whether this correlates with the stimulation of AVF activity.

A tentative model for AVF stimulation emerges from these experiments (Fig. 6), according to which the N(n)-gene determines an enzyme, which participates in the processing of pre-AVF, where the N-enzyme is much more efficient than the n-enzyme.

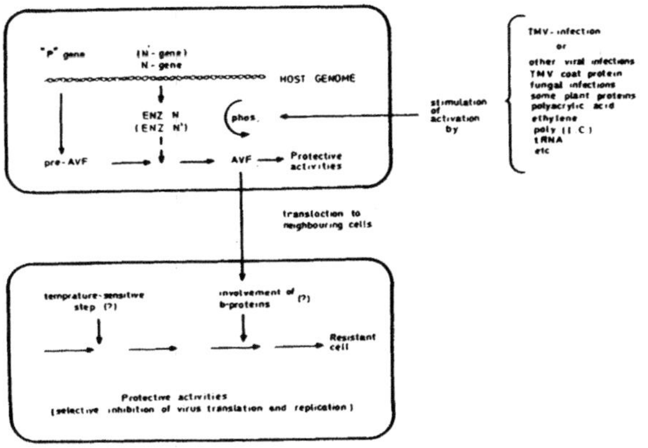

FIGURE 6. An illustration of a tentative model of the chain of events leading to protection from virus infection (after Sela, Adv. Virus Res. 26:201-237, 1981).

The activation of this inactive or slow N(n) enzyme is coupled with cyclic nucleotide-dependent phosphorylation. Hence, any stimulus that brings about the proper phosphorylation will trigger AVF processing. TMV infection is an effective stimulus, very probably through the formation of its double-stranded, replicative-form RNA. This may account for the relatively late appearance of AVF (and TMV localization) and also for the fact that there is not a single TMV mutation that can escape localization in an N-gene-carrying plant, since any mutation that is prevented from forming dsRNA is inevitably lethal.

In vitro AVF activation requires cyclic nucleotides but not dsRNA, while in vivo, dsRNA alone (and, to a certain extent, also the cyclic nucleotides by themselves) is sufficient. This suggests that the release of the cyclic nucleotides depends on dsRNA, and hence their exogenous supplementation in the in vitro studies bypasses the need for dsRNA or TMV infection.

However, until recently, the presence and role of cyclic nucleotides in higher plants was debatable (36). To complete the cycle of evidence, a direct detection of these nucleotides in plants was required. By adopting the newly-developed radio-immunoassay for cAMP we demonstrated not only the presence of basal levels of cAMP in N. glutinosa but also its stimulation, in a pulse manner, detected in various experiments between 20-40 hr after TMV inoculation (Fig. 7; ref 37).

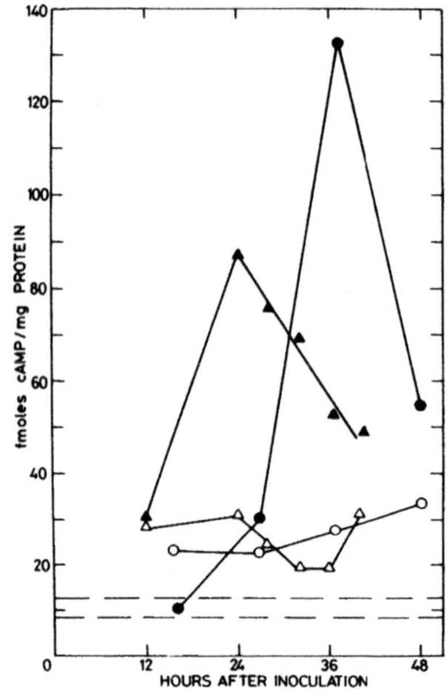

FIGURE 7. Levels of cyclic AMP in N. glutinosa leaves at various times after mock-inoculation (o;Δ) or TMV inoculation (●;▲). Basal levels in untreated leaves fell between the two horizontal lines (from Rosenberg et al., FEBS-Letters 137:105-107, 1982). (Reprinted with permission).

AVF-INDUCED ACTIVITIES IN PLANT TISSUES

The discharge of histidinyl TMV-RNA

TMV RNA can be acylated with histidine at its 3'-terminus by any eukaryotic histidinyl tRNA synthetase (38,39). This aminoacylation was inhibited by crude preparations of both AVF and interferon and by extracts of interferon-treated cells (40). When histidinyl tRNA synthetase (from pig liver) was purified to a degree where it charged, but not degraded TMV RNA, it became possible to show that the inhibiting factor in crude AVF preparations discharges histidinyl-TMV RNA rather than interfering with its formation (15). It was hence designated "discharging factor" (DF).

Upon chromatography of crude AVF on DEAE-cellulose, the antiviral and DF activities separated (Fig. 1). At this stage it was possible to show that DF is not an unrelated activity co-induced with AVF by TMV infection, but rather an activity induced by AVF itself. It was also possible to induce DF activity by AVF in the n-gene-carrying tobacco "Samsun", which does not support efficient AVF production.

The exact mechanism of DF activity is as yet unknown. One possibility is de-esterification of the aminoacyl bond; another, a restricted nuclease activity, cleaving the 3'-fragment of histidinyl TMV RNA, thus rendering it acid-soluble. The experimental answer would be to discharge histidinyl-TMV RNA and then find out if it can be re-acylated. However, any attempt, hitherto, to purify the nuclease-containing DF fraction beyond the DE-0.3 preparation, caused the loss of DF activity. Hence, the discharge-recharge experiment must await successful DF purification.

The regulation of DF activity by polymerized ATP

DEAE-cellulose-eluted DF, stored at -20°, gradually lost its activity. A fraction, soluble in trichloroacetic acid (TCA), was extracted from TMV-infected *N. glutinosa* and the TCA removed by ether extraction. This low-molecular-weight fraction was adsorbed onto DEAE-cellulose and eluted with 0.5 M NaCl. Inactive DF was reactivated by the addition of a few ml of this TCA-soluble fraction, providing both fractions were preincubated for 1 hr prior to DF assay (15).

One of the interferon-induced activities in animal cells is the induction of an enzyme which, by a dsRNA-dependent reaction, polymerizes ATP to a series of short adenosine oligomers (general formula $ppp(A^2P_5A)_n$, commulatively known as 2-5A (for a short review see ref 41). Naturally, this situation came to mind when the DF-reviving activity of the low molecular weight fraction from TMV-infected *N. glutinosa* leaves was found. Hence, another activity, namely, dsRNA-dependent ATP polymerization, was sought in DEAE-cellulose chromatograms of TMV-infected or AVF-treated plant tissues. The fraction DE-0 (Fig. 1), from TMV-infected *N. glutinosa* or AVF-treated tobacco

"Samsun" was chromatographed on a micro-column of agarose to which poly I:C was attached. The beads, with the plant material affinity-bound to them, were then incubated with ATP. A few microliters of this reaction mixture did, indeed, revive inactive DF preparations, as did the TCA-soluble plant extract. ATP polymerization was demonstrated when radioactive ATP was added to the reaction mixture and the products chromatographed on DEAE-cellulose (15). Thin-layer chromatography analyses of radioactive polymerized ATP were inconclusive. In some cases, when not masked by other plant products, spots, chromatographed as 2-5A, were observed. Charge separations demonstrated a net charge of -6 to -7 for these spots. However, at least one AVF-induced spot of a higher negative charge was always observed. Synthetic 2-5A revived DF activity as did the plant-derived polymerized ATP (42), nevertheless, the 2-5A nature of plant oligonucleotides still requires verification.

Antiviral activity of low molecular-weight plant fractions and of 2-5A

As described above, both the TCA-soluble fraction from TMV-infected *N. glutinosa* leaves, and polymerized ATP (obtained by incubating the DE-O fraction, from AVF-treated or TMV-infected plant tissues, with ATP in the presence of dsRNA) revived DF activity and exhibited resemblance to 2-5A. The introduction of 2-5A, or its core (lacking the 5'-terminal phosphates) into animal cells activated a specific nuclease which, in turn, inhibited protein synthesis (41). This well documented phenomenon also renders the treated animal cell protected from both viral infection and transformation by the Epstein-Barr virus (43;44;45;46;47). We therefore checked the antiviral potency of TCA-soluble fractions from TMV-infected *N. glutinosa* leaves, AVF-treated tobacco leaves and leaves treated with human interferon (1 unit/ml of pure γ_3 subspecies of human leukocyte interferon, as described in more detail below), against TMV in tobacco leaf-discs. We also prepared polymerized ATP from DE-O fractions derived from leaves infected with TMV or treated with AVF or interferon as described above. The results, based on the work of Reichman *et al.* (48), are briefly summarized in Table 1.

Table 1. Antiviral activity of nucleotides

Type of nucleotide	% protection
ppp 2-5A, synthetic, 200 nM	76
2-5A, core, synthetic, 200 nM	75
TCA-soluble fraction from TMV-infected *N. glutinosa* (0.005 O.D_{260})	90
TCA-soluble fraction from AVF-treated *N. glutinosa* (0.005 O.D.$_{260}$)	0
TCA-soluble fraction from interferon-treated *N. glutinosa* (0.005	0
dsRNA-dependent polymerized ATP made with synthetase (DE-O) fraction from 5 g of tissue and diluted 1:1000. Enzyme was made from:	
Untreated tobacco callus	33
AVF-treated tobacco callus	100
Interferon-treated tobacco callus	100

The TCA-soluble fraction exhibited antiviral activity only when extracted from TMV-infected leaves (but not when extracted from AVF- or interferon-treated tissue). All treated DE-O fractions, however, whether derived from TMV-infected or AVF- or interferon-treated leaves, were able to produce antiviral nucleotides upon incubation with ATP and dsRNA. The inability of AVF and interferon, by themselves, to directly induce these antiviral agents in treated, non-infected plant tissue, strongly supports the notion that dsRNA (provided by the viral infection) is essential, especially since upon exogenous supplementation of dsRNA these tissues did produce antivirally active nucleotides.

In accordance with these results, and since 2-5A synthesis in plants could not be proven beyond doubt, we decided to do the reciprocal test, i.e., to find out whether synthetic 2-5A protects plant cells from TMV infection (11). The core of 2-5A, when introduced to tobacco leaf-discs, did indeed inhibit TMV multiplication in a concentration-dependent manner (Fig. 8). Maximum protection of the tissue (80-90%) was obtained, in various experiments, by 100-200 nM of the oligonucleotide. The fact that, for maximum activity, 2-5A must be applied to the tissue within 6-9 hr of inoculation (Fig. 9), indicates its effect on an early event in TMV multiplication.

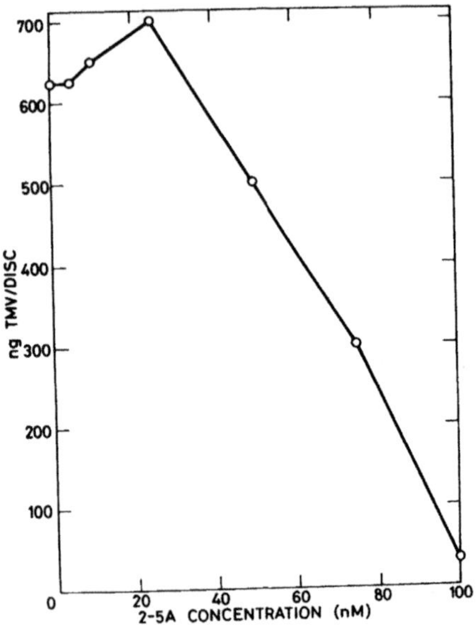

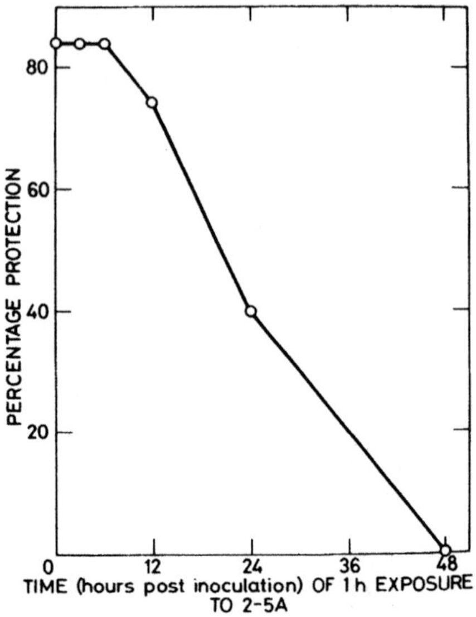

FIGURE 8. The effect, on TMV multiplication, of introducing the core of 2-5A to tobacco leaf discs (from Devash et al., Science 216:1415-1416, 1982).

FIGURE 9. The inhibition of TMV multiplication in tobacco leaf-discs, as a function of the time of application of the 2-5A core (from Devash et al., Science 216:1415- 1416, 1982).

(Reprinted with permission).

Human interferon(s) protect plants from viral infection

Interferons, in most cases, are species specific agents, in other words, they affect the animal species in whose cells they were produced (49). Hence, we were not surprised to find, time and again, that AVF is not active in animal cells. The idea to check the activity of animal interferon in plants came quite accidentally, encouraged by preliminary results which indicated the antiviral activity of 2-5A in plants.

Human leukocyte interferon can be resolved by HPLC to a number of subspecies. One, designated γ_3, seems to be less stringent regarding species specificity (21). A pure preparation of this γ_3 human leukocyte interferon (2×10^8 units/mg protein) protected tobacco leaf-discs from TMV infection (Fig. 10; ref 10). Surprisingly, 50% inhibition of virus multiplication was obtained with 0.001-0.01 units of interferon/ml. If a direct correlation with the animal system is permitted, this means that the tobacco tissue is much more

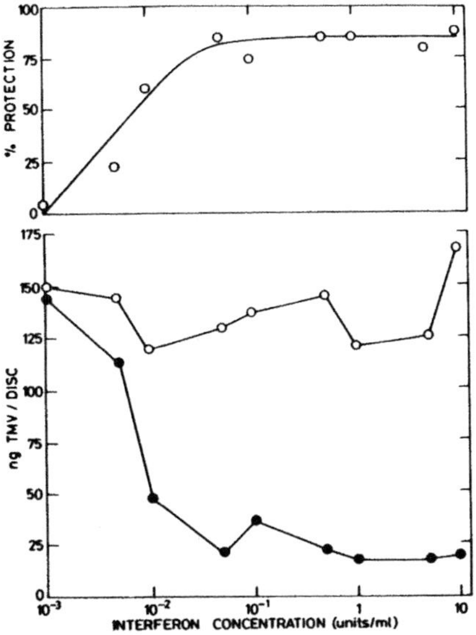

FIGURE 10. The effect of various concentrations of human leukocyte interferon (subspecies γ_3 1×10^8 units/mg protein) on TMV multiplicability. TMV content in discs treated with buffer (o) and interferon (●) is shown in the lower frame, while the upper frame shows the percentage of protection, calculated from data in the lower frame (from Orchansky et al., Proc. Nat'l. Acad. Sci. USA 79: 2278-2280, 1982).

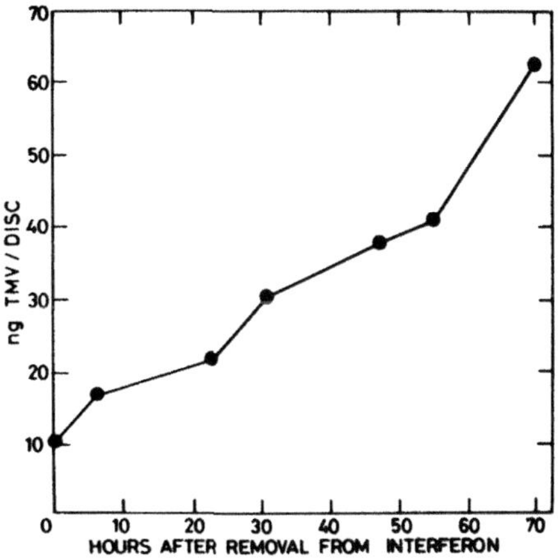

FIGURE 11. TMV-multiplicability in infected tobacco leaf-discs treated with human leukocyte interferon (γ_3; 1 unit/ml) at various times after the removal of interferon. In untreated discs, TMV level was 74 mg/disc (from Orchansky et al., Proc. Nat'l Acad. Sci. USA 79:2278-2280, 1982).

(Reprinted with permission).

sensitive to the antiviral activity of interferon than the reference animal cells. Under the above-described conditions, TMV multiplication was halted, but not abolished, by interferon. When the leaf discs were removed from the interferon, TMV multiplication gradually resumed (Fig. 11). Other subspecies of human leukocyte interferon also demonstrated antiviral activity, but, at the time of writing, γ_3 was the most consistent subspecies invariably active against TMV.

Crude human fibroblast interferon exhibited only a limited activity, often none, in the plant system. However, when treated with a mixture of glycosidases (bound to Sepharose beads as previously described for AVF deglycosidation) it became as active as the γ_3 interferon (10).

AVF AND INTERFERON

Interferons are proteins or glycoproteins, ranging from about 17-28,000 daltons, which withstand harsh treatments, such as exposure to pH 2.0 or contact with SDS. They are induced in animal tissues by viral infections or artificial agents (notable poly I:C for human fibroblast interferon). Interferons, at less than pM levels, protect animal cells from viral infections and stimulate a number of activities in cells, among them 2-5A synthesis (50).

The AVF species described in this paper is a glycoprotein (21-22,000 daltons) which remains active after lowering the pH to 2.0 or following treatment with SDS. It is induced in N. glutinosa by poly I:C and is active at pM levels or less. AVF induces the polymerization of ATP to an oligonucleotide which is antivirally active. This nucleotide resembles 2-5A, though its identification is somewhat incomplete. Furthermore, synthetic 2-5A is very active in inhibiting TMV multiplication in tobacco tissue.

Alongside these striking similarities, there are, however, some differences. Although at least some interferon species and subspecies are active in plants, a number of attepts to protect animal cells with AVF, including deglycosylated AVF, were unsuccessful. Another major difference is the process of their release. While interferon is stimulated, in animal cells, by the release of new genomic activities leading directly to its production, AVF activity, at least in part, is released by the activation of a pre-existing precursor involving cyclic nucleotides. Beside some other minor differences (51), the question of the species specificity of AVF among higher plants is still unsettled. In the limited number of plant-virus combinations tested, the antiviral substance (be it AVF of a similar substance) isolated from one such combination was active against other viral infections in the same or other plants. Attempted quantification of the antiviral activity indicates a possible quantitative difference among the various virus-host combinations. This specificity is, however, basically not as stringent as in the animal system.

REFERENCES

1. Loebenstein, G. Annu. Rev. Phytopathol. 10:177-208, 1972.
2. Sela, I. Adv. Virus Res. 26:201-237, 1981.
3. Sela, I. Trends in Biochemical Sciences 6:31-33, 1981.
4. Fulton, R.W. Virology 9:522-535, 1959.
5. Sela, I. and Applebaum, S.W. Virology 17:543-548, 1962.
6. Holmes, F.V. Adv. Virus Res. 2:1-57, 1954.
7. Samuel, G. and Bald, J.G. Ann. Appl. Biol. 20:70-99, 1933.
8. Antignus, Y., Sela, I. and Harpaz, I. J. Gen. Virol. 35: 107-116, 1977.
9. Loebenstein, G., Gera, A., Barnett, A., Shabtai, S. and Cohen, J. Virology 100:110-115, 1980.
10. Orchansky, P., Rubinstein, M. and Sela, I. Proc. Nat'l. Acad. Sci. USA 79:2278-2280, 1982.
11. Devash, Y., Biggs, S. and Sela, I. Science 216:1415-1416, 1982.
12. Gat (Eidelbaum), O. M.Sc. Thesis, 1980.
13. Clark, M.F. and Adams, A.M. J. Gen. Virol. 34:475-483, 1977.
14. Antignus, Y., Sela, I. and Harpaz, I. Physiol. Plant Pathol. 6:159-168,
15. Devash, Y., Hauschner, A., Sela, I. and Chakraburty, K. Virology 111:103-112, 1981.
16. Sela, I., Harpaz, I. and Birk, Y. Virology 22:446-451, 1964.
17. Ratner, I. M.Sc. Thesis, 1981.
18. Mozes, R., Antignus, Y., Sela, I. and Harpaz, I. J. Gen. Virol. 38:241-249, 1978.
19. Orchansky, P. M.Sc. Thesis, 1982.
20. Rubinstein, M., Rubinstein, S., Familletti, P.C., Miller, R.S., Waldman, A.A. and Pestka, S. Proc. Nat'l. Acad. Sci. USA 76:640-644, 1979.
21. Rubinstein, M., Levy, W.P., Moschera, J.A., Lai, C.Y., Hershber, R.D., Bartlett, R.T. and Pestka, S. Arch. Biochem. Biophys. 210:307-318.
22. Jacobi, D., in preparation.
23. Sela, I., Harpaz, I. and Birk, Y. Virology 28:71-78, 1966.
24. Jankowski, W.J., Davey, M.W., O'Malley, J.A., Sulkowski, E and Carter, W.A. J. Virol. 16:1124-1130, 1975.

25. Grob, P.M. and Chadha, K.C. Biochemistry 18:5782-5786, 1979.
26. Yarwood, C.E. Phytopathology 50:741-744, 1960.
27. Ross, A.F. Virology 14:329-339, 1961.
28. Ross, A.F. Virology 14:340-358, 1961.
29. Holmes, F.O. Adv. Virus Res. 11:139-161, 1965.
30. Melchers, G. Theor. Appl. Genet. 38:275-279, 1968.
31. Loebenstein, G. and Gera, A. Virology 114:132-139, 1981.
32. Atkinson, P.H. and Matthews, R.E.F. Virology 40:344-356, 1970.
33. Murakishi, H.H. and Carlson, P.S. Phytopathology 66:931-932, 1976.
34. Sela, I., Hauschner, A. and Mozes, R. Virology 89:1-6, 1978.
35. Stein, A. and Loebenstein, G. Nature 226:363-364, 1970.
36. Amrhein, N. Annu. Rev. Plant Physiol. 28:123-132, 1977.
37. Rosenberg, N., Pines, M. and Sela, I. FEBS Letters 137: 105-107, 1982.
38. Öberg, B. and Philipson, L. Biochem. Biophys. Res. Commun. 927-932, 1972.
39. Salomon, R., Sela, I., Giveon, D. and Littauer, U.Z. Virology 71:78-84, 1976.
40. Sela, I., Grossberg, S.E., Sedmak, J.J. and Mehler, A.H. Science 194:527-529, 1976.
41. Williams, R.B.G. and Kerr, I.M. Trends in Biochemical Sciences 4:138-140, 1980.
42. Devash, Y. Ph.D. Thesis, 1982.
43. Williams, R.B.G., Colgher, R.R. and Kerr, I.M. FEBS Letters 105:47-51, 1979.
44. Hovanessian, A.G., Wood, J., Meurs, E. and Montagnier, L. Proc. Nat'l. Acad. Sci. USA 76:3261-3265, 1979.
45. Hovanessian, A.G. and Wood, J.N. Virology 101:81-90, 1980.
46. Baglioni, C., D'allesandro, S.B., Nilsen, T.W., den Hartog, J.A.J., Crea, R. and Van Bovin, J.H. J. Biol. Chem. 256:3253-3257, 1981.
47. Doetsch, P.W., Suhadolnik, J.R., Sawada, Y., Mosca, J.D., Flick, M.B., Reichenbach, N.L., Dang, A.Q., Wu, J.M., Charubala, R., Pfleidener, W. and Henderson, E.E. Proc. Nat'l. Acad. Sci. USA 78:6699-6703, 1981.
48. Reichman, M., Devash, Y. and Sela, I., In preparation.
49. Lockart, R.Z. Jr. In: Interferon and Interferon Inducers (ed. N.B. Finter). North Holland-Amsterdam, London; American Elsevier, New York, pp 11-27, 1973.
50. Revel, M. In: Interferon 1 (ed. I. Gresser). Academic Press, London, New York, Toronto, Sydney, San Francisco. pp 101-163, 1979.
51. Sela, I. In: Perspectives in Virology 11:129-139, 1981.

21

THE INTERFERON SYSTEM IN MAN: NATURE OF THE INTERFERON MOLECULES AND MODE OF ACTION

MICHEL REVEL
Department of Virology, The Weizmann Institute of Science, Rehovot, Israel

I. The Various Types of Human Interferons

The task of describing the human IFN system in 1983 has been simultaneously enormously simplified and enormously complicated. The results of the past 3 years on the cloning of IFN cDNA and genes have told us the aminoacid sequence of human IFN-α, β and γ, but have also indicated the existence of multiple IFN subspecies, which are synthesized by various cells under various conditions. Even though the sequence of many IFN is known, the basis for the nomenclature of the 3 IFN classes α (leucocyte), β (fibroblast) and γ (immune), is still the immunoneutralization by specific antibodies.

A) The Human IFN Genes

The main cluster of virus-inducible IFN genes in man is on chromosome 9. *In situ* hybridization of cloned DNA probes to metaphase chromosomes, shows that the IFN-α gene family is on the distal segment of the short arm of chromosome 9 (p13-pter), while the major fibroblast IFN-β_1 gene maps even more distally (p21-pter)(1). The linkage of type I IFN genes to chromosome 9 had been established earlier by somatic cell genetics (2)and DNA hybridization (3). IFN genes can, however, be found also on other chromosomes. Thus the gene for the type II (antigen-induced) IFN-γ is on the long arm of chromosome 1 (q23-q241)(1). In addition there is somatic cell genetic evidence for the existence of genes coding for IFN species made by fibroblasts, on chromosomes 2, 5 or both (4-6). These genes may correspond to IFN-β species different from IFN-β_1, which like IFN-β_2 or others (6-8) have been detected but remain to be fully characterized. An interesting observation is that the various IFN-αs and the IFN-β_1 gene on chromosome 9 have no introns, while introns are found in the IFN-γ and also IFN-β_2 genes located on other chromosomes (see below).

B) Human Leucocyte IFN-αs

1) Multiple IFN-α subspecies. The existence of multiple IFN-α species in man is firmly based on two lines of evidence: 1) the cloning of more than 10 different cDNAs and additional genomic DNA clones and their expression in *E. coli* (9-11) and 2) the purification to homogeneity of several IFN-α subspecies by High Performance Liquid Chromatography (HPLC)(12,13), or by other methods (14).

At least 8 different polypeptides with IFN-α activity were separated by HPLC from crude preparations of IFN induced by Sendai virus in leucocytes from healthy donors or from patients with myelogenous leukemias (13). Leukemic cells from such AML or CML patients can yield up to thousands times more IFN than normal leucocytes or cells from lymphatic leukemia patients (15,16). Which cells among leucocytes (lymphocytes or monocytes) are the main producers of IFN-α is still debated (17). Since leukemia cells from patients or in cultures (e.g. KGI cells (18)) can be considered as cloned populations, their use shows that a single cell type is able to synthesize over 10 different IFN-α species.

Most of our knowledge on the amino acid sequences of IFN-α comes from sequencing cDNA clones derived from virus-induced leucocytes mRNAs (9,11,19). Fig. 1A shows the sequences deduced from 8 IFN-α cDNAs (A-H)(11). These sequences of 166 amino acids are shown starting from the aminoterminal residue of the processed mature IFN, whose position was determined by partial sequencing of purified IFN-α proteins (20,21). The signal peptide of pre-IFN-α is shown in Fig. 1B. The homology between the subspecies is extensive and fluctuates around 80% for amino acids and 90% for nucleotide sequences. A yet uncompleted task is to correlate the protein isolated to the DNA sequences (14,22,23). Table 1 gives some of these correlations. A cDNA may actually correspond to more than one polypeptide since it was found that some of the purified proteins (as "α_1, α_2, β_1 and β_2" of Rubinstein *et al*. (13), all related to the IFN-αA cDNA sequence) are shorter at their carboxyterminal end by about 10 residues (22). The 13 C-terminal residues of the IFN-α molecules may not be essential for antiviral activity (24,25) and one gene (IFN-αC2 of Fig.1B) which is actively expressed in *E. coli*, has a termination signal 17 codons before the usual position. The IFN-α species migrate on dodecylsulfate polyacrylamide gels as a series of 18-20,000-Mr proteins, but some (as "β") appear as 26,000-Mr (13,26). Differences in apparent Mr are also seen in the *E. coli* products of different IFN-α cDNAs all coding

A

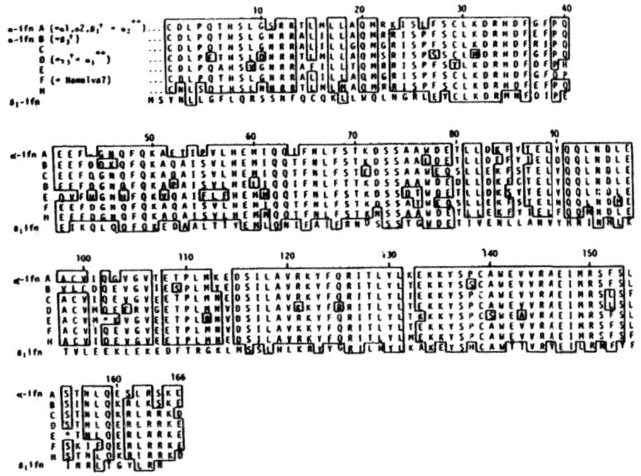

B

AMINOACID SEQUENCE OF HUMAN INTERFERON ALPHA FAMILY

```
consensus malpfsllmalvvlsckssc slgCDLPQTHSLGNRRTLMLLAQMRRISPFSCLKDRHDFGFPQEEFDGNQFQKAQA
A         ...t.a..v......c.....v...........S..T........K..L................. .......ET
C,αC*     ...s......vl...y..i..................A.I..B..B..............RI..............
psL,ps10  ...s......vl...y..ix..........T.R...A.I..B..B..............RI..............
J,α7,αC3..rs.....vvl...y..i..............R...A.I.....B............E.R..E.....H....T..
α1-L1     .v.llp..v..plch.gpvg..sw....N.G.LS.N..A..B..C...T.L.....R..R..L.IV..S.L...P.

consensus ISVLHEMIQQTFNLFSTKDSSAAWDETLLDKFYTELYQQLNDLEACVIQEVGVEETPLMNEDSILAVRKYFQRITL
A         .P.......I....................................G...T.....K...............
C,αC*     ..................E......EQS..E..S.........................................
psL,ps10  ..................E......EQS..E..S..I......................................
J,α7      ..................E......EQS..E..S............................F............
αC2                                EQS..E..S........N..........M......Y................
α1-L1     L......L..I....P.EC.....NM....QLH...HL..EC..S.LG.TKRE..SVGVI?GPT..L.T...B?MBIQBI
M                                                                               L.R..?QSMGIQRI

consensus YLTEKKYSPCAWEVVRAEIMRSFSLSTNLQKRLRRKEx
A         ..K..........................ES..S..x
C,αC*     ..I.R..................L.F...........Dx
psL,ps10  ..I.R..................L.F...........Dx
J,α7      ..N......................F....K.B....Dx
αC2       .....................Tx.F..f.........Dx
α1-L1     ..E.....D........V.Sx
M         ..K.....D........V.Sx
```

Figure 1. Comparison of amino acid sequences in human IFN-α family.

(A) The IFN-α-A to H sequences are compared; from Goeddel et al. (11). The nomenclature of Rubinstein et al. (13) is indicated. IFN-β_1 is shown in the lower line.

(B) Aminoacid replacement map of IFN-α subspecies of subfamily I (αA) and II (αC). For nomenclature see Table 1. al is α-like. Adapted from Weissmann et al. (27). Dots indicate identical amino acids.

TABLE 1

CLASSIFICATION OF HUMAN IFN-α GENES

	cDNA and GENES W	G	R	Proteins	cDNA frequency
sub family I	α_1 ···	D		γ_3 (19K)	27%
	α_2 ···	A		$\alpha_1, \alpha_2, \beta_1$	42%
	α_5 ···	G			3%
	α_6 ···	(K)			
	α_8 ···	B		β_3 (26K)	6%
sub family II (C)		C ···	αC^*		6%
		$C_1, \lambda c_1$			
	$\psi\alpha_{10}$ ···	L			
	α_7 ···	J ···	αC_3, αC_4		
		F		[Namalva]	6%
		H, λ_{2H}			6%
	α_{4a}		αC_2		
	α_{4b}				
α-like	$G_{chr_{10L}}$ ··	M ···	αI-L_2		
	$G_{chr_{26r}}$		αI-L_1		

Nomenclature of human IFN-α genes of subfamilies I and II (adapted from 27) with probable correspondence in sequences (see Fig. 1). W, G, R refer to work of the groups of Weissmann, Goeddel and Revel respectively. The Protein Nomenclature is that of Rubinstein et al. (13). Frequency of cDNA is from Goeddel et al (11).

for 166 amino acids and may reflect differences in structure (23). Most of the IFN-α species are not glycosylated as shown by the absence of glycosylation sites (N-S or N-T) in their predicted sequences (Fig.1 A,B) and by the absence of amino sugars in digests of the proteins (13,14). An exception may be species αH and αl-Ll (Fig.1 A,B) and possibly one species from Namalva cells (23).

2) The IFN-α Gene Family. Over 20 IFN-α genes are known, but have not all been correlated to the cDNAs and proteins. A classification of about 15 IFN-α genes in two large subfamilies was proposed (Table 1)(27). This classification is based on amino acid replacements as illustrated in Fig. 1B. The study of human genomic DNA clones of 15 to 20 kb showed that many contain several IFN-α genes. Reconstruction of the chromosomal DNA suggests that over 10 different genes belonging to subfamily II (IFN-αC subfamily) follow each other, separated by 4-5 kb DNA (Fig. 2). These may represent DNA repeat units (28), duplicated relatively recently in evolution. Sequences of what appears to be a separate family of α-like genes and pseudogenes not hybridizing to α genes (αl of Fig. 1B and 2) are curiously found intermingled between the genes of the IFN-αc subfamily (unpublished results). Such clusters of genes were described by several groups (27-32) and an attempt to correlate this information is shown in Table 1 and Fig. 2B. The genes of subfamily I seem to form a much less dense cluster, and can appear alone on 15-20 kb genomic DNA segments (27,33). Their duplication could be more ancient. Presumptive evolutionary trees of the IFN-α genes were published (27,34).

3) Different Biological Properties of the IFN-α Subspecies. An intriguing question is why so many functional IFN-α genes were conserved. Each subspecies may have specific properties. IFN-αA, B, C, D, and F produced by cloned cDNAs in E. coli, showed differences in their antiviral activity on various human and animal cells and on different viruses (35,36). IFN-αA and D were more active on VSV than on EMC virus in monkey Vero cells, while F showed the opposite. IFN-αD (or α_1) differs from the other subspecies by its 10-100 times lower specific activity on human cells than on bovine cells (35,37) even when purified on monoclonal antibodies (38). Similar differences in the activities of natural IFN-α subspecies purified by HPLC from human leucocytes were found (13). Recently, Goren et al. (39) compared the natural subspecies "γ3" (19K), corresponding to the

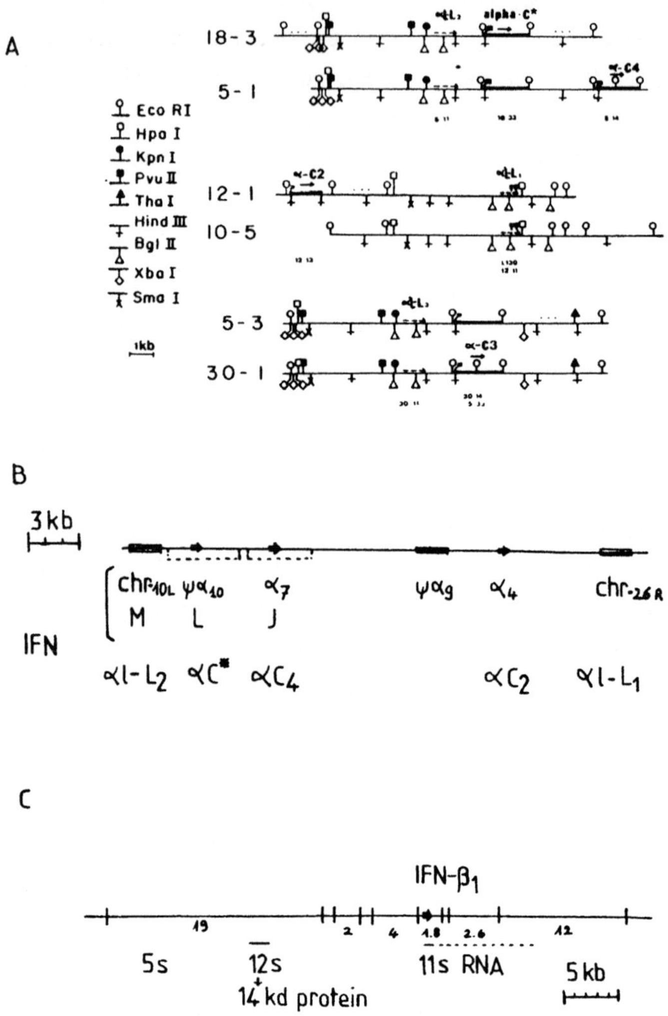

Figure 2. Maps of human genomic DNA segments containing IFN-α genes.

(A) Restriction maps of λ charon 4A clones with IFN-αC genes (arrows) and α-like genes (dotted arrows). For details see text. Section IB2.

(B) Reconstruction of a long DNA segment with 6 genes. Nomenclature is explained in Table 1. Dotted lines indicate repeat units (28). Adapted from (27).

(C) Reconstruction of DNA segment carrying human IFN-β_1 gene (arrow). Position of genes coding for poly (rI):(rC)-induced 5S and 12S RNA are shown. Data from Gross et al. (68,72) and Mory et al. (69). (See section IC1).

recombinant αD, with subspecies "β3" (26K), related to recombinant αB. Both have the same specific activity on bovine cells, but "β3" has a 100-fold higher antiviral activity on human cells ($2x10^9$ U/mg protein) and is the most active IFN-α subspecies on human cells. The same 100-fold difference in activity per protein weight is seen between "β3 and γ3" for the antigrowth effect on human cells, and for induction of HLA proteins, as well as for binding to the same IFN-receptor on human cells (39,40). Surprisingly, however, "γ3 and β3" are in protein weight equally active to induce the (2'-5') oligo A synthetase in human cells; thus, for the same antiviral, antigrowth, and HLA induction activity, subspecies "γ3" is a much better inducer of the (2'-5') oligo A synthetase than other IFN-α species (39). The striking differences between IFN-α subspecies could suggest that each performs a somewhat different function. In some cases, differences between the antiviral versus antigrowth activities of different IFN-α subspecies were seen (38,41). Similarly, differences in the efficiency of various IFN-α subspecies to activate natural killer cells were reported (42).

Possibly as a result of their different activities, the natural subspecies IFN-αD and αA were found to potentiate each other, giving when mixed a 20-fold synergistic antiviral effect (43). This potentiating effect on another α subspecies, could explain the apparent paradox that IFN-αD (α_1, "γ3") is quantitatively a major component of crude leucocyte IFN preparations (10,11), although its antiviral activity on human cells is low. In the A-H group of cDNA clones obtained by Goeddel et al. (11), D represented 30% of the total, A was 40% while the 6 other species represented each only 3-6% of the total (Table 1). A similar abundance of αD can be deduced from the purification data reported for virus-induced KGI (18) or CML cell IFN (13). Purification of IFN-α on monoclonal antibodies (44) gives mixture with variable abundance of each species. The relatively high activity of αD on cells of lower mammals (37,38) suggests that this gene may be an archaic form of human IFN-α genes.

There may still be additional IFN-α subspecies remaining to be characterized. In serum of patients with autoimmune disease, systemic lupus erythematosus, or acquired immunodeficiency syndromes (AIDS), an acid-sensitive IFN neutralized specifically by anti-IFN-α antibodies is detectable (45). This species appears also spontaneously upon incubation of a subfraction of large granular lymphocytes from healthy humans (46).

IFN-α can be produced by cell and glycoproteins interaction with lymphocytes (47-49), and even spontaneously (50), but the species produced have still to be analyzed. The existence of various forms of IFN-α mRNAs in human cells has been interpreted as indicating the existence of more IFN-α genes than presently known (51,52). Hybrid IFN-α molecules will be discussed in Section IID.

C) Fibroblast IFN-β

1). The major IFN-β_1. While leucocytes induced by viruses produce mainly IFN-α, they also make the immunologically distinct IFN-β (53) which is the main IFN species produced by cells of solid tissues. Among the cells producing IFN-β, fibroblasts of various tissues have been best studied and IFN-β was commonly named fibroblast IFN. Virus-infected fibroblasts can make some IFN-α in addition to IFN-β (54), but when these cells are induced by double-stranded (ds) RNA, the major product is IFN-β. There may be several IFN-β species (see below), but one species, IFN-β_1, represents 90% of the IFN activity produced by the fibroblasts. IFN-β_1 was obtained pure by chromatography on Cibacron-Blue Sepharose (55) or by HPLC (56) and its aminoterminal end was sequenced (56-58). The complete amino acid sequence of IFN-β_1 was deduced from sequencing the cloned cDNA (59-62), and shows about 30% homology to IFN-α at the aminoacid level (Fig. 1A) and 60% homology at the nucleotide level (63). Like in IFN-α, the 166 amino acid sequence of the mature IFN is preceded by a 21 amino-acid-long leader which is cleaved by the cell prior to secretion. IFN-β_1 is a glycoprotein (55,64), although the non-glycosylated form made by E. coli is active, but less stable (60). Study of genomic DNA clones (65-70), allows to draw a map of almost 50 kb of human DNA in which the intron-less IFN-β_1 gene is located. The gene can be directly expressed in E. coli (69). In contrast to IFN-α, the IFN-β_1 gene seems unique, with no other IFN gene in its vicinity (Fig. 2C). However, in addition to the 0.9 kb (11S) IFN-β_1 RNA, other poly (rI)(rC)-induced transcripts were described (71), some of which originate from DNA neighbouring the IFN-β_1 gene (72,73). Among the latter transcripts are 5S structural RNAs and mRNA for a 14,500-Mr protein (73). Read-through transcription of the IFN-β_1 gene into downstream DNA was also observed (74,75), and accounts for a 1.4 and a 3.8 kb RNA hybridizing to IFN-β_1 cDNA.

2). <u>Other IFN-β Species in Fibroblasts.</u> Although the IFN-β_1 gene on chromosome 9 appears unique, induced fibroblasts produce additional IFN species. About 10% of the IFN activity can be separated from IFN-β_1, and a second IFN species (IFN-β_2) which does not bind to Blue-Sepharose was purified (76). The antiviral activity of this species can be cross-neutralized by antibodies against IFN-β_1, but the IFN-β_2 protein is not immunoprecipitated by these antibodies (and vice-versa). Its immunological and chromatographic properties, and its slightly larger size (21,000-Mr versus 20,000 for β_1) show it is a distinct protein. IFN-β_2 is produced in low levels by fibroblasts treated with cycloheximide without ds RNA, conditions under which IFN-β_1 is not produced. IFN-β_2 is immunologically related to the 26,000-Mr primary translation product of a 14S IFN mRNA (1.3 kb) induced in fibroblasts along with the 11S (0.9 kb) IFN-β_1 mRNA. The 14S mRNA injected to xenopus oocytes produces IFN activity (7,8) which allowed to clone its cDNA and isolate two intron-containing genomic clones (77). The IFN-β_2 genes are not on chromosome 9, and the partial sequence data available, indicate 56% nucleotide homology with IFN-α and β_1 cDNAs, with some clearly conserved amino acid residues (unpublished). The promoter of one IFN-β_2 gene resembles that of IFN-β_1 (78). When transfected into hamster cells, the IFN-β_2 genes produce antivirally active IFN (77).

A more detailed analysis of human fibroblast RNAs which code for IFN-β activity when injected into oocytes, revealed the existence of up to 4 IFN-β mRNAs in addition to IFN-β_1. The products of these IFN-$\beta_{2,3,4,5}$ RNAs, recently reviewed by Sehgal (52) were not yet characterized. From study of mouse-human hybrid cells, it was suggested that some of these mRNAs are transcribed from genes unlinked to the IFN-β_1 gene of chromosome 9, and could represent the products of the IFN-β genes originally ascribed to chromosome 2 and 5 (4-6, 79).

D) <u>Minor IFN-β Activities Produced by Non-fibroblastic Cells</u>

White blood cells produce some IFN-β_1 when infected with viruses (54), but after induction with mitogens or lectins these cells appear to produce another IFN-β activity along with IFN-γ (80,81). The producers may be T cells, since B cell mitogens induce IFN-α (82). This activity is neutralized by anti-IFN-β_1 antibodies but is not retained on columns of these antibodies. It has not been formally excluded that this activity may be an inducer of IFN-β_1. A 28S IFN-β mRNA was detected in human lymphocytes, by injection to frog oocytes (83).

Hematopoietic cells, such as mouse Friend erythroleukemic cells undergoing differentiation in culture, secrete small amounts of an IFN-β, distinct from the virus-induced IFN-β in the same cells (84). Similarly, human U937 histiocytic lymphoma cells, which differentiate to macrophages in cultures, secrete an IFN-β activity that is cross-neutralized by anti-IFN-β_1 antibodies, but may not be IFN-β_1 itself (271). The biological significance of these minor species of IFN secreted during cell differentiation, is discussed in Section IVB. These minor IFN species, made in the absence of exogenous virus infection during physiologically important phases of the cell life, could be related to the presence of IFN genes outside the main IFN gene cluster. Expression of the IFN genes on chromosome 9 may be a response to viral infection, while the other genes may be activated by various physiological stimuli.

Unusual forms of IFN-β and α were characterized recently in human amniotic fluid which may be produced by amniotic membranes (85).

E) Immune IFN-γ

This species of human IFN is induced by cell antigens, mitogens or lectins probably in T lymphocytes and natural killer (NK) lymphocytes (for review: 17,87). For example, sensitized T cells exposed to flu virus-infected cells produce IFN-γ (88,89) but also acid-labile IFN-α (86). The recent cloning of IFN-γ cDNA, from mRNA of Staphylococcal enterotoxin B-stimulated blood cells (90,91) has revealed that this IFN of 146 aminoacids, bears little homology to the IFN-α and β species (92), yet shares many of their biological functions. There are more basic aminoacids in IFN-γ than in IFN-α and β, which may explain the instability of acid pH. The apparently unique IFN-γ gene located on chrom 12 (1), contains 3 introns (93). Its promoter differs from that of the IFN-α and β genes, as may be expected from the fact that different inducers activate these genes.

The structure of the IFN-γ protein isolated from induced blood cells was not yet determined, although an extensively purified preparation was obtained from phorbol myristate and phytohemagglutinine-induced blood cells by Yip et al. (94,95). Polyacrylamide gel electrophoresis, showed two proteins of 20,000 and 25,000-Mr, both glycoproteins. By HPLC, Rubinstein has (96) isolated under non-denaturating conditions, four peaks of activity from such IFN-γ preparations. Two of these peaks represent pure proteins of 21 and 26,000-Mr, while the other fractions may be dimers of IFN-γ (accounting for the old 40,000-Mr forms). The specific antiviral activity of IFN-γ appears lower than that of IFN-α and β_1 and is about 0.5-1×10^8 U/mg when measured with VSV on Wish cells (96). The difference between the two forms of IFN-γ is unclear at present, and it will be very important to determine the amino- and carboxyterminal ends of these molecules. Expression of cloned cDNA in E. coli (90), was done assuming that the protein starts at a cysteine residue, as in IFN-α, but this may not be the case for the naturally processed IFN-γ.

IFN-γ is species specific, but there is much variability in the sensitivity of human cells to the antiviral effect of IFN-γ (87,97): Daudi lymphoblastoid cells are not sensitive (98,99) while the amniotic Wish cell line is among the most sensitive (97,100). Purified IFN-γ (in contrast to IFN-αD) induces HLA proteins (100) and target resistance to NK cells (101) more efficiently than the antiviral state or the (2'-5) oligo A synthetase (100) in the same cells (see Section IIIB3). Purified IFN-γ may not inhibit cell growth more than other IFNs, but can induce cytolysis in some tumor cells (95,104). IFN-γ often has synergistic effects with type I IFN (102,103). A possible functional heterogeneity among IFN-γ subforms is still under study.

F) Control of IFN Gene Expression

DNA transfer experiments in which cloned IFN-genes are introduced into various heterologous cells have shown that the inducibility of the IFN-α and β genes by viruses or ds RNA, is a property of the genes themselves (105-111). Except in one case (112) it was shown that some 5' promoter DNA sequences are responsible for the induction of correctly initiated mRNA transcription and for the absence of expression in uninduced cells (73,111,113-115). IFN mRNA accumulation after poly (rI:rC) induction (116) is probably controlled at the transcription level but also at the post-transcriptional level (e.g. stabilisation)(117), which may explain superinduction by cycloheximide and similar drugs (118,119). In some cases, cycloheximide alone can induce IFN synthesis (120,121). Nevertheless, the molecular mechanisms of induction by ds RNA are still not understood (122). Genes controlling IFN induction by specific viruses were identified in the mouse (123,124), and may exist in man as well.

II. Interaction of IFN and Cells: The IFN Receptors

The different polypeptides endowed with IFN-α, β and γ activities probably all act in a similar way by first binding to cell surface receptors and subsequently inducing an intracellular response.

A) High-affinity IFN Receptors

In somatic cell hybrids, the sensitivity to human type I IFN can be transferred to mouse cells by human chromosome 21 (125-127). Since antibodies to surface determinants coded by chromosome 21 block IFN action on intact cells (128,129), the existence of a 21-coded receptor was deduced The gene(s) involved appears to be on the distal segment of the long arm of 21, close to the q22 region whose trisomy causes Down's syndrome (130). Trisomic 21 cells are more sensitive to low concentrations of IFN for the antiviral (128,130,131) and anti-mitogenic effects (132-134) and bind more type I IFN than monosomic 21 cells (135,136). The binding of IFN-α and β, but not IFN-γ, is inhibited by anti-21 antibodies (137). Monoclonal antibodies against 21-coded antigens block IFN action (138) but fail to stain human cells, indicating a low number of presumed receptors (unpublished data).

With purified mouse IFN-α, β labeled by radioactive iodine, binding to about 1000 high-affinity receptors on mouse L1210 cells, could be demonstrated (139,140). With human IFN-α, about 5,000 receptors were found on human Daudi cells (highly sensitive to IFN's antigrowth effect) (141,142) and 1,000 on bovine MDBK cells (143). These figures are much lower than for other protein hormone receptors (144,145). On the other hand, human IFN's affinity for the receptor is high (Kd of 10^{-10} to 10^{-11}) (141,142) and the binding is non-cooperative. This may explain the very high specific activity of type I IFN (13), explain why a few IFN molecules per cell suffice to produce the antiviral effect (38) and why the concentration of IFN is the critical factor in its activity (146). By chemical cross-linking to radioactive IFN, a complex of a 130,000-Mr protein (in SDS) with IFN-α, displaced by IFN-β, was observed in human cells (147). By digitonin solubilization, the complex appears as 230,000-Mr (148). IFN-γ has probably a different receptor (see section C). IFN-α and β seems to be bound to a trypsin-sensitive protein or glycoprotein, which disappears rapidly from the cell surface ($t_{½}$=5 hours) if not resynthesized (140,141,147). IFN-α binds to the receptor through the N-terminal region of the IFN molecule (149).

B) Non-Specific Binding of IFN to Cells

IFN may bind components of the cell surface without relation to its activity, as in the case of non-homologous animal cells (150). Ubiquitous gangliosides bind IFN (151-153) and inhibit its action (153,154), an effect to which the (2'-5') oligo A synthetase could be more sensitive (155) and the antigrowth effect less sensitive (156) than the antiviral effect. Glycoprotein hormones (HCG,TSH) and cholera toxin (CT), which binds to GM1 (157) can inhibit many effects of IFN (150,158-161). CT does not interfere with IFN binding to its high-affinity receptor on mouse cells (162) and may in fact alter the membrane glycolipids structure. Pure mouse IFN does not inhibit CT binding to cells (162) in contrast to previous reports (150). Mouse L1210 cells lacking GM1 (162) and other ganglioside-deficient cells (164,165) are sensitive to IFN, but gangliosides can stimulate in some cells (163). Gangliosides do not appear to be part of the high affinity receptor (166). The original proposal by Chany (146,167) that IFN binding involves two laterally-moving components: non-specific sites (gangliosides) and a species-specific site (high-affinity receptor) may explain many of the results. Presence of a sequence homologous to CT residues 72-88, in IFN-α and β (residues 128-143, Fig.1), would favor such a model. The sequence analogy is absent from IFN-γ, which does not bind GM1 (154,155) although it may bind GM2. Since in receptor-bound IFN, the C-terminal 10-15 amino acids (near the "CT-like" region) are free (149), this part may be able to bind with gangliosides.

Because of these non-specific interactions, the specific binding of IFN to its receptor must be documented (139), by showing saturation of the cellular binding sites and competitive displacement by homologous unlabeled IFN. Under these conditions, no specific binding should be found on heterologous cells or IFN-resistant cells such as $L1210^R$ (140).

C) Different Receptors for IFN-γ and for IFN-α,β

The selectivity of the IFN-receptors seems greater than expected and IFN-γ appears to have a receptor system separate from that of type I IFN. Mouse $L1210^R$ cells, which do not bind IFN-α,β are still sensitive to IFN-γ (153). Human IFN-γ does not compete against the specific binding of IFN-α to Daudi cells, while IFN-β does (141). IFN-γ binds to about 5,000 high-affinity receptors on human GM258 cells, and cannot be displaced by IFN-α (168,169), although some competition was observed with IFN-β, suggesting that IFN-β also interacts with the IFN-γ receptor. DeGrado et al. (92) noted that human IFN-γ could have more homology to IFN-β than to IFN-α. The homology is, however, small and in fact competition by IFN-β against IFN-γ binding is not seen in several human cells (170). In mouse cells, the existence of an IFN-γ receptor different from that of IFN-α,β is shown by IFN binding experiments (162) and by the study of IFN-β resistant-mutants sensitive to IFN-γ (153,171).

Is the separate IFN-γ receptor also coded by a gene on chromosome 21? One study (172) indicates that human trisomic 21 cells are not more sensitive than most diploid cells to purified IFN-γ, in contradiction with what was reported before (130,173). Furthermore, human IFN-γ was unable to induce H_2 antigen synthesis in mouse-human hybrids containing only human chromosome 21, while mouse IFN-γ was highly effective (unpublished data). Finally, antibodies against 21-containing hybrid cells fail to inhibit IFN-γ binding (137). Therefore, it remains possible that the IFN-γ receptor gene is on another chromosome than 21.

The existence of separate receptors for IFN-α,β and for IFN-γ may explain the 5-20 fold synergistic antiviral (102,103) and antitumor effects (174-176) of IFN-γ given together with IFN-β or α. If both receptors are occupied and triggered simultaneously the response of the cell could be much greater than if only one receptor type is triggered. Synergism has, however, been observed also between two human IFN-α subspecies "γ_3" (αD) and "α_2" (αA), which were shown to compete for the same receptor (43). When the binding of these biotinated IFN species was quantitated with iodinated avidin, IFN-αD was found to have a 100-fold lower affinity for Wish cells than IFN-αA, in line with the lower antiviral activity of IFN-αD on human cells. A 20-fold synergistic antiviral effect was seen when appropriate concentrations of the 2 IFN-α subspecies were added together to human cells, no synergism being observed on Bovine cells in which the 2 species have the same antiviral activity. Synergism in this case, may not be due to the triggering of different receptors but rather to a cooperative effect on intracellular events leading to the antiviral state. For example, IFN-αD appears more efficient than other IFN-α species for inducing the (2'-5') oligo A synthetase in human cells (39).

Many cells sensitive to IFN-α,β are poorly sensitive to IFN-γ (87, 97), human lymphoblastoid Daudi cells being a good example (98,99). However, one should be aware that while IFN-α may have a low antiviral effect (and hence low unitage) or induce poorly the (2'-5') oligo A synthetase, it may cause in the same cells a much more efficient HLA antigen induction (100) or anti-NK effect (101). Reovirus and vaccinia may be more sensitive to IFN-γ than VSV or EMC (177). It would be simplistic to ascribe differences in sensitivity only to the receptors without taking into account the subsequent intracellular events. While the separate IFN-γ receptor seems now on firm grounds, further heterogeneity among IFN receptors, based on the marked variations between cells in their sensitivity to various IFN species, is still speculative. In favor of two IFN-β receptors instead of one for IFN-α (169), there are cells which are more sensitive to IFN-β than IFN-α (178). However, these differences may not be related to the receptor, since in rat embryo fibroblasts which are sensitive to human IFN-β but not α, the latter binds to the rat cells (167). IFN-α is often less species-specific than IFN-β: bovine and murine cells are sensitive to human IFN-α but not β (179,180), and IFN-β competes poorly against binding of IFN-α to bovine cells (143). The IFN-α receptor of various species may fit better to some human IFN-α subspecies (as IFN-αD) than to other IFNs, as shown also by the study of hybrid IFN-α molecules (next section), which illustrate the complex pattern of interactions between IFN molecules and cells.

D) Breaking the Species Specificity: Hybrid IFN Molecules

By one of the wonders of genetic engineering, it has been possible to construct hybrid molecules between IFN-αD and IFN-αA, and study the molecular basis for their differing species specificity. This approach showed that the first 63 residues of A confer the high activity on human cells, the same region of D conferring low activity on human cells (37,38,181). More surprising, the A_{1-63}/D_{63-165} hybrid molecule became active on mouse cells, a species on which A is inactive and D very poorly active. Clearly, the structural change introduced in this hybrid IFN (called A/D Bgl) may allow its interaction with the mouse cell receptor. Studies on more hybrids (38,182) confirmed that the N-terminal part of IFN-α or idiotype A (27,37) is crucial for activity on human cells, while the C-terminal portion (idiotype C) seems more important for antiviral and antitumor action on mouse cells. However, the C-terminal of IFN-α is not silent on human cells, since the last 15 residues of αD conferred a higher activity on human cells when spliced in a α-AD' hybrid (25), although these residues are not essential for activity and are not bound to the receptor in α-A (149). In further studies with hybrid molecules, it was found that replacement of

Ser_{68}, Thr_{79} and Tyr_{85} of α-A, by respectively, Thr, Asp, Lys, leads to a decreased activity on human cells and to a dramatically increased activity on mouse cells and on EMC virus-infected mice (182). These replacements are in non-conserved position of the 64-92 region of IFN-αs, which has the most homology between IFN-α, β and γ (92). From all these results it is difficult to define an active site on IFN-α molecules, and indeed active small fragments of IFN-α could not be obtained (24). The tertiary structure maintained by the 29-138 disulfide bridge seems essential for IFN-α activity, although the 1-98 S-S bridge may not be essential. In human IFN-β, changing the Cys_{141} by Tyr abolished activity (183). In mouse IFN, the role of a histidine hydrophobic site was proposed (184). Chymotrypsin digestion of human IFN-α leads to a more rapid loss of activity on human than on bovine cells (185) and some antibodies block IFN-α action on human but not rabbit cells (186), but these results were obtained on mixtures of α species and probably reflects inactivation of individual species.

Information on the relative affinity of hybrid IFNs for the receptors is not available, but the number of IFN molecules needed per cell to give a half-maximal antiviral or antigrowth effect was computed (38). This calculation makes differences in activity on human, mouse, rat and cat cells of hybrid IFN-αs very apparent, and may have some advantage over the classical expression of specific activity (units per ml per mg protein). On the other hand, this calculation maybe misleading since the IFN effects depend on its concentration (146) and not on the number of cells used. Another important finding is that between hybrid IFN-αs, the ratio of antigrowth to antiviral effect varied markedly (38). This suggests that each IFN molecule could interact with more than one receptor site (two-idiotypes receptor model), and that the intracellular response varies according to which receptor site is triggered most by a given IFN. Such a multisite receptor model would explain how IFN-αD (α_2) can be a good inducer of the (2'-5') oligo A synthetase while being a poor inducer of the antiviral effect and having a low overall affinity for the receptor (39,43). It could also explain why some IFN-α species or hybrids are more active on VSV while others are more active on EMC (35,182) or reovirus (187) in the same cells. Evidence for two sites in IFN-α binding to human lymphoblastoid cells was reported (142).

E) Fate of the IFN-Receptor Complex After Binding

At 37°C, but not in the cold, part of the human IFN-α bound to Daudi cells appears to be internalized by the cells, and cannot be washed away by mild acid (188). After 1-2 hours, half of the bound IFN is in the cell and is being degraded, as evidenced by the release of acid-soluble IFN fragments into the medium (188). In bovine MDBK cells, up to 75% of the bound IFN is recovered in degraded form after 1 hour (143,189). The higher amount of cell-bound IFN at 37°C than at 4°C (139,141,168), could reflect recycling of receptors after internalization and IFN degradation. Slow internalization during 8 hours, followed by degradation, also occurs with human IFN-γ in GM258 fibroblasts (169,190).

IFN bound to its receptor seems to follow the fate of other protein hormones, as insulin or epidermal growth factor (EGF), and after clustering in the clathrin-coated pits, enters the cell via the receptosome system (191). Electron microscopy shows gold-stained IFN-α in coated pits (192). Dansyl-cadaverin and other transglutaminase inhibitory amines, which block clustering and endocytosis of protein-receptor complexes (191), interfere with the action of IFN-α, β and γ on human fibroblasts (193). Such inhibitors are, however, not specific enough to prove that internalization is required for IFN action. For protein hormones, it is still debated whether internalization is a prerequisite for biological action, or else is a process which removes the ligands with the receptors from the cell surface. Degradation probably occurs after receptosomes fuse to lysosomes, and IFN degradation is inhibited by lysomotropic drugs (amines, chloroquine). This leads to down-regulation of the IFN receptor, with a temporary loss of cell sensitivity to additional IFN (194). Thus, it is possible that internalization of IFN is related to recycling of receptors to allow the cell new reactions with IFN, and not to IFN action proper. In mouse cells, no evidence for internalization and degradation of bound IFN-α,β was seen during its action (140). In this case, the occupancy of receptors was correlated with biological activity and IFN was released intact from the receptor after action. The receptors alone disappear rapidly from the cells if protein synthesis is impaired and their turnover rate has a $t\frac{1}{2}$ of 5 hours (140,194).

Whether IFN functions by 1) activating a cell membrane receptor-mediated process which transfers the message intracellularly or 2) by acting directly from within the cytoplasm or nucleus, is still an unresolved question. Even in receptosomes, the hormone-receptor complexes are still functionally "extracellular" and membranal (191). Some

IFN could be delivered via the vesicles into the nucleus rather than to lysosomes, but no IFN was ever found in the nucleus. It was reported that in heterokaryons, a functioning nucleus of the same animal species as the IFN used was needed for the antiviral response (195) but this was not found by other investigators (196,127). In the case of EGF, or other hormones, a strong argument in favor of receptor activation is the fact that clustering of the receptors by polyvalent antibodies mimicks many of the hormone's action (197) showing that the hormone does not have to act directly in the cell. This phenomenon was not yet described for the IFN receptor. Activation of the receptor could lead to the formation of a factor, which transferred to the nucleus stimulates transcription, as suggested by experiments with prolactin (198). The activated receptor could also have enzymatic functions, as the protein kinase activity of the EGF receptor (197). A cytoskeleton function may also be needed for IFN action (146).

F) Time-course of Cell Response to IFN

In analogy to other hormones, the cell response to IFN comprises both rapid and slow processes. A rise in cGMP (199,200) and membrane lipid changes (201-204)(section IIIA) appear already within 5-30 minutes and could accompany receptor activation. On the other hand, induction of specific proteins and establishment of the antiviral state usually takes several hours to develop (205). IFN-γ acts more slowly than IFN-α,β (206,207). Some accumulation of IFN-induced mRNAs (e.g. C56, HLA; see section IIIB) is observed already 1-2 hours after IFN-α,β (208,209). Induction of (2'-5') oligo A synthetase and its mRNA, show a lag period of about 3 hours (210,211). During these first 3 hours, addition of anti-IFN antibodies to the cells is still able to inhibit induction of the synthetase and of the antiviral state, indicating that the functional molecules of IFN are still accessible on the cell surface, or in endocytized vesicles (210). In contrast to the rapid initial binding of IFN, a prolonged interaction of IFN with cell membranes seems needed to induce the biological response. During this time, clustering and activation of the receptors could take place. These events are reversible and, upon removal of IFN, the cells loose in a few hours their capacity to induce the synthetase (207). The availability of cloned DNA probes for the (2'-5') oligo A synthetase (211) and for other IFN-induced mRNAs (208,209,212,213) should now allow to study in more detail the relation between IFN binding and the activation of cellular genes which characterizes the response of cells to IFN.

III. The Intracellular Biochemical Responses to IFN

The mechanisms of the various biological effects of IFN are still not understood precisely, despite the fact that a number of molecular changes produced by IFNs in cells are known. We shall first review these changes and examine their possible functions in section IV.

A) Rapid and Transient Biochemical Changes Produced by IFN

A 4-fold increase in cGMP is observed 1-5 minutes after addition of IFN to human Daudi or mouse L1210 lymphoma cells (199,214). The maximal effect requires high IFN concentrations (10^3-10^4 U/ml)(200), and cGMP is thereafter rapidly degraded within 15 minutes. In IFN-treated Daudi cells, the guanylate cyclase activity is increased, but IFN does not act directly on the enzyme (214). The cGMP phosphodiesterase is unchanged. The cGMP increase is not seen in mouse L1210^R or human Daudi^R cells, resistant to IFN action, but is also absent in several other cell types which are sensitive to IFN's antiviral and antigrowth effects (215). IFN can lead to increased levels of cAMP as well, but this occurs only 24 hours after IFN and may be the indirect result of other cellular changes (see section IV, B and C). For both cGMP and cAMP, Tovey (215) concluded to the absence of a direct correlation with IFN's action. Nevertheless, the rapid cGMP rise may shed some light on early biochemical events triggered by the activation of IFN receptors.

The IFN-mediated cGMP increase requires calcium, and is prevented by inhibitors of prostaglandin (PG) synthesis (200). Interestingly, human fibroblasts treated 8 hours with IFN, have a 40-fold increase level of PG-E (216). This increase in PG-E, and PG-$F_{2\alpha}$, appears mainly to result from the release of arachidonic acid from membrane phospholipids, through a corticosteroid-inhibited phospholipase activation, protein synthesis being required for IFN to stimulate arachidonic acid release (217). IFN seems to induce the calcium-dependent phospholipase A_2 activity (218), and this activity linked to calcium fluxes, could trigger the guanylate cyclase (219). However, since the cGMP surge is prevented by aspirin and indomethacin (inhibitors of arachidonic acid cyclooxygenease, the first key enzyme of PG synthesis), it is more likely that the cGMP rise is linked to PG synthesis: the endoperoxyde PGG2, produced by cyclooxygenase is converted to PGH2 (a precursor of PG-E) by a hydroxyperoxidase, which releases hydroxyl radicals thought to be activators of the guanylate cyclase (220).

Another interesting connection may exist between the IFN-mediated increase in PG and cGMP synthesis and the change in membrane phospholipids observed by Chandrabose et al. (203,204). Within 20-30 minutes after IFN, there is a decrease in unsaturated (but not saturated) fatty acids extractable from phospholipids, an increase in free fatty acids and in cholesterol and the phospholipids become less polar. These changes are prevented by cyclooxygenease inhibitors and by an unspecific inhibitor of superoxide dismutase (diethyl dithiocarbamate, DDC)(221), suggesting that the IFN-induced lipid changes may also be driven by the oxydative reactions of PG synthesis which follow phospholipid hydrolysis.

IFN-resistant $L1210^{R}$, lacking IFN receptors, do not show the IFN-induced lipid changes and, furthermore, these cells which were selected for growth in the prolonged presence of IFN (222), seem to have lost most of their fatty acid cyclooxygenease activity (223). Nevertheless, there is no clear evidence for a role of PG, cGMP syntheses or of lipid changes in IFN's action. Pottathil et al (221,224) have claimed that the super-oxyde dismutase inhibitor DDC, as well as cyclooxygenase inhibitors oxyphenylbutazone and aspirin, can partially reduce the inhibition of VSV replication if given to L cells prior to IFN. The effects of most of these inhibitors (or of calcium deprivation) is, however, not reproducibly demonstratable on virus and cell growth (200,215,225), although partial effects of phospholipase and cyclooxygenase inhibitors on cell growth were reported (217). Despite the fact that oxygen radical intermediates appear to be required for activation of human NK cells by IFN (226), cyclooxygenase inhibitors, preventing cGMP and PG synthesis, failed to abolish mouse NK activation by IFN (227). IFN reduces the calcium requirement of NK cells, suggesting that IFN could act on calcium fluxes, possibly through the calcium-binding protein calmodulin (215). Neither cGMP, nor PG, appear to be intermediates or "second-messengers" of IFN action. PG rather inhibit IFN's antiviral action (203), the IFN-enhanced lymphocyte cytotoxicity and calcium-dependent release of lymphotoxins (228,229). PG synthesis could be a feed-back mechanism which protects the cells against excessive membrane damage by IFN, as suggested in the case of the IFN-induced cytotoxic effect of ds RNA (230), a phenomenon probably related to IFN production in which PG (231) and calmodulin (232) play a role. Similarly, cGMP could also provide a feed-back protection against free radical formation (220).

The rapid IFN-induced, calcium-dependent, lipid hydrolysis and oxydation, accompanied by free radical release, cGMP and PG synthesis, is reminiscent of rapid membrane effects caused by other hormones, and should not be neglected. Tamm et al. (201,202) described a reduction in the fluidity of the cell membrane, 30 minutes after IFN addition, which could be attributed to the loss of unsaturated fatty acids or to the cGMP surge. This membrane rigidity disappears after 1 hour and reappears at 24 hours, when the cAMP levels increase (215), and when major cytoskeletal modifications develop (202)(see section IVC). The early, transient, rigidity could regulate the movement of the IFN-receptors, their clustering and disappearance from the outer cell surface, and reflect the onset of the IFN response. Drugs which disturb the cytoskeleton were reported to inhibit IFN's antiviral effect (146,233) and similar effects were reported for ouabain which affects the Na^+/K^+ pump (146). However, changes in ion concentrations can also alter the virus sensitivity to IFN (234), and we were unable to show that vinblastin, colchicin or ouabain inhibit the later induction of (2'-5') oligo A synthetase (235). Thus the real significance of the early biochemical events produced by IFN, for the late biological response of the cells, remains largely unknown.

B) Induction of Specific Gene Products by IFNs

Since establishment of the antiviral state in IFN-treated cells is inhibited by drugs which block RNA and protein synthesis, or by enucleation (236), the prevalent model of IFN action is that it induces a number of cellular proteins which mediate the biological effects. The IFN-induced proteins identified can be divided into several groups: translation-regulatory enzymes, cell surface antigens, and intracellular proteins detected by gel electrophoresis. Their induction by IFN will be first considered, their biological function being discussed in section IV.

1) Translation regulatory enzymes

These proteins were isolated from extracts of IFN-treated cells by their ability to inhibit protein synthesis through several pathways (for review 205,237-239)(Fig. 3).

a) The (2'-5') oligo A synthetase and its pathway

This enzyme (E) polymerizes ATP into $ppp(A2'p)_nA$ dimers, trimers, and longer oligomers of up to 15 residues (240), whose only recognized function is to activate a latent ribonuclease, RNase F (241-243). Enzyme E requires ds RNA (60-80 bp per enzyme molecule)(244,245). Ds RNA may be of viral origin but cellular RNAs, as heterologous nuclear RNA (246) and some cytoplasmic $polyA^+$-RNA (247), can also activate the enzyme. The

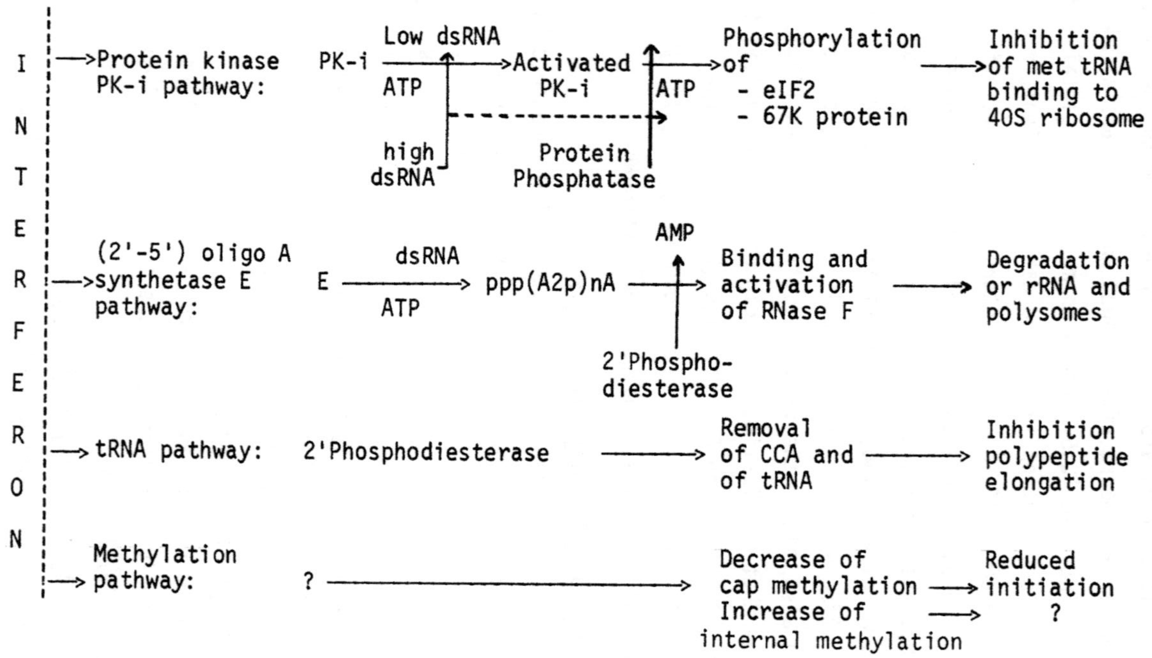

Figure 3. Translation-regulatory pathways activated by interferon treatment of cells. (Adapted from 205).

(2'-5') oligo A nucleotides are detectable in IFN-treated cells which are virus-infected (248-250), or even uninfected (248,251-253). Treatment of most vertebrate cells with IFN-α, β or γ causes 10-1,000 fold increases in the (2'-5') oligo A synthetase E, which peaks usually at 12-24 h after IFN, and slowly returns to normal even in the presence of IFN (205,238,254). Increased enzyme E was demonstrated in peripheral white blood cells of humans treated by IFN-α or β (255,256), as well as in virus-infected patients (256-258) and laboratory animals (259-261).

IFN regulates the expression of the (2'-5') oligo A synthetase E gene. Using an assay for enzyme E mRNA by translation in oocytes (210), the cDNA of human E mRNA was cloned in E. coli (211). In human fibroblasts, virtually no E mRNA is present before IFN treatment, and the E mRNA starts to accumulate after a lag of 2-3 hours, is maximal at 8-12 hours and decreases again until 24 hours. In these cells, three E mRNAs of 3.6, 1.85 and 1.65 kb are coinduced by IFN, and probably code for the two different forms of enzyme E (60-80,000-Mr and 30,000-Mr) previously identified (262). The large form, identified as a 100,000-Mr polypeptide, was purified from HeLa cells (263), while the small form was purified from human leukemic cells (264). The two forms of enzyme E are not always expressed to the same levels: human lymphoblastoid cells contain mainly the small enzyme and express the 1.85 kb E mRNA predominantly (211,262). Study of a cloned human E gene, indicates that the same gene codes for both large and small E mRNAs and enzyme, the large enzyme being longer at its N-terminal end (unpublished). The large and small E mRNAs and enzymes were found also in mouse cells, and it was shown that the small enzyme is predominant in the nucleus while the large enzyme is cytoplasmic (265). It is tempting to speculate that the nuclear IFN-induced enzyme E, plays a role in the regulation of gene expression and of cell growth (246,251), while the cytoplasmic enzyme is related to the antiviral effect.

Is the level of (2'-5') oligo A synthetase exclusively controlled by IFN? An increase in the enzyme was seen in chick oviduct after hormone withdrawal (266) and in glucocorticoid-treated human lymphoblastoid cells (267), suggesting that other controls might operate. On the other hand, the elevated levels of enzyme E in reticulocytes (268) and in mature T lymphocytes (269) [in contrast to blast cells (256)], could result from IFN which is spontaneously produced during the differentiation process, as demonstrated for Friend erythroleukemic cells (270) and for human U937 lymphoma cells (271) by the use of anti-IFN antibodies (see section IV C1). Growth-arrested cells have an elevated E enzyme as compared to growing cells (266,272-275), and it was recently shown that this could also result from spontaneous IFN secretion (275). Several cell lines have higher E enzymes than others (239); in K/Balb cells, this again is due to constitutive production of IFN (276). Thus, IFN appears to be the main regulator of (2'-5') oligo A synthetase.

The (2'-5') oligo A-activated RNase F (second enzyme of this pathway, Fig. 3) is constitutive in most cells, and IFN produces only little (2-fold) or no change in this enzyme (238,242). The 80,000-Mr RNase F can be conveniently assayed by radio binding of (2'-5') oligo A (248,277). The 2-fold increase was not seen in an IFN-resistant Daudi cell mutant (278) suggesting that the increase maybe significant. A 10-20 fold increase in RNase F induced by IFN was found in mouse JLSV9-R cells (IFN resistant for EMC but sensitive for VSV)(279,415). The increase was dependent on RNA synthesis. The same cells showed also an increase in RNase F when growth-arrested. In addition, in teratocarcinoma cells (IFN-resistant for certain viruses, see section IV A2), RNase F increased during retinoic acid-induced differentiation when the cells became IFN-sensitive. In teratoma line PSA5A, RNase F remained low, but became inducible 10-fold by IFN (279). Inhibition of protein synthesis by the (2'-5') oligo A pathway, depends both on the level of synthetase and on that of RNase F: IFN regulates mainly the first enzyme, but can induce the second enzyme when it is low. RNase F is decreased in cells infected by EMC virus, through a yet unknown mechanism (280).

The third enzyme which regulates the (2'-5') oligo A pathway is the 2'-phosphodiesterase (2'-PDi), a 35-40,000-Mr enzyme which degrades the ppp(A2'p)$_n$A oligonucleotides back to AMP and ATP (281). This enzyme may have a role in protecting the cells against excess (2'-5') oligo A, which can completely block protein RNA and DNA synthesis when more than the physiological nanomolar concentrations are introduced in cells (282). The 2'-PDi is increased 2-3 fold by IFN in mouse L cells (281); in human Daudi cells, an increase in 2'-PDi was seen with 1,000 U/ml of pure IFN-α, a concentration higher than that needed to induce the (2'-5') oligo A synthetase (39). Under other conditions, 2'-PDi may not be increased by IFN (283). In line with a role of the (2'-5') oligo A pathway in the inhibition of cell growth (see also section IV B2), the 2'-PDi is increased in growth-stimulated cells (272), the balance between synthesis and degradation being 10 times higher in resting than growing cells.

The (2'-5') oligo A pathway leads to inhibition of protein synthesis through activation of RNase F, which remains active as long as ppp(A2'p)$_n$A trimers or longer chains are bound to the enzyme (242,243); dimers (249) or monophosphates (284) are inactive. RNase F has no obvious specificity for viral mRNA (238), although it cleaves preferentially ss-mRNA linked to ds RNA structures (285). Alternatively, RNase F may cleave the RNA of cytoplasmic ribosomes (247,286). In the cell nucleus, there are other proteins in addition to the RNase, which bind ppp(A2'p)$_n$A (251,265). Some of these proteins could be increased by IFN treatment (265), and their function in RNA or DNA synthesis will be interesting to investigate.

b) tRNA deficiency and 2'-phosphodiesterase

The increase in 2'-PDi observed at high IFN concentrations in some cells (39,281) could mediate another pathway of translational inhibition (Fig. 3), characterized by the loss of some minor species of tRNA, mainly leu-tRNAs, in extracts of IFN-treated cells (287). While RNase F does not degrade tRNA, the 2'-PDi selectively removes the -CCA terminus of tRNA (281), and since minor tRNA species are likely to become rate-limiting, this could produce codon-specific blocks in polypeptide chain elongation (287). Accelerated loss of tRNA in IFN-treated cell extracts have been observed in several systems (287-289), and a leu-tRNA deficiency caused by an increased degradation of the -CCA terminus was recently demonstrated (290).

c) IFN-induced protein kinase and its pathway

A ds RNA-activated protein kinase which phosphorylates the small 35,000-Mr subunit of initiation factor eIF-2, a 67,000-Mr protein (in mouse cells; 70,000-Mr in human cells) and histones, is induced by IFN in several cell systems (for review 205,237-239). This enzyme, PK-i, increases usually 5-10 fold and starts to accumulate 2-4 hours after IFN addition; its induction is blocked by inhibitors of RNA and protein synthesis. The 67,000-Mr phosphoprotein binds to ds RNA, copurifies with the protein kinase activity (291-293), and is often considered as an autophosphorylated form of the enzyme, but could as well be a separate substrate bound to ribosomes (205,294). PK-i activity, when assayed for phosphorylation of eIF-2 appears smaller than the phosphoprotein (50-60,000-Mr), and a 70,000-Mr substrate, which can be cross-phosphorylated by mouse PK-i, is present in human cells already before IFN treatment (205,235). The identification of the IFN-induced PK-i is further complicated by the existence of another 67,000-Mr protein induced by IFNs, mostly by IFN-γ (295,296), which has a guanylate binding activity (297). The ds RNA-activated, eIF-2 protein kinase PKi is induced by all IFN-α, β and γ (298,299), and its mRNA remains to be identified. A similar enzyme is present in reticulocytes (300), in analogy to the (2'-5') oligo A synthetase. When ds RNA and ATP activated, the IFN-induced PKi inhibits initiation of protein synthesis (205,294), the inhibition being reversed by eIF-2 (301). When phosphorylated by enzymes similar to PKi, eIF-2 forms a stable complex with GDP, which prevents GTP binding and the reutilization of eIF-2 for further rounds of initiation (302).

Increase in ds RNA-dependent protein kinase PKi was demonstrated in virus-infected and IFN-treated animal and human tissues (303), as well as the presence of a similar enzyme in platelets from human plasma, where it could phosphorylate fibrinogen (304). There are probably additional protein kinases which are increased by IFN-treatment but not ds RNA-dependent: an IFN-induced enzyme which is activated by basic proteins such as histones and protamines was separated from PKi during purification (292) and a calcium-dependent protein kinase induced by IFN-γ was identified in L cells (305). The role of these IFN-induced protein kinases remains largely unexplored. The ds RNA-dependent PKi is probably involved in the control of protein synthesis initiation. Vaccinia virus seems to produce an inhibitor of PKi (306). A protein phosphatase, which dephosphorylates eIF-2, is active in most cells (292) and could be itself inhibited by ds RNA (307), completing the regulation of the PKi pathway (Fig. 3).

2) Other intracellular proteins induced by IFN

One or two-dimensional gel electrophoresis of proteins shows that up to 12 proteins (M_r from 28,000-to 120,000) are either increased or induced de novo by all IFNs in various mouse and human cells (295,296, 298,308-311). IFN-γ induces the same proteins as type I IFNs (295,298) but, in addition, another group of 12 proteins seem to be induced only by IFN-γ (296). IFNs probably activates the expression of the genes coding for these proteins and IFN-treated cells contain increased amounts of mRNA for several induced proteins (312) among which mRNAs for a small 14,000-Mr polypeptide (311) and a 56,000-Mr protein (208,213) are prominent. There is no massive change in the mRNA population of IFN-treated cells (209) indicating that IFN's effects are rather specific.

Molecular cloning of cDNA corresponding to IFN-induced mRNA sequences is actively pursued in several laboratories. Starting from mRNA of human cells treated 12 hours by 200 U/ml of IFN-β, it was estimated that about 1% of the cDNA clones obtained are IFN-induced sequences (208). One of the most strongly induced cDNA clones originates from an 18S RNA coding for a 56,000-Mr protein (C56 mRNA), which is undetectable in untreated human cells and accumulates to 0.1% of the total cell mRNAs (208). A similar cDNA was cloned from mouse cells (213). Induction is rapid (1 hour) and transient; the C56 mRNA is maximum at 4 hours and disappears at 12-16 hours even if IFN is continuously present. Transient induction is seen for many IFN-induced mRNAs and proteins (310,313) but may not be general. The 56,000-Mr protein is present in cells-treated by IFN-α, β or γ (208, 295) but its function is unknown. Preliminary findings indicate it may bind nucleotides (GTP, ATP)(297) and ds RNA.

The role of these IFN-induced proteins which were detected by gel electrophoresis and not by their activity, remains to be investigated. In one case, mouse genetics has provided evidence for a role in IFN's antiviral effect. Mice of the Mx/Mx genotype are naturally resistant to the orthomyxovirus influenza, and their macrophages or embryonic cells treated with IFN develop a more efficient antiviral state towards influenza virus (but not VSV) than cells from Mx-negative mice (314,315). A 72,000-Mr protein is induced by type I IFN in Mx/Mx cells but is not seen in Mx-negative cells, suggesting a role in the inhibition of influenza virus.

3) Cell surface antigens

Another important group of proteins, whose synthesis and gene expression are controlled by IFNs, are the histocompatibility antigens (HLA in humans, H_2 and Ia in mice). Other functionally important surface antigens are probably also regulated by IFN.

a) Class I histocompatibility antigens: HLA-A,B,C and β_2 microglobulin

An increase in H_2 antigens after IFN treatment was first observed on mouse immune cells *in vitro* (316,317) and *in vivo* (318), but is seen as well on fibroblasts and other cells (319). HLA-A,B,C and β_2-microglobulin are similarly increased by IFN on a variety of human cells *in vitro* (320-325) and *in vivo* (326). The increases in cell surface HLA are often small (3-fold), but were confirmed with monoclonal anti-HLA and appear quite specific since many other surface antigens remain unaltered (209,325,327). The increased availability of surface HLA antigens was at first interpreted as resulting from membrane rearrangements of the type described by Tamm et al (202), but has now clearly been shown to result from an increased synthesis of the HLA-A,B,C chains (44,000-Mr) and β_2-microglobulin (11,600-Mr)(100,209,328,329). IFN activates the expression of the HLA genes and produces a 5-10 fold increase in the mRNA for HLA-A,B,C and β_2-microglobulin in human B, T and fibroblastic cells (209,329,100). The HLA mRNAs are maximum 4 hours after IFN and can remain high for 24 hours (209). However, untreated cells already contain HLA mRNA, in contrast to the (2'-5') oligo A synthetase and C56 mRNAs (208,211). Nevertheless, the rate of HLA protein synthesis can increase 20-fold in melanoma cells (328) or fibroblasts (100), exceeding the increase in surface HLA antigens. It is likely that not all the newly synthesized chains are assembled on the membrane, and secretion of β_2-microglobulin from the cells is markedly increased by IFN-α and β in vitro (323). In patients, the serum level of β_2-microglobulin is increased by IFN-α injection (326).

Immune IFN-γ is relatively much more active than type I IFN for the induction of HLA class I mRNAs and proteins. Concentrations of human IFN-γ which are 100 times lower than those needed for inhibiting VSV growth or inducing the (2'-5') oligo A synthetase, are sufficient to stimulate HLA-A,B,C and β_2-microglobulin mRNAs and protein synthesis in the same cells (100). In contrast, the concentrations of IFN-α and β required to induce HLA are as high as those needed to induce the antiviral effect and the synthetase. With IFN-αD (α-1), the (2'-5') oligo A synthetase is even induced at lower concentrations than those which produce virus inhibition and HLA induction (39). Thus, induction of the HLA mRNAs and of the (2'-5') oligo A synthetase mRNAs may be separate effects of IFNs possibly mediated by different receptor sites on the cell membrane (see section II). IFN-γ appears "specialized" in HLA induction, suggesting that its immunoregulatory functions (87,101,330) may be more important than its antiviral action (see section IV 3a).Another protein of 28,000-Mr, probably associated with β_2-microglobulin, is also preferentially induced by IFN-γ as are the HLA proteins (331).

Further evidence that IFN activates the expression of the HLA class I genes comes from experiments in which a cloned human HLA-A_3 gene, introduced by co-transformation into mouse L-TK$^-$ cells, responds to treatment of the cells by mouse IFN but not human IFN. The amount of HLA-A_3 in the mouse plasma membrane increases and 4 times more human HLA mRNA accumulates in the mouse IFN-treated, HLA-transformed cells (332). The IFN-mediated activation of the HLA gene promoter is now under study. Similar experiments were reported by the groups of F.H. Ruddle and P. Lengyel. Activation of HLA class I gene expression does not depend on the translation of the proteins or their presence on the cell surface. This can be shown in Daudi cells which lack HLA class I antigens on their surface, and in which no β_2-microglobulin is synthesized, as a result of a mutation in the initiation codon of β_2-microglobulin mRNA (333). Nevertheless, HLA mRNA and the inactive β_2-microglobulin mRNA are increased in Daudi cells treated by IFN (209,334). Thus, IFN seems to activate the expression of genes located on at least 3 different human chromosomes: 6 for HLA, 15 for β_2-microglobulin and 11 for the (2'-5') oligo A synthetase.

Stimulation of HLA synthesis is not seen in melanoma cells resistant to the antiviral and antigrowth effects of IFN, and in sensitive melanoma cells, the HLA-stimulation is maximum in Go/G_1-arrested cells (328). IFN also failed to induce HLA-mRNA in Namalva cells and in an IFN-resistant Daudi cell (209,334). Embryonic K562 cells, which still lack HLA antigens, did not show HLA-mRNA synthesis after type I IFN treatment (209), but did after IFN-γ (Fellous M., unpublished). The defective expression of HLA antigens in cells of patients with certain combined immuno deficiencies, could be corrected by IFN-α injections (335). The developmental induction of HLA class I gene expression was studied in a human histiocytic lymphoma cell line, U937, which differentiate into macrophages when exposed to phorbol esters as TPA. In these cells, a 10-fold increase in HLA class I mRNA and surface antigens takes place during differentiation, and this increase is abolished if antibodies to human IFN-β are added to the culture (271). IFN secretion by the differentiating culture seems to provide the developmental signal which cause HLA class I antigens to be made. IFNs are at present the only factors known to modulate the expression of the HLA genes, and could thereby, regulate the recognition of foreign (viral) antigens by cytotoxic T lymphocytes (see section IV 3a).

b) Class II histocompatibility antigens: HLA-DR

The HLA-D region of chromosome 6 in man codes for the human equivalent of the mouse Ia antigens, in particular for HLA-DR composed of a 34,000-Mr α chain and polymorphic 28,000-Mr β chains (for review 336). HLA-DR is expressed on macrophages, B lymphocytes, vascular endothelial cells, but also on some tumor cells (melanoma, colorectal cancer). It may function for antigen presentation in some T lymphocytes and monocytes and these cells can acquire HLA-DR when immunologically stimulated, mouse thymocytes showing an increase in Ia antigens when exposed to lymphokines mixtures containing IFN-γ (337). IFN-γ is actually the most potent inducer of HLA-DR synthesis in melanoma cells (338,339) and in lymphoblastoid B cell lines (339), as shown by the use of purified or recombinant DNA-produced IFN-γ. A 5 to 10-fold increase in mRNA for HLA-DR α and β chains, is observed in cells exposed for 2 hours to IFN-γ (339), suggesting that the effect is on HLA-DR gene expression as is the case for class I HLA genes. IFN-α and β also increase HLA-DR mRNA, but to a much lower extent than does IFN-γ (339). The increase in cell surface HLA-DR antigens is more variable: it is increased 2-4 fold in some melanoma cells by IFN-γ (338,339) and to a lesser extent by IFN-β (323,339), but it is unchanged in many cells having a high basal-level of HLA-DR, even after treatment by IFN-γ (339), which may explain earlier negative results (209,322,324,327).

By the stimulation of HLA-DR antigens, IFNs most probably participate in the initiation of the immune response. As for class I HLA antigens, IFN-γ appears specialized in this immunoregulatory function. Pathological situations in which HLA-DR antigens play a central role, as autoimmune diseases [e.g. rheumatoid arthritis (340)], or other immunological disorders in which T cell functions are enhanced or suppressed, could reflect alterations in the production of IFN-γ or other IFN species.

Immune IFN-γ, by its potent effect on both HLA antigen classes, could not only control Ia^+-macrophages (336) but also helper T cells ($T4^*$, HLA-DR restricted) as well as cytotoxic or suppressor T cells ($T8^*$, HLA-A-B,-restricted)(341).

4) <u>Other surface, cytoskeletal and secreted proteins increased by IFNS</u>

Burrone and Milstein (329) have described a 16,000-Mr protein of the cell surface of T-leukemic cells MOLT 4 and B-lymphoblastoid cells Raji or Daudi, which is strongly increased after treatment by IFN-α. This protein becomes the major surface antigen of the IFN-treated cells, but its function is yet unknown. Whether this protein is formed in all cells and could explain the gross changes in cell surface charge observed in L cells (342), also remains to be examined.

Type I IFN treatment enhances the expression of IgG-Fc receptors (343,344) which, in stimulated macrophages, is accompanied by increased binding and phagocytosis of IgG-coated cells through Fc receptors (345). Phagocytosis and macrophage cytotoxicity may be increased by other IFN-mediated effects as well (for review 346,347) and are accompanied by changes in the cytoskeleton (section IV C3). Recent work shows that while Fc receptors increase, another antigen of early macrophage differentiation (Mac-1) seems to decrease, and late differentiation antigens, as C3b receptors or Ia antigens, were not stimulated by type I IFN (345). Such studies need to be completed, however, with IFN-γ. The effects of IFN on surface antigens of cells is generally selective: in melanoma, tumor-associated antigens were unchanged (325,328), while in acute lymphatic leukemia, the CALLA antigen even decreased after IFN type I (348) under conditions where HLA antigens increased. Among T-cell and monocyte specific antigens, only T10 seems to increase after IFN, possibly reflecting an activation of immature thymocytes (348).

Besides surface proteins, IFN treatment may also affect the activity of genes for cytoskeletal proteins. IFN-α and β produced a transient 4-7 fold increase in α-tubulin mRNA in Ramos cells, maximum at 2-4 hours but not seen anymore at 6 hours after IFN (349). The β-tubulin mRNA was less affected and the synthesis of tubulins was not significantly changed in the cells. The effect could result from the increased polymerization of tubulin into microtubules produced by IFN (350) since free tubulin produced by colchicin inhibited tubulin mRNA synthesis (349).

The secretion of certain proteins can also be increased by IFN. Thus, the release of β_2-microglobulin is stimulated by IFN *in vivo* and *in vitro* (323,326). Plasminogen activator production is increased by IFN in macrophages (351) but is decreased in SV3T3 tumor cells (352). A lymphotoxin which causes killing of virus-infected target cells is secreted in much larger amounts by lectin-stimulated human blood lymphocytes when treated by IFN (229,353) and may explain some of the increase in lymphocyte and NK cytotoxicity (42,347)(see section IV 3a). Secretion of interleukine 1 and 2 may be increased by IFN treatment as well as the response (receptor?) to these lymphokines (104, Vilcek J., private communication). The list of IFN-induced proteins and surface receptors is likely to grow even larger and provide us with more clues about the multiple mechanisms by which IFNs exert their biological effects.

IV. Biological Functions of the IFN system

The three main groups of IFN effects are 1) the inhibition of virus multiplication or antiviral effect, 2) the inhibition of cell proliferation or antimitogenic effect and 3) cell-regulatory activities, such as effects on cell differentiation and immune functions.

A) The Antiviral Effects of IFN

1) Evidence for the role of IFN in viral infections. Injection of antibodies against type I IFN to virus-infected mice causes a more rapid development of symptoms and death under conditions where the untreated animals survive, indicating that IFN normally protects the animal against the viral disease. This was demonstrated for EMC virus (354), Herpes simplex virus (HSV) and VSV (355), NDV, polyoma and mouse hepatitis virus (356,357), Semliki forest virus (358) as well as for influenza virus (359). After inoculation of Maloney sarcoma virus, tumors developed more rapidly in mice treated by anti-IFN antibodies (355). Mice which are relatively resistant to high doses of HSV-1, such as the C57 Bl/6 strain, are killed if anti-IFN is injected but are also killed if low virus doses are given, presumably because the low virus dose does not induce enough IFN production (360). In humans, anti-interferon antibodies increase rhinovirus secretion from the nasal mucosa (361). IFN is produced during viral infections in man, and if transient interferonemia is detected in only part of the patients (362-364), the more prolonged increase in (2'-5')-oligo A synthetase of circulating white blood cells (261) is observed in almost all human patients with viral diseases (256-258,365). Deficiencies in IFN production in man are associated with increased viral infections (366,367) and life-threatening viral diseases (368,369).

Studies on the Mx gene in influenza-resistant A2G mice provide clear genetic evidence that the IFN-system is involved in host resistance to virus (315). Cells of Mx/Mx animals are specifically resistant to orthomyxovirus infection (but not to EMC or VSV) and anti-IFN antibodies make these cells and animals again sensitive to the lethal effects of influenza virus (370). Cells and Mx-animals are protected against influenza by low doses of IFN which do not protect wild-type mice (314,371), an effect probably mediated by IFN-induced proteins under the control of the Mx gene (see section III B2). Thus, a mutation affecting the response of cells to IFN determines the resistance of the animal to orthomyxovirus infections.

2) Mechanisms involved in the antiviral effects of IFN in cells

In living organisms, IFN may affect the course of viral infections by a direct action on virus multiplication in cells, and also by indirect actions (e.g. on the immune system) which inhibit virus spreading within the organism. This section deals first with the direct inhibition of virus-growth in IFN-treated cells.

a) Multiphase antiviral state for different viruses and cells

The genetic data on the Mx-mice warns us that the antiviral effect of IFN may differ from one virus to another, and this is indeed what can be concluded from the numerous studies on how IFN blocks virus replication in cell cultures. As pointed out in earlier reviews (205,372,373), IFN-treated cells can show inhibition of virus functions at different phases of the virus cycle (multiphase effect). IFN may act on one or more of the following steps: uptake and processing of virions, early transcription, DNA or RNA replication, viral protein synthesis, assembly of virions and their release from the cells. This multiphase action could explain why no IFN-resistant virus mutant was yet found. Each of these different effects of IFN was observed in some cells and with some viruses, so that the mechanism of IFN action has to be considered separately for each virus-cell system and even then, IFN may have more than one effect on a given virus.

An example of the multiphase action of IFN is provided by SV40 in monkey cells, a system that we described in detail previously (205). At least two steps in the lytic cycle of SV40 are inhibited by IFN: the onset of RNA transcription immediately upon entry of the virion and the translation of all viral RNAs (T-antigens and capsid protein in mRNAs) in the late phase of the virus cycle (after DNA replication). Because host protein synthesis is not shut-off by SV40, and the virus cycle is long, this virus offers the possibility to study the effect of IFN addition either before or after infection (with other viruses, host shut-off prevents the establishment of the antiviral state, IFN being unable to release host shut-off). Similar inhibitions of virus yield (90%) may be obtained with monkey BSC-1 cells treated by monkey IFN either 24 hours before infection or 24 hours after infection (374). When cells are pretreated by IFN, no synthesis of early T-ag mRNAs (either large T or small tRNA) is observed (375-377), and most viral functions are inhibited including the cytopathic effect (378,379). However, IFN does not inhibit SV40 RNA transcription *per se*, since if IFN is given to the cells immediately after infection, the T-ag mRNA synthesis in the early viral phase is not inhibited (205,379,380). Some event preceding transcription must, therefore, be sensitive to IFN. This could be uncoating or the onset of transcription itself; it is unlikely to be virion transport to the nucleus since in Cos-1 cells containing SV40 T-antigens, there is SV40 DNA replication even after IFN pretreatment (at 20% of the normal rate) indicating that virions reach the nucleus and are uncoated enough for DNA replication (379). Transfection with naked SV40 DNA allows T-ag mRNA transcription in IFN-pretreated cells (376,379), although later steps are inhibited (378,380). For transcription, RNA polymerase probably enters the SV40 DNA in the enhancer region which forms a nucleosome free gap on the SV40 minichromosome (381). A viral or cellular protein may either have to be removed from or bound to the enhancer DNA before transcription can be initiated. This event, which activates the transcription of a small portion of the SV40 DNA molecules reaching the cell nucleus, may be the one controlled by IFN pretreatment. New protein synthesis is not required for the onset of SV40 transcription (380) but removal of the VP2, VP3 seems essential (379).

The second step at which SV40 is sensitive to IFN is the translation of viral mRNAs. When BSC-1 monkey cells are treated by IFN after SV40 infection, the amount and rate of synthesis of the T-ag mRNAs and of the VP1, VP2, VP3 mRNAs is normal but these mRNAs are not in polysomes and the synthesis of the proteins is inhibited by 80% (377). The SV40 mRNA are overmethylated at internal positions (m^6A) but are translatable in cell-free systems (382). Host protein synthesis is not inhibited, with the exception of histones and of a 80,000-Mr nuclear protein. The mechanism of this specific translational inhibition is not certain; it starts 2-3 hours after IFN addition and is parallel to the increase in (2'-5') oligo A synthetase (205,247). A trans-acting inhibitor of protein synthesis is present in extracts of IFN-treated SV40-infected cells. While viral mRNAs are not degraded, a cleavage of ribosomal RNAs characteristic of RNase F activation (sectionIII B1 a)is observed (247). In addition to (2'-5')ppp$(Ap)_n$A, other (2'-5') oligoadenylates which do not activate RNase F accumulate in the SV40-infected IFN-treated cells, indicating that enough ds RNA must be made to active the (2'-5') oligo A synthetase (383).

Translational inhibition of SV40 mRNAs by IFN given a few hours after infection, develops only after SV40 replication is underway: if DNA synthesis is prevented by cytosine arabinoside or by the use of a tsA mutant of SV40, the T-ag mRNAs are synthesized and translated normally; when the drug is removed or the permissive temperature is restored, the translation block appears (205,380). This situation is similar to that of SV40-transformed rodent cells in which IFN does not inhibit T-ag synthesis from the integrated SV40 genome (384). Use of permissive cells transformed by SV40 (385), superinfected by SV40, suggests that the translational inhibition which takes place in the late phase of the lytic cycle does not operate on the early T-ag mRNAs (205,380). It is very interesting that the T-ag mRNAs made prior to DNA replication start at a different site on the SV40 DNA than the T-ag mRNAs made in the late phase (386). The different 5'-end of these mRNAs, which map on either side of the origin of replication, could explain their different sensitivity to the IFN-induced transcriptional block. This resistance to IFN in the early phase, appliesonly to the translational inhibition seen when IFN is given after infection; if cells are treated by IFN prior to infection, no transcription is obtained in the early phase prior to DNA synthesis in vivo (387), and in vitro (388), as a result of the unsuccessful onset of transcription.

The antiviral action of IFN on SV40 is different according to the state of the genome (integrated in the host chromosomes or episomal) and according to the phase of the virus cycle. The two steps at which SV40 is inhibited may insure that if a few viruses have escaped the first block, their multiplication will still be strongly inhibited, although the cells can no longer be saved from the viral cytopathic effect. Because SV40 is a transforming virus, it may have acquired the ability to escape IFN action during the early phase (which is the only one expressed in the transformed cell), provided it is already in the cell. The same could apply to integrated retroviruses whose transcription and translation is not inhibited by IFN, while entry and exit of the virus from the cell is blocked by IFN (see below). Escape from IFN may also characterize slow neurotrope viruses. In these cases, maintenance of the chronical viral infection of the cell is insured by integration of the virus, in the host genome, replication occurring at cell division by the host machinery. In contrast, a virus which is maintained in the cell episomally and replicates autonomously, as Bovine Papilloma Virus in transformed mouse cells, remains sensitive to IFN and is eliminated from the cell culture by IFN treatment (389).

Multiphase action of IFN seems to be the common rule. For many viruses more than one effect of IFN was observed. Thus, for VSV (most commonly used for IFN assays), a 65% inhibition of virus penetration by endocytosis is seen with relatively high levels of IFN (390); this could explain in part the inhibition of primary VSV RNA transcription which occurs in the absence of protein synthesis (391,392). All VSV protein synthesis is inhibited by 90% and mRNA is not in polysomes (393), which in turn prevents RNA replication (394). In addition, the 10% VSV particles produced despite the IFN treatment,are non-infectious (395) possibly as a result of a deficiency in glycocoprotein G and membrane protein M (396,397) or because (2'-5') oligo A synthetase is incorporated into the virions (398). Formation of non-infectious virions may be an important part of IFN action on other types of viruses, as in retroviruses. In transformed cells, chronically producing RNA tumor viruses, IFN acting on the already integrated virus, does not inhibit significantly the bulk of viral mRNA and protein synthesis, as found also for SV40 transformed cells. However, IFN often decreases the number of viral particles released from the cells and always produces a larger decrease in infectious virions (reviewed in 205,373,399, 400). As for VSV, there seems to be a defect in the late assembly of the retrovirus which takes place on the cell membrane during the budding process: abnormal virion structures (empty capsids) are seen by EM (401), or protein analysis (402), uncleaved envelope glycoprotein (403) or gag polyprotein precursors (404) are present, and (2'-5') oligo A synthetase is found in the virions (398). In addition, retrovirus infection of virgin cells is inhibited by IFN (405), uncoating and onset of cDNA synthesis is delayed (406,407), while cell transformation can be blocked (408). Clearly again multiple steps of the virus infection are affected by IFN.

The few examples summarized here suffice to illustrate the multiplicity of IFN actions on different viruses. These effects must be mediated by distinct mechanisms since a number of variant or mutant cells have been found in which IFN action on different viruses are dissociated. Thus, the anti-retrovirus action (assembly inhibition) is often dissociated from the inhibition of picornavirus EMC or from that of VSV. In NIH 3T3 clone 1 cells, which are deficient in RNase F (409), IFN inhibits murine leukemia virus production, but not EMC virus and only partially VSV; cell growth is also not inhibited by IFN in these cells (410). Another NIH 3T3 clone, A10, with similar properties but deficient in the induction of (2'-5') oligo A synthetase was reported (411). In RD114 cells, (clone C1), which are human rhabdosarcoma cells infected by a feline retrovirus, human IFN inhibits the retrovirus production (412), but fails partially to inhibit VSV and totally to inhibit EMC virus or cell growth (413); in RD114, IFN also failed to induce the (2'-5') oligo A synthetase and protein kinase PK-i, but inhibited Sendai virus-induced cell fusion in line with a retrovirus block at the membrane budding stage (413). Cells can become resistant to IFN action on the retrovirus while sensitive for VSV (414); JLSV9-R cells (mouse bone marrow cells transformed by Rauscher Leukemia virus) are in addition resistant to EMC virus (although sensitive for VSV): RNase F is very low in these cells, but the RNase F levels increase 10-fold during IFN treatment (415). The embryonic teratocarcinoma (EC) cells present an interesting case: when undifferentiated, EC cells are resistant to IFN effect on Sindbis and on wild type VSV, but sensitive for two VSV-ts mutants (accumulating ds RNA in contrast to wild type)(416); the picornavirus EMC was found resistant in one report (417), while EMC and Mengo virus were sensitive in another (416). In undifferentiated EC cells, the protein kinase PK-i is not induced (418) and RNase F is low (279): these two defects are corrected after the cells differentiate under the influence of retinoic acid and the cells become sensitive to IFN action on EMC and wild type VSV. A more complete review of variants showing IFN effects on some viruses but not others was published by Lebleu and Content (239). The differential effects of human IFN-α subspecies on different viruses was discussed in section I B3.

The unescapable conclusion of these studies is that there must be different IFN-mediated biochemical mechanisms of virus inhibition. Knowing only a few of IFN's biochemical effects, it would be premature to hope for a complete correlation between these and the antiviral effects. The best one can do is to indicate some of the possible mechanisms involved.

b) The inhibition of viral mRNA translation in IFN-treated cells

Among the mechanisms by which virus growth is inhibited in IFN-treated cells, the most ubiquitous remains the inhibition of viral mRNA translation (with the exception of transforming retrovirus as discussed above). This may be because the need for rapid synthesis of viral proteins in infected cells is a common feature of lytic viruses, while transcription and replication differ fundamentally between plus-strand and minus-strand RNA and DNA viruses. The translational inhibition is also, by far, the best studied effect of IFN because it is the only one that could be demonstrated in cell-free extracts of IFN-treated cells. It is not, however, a simple phenomenon andis mediated by several pathways (extensively reviewed recently in 238,239)(Fig. 3, section III B1). Two of these pathways are activated by ds RNA, and may be stimulated in the cells by viral infection if ds RNA accumulates. In most cell-virus systems, it is difficult to ascertain whether the translational inhibition is selective for viral protein synthesis because, host protein synthesis is shut-off after viral infection. However, in SV40-infected permissive cells, selectivity against viral mRNA can be demonstrated (205). The IFN-treated infected cell may die, preventing further virus spreading. In IFN-treated, Mengo virus-infected cells, the shut-off of host protein synthesis occurs as in untreated cells (419,420) presumably by inactivation of the cap binding factor (421), but the cellular RNAs are not degraded and cellular protein synthesis can recover after virus replication has aborted, if the multiplicity of infection is low enough (419,422).

One of the mechanisms most frequently considered for inhibition of protein synthesis in IFN-treated cells is formation of $(2'\text{-}5')ppp(Ap)_nA$ by the IFN-induced synthetase and activation of RNase F (Fig. 3). Accumulation of the (2'-5') oligo A nucleotides was demonstrated in IFN-treated cells infected by EMC (249,250), by reovirus (251) and by SV40 (424); in the latter case inactive cores and other products also accumulate (383). Ds RNA alone (poly rI:rC) causes the accumulation of high levels of (2'-5') oligo A in IFN-treated HeLa cells, and activates cellular mRNA and polyribosomes degradation (425). Artificial introduction of (2'-5') oligo A into cells (e.g. by $CaCl_2$ precipitation) also causes extensive mRNA degradation, inhibition of protein synthesis and, if the cells were infected, an antiviral effect (282,426). Viral mRNA degradation can be seen, however, only in certain virus-infected IFN-treated cells. A degradation of reovirus mRNA was seen in intact IFN-treated cells infected at high multiplicity of virus (423). This is in line with the shorter half-life of reovirus mRNA (238) and with the endonuclease activity of subviral particles from IFN-treated cells (427). In EMC or Mengo-infected cells, it is presumed that newly transcribed viral RNA is degraded since an enhanced nuclease activity is found in a viral complex (428), but in fact only a typical pattern of ribosomal RNA degradation by RNase F has actually been observed (250,286). In VSV-infected HeLa cells (393) and in SV40-infected

BSC-1 cells (377), viral mRNA is not degraded at all, although it is not in polyribosomes and is not translated in the cells. In the case of SV40, however, ribosomal RNA cleavage was clearly seen in the monosomes, but not in polyribosomes still translating the host mRNAs (247, unpublished data), suggesting that an RNase F cleavage may selectively inactivate the ribosomes engaged in SV40 mRNA translation. The SV40 mRNAs from IFN-treated cells can be translated in cell-free systems (382). On the other hand, cell-free extracts of IFN-treated cells show a complete block in all protein synthesis (205,247) suggesting that the selective inhibition is lost upon disruption of the cell. Since RNase F has little sequence specificity (429,430), it is assumed that its action is localized through a local activation of (2'-5') oligo A synthesis, dependent on the presence of a ds RNA structure (as EMC virus replicative intermediate) on or near the mRNA-ribosome target (285). Diffusion of the nucleotide could be prevented by its binding to a ribosome-associated RNase F or by rapid degradation through the 2'-PDi (281). The cell may have several ways of maintaining its own protein synthesis: in reovirus-infected cells, although reovirus mRNAs are degraded and rRNA cleavage occurs, protein synthesis is maintained by a more efficient synthesis of cellular mRNAs (431).

The (2'-5') oligo A pathway cannot, by itself, account for all the IFN-mediated translational inhibition. As extensively reviewed by Lebleu and Content (239), there are for example cells in which (2'-5') oligo A synthetase is present, with or without IFN, and some viruses are not inhibited (98,432,433), and cells in which the synthetase is not induced while viruses are inhibited (434). The absolute level of one enzyme in the pathway may not be always significant. The final result depends on the relative ratio of all enzymes involved which can regulate positively or negatively the pathway (Fig. 3). For example, in HeLa cells, which naturally contain (2'-5') oligo A, it was found that RNase F is inhibited after EMC virus infection (250,280) and that IFN increases the RNase activity in the infected cells. Nevertheless, it is clear that viral protein synthesis can be inhibited by other mechanisms. The phosphorylation of eIF-2 by IFN-induced PK-i could inhibit the initiation of mRNA translation. This phosphorylation was not demonstrated in intact cells, but increased phosphorylation of the 67,000-Mr protein associated with PK-i (see section III B1c) was seen in intact cells exposed to ds RNA (205) or infected by reovirus (435,436). An inhibition of PK-i was observed after vaccinia virus infection (306), suggesting that the enzyme is interacting with the virus cycle. Activation of PK-i is, however, not always correlated with the antiviral effect (239) or with inhibition of protein synthesis (437). Even less is known of the role of the tRNA deficiency seen in IFN-treated cell extracts, for the inhibition of viral protein synthesis. However, even a small reduction in the rate of protein elongation by cycloheximide can produce a selective inhibition of viral protein synthesis (438).

Modification of the viral mRNA structure is another mechanism by which IFN may inhibit translation. In VSV-infected cells, the non-polysomal viral mRNA was found to be partly (60%) missing the 7-methyl residue on the 5' terminal capping guanine (439). The VSV mRNA with the unmethylated caps binds poorly to ribosomes and is not well translated (440). Whether this is a cause or consequence of the translation inhibition is unclear. In reovirus (441) and vaccinia (442) mRNAs from IFN-treated cells, the 2'-O ribose methylation of the cap is deficient. Methylation by the viral enzymes could be inhibited and a labile inhibitor of reovirus mRNA methylation was detected but could not be isolated (443). It could be an IFN-induced protein yet unidentified. Cell-free extracts do not show inhibition of VSV RNA cap methylation (444), and the undermethylation could as well result from a decrease in S-adenosylmethionine. An inhibition of vaccinia cap methylation by (2'-5') oligo A nucleotides was reported (445) but remains to be confirmed. Inhibition of methylation is not a general mechanism: picornavirus RNA are not capped (440). In SV40 mRNA, there is no decrease in cap methylation, but the untranslated viral mRNA is more fully methylated on the internal 6^{m}A positions (382). This modification does not impair translation in cell-free systems but could stabilize the SV40 mRNAs against RNase. Cellular mRNAs were also partially overmethylated. Modifications in mRNA processing should clearly be further investigated in IFN-treated cells.

c) Other effects of IFN on virus replication

What is known of the mechanisms by which IFN inhibits virus replication occurs at other steps than protein synthesis? Endocytosis of viruses and parasites (see section IV A2a) could be inhibited through alterations of the cell membrane (201,446) or cytoskeleton (202,447), which also inhibit Sendai virus-induced cell fusion (413,448). Increase in membrane rigidity and in actin-cables may inhibit fusion of enveloped viruses with the cell membrane and also the release of budding viruses from the cell membrane, as observed for retroviruses.

An IFN-induced impairment of protein glycosylation, leading to a reduced synthesis of sugar-dolichol intermediates (449), could explain the glycoprotein deficiency of non-infectious VSV and retrovirus virions produced in IFN-treated cells (see section IV A2a). Impairment of proper processing and glycosylation through the golgi vesicles, could prevent correct assembly of the virions and release from the cell membrane. It is not known whether this glycosylation-processing impairment is specific for viral proteins.

The inhibition of viral RNA and DNA synthesis observed in IFN-treated cells, have not been shown to result from a direct effect of IFN and are usually considered as resulting from inhibition of the synthesis of early proteins required for replication. Cell transformation by transfected DNA (e.g. TK^- cell transformation by the thymidine kinase gene) is inhibited by IFN (450). It is not known if this effect is on expression of the transfected gene and growth of the transformed colonies, or if it really represents an inhibition of DNA transport and integration in the host genome. Obviously, inhibition of cell transformation by oncogenic viruses (408) would be an important function of IFNs.

A recent report (451) indicated that thymidine kinase negative cells may not respond to IFN (no VSV inhibition nor (2'-5') oligo A synthetase induction). Transformation with a TK gene restored the IFN response. This potentially important observation seems, however, restricted to certain cell lines and its significance for an effect of IFN on nucleotide metabolism remains uncertain.

3) Indirect antiviral mechanisms mediated by IFN

Not less important than the direct inhibition of virus growth in infected cells, are the indirect effects which, in the whole organism, may eliminate virus spreading.

a) Killing of virus-infected cells

It is not enough for the organism to reduce the number of virions emerging from cells, it must also destroy infected-cells which can serve as reservoir for latent infections or for continuous spreading of the virus. This need is probably greatest for viruses which cross from cell to cell and can be hidden from antibodies. Herpes-virus infected cells are killed by such cell-mediated immunity (452).

Virus-infected cells often have viral antigens on their surface which are recognized as non-self antigens. Presentation of these viral antigens to the immune T cells requires the formation of a complex with HLA-class I proteins (453). The cytotoxic effector T cells are HLA-A,B restricted and, after sensitization acquire a receptor for the specific HLA-viral antigen complex. T-cell-mediated lysis of influenza virus-infected human cells requires the matching HLA class I antigens (454). Since IFN-α, β and γ stimulate HLA class I antigen synthesis and expression on the cell surface (see section III B3a), it is possible that this enhances the ability of the immune cells to recognize the virus-infected cells and destroy them. IFN are the only factors known to stimulate HLA synthesis. IFN-γ stimulates strongly HLA class I expression, at concentrations much lower than those needed to inhibit virus proliferation (100). By not inhibiting viral functions, IFN-γ may favor presentation of the viral antigen-HLA complex and reduce the chances of virus carrier cells to escape killing. Although this effect of IFN on recognition of non-self antigens on the target cell was not demonstrated _in vitro_, immune deficient patients with low HLA-A or B treated with IFN showed an increased ability of their cells to be recognized by specific cytotoxic T cells (335). This phenomenon may be difficult to observe in cells saturated with HLA antigens, where even inhibitory effects may appear (see Fig. 4).

IFNs stimulate cell killing also by a direct action on immune cells. The cytotoxicity of T lymphocytes (455-458), of macrophages (459-461), of monocytes (462) and of NK cells (42,463-465) against viral-infected or tumor cells, is enhanced after exposure of these cells to IFN-α, β or γ. The mechanism is unknown, but could be related to the increased synthesis and expression of membrane antigens (see section III B3,4). HLA-DR of class II histocompatibility antigens play a major role in macrophage-lymphocytes communication and in the regulation by helper T cells. HLA-DR is strongly increased by IFN-γ (338,339) on various immune cells (B and T) lymphocytes, macrophages) but also on potential target cells (melanoma). The increased cytotoxicity may reflect differentiation of the immune cells, under the influence of IFN. An increase in HLA-DR expressing lymphocytes and macrophages was reported in rheumatoid arthritis (340) with a delayed-type hypersensitivity picture. Suppressor T cell activity is depressed by IFN (466).

In patients treated with IFN, the NK cell activity is first increased but may decrease later, even below normal (467). This does not occur

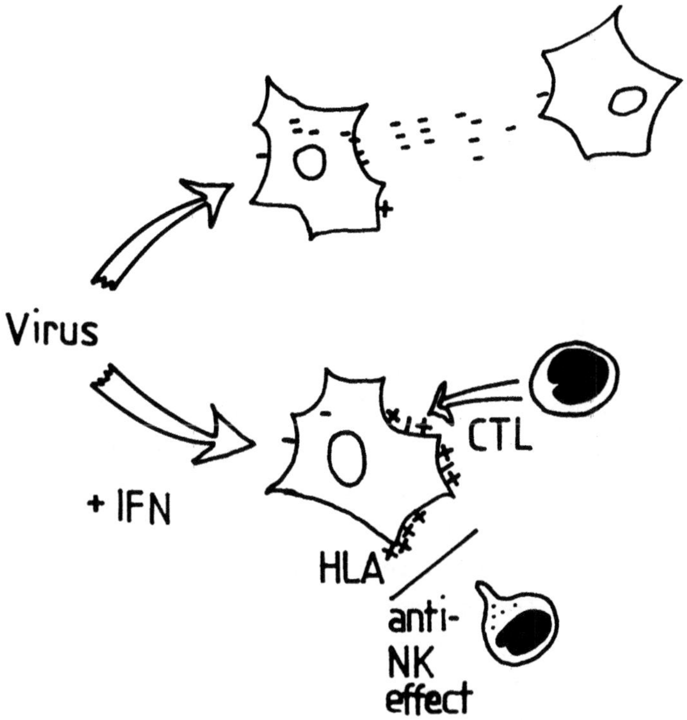

Figure 4. Interferon effect on the killing of virus-infected cells.

Symbols: +, HLA antigens; ', virus or viral antigens; CTL, cytotoxic T lymphocyte; NK, natural killer lymphocytes. IFN would increase presentation of viral antigens complexed with HLA to the CTL. IFN also makes the cell resistant to NK killing. (For details see section IV A3a).

with $3x10^6$ U of IFN-α daily (Strander H., personal communication) and also not in patients with low NK; much variability being observed in the basal NK level and its change after IFN (468). The secondary decrease in NK cells may result from a depletion of pre-NK cells, which differentiate after IFN (469). IFN has also a protective effect in vitro on the target cell, against killing by NK cells (464). This effect is exhibited by cloned IFN-γ at the very low concentrations, which stimulate HLA antigens, but do not inhibit virus growth nor stimulate NK cells (101). Type I IFN produces both protection and stimulation at the same concentration. Protection may involve changes in membrane antigens, but could also result from an increase in prostaglandin by IFN. The anti-NK effect requires protein synthesis and most virus-infections prevent the protective effect of IFN. This could endow the NK-target couple with specificity: only cells able to induce IFN-γ and to respond to it, escaping the killing (470). Some tumor cells may escape killing by inducing IFN-γ efficiently in the NK cells (471). IFN does not protect the target against cytotoxic lymphocytes, the selectivity resulting in this case from a specific recognition of some foreign antigen (see Fig. 4).

Killing of a target by an immune cell proceeds in several steps: i) formation of cell conjugates, ii) triggering of the killer and iii) cytolysis of the target. While the positive and negative effects of IFN discussed above seem to belong to the first two steps, IFN participates also in the last step. Target cytolysis probably results from the release of toxic molecules by killer cells as oxygen radicals (226) or lymphotoxins. The activity of one of the known lymphotoxins (472) (α of 75,000-Mr) released by lectin-triggered human blood lymphocytes, increases strongly if lymphocytes are treated by IFN-β (229). The release requires calcium but not protein synthesis. The same occurs with enriched NK cells treated with IFN-α (473). Prolonged lectin stimulation induces IFN-γ and anti-IFN-γ was found to inhibit the production of lymphotoxin activity. In addition, cytolysis of human fibroblasts by the IFN-stimulated lymphotoxin, occurs only if the cells are either virus-infected, or treated with cycloheximide, or treated with IFN (353). IFN-γ and VSV-infection have a strong synergism in enhancing cytolysis by the lymphotoxin; in contrast IFN-α or β rather inhibit, probably by preventing VSV replication. Thus IFN-γ, in this system, enhances selectively the destruction of a virus-infected cell. IFN-γ by itself may also cause the death of tumor cells at high concentration

(Vilcek J., personal communication), and IFNs are known to enhance cell killing by ds RNA (474,230) with characteristics (dexamethasone, prostaglandin inhibition) similar to the lymphotoxin system. IFNs can also enhance the cytotoxic effects of vaccinia virus in cell cultures (475). It is likely that in vivo, the indirect effects of IFN on virus-infected cell killing by the complex cellular immunity system are the most important for the overall antiviral effect. The implications of tumor cell killing for the antitumor effect are similarly obvious (42,467).

b) Lymphocyte mobilisation

Lymphopenia often accompanies the early stages of viral diseases. By the use of anti-IFN-α,β antibodies in VSV-infected mice, it was recently shown (476) that the virus-induced lymphocyte decrease in peripheral blood, is mediated by IFN. Injection of IFN produces a rapid lymphopenia (476,477) which, in the mouse, is accompanied by a reduction of lymphocytes in the thoracic duct (477). Vaccinia virus or poly (rI:rC) produce the same effects and these are blocked again by anti-IFN (478). This early lymphopenia is not due to bone marrow depression, it is transient and reflects a redistribution of the lymphocytes. Lymph nodes are sometimes enlarged (477), but translocation of lymphocytes is seen also in the spleen where lymphocytes are depleted in several areas (476). Mobilised lymphocytes are seen around blood vessels and in perivascular inuses of the spleen. It may be speculated that lymphocyte migrate from peripheral blood and thoracic duct to the foci of infection, and that this IFN-mediated effect regulates the contact of lymphocytes with antigen-presenting infected cells. This IFN effect could, therefore, be part of the indirect antiviral effect. In some cases, it may, however, decrease the immune inflammatory response (478). It is also tempting to speculate that changes in histocompatibility antigens (H_2 or Ia) induced by IFN, either in lymphocytes or in endothelial cells, with which they interact to cross the vascular barrier, cause the migration of lymphocytes (479). Variable expression of Ia antigens was observed in vascular endothelium (480) and could be IFN-mediated.

c) Rise in body temperature

Human IFN injections are invariably pyrogenic to man. With pure natural or recombinant IFN-α, β or γ, a raise of body temperature to 38.5 - 40°C is observed. High temperature inhibits virus replication in cells and has been proposed by Lwoff (481) as a defence mechanism against viral diseases. IFN-induced hyperthermia could result from prostaglandin

E2 synthesis, which is increased by IFN (216) or from a stimulation in the production of interleukin-1 (the endogenous pyrogene lymphokine) by IFN (104).

B) The Antimitogenic Effect of IFN on Cells

1) Effects of IFN on the cell cycle and on mitogenesis. Control of normal cell proliferation occurs primarily during the G_1 phase of the cell cycle. Biochemical changes and synthesis of proteins, required for the onset of DNA replication, take place during G_1 under the influence of hormone-like serum growth factors (482). Commitment to divide is defined as reaching a restriction point (R) during G_1 (483), after which cells are no longer dependent on growth factors and go through the S, G_2 and M phases at a rate relatively independent of external factors. The rates of protein, mRNA and rRNA synthesis increase in G_1 and if protein synthesis is inhibited, less cells pass the R-point and remain in a Go/G_1 state (484). The identity of the unstable proteins restricting transition from G_1 to S is still unknown (485) but enzymes for the synthesis of nucleotides, DNA and polyamines (e.g. ODC) are probably synthesized during G_1. Tumor cells are not readily arrested at the Go/G_1 restriction point (486).

Inhibition of normal and tumor cell growth is a genuine function of the IFN-α, β and γ molecules (for review 487). Addition of IFN to growing cell populations can prolong all phases of the cell cycle (488) although S and M are usually less affected than G_1 and G_2 (489). In cells still restricted at R, IFN decreases the transition from Go/G_1 to S (490,491) and Go/G_1 quiescent cells appear more sensitive to IFN's antigrowth effect. IFN seems to act primarily on the Go/G_1 to S transition, but certainly has additional effects on preparation to mitosis, which can abort and produce binucleated cells (202).

The effect of IFN on the Go/G_1 to S transition can be best studied in growth-arrested cells which are stimulated synchroniously to resume growth. In mature T lymphocytes or spleen cells stimulated by lectins, IFN treatment inhibits the onset of DNA synthesis which normally occurs after 48 hours (492-494). In serum-starved quiescent 3T3 fibroblasts restimulated by serum, IFN inhibits the onset of DNA synthesis at 24 hours and the subsequent growth (495-497). IFN inhibits similarly the mitogenic stimulus of insulin, vasopressin, or of peptide growth factors, as epidermal growth factor (EGF)(498,498a). An antagonism between IFN action and growth factors such as platelet-derived growth factor (PDGF), has been documented by Inglot et al. (499); IFN antigrowth and even antiviral effects can also be inhibited by PDGF (500,501). Since the product of certain oncogenes, as sis of simian sarcoma virus, has strong structural homologies to PDGF (502), while other transforming growth factors, which allow retrovirus-transformed fibroblast to escape the cell cycle regulation, are homologous to EGF (503), the antagonistic action of IFN and these GF may have profound implications for the antitumor effect of IFN (498a-501).

2) IFN-induced enzymes and the antimitogenic effect. In IFN-treated cells, the inhibition of Go/G_1 to S transition is accompanied by an inhibition of the large increase in the rate of protein synthesis which normally occurs before the onset of DNA synthesis; protein synthesis in resting cells, and during the first hours after mitogen stimulation, is only marginally inhibited by IFN (496,497). Although this may not be the only antimitogenic action of IFN (see next section), it was tempting to correlate this effect with the induction of translation-regulatory enzymes by IFN. Resting lymphocytes have a high level of (2'-5') oligo A synthetase in mice (269,494, 504) and humans (257). In proliferating thymocytes (269) and in human leukemic lymphoid blasts (256), the enzyme level is much lower. After mitogenic stimulation, lymphocyte extracts synthesize less (2'-5') oligo A, partly as a result of a 5-fold increase in 2'-PDi activity (272). In fibroblasts, a decrease in synthetase coupled with an increase in ?'-PDi, produce after growth-stimulation a 10-fold decrease in the ratio between synthesis and degradation of (2'-5') oligo A (272). These early observations were confirmed: many growth-arrested cells have increased (2'-5') oligo A synthetase levels (254,266,267,270-273,505) and an inverse correlation between synthetase levels and number of cells in S phase was shown in melanoma tumor cells (275). Moreover, it appears that a spontaneous secretion of some IFN-β is the cause of the synthetase increase in growth-arrested cells (270,271,275) and anti-IFN-β antibodies lead to a more rapid entry of cells in S phase (275). Uninfected cells contain RNA capable of replacing viral ds RNA in the activation of the (2'-5') oligo A synthetase (246,247). Indeed, a radioimmunoassay of the (2'-5') oligo A nucleotides, shows that their level varies during cell cycle (253); interestingly, dephosphorylation of $ppp(A2'p)_nA$ seems to lead to core nucleotide accumulation at the onset of the S phase. A transient 4-fold decrease in synthetase and 10-fold decrease in (2'-5') oligo A nucleotides, could be corelated with the onset of DNA synthesis in liver, 24 hours after partial

hepatectomy (252,274). Finally, increase in RNase F was also observed in confluent JLSV9-R cells (279); NIH 3T3 clone 1 cells, deficient in RNase F, are resistant to the antimitogenic effects of IFN (410,497,506).

The use of chemically synthesized (2'-5') oligo A nucleotides allowed to test more directly their role in the antimitogenic effect of IFN. The phosphorylated ppp(A2'p)$_n$A produced by the synthetase do not penetrate in intact cells, but require permeabilization of the cells. When introduced by calcium-coprecipitation, there is a profound inhibition of protein synthesis and also DNA synthesis (282). To avoid the permeability problem, Kimchi et al. (272,494,497) used chemically synthesized (A2'p)$_n$A trimer or tetramer cores. In mouse splenocytes (499) or in Balb/C 3T3 fibroblasts (497), stimulated to undergo a Go/G$_1$ to S transition, a delay or an inhibition in the onset of DNA synthesis was observed at 2-20 μM nucleotide concentrations. This concentration is critical, since nM amounts (as found _in vivo_) are insufficient, and higher amounts can produce unspecific effects: adenosine can itself inhibit DNA synthesis at 50 μM (497), and analogs (cordycepin, xylose) could be even more toxic. In the correct concentration range, (3'-5') oligo A does not inhibit mitogenesis. The specific antimitogenic effect of the cores, requires a chain-length of 3 adenines and an unblocked 5'-OH (497). It is not known if the cores become phosphorylated on the 5'-OH in the cell to form pp(A2'p)$_n$A capable of RNase F activation, or if they have a sparing effect on endogenous ppp(A2'p)nA. Cores exist in intact cells (253), as well as other (2'-5') oligo A nucleotides unable to activate RNase F (383), and the RNase may not be the unique target of (2'-5') oligo A (265). However, in support of an effect mediated by RNase F, it was found that p(A2'p)$_n$A at 5 μM are more potent antimitogens than cores in synchronized Balb/C 3T3 cells (507) and that NIH 3T3 clone 1 cells (deficient in RNase F) are much more resistant to the inhibition of DNA synthesis by (2'-5') oligo A cores (497,508). The inhibition of clone 1 cells at high concentrations of cores may indicate an additional effect or an unspecific toxicity (508). The specific antimitogenic effect of (2'-5') oligo A cores is preceded by a partial inhibition of the increase in the rate of protein synthesis and of RNA synthesis late in G$_1$, while protein synthesis is not inhibited in early G$_1$ (272,497). This suggests that (2'-5') oligo A interferes with some event required for the onset of DNA synthesis which occurs late in G$_1$ (see next section). Indeed, addition of (2'-5') oligo A or of IFN during late G$_1$ fails to inhibit the entry of cells in S phase (497). Therefore, the model proposed in Fig. 5 (272), remains a useful working hypothesis.

An antimitogenic or antiviral effect of exogenously added (2'-5') oligo A is hard to see in exponentially growing cells, but a 1 log reduction of VSV was observed in 3T3 cells undergoing the Go/G_1 to S transition (497). These nucleotides inhibit transformation of human lymphocytes by Epstein-Barr virus (EBV)(509). In growing Daudi cells, which are very sensitive to IFN antigrowth effect, no evidence of ppp(A2'p)$_n$A involvement or typical rRNA cleavage was found (278). However, Daudi cells resistant to the antigrowth effect of IFN, did not show the usual two-fold increase in RNase F (278). Thus, the rRNA cleavage could be small, selective or masked by active resynthesis of rRNA. Nevertheless, it is most likely that the (2'-5') oligo A-mediated effects of IFN are not the only one involved in antigrowth action.

3) Possible antimitogenic mechanisms. As indicated in section 2, one of the effects of IFN on mitogenesis could be due to induction of a translational regulation which prevents the synthesis of critical proteins in G_1 and the increase in the rate of protein synthesis late in the G_1 phase of mitogen-stimulated cells. A preferential inhibition of the synthesis of 4 proteins, whose synthesis is stimulated by mitogens, was seen by 2D-gel electrophoresis in (2'-5') oligo A or IFN-treated splenocytes and, in 3T3 cells, the synthesis of two proteins (Mr = 69,000 and 76,000) was decreased prior to the reduction in the overall rate of protein synthesis (497 and unpublished data). Strikingly, a 68,000-Mr protein was recently proposed as the restriction point R-protein (485). The decrease in the synthesis of these mitogen-stimulated proteins was more pronounced than the overall decrease in protein synthesis during late G_1. Since in growing cells, IFN has little effect on overall protein synthesis, a selective mechanism must be at work in mitogen-stimulated cells. One possible mechanism is an effect on ribosome metabolism. Transition from quiescence to growth is accompanied by a 2-3 fold increase in ribosomes, which may be impaired by IFN-induced breakdown of rRNA. Maroun et al. (510,511) have pointed out that IFN affects rRNA metabolism in a way reminiscent of the transition from growth to quiescence. A role for the (2'-5') oligo A system in the processing of heterogenous nuclear RNA was suggested (241) and pre-ribosomal RNA processing may also be affected. This may be a function of the nuclear form of the (2'-5') oligo A synthetase (251,265).

However, neither additional effects of (2'-5') oligo A, nor selective translational inhibition by the IFN-induced mechanisms (see section III B1) are excluded.

Among late G_1 proteins which could be under IFN control, are several proteins known to be coupled to DNA synthesis. Ornithin decarboxylase (ODC), a rate-limiting enzyme of polyamine synthesis, increases in late G_1 and this increase is blocked by IFN in 3T3 cells (496). ODC synthesis may be prevented, but ODC could in addition be phosphorylated, for example, in cells (e.g. macrophages) in which cAMP inhibits growth (512,513). However, the role of cAMP in IFN action is controversial (215) and in cells like 3T3 fibroblasts, cAMP stimulates growth! (512). An IFN-induced protein kinase, different from PK-i, and activated by basic proteins but not by ds RNA (292, 514), could be another candidate for ODC phosphorylation. In 3T3 cells, the increase in uptake of sugars, uridine and ions, which occurs both in early and late G_1, is not inhibited by IFN (496). Uptake of thymidine and its phosphorylation may be inhibited by IFN (515). This may result at least in part from a lower rate of thymidine kinase increase in some IFN-treated cells, an effect also seen for DNA polymerases (516). Histone synthesis is preferentially inhibited in IFN-treated (247) and (2'-5') oligo A-treated cells (272). More important may be the effect of IFN on oncogene products: Rous sarcoma virus pp60 src is inhibited 50% by IFN (517) and the cellular p53 protein, which accumulates in transformed and growing cells (518), decreases more in Friend cells induced to differentiate and to stop growing, when IFN is added (519). Non-growing cells have more HLA on their surface (520) and the stimulation of HLA synthesis by IFN is higher in Go/G_1 cells (328). Finally, the increase in calmodulin preceding the S phase is inhibited by IFN (521). These are just examples of proteins influenced by IFN which may control the progression of cells through G_1. How many of these changes are just the result of yet unknown primary action, remains to be seen.

Selection of cell mutants resistant to IFN's antiproliferative effects, should become the best tool to understand this action. In some cell variants (e.g. IF^R cells of Kuwata or macrophages J774.2 of Bloom and Rosen) the antiviral and antiproliferative effects are dissociated, while in others, the inductions of the (2'-5') oligo A and PK-i systems appear dissociated from the antiproliferative effect (reviewed in 239). We have already warned against simplistic conclusions, the IFN-induced pathways being regulated by more than one enzyme. However, it is clear that IFN, in its complexity, affects cell division in more than one way. For example, changes in the cytoskeleton are numerous (202) and could lead to abnormal control of cell size and shape, as well as to abortive mitosis. What is certainis thatthe role of IFN as a negative growth-factor should receive more attention in the future, as it may be the most fundamental aspect of IFN action, of which the antiviral effect is only an episode.

C) Cell Regulatory Effects of IFN

1) IFN and cell differentiation. Since IFN inhibits the synthesis of many proteins coded by viral genomes or stimulated by mitogens, it is important to examine if this inhibition is selective and whether the synthesis of proteins related to differentiated functions of the cell are also affected. Non-infected cells treated with IFN do not die, but can cells still express a new differentiation program? Friend erythroleukemic cells, induced to differentiate by DMSO, synthesize, in fact, more hemoglobin when pretreated by amounts of mouse IFN (100 U/ml) which inhibit Friend virus production (522). Both the number of hemoglobin-containing cells and the amount of globin per cell is increased by IFN addition (401). High concentrations of mouse IFN-α,β (>1,000 U/ml), however, were reported to inhibit globin mRNA and protein synthesis (523,524) while, at up to 500 U/ml, the enhancement of differentiation was confirmed (524,525). In human K562 cells, which differentiate if treated by butyric acid or hemin, IFN-α (up to 5,000 U/ml) enhanced 3-5 fold the number of hemoglobin containing cells with or without inducer (524) while doses of 100,000 U/ml were needed to inhibit differentiation. The significance of such high IFN concentrations is difficult to ascertain, as secondary effects may operate. Stimulation of cell differentiation by IFN was seen in other systems: mouse myeloid leukemic cells induced to differentiate into granulocytes (526) or macrophages (527) were stimulated by IFN, more lysozyme being produced; the differentiation of promyelocytic HL60 human cells induced by retinoic acid or TPA was stimulated by purified IFN-α or β (528); pre NK cells are triggered to differentiate in active NK cells by IFN (467); IFN-treated neuroblastoma cells differentiate more actively into firing neuronal cells after DMSO induction (235); cloned IFN-αA (100-5,000 U/ml) accelerates myotube formation and muscle creatine kinase synthesis in human myoblast cultures (529). In other cases, only the inhibitory effect of IFN was reported as in steroid-dependent glutamine synthetase formation in retina

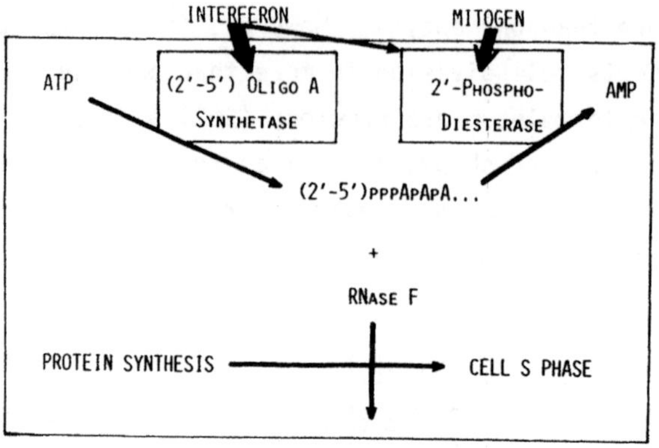

Figure 5. Schematic view of the regulation of mitogenesis by (2'-5') oligo A synthetase E and 2'-phosphodiesterase.

The rectangle symbolizes the cell. Black arrows indicate that interferon induces the (2'-5') oligo A synthetase and to a lesser extent the 2'-phosphodiesterase. A mitogen stimulus such as concanavalian A in lymphocytes or serum in starved fibroblasts leads to an increase in 2'-phosphodiesterase activity (see section IV B2). Adapted from Revel et al. (247).

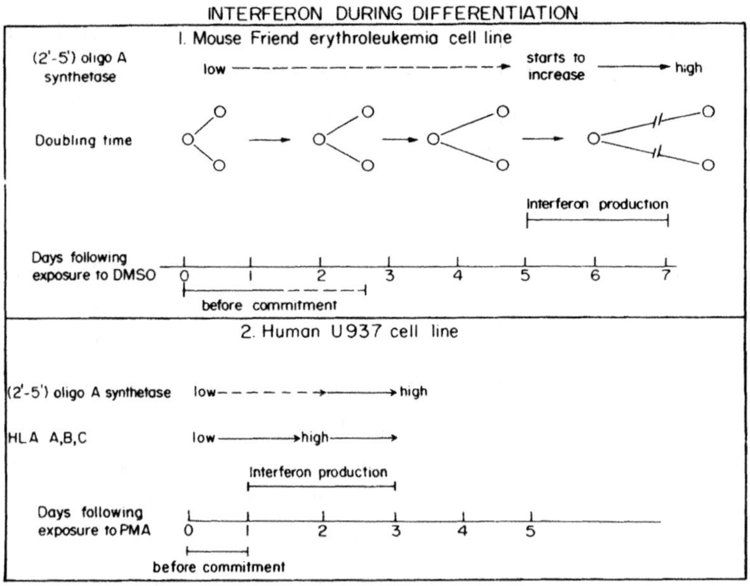

Figure 6. Interferon in cell differentiation.

Spontaneous IFN production correlated with increase in (2'-5') oligo A synthetase and HLA antigens, was observed in two differentiating cell systems. For details see section IV C1. From Kimchi et al. (271).

(530), melanin production in melanoma cells (531) or insulin-induced formation of adipocytes from 3T3 fibroblasts (532). The detoxification enzymes of cytochrome P-450 are decreased in liver of IFN-treated animals (532a). Whether these are true differences or dose effects are unknown, but the general rule seems to be that IFN stimulates differentiation and enhances the synthesis of specific differentiation proteins.

IFN appears fundamentally involved in cell differentiation: two different cell systems have been found to secrete IFN during their differentiation program. The clue to this finding was a 30-fold increase in (2'-5') oligo A synthetase observed in Friend cells 5-7 days after DMSO (269,270) which can be blocked if anti-IFN-β is added to the culture. This increase is much higher than what is seen when cells stop growing without differentiation, and small amounts of a mouse IFN-β are detectable in the culture medium. This IFN-β differs in its molecular size, from Sendai-virus IFN-β produced in the same cells, and is 50 times more efficient to inhibit mitogenesis in Balb/C 3T3 cells than viral-induced IFN for the same antiviral action (270). Cultures with anti-IFN-β have 3-4 times less globin per cell than Friend cells exposed to IFN or even to (2'-5') oligo A cores (269). In addition, less cells become enucleated, suggesting an inhibition of the late stages of differentiation (unpublished data). A similar phenomenon was found in human histiocytic lymphoma cells U937 induced to differentiate into macrophages by TPA (271). The increase in (2'-5') oligo A synthetase is inhibited by antibodies to human IFN-β and, more significantly, HLA class I mRNA and surface antigens do not appear on the differentiating macrophages, although morphological cell changes occur (Fig. 6).

The secretion of small amounts of particular minor species of IFN could be part of the differentiation program in many cells. Terminal differentiation is not only accompanied by cell-growth arrest, but also by destablization of unwanted mRNAs (533) which may be a function of the (2'-5') oligo A system. Differentiated cells often have a high (2'-5') oligo A synthetase levels (269). IFN secretion during pregnancy (86,258, 534) may have this origin. The effect of IFN on cell differentiation may be relevant to its antitumor action: IFN not only decreases tumor growth, but would force the malignant cells to differentiate more normally. Thus, oestrogen receptors are increased in breast cancer cells of patients receiving IFN-β therapy (535). Inhibitory effects on differentiation of bone marrow cells observed at high IFN doses (524) may explain some side effects of high-dose IFN therapy (leucopenia), but are probably not physiological and not typical of IFN action, if properly used.

2) Reversal of the transformed phenotype. Prolonged treatment of tumor cells with IFN can progressively reverse the transformed phenotype of murine sarcoma virus-transformed (536) or x-ray transformed malignant cells (537,538). The morphology and anchorless growth in agar of the cancer cell can be reversed. There are profound changes in the cytoskeleton (538): increase of actin cables (microfilament bundles) and of vinculin adhesion plaques [vinculin being a target of sarc protein kinase (539)]; changes in fibronectin and microtubules (see next section) appear less characteristic of the reversion from malignancy (540). There are also changes in cytoskeletal-associated proteins. Release of plasminogen activator by the transformed cells may be inhibited by IFN (352). Reversal of the malignant phenotype occurs slowly over 30 passages in the constant presence of IFN; the malignancy can reappear if IFN is withdrawn, suggesting that there is no genetic cure. Since retrovirus products are still made, the viral genome is not eliminated during reversal (539). This contrasts with the cure and loss of virus from mouse cells transformed by the non-integrated bovine papilloma virus (389).

The mechanism of the slow reversal of the malignant cell phenotype in cultures could involve a partial inhibition of the expression of activated integrated oncogenes, as was observed for $pp60^{sac}$ of Rous sarcoma-transformed cells (517). Oncogenes are cellular genes which by recombination with retroviruses and their promoters, by mutation or translocation to actively transcribed regions of the genome, become activated (for review 541). If these events occur continuously in cultured cells, they may be inhibited by IFN as is the primary transformation by retroviruses (408) or by DNA (450). Membranal oncogene products may possess growth factor activity (502,503) to which IFN acts as an antagonist (498-501), the slow action being explained by the continued action of the oncogene product (sis, ras) inserted in the cell membrane. Other oncogene products are nuclear and immortalize the cells (e.g. myc). The effect of IFN on the cell cycle and on differentiation (see above) could inhibit these oncogene effects. Finally, IFN may act by reversing the cytoskeletal alterations caused by oncogenes or growth factors.

3) Change in cytoskeleton and cell behavior. Exposure of cultured HeLa or L-929 cells, which have little cytoskeletal organization, to 500-1,000 U/ml IFN for 2-3 days causes multiple changes in cell structure and behavior. Actin-containing microfilaments become more dense and form meshworks under the membrane and also around the nucleus, with actin fibers radiating to the periphery of the cell, vaguely similar to what is seen in normal fibroblasts (202,350,44). Numerous microbubules can also be seen near the membrane, as is found in IFN-reverted cancer cells (540). Polyribosomes are no more seen near the plasma membrane, being excluded by the microfilaments. Binucleated cells and figures of abortive mitosis, in which daughter cells cannot separate being still linked in telophase by abnormal microfilament bridges, appear frequently after 3 days of IFN treatment (202,447,542) and contribute to the arrest of cell growth which becomes maximum at that time. During the first days, the antiproliferative effect of IFN is probably mainly on the G_1/S transition (section IV B1) and is far from complete in these tumor cells. After 3 days, cell death may be seen. Changes in actin microfilaments also impair cell surface receptor movement and capping of concanavalin A receptors is inhibited (447,543). Membrane lipid fluidity, which is transiently decreased during the first hours after IFN (section III A), is again reduced on the 2nd day after IFN (201,202). Cell size is increased up to two-fold. Cell motility is decreased and fibroblasts or epithelial cells migrate at a much reduced rate in wound-repair experiments on cell cultures (447,544,545); this contrasts with the increased lymphocyte migration *in vivo* (section IV A3b), and may have clinical implications by retarding reepithelization of viral skin or cornea lesions. How many of these late effects are due to IFN action directly and how many result from primary antimitogenic effects, or antagonism to growth factors, has not been clarified. IFN does derepress the tubulin genes (349) but no stimulation of actin synthesis is seen. Some of the actin and tubulin polymerization could result from cAMP action (542), which is increased at these late times (215), but larger cell size and binucleated cells are not produced by cAMP (202). A more specific relation between IFN and cytoskeleton structure may exist: butyrate, which enhances cytoskeleton organization, increases tumor cell response to IFN (233,546) while colchicine may inhibit it, suggesting that cytoskeleton organization plays a role in the response of the cell to IFN (146).

Blood and peritoneal macrophages (induced by irritants but not residents) show an increased phagocytic activity when exposed to IFN _in vivo_ and _in vitro_ (for review 346,347). When treated by 5,000 U/ml IFN for 24 hours, the peritoneal macrophages show a large increase in actin microfilaments near phagocytic cups (202,547) which may explain the increased phagocytosis mediated by Fc-receptors. The microtubules of these IFN-treated cells, aggregate around the nucleus and their disorganization, with that of the intermediate filaments, may explain the decreased pinocytosis (202). IFN can cause structural changes in kidney and liver cells of newborn mice, in which IFN is lethal (548): tubular aggregates can be seen in liver endoplasmic reticulum (549) and fibrillar material thickens the renal glomerular basal membrane (550), but the basis for these changes is yet unknown. In neuronal cells, in which differentiation is stimulated by IFN (235), microtubule assembly may participate in the densification of nerve projections leading to more efficient neurone firing (551). Beat frequency of myocardial cells is also increased by IFN (552). Some of these changes in cell structure and behavior may explain toxic side-effects of IFN, for example, nausea, muscle pain, fatigue, central nervous system disorders and cardiotoxicity which may appear during very high dosage ($>30x10^6$ units) IFN-therapy in man.

D) Immunoregulatory Effects of IFN

Many effects of IFN on the immune system which we have discussed in preceding sections (see III B3 and 4, IV A 3a and b) can be seen as resulting from cell-regulatory actions such as inhibition of proliferation, stimulation or inhibition of differentiation, and changes in cytoskeleton. IFNs act on immune cells as on other cells, but the outcome is often complex and contradictory because of the cell-cell and cell-lymphokines network of the immune system. Antibody formation is inhibited if IFN is given with or before the antigen (inhibition of clonal proliferation? protein kinase activation? (553)), but enhanced if antibody-secreting lymphocytes are already formed (increased differentiation, helper cells, HLA-DR?)(554). NK cell activity is increased by IFN _in vitro_, (differentiation of preNK?), but may be decreased _in vivo_ if excess IFN is given (depletion of preNK?)(467). CTL activity is increased (differentiation, HLA, lymphotoxin?), ADCC is less sensitive. Monocyte and macrophage cytotoxicity is increased (differentiation, receptors?)(347). Lymphocyte migration from the central circulation to the periphery is triggered (surface antigens, motility?) but this may depress part of the immune response (478). Suppressor T cells are depressed (surface antigen changes, proliferation decrease?)(466,555). Delayed type hypersensitivity is inhibited if IFN is given before sensitization (inhibition of clonal proliferation?) but enhanced if given to sensitized animals (decreased suppressor cells? enhanced differentiation, HLA-DR?)(347). Ig E-mediated histamine release by basophils is increased (556). Macrophage, monocyte phagocytosis is enhanced (differentiation, cytoskeleton)(346). These are only part of the described effects of IFN on the immune cells, (and mainly reviews were quoted), but this list should indicate the need to search for common principles in the diversified action of IFNs.

V. Remarks on Some Clinical Uses of IFN and Perspective

Interferon was studied for the first 20 years following its discovery by Isaacs and Lindenmann in 1957, as an antiviral agent. Today, we have a multitude of IFN molecules α, β and γ and IFNs appear more as cell-regulatory factors. Can IFN still justify the hopes of many for clinical use against viral infections? First, it should be remembered that the IFN system (including induction of the various types and response of the cells to them) is the first line of defence of the higher organisms to viral infections. It is activated over its basal level during early stages of viral diseases in man. It mediates many of the symptoms of viral diseases (fever, leucopenia, inflammation) and, therefore, cannot alleviate these symptoms. However, it is beyond doubt that it can limit the proliferation of lytic viruses and protects the organism from death (section IV A1). Second, the proper way of using IFNs in human therapy, is not always apparent. Should it be given to further strengthen the endogenous IFN response to viral infections or should it be used only as replacement therapy in case of endogenous IFN deficiencies? One could argue that the mere fact that a viral disease becomes clinically active, indicates that viral infection has overpowered the body defence capacity. On the other hand, patients dying from hemorrhagic fever have often 50,000 U/ml of IFN in their blood: injection of IFN seems then useless and could even be detrimental because part of the immune response can be inhibited and side effects worsened by excess dosage of IFN (Falcoff R., private communication). In lupus (SLE), autoimmune diseases and AIDS, a mysterious acid-labile IFN-α is present in the blood (45,169): is this the result or the cause of the immune disorder? would injection of IFN be good or bad?. In fact, AIDS and Kaposi sarcoma patients seem to benefit from IFN-therapy (Krown S., private communication). Which type of IFN should be given and what route should be used (topical, systemic, intrathecal or cerebral) are questions to explore for each disease, according to its tissue site: IFN may be produced by the body but not reach critical tissues (e.g. in rabies). Replacement therapy in IFN deficient or immunodepressed patients sounds more logical but applies to only few cases. Levin et al. (368,369) have obtained good clinical responses in patients with no endogenous IFN response and emphasize the relation between this deficiency and the life-threatening character of hepatitis B, encephalitis or herpes simplex (HSV) infections.

To alleviate uncertainties of IFN use, monitoring the IFN system of patients before, and during IFN therapy, could be extremely useful. This is one of the justifications for continued research on the molecular and cellular aspects of IFN action, and for our extensive review of these. Measuring (2'-5') oligo A synthetase in white blood cells (257,365), β_2-microglobulin in serum (326,535) or NK cell activity (467) can tell us if the IFN system is active, increased or depressed. This is also useful in chronic diseases: the (2'-5) synthetase level is increased in persistent chronic hepatitis, but not in chronic active hepatitis, suggesting that the bad prognostic of CAH may result from a deficient IFN system and would justify IFN therapy (557).

Human IFNs from cell cultures or from cloning technology, have proven effective in controlled clinical trials for a number of viral diseases. A review of these results is outside the scope of this chapter, and the reader is referred to two recent reviews (558,559). Most of the results pertain to type I IFNs, which proved beneficiary in early therapy of systemic herpes, varicella zoster of immunodepressed patients, in prophylaxis of cytomegalovirus infection in renal transplant patients, against eruption of facial herpes in trigeminal nerve root surgery, in decreasing Dane particles in chronic active hepatitis especially when combined with adenosine arabinoside, in preventing rhinovirus infection of volunteers, in eliminating skin warts intralesionally and juvenile laryngal papillomatosis systemically, and in hasting healing of dendritic herpetic keratitis when associated topically with antiviral chemotherapy. In our own experience with IFN-β from InterYeda Ltd (Rehovot, Israel), controlled trials have shown that IFN is efficient to shorten and prevent epidemic adenovirus conjunctivo-keratitis when given as eye drops, in preventing recurrences of herpes virus eruption and alleviating their symptoms in genital and facial herpes when used as an ointment, and in eliminating condyloma acuminata (genital papillomavirus warts) when given by intramuscular injections. Positive effects on life-threatening acute hepatitis and encephalitis as well as on progression of multiple sclerosis were reported in a few cases (560). These results due to many groups around the world are encouraging, since other effective treatment or immunization is not available for most of the diseases where IFN is active. The wide-spectrum IFNs, being now mass-produced, will certainly have their place next to specific chemical antiviral drugs. Cancer therapy by IFNs, which has sometimes pushed the antiviral applications to the background, gives encouraging results in hairy cell leukemia, chronic myelocytic leukemia, IgA myelomas, renal cancer, skin cancers due to papillomavirus, melanomas and breast cancer metastasis. IFN specialists are convinced that IFNs will be an important adjuvant tumor therapy and that much can be learned about the role of the IFN system by studying the mechanism of these non-antiviral effects. In many cancers, only a 1/3-1/4 of the patients show tumor regression, and studying whether these individuals are deficient in some of the IFN species should be gratifying. As for viral diseases, IFN therapy has to be evaluated in the framework of the entire functions of the IFN system (induction of each species, their circulation through blood, lympha, tissues, and their action on specific cells) to allow the most useful modulation of this system to maintain or restore health.

Acknowledgements: Work from our laboratory supported in part by InterYeda Ltd., Israel.
It is a pleasure to thank Drs. M. Rubinstein, M. Fellous, A. Kimchi, D. Wallach, Y. Mory and J. Chebath for helpful discussions, and Mrs. P. Wynick for typing.

REFERENCES

1. Trent, J.M., Olson, S. and Lawn, R.M. Proc.Natl.Acad.Sci. USA 79: 7809-7813, 1982.
2. Meager, A., Graves, H., Burke, D.C. and Swallow, D.M. Nature 280: 493-495, 1979.
3. Owerbach, D., Rutter, W.J., Shows, T.B., Gray, P., Goeddel, D. and Lawn, R.M. Proc.Natl.Acad.Sci. USA 78:3123-3127, 1981.
4. Tan, Y.H., Creagan, R.P. and Ruddle, F.H. Proc.Natl.Acad.Sci. USA 71:2251-2255, 1974.
5. Slate, D.L. and Ruddle, F.H. Cell 16:171-180, 1979.
6. Sagar, A.D., Sehgal, P.B., Slate, D.L. and Ruddle, F.H. J. Exp. Med. 156: 744-755, 1982.
7. Weissenbach, J., Chernajovsky, Y., Zeevi, M., Shulman, L., Soreq, H., Nir, U., Wallach, D., Perricaudet, M., Tiollais, P. and Revel, M. Proc.Natl.Acad.Sci. USA 77:7152-7156, 1980.
8. Sehgal, P.B. and Sagar, A.D. Nature 288:95-97, 1980.
9. Streuli, M., Nagata, S. and Weissmann, C. Science 209:1343-1347, 1980.
10. Nagata, S., Mantei, N. and Weissmann, C. Nature 287:401-408, 1980.
11. Goeddel, D.V., Leung, D.W., Dull, T.J., Gross, M., Lawn, R.M., McCandliss, R., Seeburg, P.H., Ullrich, A., Yelverton, E. and Gray, P.W. Nature 290:20-26, 1981.
12. Rubinstein, M., Rubinstein, S., Familletti, P.C., Miller, R.S., Waldman, A.A. and Pestka, S. Proc.Natl.Acad.Sci. USA 76: 640-644,
13. Rubinstein, M., Levy, W.P., Moschera, J.A., Lai, C.Y., Hershberg, R.D., Bartlett, R.T. and Pestka, S. Arch.Biochem.Biophys. 210:307-318, 1981.
14. Allen, G. and Fantes, K.H. Nature 287: 408-411, 1980.
15. Rubinstein, M., Rubinstein, S., Familletti, P.C., Gutterman, J., Hester, J. and Pestka, S. In: Proteins, Structure and Biological Functions (Eds. E. Gross and J. Meinhofer), Pierce Chemical Company, Rockford, Il, 1980, pp. 99-103.
16. Plan, D. M.D. Thesis, University of Lyon, France, 1983.
17. Stanton, G.J., Langford, M.P. and Weigert, D.A. Texas Rep. Biol. Med. 41:84-88, 1982.
18. Familetti, P.C. and Pestka, S. Methods Enzymol. 78:83-86, 1981.
19. Mantei, N., Schwarzstein, M., Streuli, M., Panem, S., Nagata, S. and Weissmann, C. Gene 10:1-10, 1980.
20. Zoon, K.C., Smith, M.E., Bridgen, P.J., Anfinsen, C.B., Hunkapillar, M.W. and Hood, L.E. Science 207:527-528, 1980.
21. Levy, W.P., Shively, J., Rubinstein, M., Del Valle, U. and Pestka, S. Proc.Natl.Acad.Sci. USA 77:5102-5104, 1980.
22. Levy, W.P., Rubinstein, M., Shively, J., Del Valle, U., Lay, C.Y., Moschera, J., Brink, L., Gerber, L., Stein, S. and Pestka, S. Proc.Natl.Acad.Sci. USA 78:6186-6190, 1981.
23. Rubinstein, M. Biochim. Biophys. Acta 695:5-16, 1982.
24. Wetzel, R., Levine, H.L., Estell, D.A., Shire, S., Finer-Moore, J., Stroud, R.M. and Bewley, P. In: Interferons (Eds. T. Merigan, R.M. Friedmann and C. Fox) Academic Press, New York, 1982. pp.365-376.
25. Franke, A.E., Shepard, H.M., Houck, C.M., Leung, D.W., Goeddel, D.V. and Lawn, R.M. DNA 1:223-230, 1982.
26. Berg, K. and Heron, I. J. Gen. Virol. 50:441-446, 1980.

27. Weissmann, C., Nagata, S., Boll, W., Fountoulakis, M., Fujisawa, A., Fujisawa, J-I., Haynes, J., Henco, K., Mantei, N., Ragg, H., Schein, C., Schmid, J., Shaw, G., Streuli, M., Taira, H., Todokoro, K. and Weidle, U. In: Interferons (Eds. T. Merigan, R.M. Friedmann, and C. Fox), Academic Press, New York, 1982, pp. 295-326.
28. Ullrich, A., Gray, A. Goeddel, D.V. and Dull, T.J. J. Mol. Biol. 156: 467-486, 1982.
29. Brack, C., Nagata, S., Mantei, N. and Weissmann, C. Gene 15:379-394, 1981.
30. Nagata, S., Brack, C., Henco, A., Schambock, A. and Weissmann, C. J. Interferon Res. 1:333-336, 1981.
31. Lawn, R.M., Adelman, J., Dull, T.J., Gross, M., Goeddel, D.V. and Ullrich, A. Science 212:1159-1162, 1981.
32. Shows, T.B., Sakaguchi, A.Y., Naylor, S.L., Goeddel, D.V. and Lawn, R.M. Science 218:373-374, 1982.
33. Lawn, R.M., Gross, M., Houck, C.M., Franke, A.E., Gray, P.W. and Goeddel, D.V. Proc.Natl.Acad.Sci. USA 78:5435-5439, 1981.
34. Miyata, T. and Hayashida, H. Nature 295:165-168, 1982.
35. Weck, P.K., Apperson, S., May, L. and Stebbing, N. J. Gen. Virol. 57:233-237, 1981.
36. Yelverton, E., Leung, D.W., Weck, P.K., Gray, P.W. and Goeddel, D.V. Nucl. Ac. Res. 9:731-741, 1981.
37. Streuli, M., Hall, A., Boll, C., Stewart, W.E., Nagata, S. and Weissmann, C. Proc.Natl.Acad.Sci. USA 78:2848-2852, 1981.
38. Rehberg, E., Kelder, B., Hoal, E.G. and Pestka, S. J. Biol. Chem. 257:11497-11502, 1982.
39. Goren, T., Kapitkovsky, A., Kimchi, A. and Rubinstein, M. Virology in press, 1983.
40. Hannigan, G.E., Gewert, D.R., Fish, E.N., Read, S.E. and Williams, B.R.G. Biochem. Biophys. Res. Commun. 110:537-544, 1983.
41. Evinger, M., Rubinstein, M. and Pestka, S. Arch. Biochem. Biophys. 210:319-329, 1981.
42. Herberman, R.B., Ortaldo, J.R., Timonen, T., Reynolds, C.W., Djeu, J.Y., Pestka, S. and Stanton, J. Texas Rep. Biol. Med. 41:590-595, 1982.
43. Orchansky, P., Goren, T. and Rubinstein, M. In: The Biology of the Interferon System (Eds. E. De Maeyer, G. Galasso, and H. Schellekens). Elsevier-North Holland, Amsterdam, 1983, in press.
44. Staehelin, T., Durrer, B., Schmidt, J., Takacs, B., Stocker, J., Miggiano, V., Stahli, C., Rubinstein, M., Levy, W.P., Hershberg, R.D. and Pestka, S. Proc.Natl.Acad.Sci. USA 78:1848-1852, 1981.
45. Preble, O.T., Black, R.J., Friedman, R.M., Klipper, J.H. and Vilcek, J. Science 216:429-431, 1982.
46. Fischer, D. and Rubinstein, M. Cellular Immunol., in press, 1983.
47. Trinchieri, G.D., Santoli, D., Dee, R.R. and Knowles, B.B. J. Exp. Med. 147:1299-1313, 1978.
48. Blalock, J.E., Langford, M.P., Georgiades, J. and Stanton, G.J. Cell.Immunol. 43:197-210, 1979.
49. Smith, E.M., Hughes, T.K. and Blalock, J.E. Infect.Immun. 39:220-224, 1983.
50. Pickering, L.A., Kronenberg, L.H. and Stewart, W.E. Proc.Natl.Acad. Sci. USA 77:5938-5942, 1980.
51. Sehgal, P.B., Sagar, A.D. and Braude, I.A. Science 214:803-805, 1981.
52. Sehgal, P.B. In: Interferon 4 (Ed. I.Gresser), Academic Press, New York, 1982, pp. 1-22.

53. Havell, E.A., Berman, B., Ogburn, C.A., Berg, K., Paucker, K. and Vilcek, J. Proc.Natl.Acad.Sci. USA 72:2185-2190, 1975.
54. Hayes, T.G., Yip, Y.K. and Vilcek, J. Virology 98:351-363, 1979.
55. Knight, E. Jr. and Fahey, D. J.Biol.Chem. 256:3609-3611, 1981.
56. Friesen, H.J., Stein, S., Evinger, M., Familletti, F.C., Moschera, J., Meienhoffer, S., Shively, J. and Pestka, S. Arch.Biochem.Biophys. 206:432-450, 1981.
57. Knight, E., Jr., Hunkapillar, M.W., Korant, B.D., Hardy, R.W.F. and Hood, L.E. Science 207:525-526, 1980.
58. Okamura, H., Berthold, W., Hood, L., Hunkapillar, M., Inoue, M., Smithson, H. and Tan, Y.H. Biochem.Biophys.Res.Commun. 19:3831-3835, 1980.
59. Taniguchi, T., Ohno, S., Fuji-Kuriyama, Y. and Muramatsu, M. Gene 10: 11-15, 1980.
60. Derynck, R., Content, Y., DeClercq, E., Volckaer, G., Tavernier, J., Devos, R. and Fiers, W. Nature 285:542-547, 1980.
61. Goeddel, D.V., Shepard, H.M., Yelverton, E., Leung, D., Crea, P., Sloma, A. and Pestka, S. Nucl. Acids Res. 8:4057-4074, 1980.
62. Houghton, M., Eaton, M.A.W., Stewart, A.G., Smith, J.C., Doel, S.M., Catlin, G.H., Lewis, H.M., Patel, T.P., Emtage, J.S. and Carey, N.H. Nucl. Acid Res. 8:2885-2894, 1980.
63. Taniguchi, T., Mantei, N., Schwarzstein, M., Nagata, S., Muramatsu, M., and Weissmann, C. Nature 285:547-549, 1980.
64. Tan, Y.H., Barakat, F., Berthold, W., Smith-Johansen, H. and Tan, C. J.Biol.Chem. 254:8067-8073, 1979.
65. Lawn, R.M., Adelman, J., Franke, A.E., Houck, C.M., Gross, M., Najarian, R. and Goeddel, D.V. Nucl. Acid Res. 9:1045-1052, 1981.
66. Ohno, S. and Taniguchi, T. Proc.Natl.Acad.Sci. USA 78:5305-5309, 1981.
67. Degrave, W., Derynck, R., Tavernier, J., Haegeman, G. and Fiers, W. Gene 14:137-143, 1981.
68. Gross, G., Mayr, U., Bruno, W., Grosveld, F., Dahl, M.H.M. and Collins, J. Nucl. Acid Res. 9:2495-2507, 1981.
69. Mory, Y., Chernajovsky, Y., Feinstein, S.I., Chen, L., Nir, U., Weissenbach, J., Malpiece, Y., Tiollais, P., Marks, D., Ladner, M., Colby, C. and Revel, M. Eur.J.Biochem. 120:197-202, 1981.
70. Houghton, M., Jackson, I.J., Porter, A.G., Doel, S.M., Catlin, G.H., Barber, C. and Carey, N.H. Nucl. Acid Res. 9:247-265, 1981.
71. Raj, N.B.K. and Pitha, P.M. Proc.Natl.Acad.Sci. USA 77:4918-4922, 1980.
72. Gross, G., Mayr, V. and Collins, J. In: "The Biology of the Interferon System" (Eds., E. DeMaeyer, G. Galasso and H. Schellekens), Elsevier-North Holland, Amsterdam, pp. 85-90, 1981.
73. Dinter, H., Hauser, H., Mayr, U., Lammers, R., Bruns, W., Gross, G. and Collins, J. In "The Biology of the Interferon System" (Eds., E. DeMaeyer andH. Schellekens) Elsevier, Amsterdam, in press, 1983.
74. Content, J., DeWitt, L., Tavernier, J. and Fiers, W. Nucl.Acid. Res. 11:2677-2687, 1983.
75. Nir, U., Maroteaux, L., Mory, Y. and Revel, M. in press, 1983.
76. Zilberstein, A., Nissim, A. and Revel, M. in press, 1983.
77. Zilberstein, A., Chernajovsky, Y., Mory, Y., Nir, U., Weissenbach, J. and Revel, M. in press, 1983.
78. Maroteaux, L., Kahana, C., Mory, Y., Groner, Y. and Revel, M. EMBO J. 2:325-332, 1983.

79. Pitha, P.M., Slate, D.L., Raj, N.B.K. and Ruddle, F.H. Mol.Cell Biol. 2:564-570, 1982.
80. Van Damme, J., DeLey, M., Claeys, H., Hilliau, A., Vermylen, C. and DeSomer, P. In: "The Biology of the Interferon System" (Eds. E.DeMayer, G.Galasso, and H. Schellekens), Elsevier-North Holland, Amsterdam, 1981, pp. 331-334.
81. Van Damme, J., Billiau, A., DeLey, M. and DeSomer, P. Antiviral Res. ABSTR. 1, No.2, 54, 1983.
82. Wietzerbin, J., Stefanos, S., Falcoff, R. and Falcoff, E. Texas Rep. Biol. Med. 35:205-210, 1977.
83. Vaquero, C., Sanceau, J., Catinot, L., Andreu, J. and Falcoff, R. J. Interferon Res. 2, 217-228, 1982.
84. Friedman-Einat, M., Revel, M. and Kimchi, A. Mol.Cell.Biol.2:1472-1480, 1982.
85. Duc-Goiran, P., Robert-Gallio, B., Chudzio, T. and Chany, C. Proc. Natl.Acad.Sci. USA 80:2628-2631, 1983.
86. Balkwill, F.R., Griffin, D.B., Band, H.A. and Beverley, P.C. J.Exp. Med. 157:1059-1063, 1983.
87. Epstein, L.B. In: Interferon 3 (Ed. I.Gresser), Academic Press, New York, 1981, pp. 13-44.
88. Ennis, F.A. and Meager, A. J.Exp.Med. 154:1279-1289, 1981.
89. Morris, A.G., Lin, Y.L. and Askonas, B.A. Nature 295:150-152, 1982.
90. Gray, P.W., Leung, D.W., Pennica, D., Yelverton, E., Najarian, R., Simonsen, C.C., Derynck, R., Sherwood, P.J., Wallace, D.M., Berger, S.L., Levinson, A.D. and Goeddel, D.V. Nature 295:503-508, 1982.
91. Devos, R., Cheroutre, H., Taya, Y., Degrave, W., Van Heuwerszwyn, H. and Fiers, W. Nucl.Acid Res. 10:2487-2501, 1982.
92. De Grado, W.F., Wasserman, Z.R. and Chowdry, V. Nature 300:379-381, 1982.
93. Gray, P.W. and Goeddel, D.V. Nature 298:859-863, 1982.
94. Yip, Y.K., Barrowclough, B.S., Urban, C. and Vilcek, J. Science 215: 411-413, 1982.
95. Yip, Y.K., Barowclough, B.S., Urban, C. and Vilcek, J. Proc.Natl. Acad. Sci. USA 79:1820-1824, 1982.
96. Novick, D., Eshhar, Z., Fischer, D.G., Friedlander, J. and Rubinstein, H. EMBO J. 2: 1527-1530, 1983.
97. Vilcek, J., Yip, Y.K. and Pang, R.H.L. In: Interferon, Properties, Modes of Action, Production, Clinical Application (Eds. K. Munk, and H. Kirchner), Karger, Basel, pp.150-158, 1982.
98. Tomita, Y., Cantell, K. and Kuwata, T. Int. J. Cancer 30:161-165, 1982.
99. DeLey, M., Van Damme, J., Claeys, H., Weening, H., Heine, J., Billiau, A., Vermylen, C. and DeSomer, P. Eur.J.Immunol. 10:877-883, 1980.
100. Wallach, D., Fellous, M. and Revel, M. Nature 299:833-836, 1982.
101. Wallach, D. Cell. Immunol. 75:390-395, 1983.
102. Fleishmann, W.R., Georgiades, J.A., Osborne, L.C. and Johnson, H.M. Infect. Immun. 26:248-253, 1979.
103. Zerial, A., Hovanessian, A.G., Stefanos, S., Huygen, K., Werner, G.H. and Falcoff, E. Antiviral Res., 2:227-239, 1982.
104. Vilcek, J. et al., In: "The Biology of the Interferon System", (Eds. E. DeMaeyer and H. Schellekens), Elsevier, Amsterdam, 1983, in press.
105. Ohno, S. and Taniguichi, T. Nucl.Acid Res., 10:967-977, 1982.
106. Mantei, N. and Weissmann, C. Nature 297:128-132, 1982.
107. Hauser, H., Gross, G., Burns, W., Hochkeppel, H.K., Mayr, U. and Collins, J. Nature 297:650-654, 1982.
108. Canaani, D. and Berg, P. Proc.Natl.Acad.Sci. USA 79:5166-5170, 1982.

109. Zinn, K., Mellon, S., Ptashne, M. and Maniatis, T. Proc.Natl.Acad. Sci. USA 79:4897-4901, 1982.
110. Mitrani-Rosenbaum, S., Maroteaux, L., Mory, Y., Revel, M. and Howley,P.M. Mol.Cell.Biol. 3:233-240, 1983.
111. Maroteaux, L., Kahana, C., Mory, Y., Groner, Y. and Revel, M. EMBO. J. 2:325-332, 1983.
112. Pitha, P.M., Ciufo, D.M., Kellum, M.,Raj, N.B.K. and Reyes, G.R. and Hayward, G.S. Proc.Natl.Acad.Sci. USA 79:4337-4341, 1982.
113. Tavernier, J., Gheysen, D., Duerinck, F., Van der Heyden, J. and Fiers, W. Nature 301:634-636, 1983.
114. Ragg, H. and Weissmann, C. Nature 303: 339-442, 1983.
115. Weidle, U. and Weissmann, C. Nature 303:442-446, 1983.
116. Raj, N.B.K. and Pitha, P.M. Proc.Natl.Acad.Sci. USA 78:7426-7430,1981.
117. Nir, U., Cohen, B., Chen, L. and Revel, in press, 1983.
118. Cavalieri, R.L., Havell, E.A., Vilcek, J. and Pestka, S. Proc.Natl. Acad.Sci. USA 74:4415-4419, 1977.
119. Sehgal, P.B., Lyles, D.S. and Tamm, I. Virology 89:186-198, 1978.
120. Tan, Y.H. and Berthold, W. J.Gen.Virol. 34:401-411, 1977.
121. Maroteaux, L., Chen, L., Mitrani-Rosenbaum, S., Howley, P.M. and Revel, M. J.Virol. 47:89-95, 1983.
122. Borden, E.C. Texas Rep. Biol. Med. 41:122-128, 1982.
123. DeMaeyer, E. and DeMaeyer, J. In: Interferon I (Ed. I.Gresser) Academic Press, New York, pp.75-100, 1979.
124. DeMayer-Guignard, J., Zavatzky, R., Dandoy, G. and DeMaeyer, E. J. Interferon Res. 3:241-252, 1983.
125. Tan, Y.H., Tischfield, J.A. and Ruddle, F.H. J.Exp.Med. 137:317-330, 1973.
126. Chany, C., Vignal, M., Couillin, P., Cong, N.V., Boue, J. and Boue, A. Proc.Natl.Acad.Sci. USA 72: 3129-3133, 1975.
127. Slate, D.L., Shulman, L., Lawrence, J.B., Revel, M. and Ruddle, F.H. J. Virol. 25:319-325, 1978.
128. Revel, M., Bash, D. and Ruddle, F.H. Nature 260:139-141, 1976.
129. Slate, D.L. and Ruddle, F.H. Cytogenet. Cell Genet. 22:265-269, 1978.
130. Epstein, L.B. and Epstein, C.J. J. Infect.Dis. 133, Suppl. A56-A62, 1976.
131. Tan, Y.H., Schneider, E.L., Tischfield, J., Epstein, C.J. and Ruddle, F.H. Science 186:61-63, 1974.
132. Cupples, C.G. and Tan, Y.H. Nature 267:165-167, 1977.
133. Gurari-Rotman, D., Revel, M., Tartakovsky, B., Sehgal, S., Hahn, T. and Levin, S. FEBS Letters 94:187-190, 1978.
134. Epstein, L.B. and Epstein, C.J. Cell Immunol. 51:303-318, 1980.
135. Wiranowska-Stewart,M. and Stewart, W.E. J.Gen.Virol. 37:629-633, 1977.
136. Epstein, C.J., McManus, N.H., Epstein, L.B., Branca, A.A., D'Alessandro, S.B. and Baglioni, C. Biochem.Biophys.Res.Comm. 107:1060-1066, 1982.
137. Shulman, L., Kamarck, M.E., Slate, D.L., Ruddle, F.H., Branca, A.A., Baglioni, C., Maxwell, B.L., Gutterman, J., Anderson, P., Nagler, C. and Vilcek, J. Nature, in press, 1983.
138. Kamarck, M.E., Slate, D.L., D'Eustachio, P., Barel, B. and Ruddle, F.H. Fedn. Proc. Fedn. Am. Socs. Exp. Biol. 40:1051, 1981.
139. Aguet, M. Nature 284:459-461, 1980.
140. Aguet, M. and Blanchard, B. Virology 115:249-261, 1981.
141. Branca, A.A. and Baglioni, C. Nature 294:768-770, 1981.
142. Mogensen, K.E., Bandu, M.T., Vignaux, F., Aguet, M. and Gresser, I. Int. J. Cancer 28:575-582, 1981.
143. Zoon, K., Zur Nedden, D. and Arnheiter, H. J.Biol.Chem.257:4695-4697.1982.

144. King A.C. and Cuatrecasas, P. New England J. Med. 305:77-88, 1981.
145. Carpenter, G. and Cohen, S. Ann.Rev. Biochem. 48:193-216, 1979.
146. Chany, C., Rousset, S. and Chany-Fournier, F. Texas Rep.Biol. Med. 41:307-312, 1982.
147. Joshi, A.R., Sarkar, F.H. and Gupta, S.L. J.Biol.Chem. 257:13884-13887, 1982.
148. Eid, P. and Morgensen, K.E. Antiviral Res. ABSTR 1, No.2, 58, 1983.
149. Arnheiter, H., Ohno, M., Smith, M., Gutte, B. and Zoon, K.C. Proc. Natl.Acad.Sci. USA 80:2539-2543, 1983.
150. Kohn, L.D., Friedman, R.M., Holmes, J.M. and Lee, G. Proc.Natl. Acad.Sci. USA 73:3695-3699, 1976.
151. Besancon, F. and Ankel, H. Nature 252:478-480, 1974.
152. Besancon, F., Ankel, H. and Basu, S. Nature 259:576-578, 1976.
153. Ankel, H., Krishnamurti, C., Besancon, F., Stefanos, S. and Falcoff, S. Proc.Natl.Acad.Sci. USA 77:2528-2532, 1980.
154. Fuse, A., Handa, S. and Kuwata, T. Antiviral Res. 2:161-166, 1982.
155. Krishnamurti, C., Besancon, F., Justesen, J., Poulsen, K. and Ankel, H. Eur. J.Biochem. 124:1-6, 1982.
156. Kuwata, T., Handa, S., Fuse, A. and Morigana, N. Biochem.Biophys. Res. Commun. 85:77-84, 1978.
157. Gill, D.M. Biochemistry 15:1242-1248, 1976.
158. Degre, M. Proc.Soc.Exp.Biol.Med. 157:253-255, 1978.
159. Fuse, A. and Kuwata, T. Infect.Immun. 26:235-239, 1979.
160. Schellekens, H., DeWilde, G.A. and DeReus, A. Antiviral Res. 1: 135-140, 1981.
161. Belardelli, F., Ausiello, C., Tomasi, M. and Rossi, G.B. Virology 107:109-120, 1980.
162. Aguet, M., Belardelli, F., Blanchard, B., Marcucci, F. and Gresser, I. Virology 117:541-544, 1982.
163. Vengris, V.E., Reynolds, F.H., Hollenberg, M.D. and Pitha, P.M. Virology 72:486-493, 1976.
164. Moss, J., Fishman, P.H., Manganiello, V.C., Vaughan, M. and Brady, R.O. Proc.Natl.Acad.Sci. USA 73:1034-1037, 1976.
165. Schiffman, D. and Koschel, K. Med.Microbiol.Immunol. 169:281-289, 1981.
166. Belardelli, F., Aliberti, A., Santurbano, B., Antonelli, G., D'Agnolo, G. and Rossi, G.B. Virology 117:391-400, 1982.
167. Chany, C. Biomedicine 24:148-157, 1976.
168. Anderson, P., Yip, Y.K. and Vilcek, J. J.Biol.Chem. 257:11301-11304, 1982.
169. Vilcek, J. In: "Interferon 4" (Ed. I.Gresser). Academic Press, New York, pp. 129-154, 1982.
170. Orchansky, P., Novick, D., Fischer, D.G. and Rubinstein, M. J. Interferon Res. in press, 1983.
171. Afflabris, E., Jemma, C., Romeo, G. and Rossi, G.B. In: "Biology of the IFN system" (Eds. E.DeMaeyer, and H. Schellekens) Elsevier, Amsterdam, in press, 1983.
172. DeLey M. and Billiau, A. Antiviral Res. 2:97-102, 1982.
173. Weil, J., Epstein, C.J. and Epstein, L.B. Nature 301:437-439, 1983.
174. Fleishmann, W.R., Kleyn, K.M. and Baron, S. J.Natl.Cancer Inst. 65:963-966, 1980.
175. Schwarz, L.A. and Fleishmann, W.R. Texas Rep.Biol.Med. 41:288-306, 1982.
176. DeClercq, E., Zhang, Z-X. and Huygen, K. Cancer Letters 15:223-228, 1982.
177. Rubin, B.Y. and Gupta, S.L. Proc.Natl.Acad.Sci. USA 77:5928-5932,1980.

178. Hayes, T.G. Archives Virol. 67:267-281, 1981.
179. Gresser, I., Bandu, M.-T., Brouty-Boyé, D. and Tovey, M.G. Nature 251:543-545, 1974.
180. Hayes, T.G., Yip, Y.K. and Vilcek, J. Virology 98:351-363, 1979.
181. Weck, P.K., Apperson, S., Stebbing, N., Gray, P.W., Leung, D., Shephard, H.M. and Goeddel, D.V. Nucl.Acid Res. 9:6153-6166, 1981.
182. Lee, S.H., Weck, P.K., Moore, J., Chen, S. and Stebbing, N. In: "Interferons" (Eds, T.Merigan, R.M. Friedman and C.Fox), UCLA Symposium Vol XXV, Academic Press, New York, pp.341-351, 1982.
183. Shepard, H.M., Leung, D., Stebbing, N. and Goeddel, D.V. Nature 294, 563-565, 1981.
184. McGray, J.W. and Weil, R. Proc.Natl.Acad.Sci. USA 79:4829-4833, 1982.
185. Braude, I.A., Lin, L. and Stewart, W.E. Biochem.Biophys.Res. Comm. 89:612-619, 1979.
186. Paucker, K., Dalton, B.J., Ogburn, C.A. and Torma, E. Proc.Natl. Acad.Sci. USA 72:4587-4591, 1975.
187. Samuel, C.E., Knutson, G.S. and Masters, P. J.Interferon Res. 2: 563-574, 1982.
188. Branca, A.A., Faltynek, C.R., D'Alessandro, S.B. and Baglioni, C. J. Biol.Chem. 257:13291-13296, 1982.
189. Baglioni, C., Branca, A.A. and D'Alessandro, S.B. Biochemistry, in press, 1983.
190. Anderson, P., Yip, Y.K. and Vilcek, J. J.Biol.Chem. 258:6497-6502, 1983.
191. Pastan, I.H. and Willingham, M.C. Science 214:504-509, 1981.
192. Zoon, K.C., Arnheiter, H., Zur Nedden, D., Fitzgerald, D.J.P and Willingham, M. Virology, in press, 1983.
193. Anderson, P., Tycko, B., Maxfield, F. and Vilcek, J. Virology 117: 510-515, 1982.
194. Branca, A.A. and Baglioni, C. J.Biol.Chem. 257:13197-13202, 1982.
195. Sakaguchi, A.Y., Stevenson, D. and Gordon, I. Virology 116:441-453, 1982.
196. Veomett, M.J.and Veomett, G.E. J.Virol. 31:785-794, 1979.
197. Schreiber, A.B., Lax, I., Yarden, Y., Eshhar, Z. and Schlessinger, J. Proc.Natl.Acad.Sci. USA 78:7535-7539, 1981.
198. Teysott, B., Houdebine, LM. and Djiane, J. Proc.Natl.Acad.Sci. USA 78:6729-6733, 1981.
199. Tovey, M.G., Rochette-Egly, C. and Castagna, M. Proc.Natl.Acad.Sci. USA 76:3890-3893, 1979.
200. Tovey, M.G. and Rochette-Egly, C. Virology 115:272-281,1981.
201. Pfeffer, L.M., Landsberger, F.R and Tamm, I. J. Interferon Res, 1: 613-620, 1981.
202. Tamm, I., Wang, E., Landsberger, F.R. and Pfeffer, L.M. In: "Interferons" (Eds., T.Merigan, R.M.Friedman and C.F.Fox), UCLA Symposium Vol. XXV, Academic Press, pp. 159-179, 1982.
203. Chandrabose, K.A., Cuatrecasas, P. and Potatthil, R. Texas Rep. Biol. Med. 41:499-505, 1982.
204. Chandrabose, K.A.,Cuatrecasas, P., Potatthil, R. Biochem. Biophys. Res. Comm. 98:661-668, 1981.
205. Revel, M. In: "Interferon 1" (Ed. I.Gresser), Academic Press, New York, pp.101-163, 1979.
206. Dianzani, F., Zucca, M., Scupham, A. and Georgiades, J.A. Nature 283:400-402, 1980.
207. Baglioni, C. and Maroney, P. J.Biol.Chem. 255:8390-8393, 1980.
208. Chebath, J., Merlin, G., Metz, R. Benech, P. and Revel, M. Nucl.Acid Res. 11:1213-1225, 1983.

209. Fellous, M., Nir, U., Wallach, D., Merlin, G., Rubinstein, M. and Revel, M. Proc.Natl.Acad.Sci. USA 79:3082-3086, 1982.
210. Shulman, L. and Revel, M. Nature 287: 98-100, 1980.
211. Merlin, G., Chebath, J., Benech, P., Metz, R. and Revel, M. Proc. Natl. Acad.Sci. USA 80, 4904-4908, 1983.
212. Rosa, F., Hatat, D., Abadie, A., Wallach, D., Revel, M. and Fellous,M. EMBO J. 2, 1585-1589, 1983.
213. Samanta, H., Dougherty, J.P., Brawner, M.E., Schmidt, H. and Lengyel, P. In: Interferons (Eds. T.Merigan, R.M. Friedman and C.F. Fox), UCLA Symposium Vol. XXV, Academic Press, New York, pp.59-72, 1982.
214. Rochette-Egly, C. and Tovey, M.G. Biochem.Biophys.Res. Commun. 107:150-156, 1982.
215. Tovey, M.G. In: "Interferon 4" (Ed. I.Gresser), Academic Press, New York, pp.23-46, 1982.
216. Yaron, M., Yaron, I., Gurari-Rotman, D., Revel, M., Lindner, H.R. and Zor. U. Nature 267:457-459, 1977.
217. Fuse, A., Mahmud, I. and Kuwata, T. Cancer Res. 42:3209-3214, 1982.
218. Bougnoux, P., Hirata, F., Timonen, T. and Hofman, T. In: Mechanisms of Lymphocyte Activation. (Eds., K.Resch and H.Kirchner), Elsevier, North Holland, Amsterdam, pp.577-580, 1981.
219. Coffey, R.G., Hadden, E.M. and Hadden, J.W. J.Biol.Chem. 256:4418-4424, 1981.
220. Murad, F., Arnold, W.P., Mittal, C.K. and Braughler, J.M. In: Advances in Cyclic Nucleotides Research (Eds., P. Greengard and G.A. Robison) 11, Raven Press, New York, pp.175-204, 1979.
221. Pottathil, R., Chandrabose, K.A., Cuatrecasas, P. and Lang, D.J. Proc.Natl.Acad.Sci. USA 78:3343-3347, 1981.
222. Gresser, I., Bandu, M.T. and Brouty-Boye, D. J.Nat.Cancer Inst. 52:553-559, 1974.
223. Chandrabose, K.A., Cuatrecasas, P., Pottathil, R. and Lang, D.J. Science 212:329-331, 1981.
224. Pottathil, R., Chandrabose, K.A., Cuatrecasas, P. and Lang, D.T. Proc.Natl.Acad.Sci. USA 77:5437-5440, 1980.
225. Bendori, R. M.Sc. Thesis (U.Zor and M.Revel, advisors), The Weizmann Institute of Science, 1979.
226. Helfand, S.L., Werkmeister, J., Pross, H. and Roder, J.C. J.Interferon Res. 3:143-151, 1983.
227. Tovey, M.G., Rochette-Egly, C. and Kolb, J.P. J.Interferon Res. 2:549-561, 1982.
228. Brunda, M.J., Herberman, R.B. and Holden, H.T. J.Immunol. 124: 2682-2688, 1980.
229. Wallach, D. and Hahn, T. Cell.Immunol. 76:390-396, 1983.
230. Wallach, D. and Revel, M. FEBS Letters 101:364-368, 1979.
231. Fitzpatrick, F.A. and Stringfellow, D.A. J.Immunol. 125:431-438,1980.
232. Gurari-Rotman, D. FEBS Letters 148:17-20,1982.
233. Bourgeade, M.F. and Chany, C. Proc.Soc.Exp.Biol.Med. 153, 501-510, 1976.
234. Garry, R.F. and Waite, M.R.F. Virology 96:121-128, 1979.
235. Shulman, L. Ph.D. Thesis (M. Revel, advisor), The Weizmann Institute of Science, 1981.
236. Radke, K.L., Colby, C., Kates, J.R., Krider, H.M. and Prescott, D.M. J. Virol. 13:623-630, 1974.
237. Baglioni, C. Cell 17:255-264, 1979.
238. Lengyel, P. Annu. Rev. Biochem. 51:251-282, 1982.
239. Lebleu, B. and Content, J. In: "Interferon 4" (Ed. I.Gresser),

Academic Press, New York, pp.47-94, 1982.
240. Kerr, I.M. and Brown, R.E. Proc.Natl.Acad.Sci. USA 75:256-260,1978.
241. Baglioni, C., Minks, M.A. and Maroney, P.A. Nature 273:684-687,1978.
242. Schmidt, A., Zilberstein, A., Shulman, L., Federman, P., Berissi, H. and Revel, M. FEBS Letters 25:257-264, 1978.
243. Slattery, E., Ghosh, N., Samanta, H. and Lengyel, P. Proc.Natl.Acad. Sci. USA 76:4778-4782, 1979.
244. Samanta, H., Dougherty, J.P. and Lengyel, P. J.Biol.Chem.255:9807-9813, 1980.
245. Minks, M.A., West, D.K., Benuin, S. and Baglioni, C. J.Biol.Chem. 254:10180-10183, 1979.
246. Nilsen, T.W., Maroney, P.A., Robertson, H.D. and Baglioni, C. Mol. Cell.Biol. 2:154-160, 1982.
247. Revel, M., Kimchi, A., Shulman, L., Fradin, A., Shuster, R., Yakobson, E., Chernajovsky, Y., Schmidt, A., Shure, H. and Bendori, R. Ann. N.Y. Acad. Sci. 350:459-472, 1980.
248. Knight, M., Cayley, P.J., Silverman, R.H., Wreschner, D.H., Gilbert, C.S., Brown, R.E. and Kerr, I.M. Nature 288:189-192, 1980.
249. Williams, B.R.G., Golgher, R.R., Brown, R.E., Gilbert, C.S. and Kerr, I.M. Nature 282:582-586, 1979.
250. Silverman, R.H., Cayley, P.J., Knight, M., Gilbert, C.S. and Kerr, I.M. Eur. J. Biochem. 124:131-138 ,1982.
251. Nilsen, T.W., Wood, D.L. and Baglioni, C. J.Biol.Chem. 257:1602-1605, 1982.
252. Etienne-Smekens, M., Vandenbussche, P., Content, J. and Dumont, J.E. Proc.Natl.Acad.Sci. USA 80:4609-4613, 1983.
253. Samuel, D. and Cailla, H. Antiviral Res. ABSTR. 1, No.2, p.83, 1983.
254. Krishnan, I. and Baglioni, C. Virology 111:660-670, 1981.
255. Schattner, A., Merlin, G., Wallach, D., Rosenberg, H., Bino, T., Hahn, T., Levin, S. and Revel, M. J.Interferon Res. 1:587-594, 1981.
256. Wallach, D., Schattner, A., Merlin, G., Kimchi, A., Fellous M. and Revel, M. In: Interferons (Eds., T.C.Merigan, R.M. Friedman and C.F.Fox), UCLA Symposia Vol. XXV, Academic Press, New York, pp.449-463, 1982.
257. Schattner, A., Wallach, D., Merlin, G., Hahn, T., Levin, S. and Revel, M. Lancet 2:497-500, 1981.
258. Williams, B.R.G., Read, S.E., Freedman, M.H., Carver, D.H. and Gelfand, E.W. In: Interferons (Eds., T.C.Merigan, R.M.Friedman and C.F.Fox), UCLA Symposia Vol XXV, Academic Press, pp.253-267, 1982.
259. Krishnan, I. and Baglioni, C. Nature 285:485-488, 1980.
260. Sokawa, Y., Ando, T. and Ishihara, Y. Infect.Immuno.28:719-723,1980.
261. Schattner, A., Merlin, G., Shapira, A., Revel, M. and Wallach, D. J. Interferon Res. 2:285-289, 1982.
262. Revel, M., Kimchi, A., Friedman, M., Wolf, D., Merlin, G., Panet, A., Rapoport, S. and Lapidot, Y. Texas Rep.Biol.Med. 41:452-462, 1982.
263. Yang, K., Samanta, H., Dougherty, J., Jayaram, B., Broeze, R. and Lengyel, P. J.Biol. Chem. 256:9324-9328, 1981.
264. Revel, M., Wallach, D., Merlin, G., Schattner, A., Schmidt, A., Wolf, D., Shulman, L. and Kimchi, A. Methods Enzymol. 79:149-161, 1981.
265. St.Laurent, G., Yoshie, O., Floyd-Smith, G., Samanta, H., Sehgal, P.B. and Lengyel, P. Cell 33:95-102, 1983.
266. Stark, G.R., Dower, W.J., Schimke, R.T., Brown, R.E. and Kerr, I.M. Nature 278:471-473, 1979.
267. Krishnan, I. and Baglioni, C. Proc.Natl.Acad.Sci. USA 77:6506-6510,1980.

268. Hovanessian, A.G. and Kerr, I.M. Eur.J.Biochem. 84:149-159, 1978.
269. Kimchi, A. J.Interferon Res. 1:559-569, 1981.
270. Friedman-Einat, M., Revel, M. and Kimchi, A. Mol.Cell.Biol. 2, 1472-1480, 1982.
271. Kimchi, A., Bendori, R., Einat, M., Yarden, A., Resnitzky, D. and Gottlieb, H. In: "Biology of the Interferon System" (Eds., E.DeMaeyer and H. Schellekens), Elsevier, Amsterdam, pp. 83-90, 1983.
272. Kimchi, A., Shure, H. and Revel, M. Eur.J.Biochem. 114:5-10, 1981.
273. Krishnan, I. and Baglioni, C. Mol.Cell.Biol. 1:932-938, 1981.
274. Etienne-Smekens, M., Goldstein, J. and Dumont, J.E. Eur.J.Biochem. 130:269-273, 1983.
275. Creasey, A.A., Eppstein, D.A.,Marsh, Y.V., Khan, Z. and Merigan, T.C. Mol.Cell.Biol. 3:780-786, 1983.
276. Hovanessian, A.G., Meurs, E. and Montagnier, L. J.Interferon Res. 1:179-190, 1981.
277. Floyd-Smith, G., Yoshie, O. and Lengyel, P. J.Biol.Chem. 257:8584-8587, 1982.
278. Silverman, R.H., Watling, D., Balkwill, F.R., Trowsdale, J. and Kerr, I.M. Eur.J.Biochem. 126:333-341, 1982.
279. Silverman, R.H., Krause, D., Jacobsen, H. and Friedman, R.M. In: "Biology of the IFN System" (Eds., E. DeMaeyer and H.Schellekens), Elsevier, Amsterdam, in press, 1983.
280. Cayley, P.J., Knight, M. and Kerr, I.M. Biochem.Biophys.Res. Commun. 104:376-382, 1982.
281. Schmidt, A., Chernajovsky, Y., Shulman, L., Federman, P., Berissi, H. and Revel, M. Proc.Natl.Acad.Sci. USA 76:4788-4792, 1979.
282. Hovanessian, A.G. and Wood, J.N. Virology 101:81-90, 1980.
283. Verhaegen-Lewalle M. and Content, J. Eur.J.Biochem. 126:639-643,1982.
284. Martin, E.M., Birdsall, W.J.M., Brown, R.E. and Kerr, I.M. Eur. J. Biochem. 95:295-307, 1979.
285. Nilsen, T.W. and Baglioni, C. Proc.Natl.Acad.Sci. USA 76:2600-2604, 1979.
286. Wreschner, D.H., James, T.C., Silverman, R.H. and Kerr, I.M. Nucl. Acid Res. 9:1571-1581, 1981.
287. Zilberstein, A., Dudock, B., Berissi, H. and Revel, M. J.Mol.Biol. 108:43-54, 1976.
288. Sen. G.C., Gupta, S.L., Brown, G.E., Lebleu, B., Rebello, M. and Lengyel, P. J.Virol. 17:191-203, 1976.
289. Falcoff, R., Falcoff, E., Sanceau, J. and Lewis, J.A. Virology 86: 507-515, 1978.
290. Keith, G., Gangloff, M.L., Falcoff, R. and Dirheimer, G. Comptes Rendus Academic des Sciences, 294-III:795-798, 1982.
291. Sen, G.C., Taira, H. and Lengyel, P. J.Biol.Chem.253:5915-5921, 1978.
292. Kimchi, A., Zilberstein, A., Schmidt, A., Shulman, L. and Revel, M. J.Biol.Chem. 254:9846-9853, 1979.
293. Hovanessian, A.G. and Kerr, I.M. Eur.J.Biochem. 93:515-526, 1979.
294. Ohtsuki, K., Nakamura, M., Koike, T., Ishida, N. and Baron, S. Nature 287:65-67, 1980.
295. Rubin, B.Y. and Gupta, S.L. Proc.Natl.Acad.Sci. USA 77:5928-5932, 1980.
296. Weil, J., Epstein, C.J., Epstein, L.B., Sedmark, J., Sabran, J.L. and Grossberg, S.E. Nature 301:437-439, 1983.
297. Cheng, Y.-S. Antiviral Res. ABSTR. 1, No.2, p. 50, 1983.
298. Hovanessian, A.G., Meurs, E., Aujean, O., Vaquero, C., Stefanos, S. and Falcoff, E. Virology 104:195-204, 1980.
299. Broeze, R.J., Dougherty, J.P., Pichon, J., Jayaram, B.M. and Lengyel, P.

J. Interferon Res. 1:191-201, 1981.
300. Levin, D.H., Petryshyn, R. and London, I.M. J.Biol.Chem. 256:7638-7641, 1981.
301. Kaempfer, R., Israeli, R., Rosen, J., Knoller, S., Zilberstein, A., Schmidt, A. and Revel, M. Virology 99:170-173, 179.
302. Clemens, M.J., Pain, V.M., Wong, S.T. and Henshaw, E.C. Nature 296: 93-96, 1982.
303. Hovanessian, A.G., Rollin, P., Riviere, Y., Pouillart, P., Sureau, P. and Montagnier, L. Biochem.Biophys.Res.Commun. 103:1371-1377, 1981.
304. Hovanessian, A.G., Galabru, J., Krust, B., Montagnier, L. and Riviere, Y. In: "Biology of the Interferon System" (Eds. E. DeMaeyer and H. Schellekens), Elsevier, Amsterdam, pp. 323-328, 1983.
305. Ohtsuki, K. and Baron, S. J.Biol.Chem. 92:967-970, 1982.
306. Rice, A.P. and Kerr, I.M. J.Virol. in press, 1983.
307. Epstein, D.A., Torrence, P.F. and Friedman, R.M. Proc.Natl.Acad.Sci. USA 77:107-111, 1980.
308. Knight, E. and Korant, B.D. Proc.Natl.Acad.Sci. USA 76:1824-1827, 1979.
309. Rubin, B.Y. and Gupta, S.L. J.Virol. 34:446-454, 1980.
310. Gupta, S.L., Rubin, B.Y. and Holmes, S.L. Virol. 111:331-340, 1981.
311. Farrell, P.J., Broeze, R.J. and Lengyel, P. Nature 279:523-525, 1979.
312. Colonno, R.J. Proc.Natl.Acad.Sci. USA 78:4763-4766, 1981.
313. Colonno, R.J. and Pang, R. J.Biol.Chem. 257:9234-9237, 1982.
314. Haller, O., Arnheiter, H., Lindenmann, J. and Gresser, I. Nature, 283: 660-662, 1980.
315. Lindenmann, J. In:"Interferon 3" (Ed. I.Gresser), pp.1-12, Academic Press, 1981.
316. Lindahl, P., Leary, P. and Gresser. I. Proc.Natl.Acad.Sci. USA 70: 2785-2788, 1973.
317. Lonai, P. and Steinman, L. Proc. Natl. Acad.Sci. USA 74:5662-5667,1977.
318. Lindahl, P., Gresser, I., Leary, P. and Tovey, M. Proc.Natl.Acad. Sci. USA 73:1284-1288, 1976.
319. Vignaux, F. and Gresser, I. Proc.Soc.Exptl.Biol.Med.157:456-460, 1978.
320. Heron, I., Hokland, M. and Berg, K. Proc.Natl.Acad.Sci. USA 75: 6215-6219, 1978.
321. Fellous, M., Kamoun, M., Gresser, I. and Bono, R. Eur. J. Immunol. 9:446-449, 1979.
322. Attalah, A.M. and Strong, D.M. Int.Arch.Allergy Appl.Immunol. 60: 101-105, 1979.
323. Dolei, A.F., Ameglio, M.R., Capobianchi, T. and Tosi, R. Antiviral Res. 1:367-373, 1981.
324. Hokland, M., Heron, I. and Berg, K. J.Interferon Res. 1:483-494, 1981.
325. Imai, K., NG A.-K., Glassy, M.C. and Ferrone, S. J.Immunol. 127: 505-509, 1981.
326. Lucero, M.A., Magdelenat, H., Fridman, W., Pouillart, P., Billardon, C., Billiau, A., Cantell, K. and Falcoff, E. Eur.J. Cancer Clin. Oncol. 18:243-251, 1982.
327. Vignaux, F. and Gresser, I. J.Immunol. 118:721-723, 1977.
328. Basham, T.Y., Bourgeade, M.F., Creasey, A.A. and Merigan, T.C. Proc.Natl.Acad.Sci. USA 79:3265-3269, 1982.
329. Burrone, O.R. and Milstein, C. EMBO J. 1:345-349, 1982.
330. Sonnefeld, G., Mandel, A.D. and Merigan, T.C. Cell.Immunol. 34: 193-206, 1977.
331. Wallach, D. Eur.J.Immunol. in press, 1983.

332. Rosa F., Le Bouteiller, P., Abadie,A., Mishal, Z., Lemonnier, F.A., Bourrel, D., Lamotte, M., Kalil, J., Jordan, B. and Fellous, M. Eur.J.Immunol. 13:495-499, 1983.
333. Rosa, F., Berissi, H., Weissenbach, J., Maroteaux, L., Fellous, M. and Revel, M. EMBO J. 2:239-243, 1983.
334. Rosa, F., Fellous, M., Bron, M., Tovey, M. and Revel, M. Immunogenetics, 17:121-131, 1983.
335. Durandy, A., Virelizier, J.L. and Griscelli, C. Clin.exp. Immunol. 52, in press, 1983.
336. Kapp, J., Pierce, C., Cullin, S., Shreffler, D. and Schwartz, B. Immun. Today 4:1-4, 1983.
337. Sonnenfeld, G., Meruelo, D., McDevitt, H.O. and Merigan, T.C. Cell. Immunol. 57:427-435, 1981.
338. Basham, T.Y. and Merigan, T.C. J. Immunol. 130:1492-1495, 1983.
339. Rosa, F., Hatat, D., Abadie, A., Wallach, D., Revel, M. and Fellous, M. EMBO J. 2:1585-1589, 1983.
340. Klareskog, L., Forsum, U., Scheynius, A., Kabelitz, D. and Wigzell, H. Proc.Natl.Acad.Sci. USA 79:3632-3636, 1982.
341. Meuer, S.C., Schlossman, S.F. and Reinherz, E.L. Proc.Natl.Acad. Sci. USA 79:4395-4399, 1982.
342. Knight, E. and Korant, B.D. Biochem.Biophys.Res.Commun. 74:707-713,1977.
343. Fridman, W.H., Gresser, I., Bandu, M.T., Aguet, M. and Neauport-Sautes, C. J.Immunol. 124:2436-2441, 1980.
344. Itoh, K., Inous, M., Katadka, S. and Kumagai, K. J.Immunol. 124: 2589-2595, 1980.
345. Vogel, S., English, K.E., Fertsch, D. and Fultz, M.J. J.Interferon Res. 3:153-160, 1983.
346. Huang, K.Y. Texas Re.Biol.Med. 41:375-380, 1982.
347. DeMaeyer, E. and DeMaeyer-Guignard, J. Texas Rep.Biol.Med. 41:420-425, 1982.
348. Hokland, M., Ritz, J. and Hokland, P. J.Interferon Res. 3:199-210,1983.
349. Fellous, A., Ginzburg, I. and Littauer, U.Z. EMBO J. 1, 835-839, 1982.
350. Wang, E., Pfeffer, L.M. and Tamm, I. Proc.Natl.Acad.Sci. USA 78: 6285-6289, 1981.
351. Jones, C.M., Varesio, L., Herberman, R.B. and Pestka, S. J.Interferon Res. 2:377-386, 1982.
352. Schroder, E.W., Chou, I.H., Jaken, S. and Black, P.H. Nature 276: 828-829, 1978.
353. Wallach, D., Aderka, D., Budilovsky, S. and Hahn, T. In: "The Biology of the Interferon System" (Eds., E.DeMaeyer, G.Galasso and H.Schellekens), Elsevier-North Holland, in press, 1983.
354. Gresser, I., Tovey, M.G., Bandu, M.T., Maury, C. and Brouty-Boye, D. J.Exp.Med. 144:1305-1315, 1976.
355. Gresser, I., Tovey, M.G., Maury, C. and Bandu, M.T. J.Exp.Med. 144: 1316-1323, 1976.
356. Gresser, I., Maury, C., Kress, C., Blangy, D. and Maunoury, M.T. Int. J.Cancer. 24:178-183, 1979.
357. Virelizier, J.L. and Gresser, I. J.Immunol. 120:1616-1617,1978.
358. Fauconnier, B. C.R. hebd. Acad.Sci. Paris D 271:1464, 1970.
359. Hoshino, A., Takenaka, H., Migushoki, O., Imanishi, J., Kishida, T. and Tovey, M.G. Antiviral Res. 3:59-65, 1983.
360. Zawatzky, R., Gresser, I., DeMaeyer, E. and Kirchner, H. J.Inf.Dis. 146:405-410, 1982.
361. Hayden, F.G. and Gwatlney, J.M. Antiviral Res. 3:67-71, 1983.
362. Matthews, T.H.J. and Lawrence, M.K. Arch.Virol. 59:35-38, 1979.

363. Parry, R.P. and Parry, J.V. Lancet 1: 506-507, 1981.
364. Levin, S. and Hahn, T. Clin.Exp.Immunol. 46:475-483, 1981.
365. Schattner, A., Wallach, D., Merlin, G., Hahn, T., Levin, S., Ramot, B. and Revel, M. J.Interferon Res. 2:355-361, 1982.
366. Virelizier, J.L., Lenoir, G. and Griscelli, C. Lancet 2:231-234, 1978.
367. Isaacs, D., Clarke, J.R., Tyrrell, D.A.J., Webster, A.D.B. and Valman, H.B. Lancet 2:950-952, 1981.
368. Levin, S. and Hahn, T. In: "Primary Immunedeficiences" (Eds., M. Seligman and W.H. Hitzig), Elsevier-North Holland, Amsterdam, pp.465-472, 1980.
369. Levin, S. and Hahn, T. Lancet 1:592-594, 1982.
370. Haller, O., Arnheiter, H., Gresser, I. and Lindenmann, J. J.Exp.Med. 149:601-612, 1979.
371. Haller, O., Arnheiter, H., Gresser, I. and Lindenmann, J. J.Exp.Med. 154:199-203, 1981.
372. Lewis, J.A. In: "Hormone Receptors" (Ed. L.D.Kohn), John Wiley and Sons, Ltd., London, pp.357-384, 1982.
373. Sen, G.C. Progress Nucl. Acid Res. 27:105-156, 1982.
374. Yakobson, E., Revel, M. and Winocour, E. Virology 80:225-228, 1977.
375. Metz, D.H., Levin, M.J. and Oxman, M.N. J.Gen.Virol. 32:227-240, 1976.
376. Yamamoto, K., Yamaguchi, N. and Oda, K. Virology 68:58-70, 1975.
377. Yakobson, E., Prives, C., Hartman, J.R., Winocour, E. and Revel, M. Cell 12:73-81, 1977.
378. Mozes, L.W. and Defendi, V. Virology 85:315-318, 1978.
379. Brennan, M.B. and Stark, G.R. Cell 33:811-816, 1983.
380. Yakobson, E. Ph.D.Thesis (M. Revel, Advisor). Weizmann Institute of Science, Rehovot, 1978.
381. Breatnach, R. and Chambon, P. Annu.Rev.Biochem. 50:349-383, 1981.
382. Kahana, C., Yakobson, E., Revel, M. and Groner, Y. Virology 112: 109-118, 1981.
383. Hersh, C.L., Brown, R.E., Roberts, W.K., Swyryd, E.A., Kerr, I.M. and Stark, G.R. J.Biol.Chem, in press, 1983.
384. Oxman, M.N., Levin, M.J. and Lewis, A.M. J.Virol. 13:322-330, 1974.
385. Gluzman, Y., Davidson, J., Oren, M. and Winocour, E. J.Virol. 22: 256-266, 1977.
386. Elder, J.T., Spritz, R.A. and Weissmann, S.M. Annu. Rev.Genet. 15: 295-340, 1981.
387. Kingsman, S.M. and Samuel, C.E. Virology 101:458-465, 1980.
388. Metz, D.H., Oxman, M.N. and Levin, M.J. Biochem.Biophys.Res.Commun. 75:172-178, 1977.
389. Turek, L., Byrne, J.C., Lowy, D.R., Dvoretsky, I., Friedman, R.M. and Howley, P.M. Proc.Natl.Acad.Sci. USA 79:7914-7918, 1982.
390. Whitaker-Dowling, P., Wilcox, D.K., Widnell, C.C. and Youngner, J.S. Proc.Natl.Acad.Sci. USA 80:1083-1086, 1983.
391. Marcus, P. and Sekellick, M.J. J.Gen.Virol. 38:391-408, 1978.
392. Thacore, H.R. J.Gen.Virol. 41:421-425, 1978.
393. Simili, M., DeFerra, F. and Baglioni, C. J.Gen.Virol. 47:373-384,1980.
394. Repik, P., Flamand, A. and Bishop, D.H.L. J.Virol. 14:1169-1178, 1974.
395. Maheshwari, R.K. and Friedman, R.M. J.Gen.Virol. 44:261-264, 1979.
396. Maheshwari, R.K. and Friedman, R.M. Virology 101:399-407, 1980.
397. Maheshwari, R.K., Demsey, A.E., Mohanti, S.B. and Friedman, R.M. Proc.Natl.Acad.Sci. USA 77:2284-2287, 1980.
398. Wallach, D. and Revel, M. Nature 287:68-70, 1980.
399. Friedman, R.M. Bact. Rev. 41:543-567, 1977.
400. Pitha, R.M., Bilello, J.A. and Riggin, C.H. Texas Rep.Biol.Med.41:603-609, 1982.

401. Luftig, R.B., Conscience, J.F., Skoultchi, A., McMillan, P., Revel, M. and Ruddle, F.H. J. Virol. 23:799-810, 1977.
402. Bandyopadhyay, A.D., Chang, E.H., Levy, C.C. and Friedman, R.M. Biochem.Biophys.Res.Commun. 87:983-990, 1979.
403. Chang, E.J. and Friedman, R.M. Biochem.Biophys.Res.Commun. 77:392-398, 1977.
404. Pitha, P.M., Fernie, B., Moldarelli, F., Haltman, T. and Wivel, N.A. J.Gen.Virol. 46:97-105, 1980.
405. Pitha, P.M., Rowe, W.P. and Oxman, M.N. Virology 70:324-338, 1976.
406. Aboud, M., Shoor, R. and Salzberg, S. J.Gen.Virol. 51:425-428, 1980.
407. Aboud, M., Moldovan-Levin, D. and Salzberg, S. J.Virol. 37:836-839, 1981.
408. Avery, R.I., Norton, J.D., Jones, J.S., Burke, D.C. and Morris, A.G. Nature 288:93-96, 1980.
409. Epstein, D.A., Czarniecki, C.W., Jacobsen, H., Friedman, R.M. and Panet, A. Eur.J.Biochem. 118:9-15, 1981.
410. Czarniecki, C.W., Sreevalsan, T., Friedman, R.M. and Panet, A. J. Virol. 37:827-831, 1981.
411. Bakhanashvili, M., Wreschner, D.H., Oberman, F., Panet, A. and Salsberg, S. Antiviral Res. ABSTR 1, No.2, p.41. 1983.
412. Salsberg, S., Heller, A., Aboud, M., Gurari-Rotman, D. and Revel, M. Virology 93:209-214, 1979.
413. Tomita, Y., Nishimaki, J., Takahashi, F. and Kuwata, T. Virology 120: 258-263, 1982.
414. Allen, P.T., Schellekens, H., Van Griesven, L.J. and Billiau, A. J.Gen.Virol. 31:429-435, 1976.
415. Jacobsen, H., Czarniecki, C.W., Friedman, R.M. and Silverman, R.H. Virology 125:496-501, 1983.
416. Nilsen, T.W., Wood, D.L. and Baglioni, C. Nature 286:178-180, 1980.
417. Aguet, M., Gresser, I., Hovanessian, A.G., Bandu, M.T., Blanchard, B. and Blangy, D. Virology 114:585-588, 1981.
418. Wood, J.N. and Hovanessian, A.G. Nature 282:74-76, 1979.
419. Falcoff, R. and Sanceau, J. Virology 98:433-438, 1979.
420. Vaquero, C., Sanceau, J. and Falcoff, R. Biochem.Biophys.Res.Commun. 107:974-980, 1982.
421. Trachsel, H., Sonenberg, N., Shatkin, A.J., Rose, J.K., Leong, K., Bergmann, J.E., Gordon, J. and Baltimore, D. Proc.Natl.Acad.Sci. USA 77:770-774, 1980.
422. Vaquero, C., Aujean-Rigand, O., Sanceau, J. and Falcoff, R. Antiviral Res. 1:123-134, 1981.
423. Nilsen, T.W., Maroney, P.A. and Baglioni, C. J.Virol. 42:1039-1045, 1982.
424. Hersh, C.L. and Stark, G.R. J.Cell Biochem. Suppl. 6:98, 1982.
425. Nilsen, T.W., Maroney, P.A. and Baglioni, C. J.Biol.Chem. 256:7806-7811, 1981.
426. Williams, B.R.G., Golgher, R.R. and Kerr, I.M. FEBS Letters 105:47-52, 1979.
427. Galster, R.L. and Lengyel, P. Nucl. Acid Res. 3:581-598, 1976.
428. Aujean, O., Sanceau, J., Falcoff, E. and Falcoff, R. Virology 92: 583-586, 1979.
429. Wreschner, D.H., McCauley, J.W., Skehel, J.J. and Kerr, I.M. Nature 289, 414-417, 1981.
430. Floyd-Smith, G., Slattery, E. and Lengyel, P. Science 212:1030-1032, 1981.
431. Nilsen, T.W., Maroney, P.G. and Baglioni, C. Mol. Cell.Biol. 3: 64-69, 1983.

432. Verhaegen-Lewalle, M., Kuwata, T., Zhang, Z.-X., DeClercq, E., Cantell, K. and Content, J. Virology 117:425-434, 1982.
433. Hovanessian, A.G., Meurs, E. and Montagnier, L. J.Interferon Res. 1:179-190, 1981.
434. Meurs, E., Hovanessian, A.G. and Montagnier, L. J.Interferon Res. 1:219-232, 1981.
435. Gupta, S.L., Holmes, S.L. and Mehra, L. Virology 120:495-499, 1982.
436. Nilsen, T.W., Maroney, P.G. and Baglioni, C. J.Biol.Chem. 257:14593-14596, 1982.
437. Jacobsen, H.J., Epstein, D.A., Friedman, R.M., Safer, B. and Torrence, P.F. Proc.Natl.Acad.Sci. USA 80:41-45, 1983.
438. Yau, P.M., Godefroy-Colburn, T., Birge, C.H., Ramabhadran, T.V. and Thach, R.E. J.Virol. 27:648-658, 1978.
439. DeFerra, F. and Baglioni, C. Virology 112:426-435, 1981.
440. Banerjee, A.K. Microbiol. Rev. 44:175-205, 1980.
441. Desrosiers, R.C. and Lengyel, P. Biochim.Biophys.Acta 562:471-480, 1979.
442. Kroath, H., Gross, H.J., Jungwirth, C. and Bodo, G. Nucl.Acid Res. 5:2441-2445, 1979.
443. Sen, G.C., Shaila, S., Lebleu, B., Brown, G.E., Desrosiers, R.C. and Lengyel, P. J. Virol. 21:69-83, 1977.
444. Ball, L.A. and White, C.N. Virology 84:496-508, 1978.
445. Sharma, O.K. and Goswami, B.B. Proc.Natl.Acad.Sci. USA 78:2221-2224, 1981.
446. Chang, E.H., Jay, F.T. and Friedman, R.M. Proc.Natl.Acad.Sci. USA 75: 1859-1863, 1978.
447. Pfeffer, L.M., Wang, E. and Tamm, I. J.Cell.Biol. 85:9-17, 1980.
448. Chatterjee, S., Cheung, H.C. and Hunter, E. Proc.Natl.Acad.Sci. USA 79:835- 839, 1982.
449. Maheshwari, R.K., Banerjee, D.K., Waechter, C.J., Olden, K. and Friedman, R.M. Nature 287:454-456, 1980.
450. Dubois, M.F., Vignal, M., Le Cunff, M. and Chany, C. Nature 303:433-435, 1983.
451. Lewis, M.F., Mengheri, E. and Esteban, M. Proc.Natl.Acad.Sci. USA 80:26-30, 1983.
452. Rager-Zisman, B. and Bloom, B. Nature 251:542-543, 1974.
453. Zinkernagel, R.M. and Doherty, P.C. Advances Immunol. 27:96, 1979.
454. Michael, A.J., Ting, A., Zweering, H.J. and Askonas, B.A. Nature 270: 524-526, 1977.
455. Lindahl, P., Leary, R. and Gresser, I. Proc.Natl.Acad.Sci. USA 69: 721-725, 1972.
456. Heron, I., Berg, K. and Cantell, K. J.Immunol. 117:1370-1374, 1976.
457. Sethi, K.K. and Brandis, H. Arch.Virol. 57:117-125, 1978.
458. Zarling, J.H., Sosman, J., Eskra, L., Borden, E.C., Horoszewicz, J.S. and Carter, W. J.Immunol. 121:2002-2008, 1978.
459. Schultz, R.M., Papamatheakis, J.D. and Chirigos, M.A. Science 197: 674-676, 1977.
460. Schultz, R.M. and Chirigos, M.A. Cell.Immunol. 48:52-60, 1979.
461. Stanwick, T.L., Campbell, D.E. and Nahmias, A.J. Cell.Immunol. 53: 413-418, 1980.
462. Jett, J.R., Mantovani, A. and Herberman, R.B. Cell.Immunol. 54:425-430, 1980.
463. Herberman, R.B., Ortaldo, J.R. and Bonnard, G.D. Nature 272:221-223, 1979.
464. Trinchieri, G. and Santoli, D. J.Exp.Med. 147:1314-1320, 1978.
465. Welsh, R.M. Antiviral Res. 1, 5-12, 1981.

466. Stremmer, R., Neumann, C., DeMaeyer, E and Macher, E. Nature 296: 757-758, 1982.
467. Herberman, R.B., Ortaldo, J.R., Riccardi, C., Timonen, T., Schmidt, A., Maluish, A. and Djeu, J. In: Interferons (Eds., T.C.Merigan, R.M. Friedman, and C.F.Fox), UCLA Symposium XXV, Academic Press, pp.287-294, 1982.
468. Lotzova, E., Savary, C.A., Gutterman, J.U. and Hersh, E.M. Cancer Res. 42:2480-2488, 1982.
469. Golub, S.H., D'Amore, P. and Rainey, M. J.Natl.Cancer Inst. 68:711-717, 1982.
470. Trinchieri, G., Granato, D. and Perussia, B. J.Immunol. 126:335-340, 1981.
471. Grönberg, A., Kiessling, R., Masucci, G., Guevara, L., Eriksson, E. and Klein, G. in press, 1983.
472. Ross, M.W., Tiangco, G.J., Horn, P., Hiserodt, J.C. and Granger, G.A. J.Immunol. 123:325-330, 1979.
473. Farram, E. and Targan, S.R. J.Immunol. 130:1252-1256, 1983.
474. Stewart, W.E., DeClercq, E., Billiau,A., Desmyter, J. and DeSomer, P. Proc.Natl.Acad.Sci. USA 69:1851-1854, 1972.
475. Harak, I., Jungwirth, C. and Bodo, G. Virology 45:456-462, 1971.
476. Schattner, A., Meshorer, A. and Wallach, D. Cell.Immunol. 79:11-25, 1983.
477. Gresser, I., Guy-Grand, D., Maury, C. and Maunoury, M.T. J.Immunol. 127:1569-1575, 1981.
478. Korngold, R., Blank, K.J. amd Murasko, D.M. J.Immunol. 130:2236-2240, 1983.
479. Degos, L., Pla, M. and Colombani, J.M. Eur.J.Immunol. 9:808-815, 1979.
480. DeWaal, R.M., Bogman, M.J., Maass, C.N., Cornelissen, L.M., Tax, W.J. and Koene, R.A. Nature 303:426-428, 1983.
481. Lwoff, A. Bact.Rev. 33:390-420, 1969.
482. Pardee, A.B., Dubrow, R., Hamlin, J.L. and Kletzien, R.F. Annu.Rev. Biochem. 47:715-750, 1978.
483. Yen, A. and Pardee, A.B. Exp.Cell.Res. 116:103-113, 1978.
484. Rossow, P.W., Riddle, V.G.H. and Pardee, A.B. Proc.Natl.Acad.Sci. USA 76:4446-4450, 1979.
485. Croy, R.G. and Pardee, A.B. Proc.Natl.Acad.Sci. USA 80:4699-4703, 1983.
486. Campisi, J., Medrano, E.E., Morreo, G. and Pardee, A.B. Proc.Natl. Acad.Sci. USA 79:436-440, 1982.
487. Taylor-Papadimitriou, J. In: "Interferon 2" (Ed. I.Gresser), Academic Press, New York, pp.13-46, 1980.
488. Balkwill, F.R. and Taylor-Papadimitriou, J. Nature 274:798-800, 1978.
489. Balkwill, F.R., Watling, D. and Taylor-Papadimitriou, J. Int.J.Canc. 22:258-265, 1978.
490. Creasey, A.A., Bartholomew, J.C. and Merigan, T.C. Proc.Natl.Acad.Sci. USA 77:1471-1475, 1980.
491. Creasey, A.A., Bartholomew, J.C. and Merigan, T.C. Exp.Cell.Res. 134: 155-160, 1981.
492. Lindahl-Magnusson, P., Leary, P. and Gresser, I. Nature New Biol. 237:120-121, 1972.
493. Weinstein, Y., Brodeur, B.R., Melman, K.L. and Merigan, T.C. Immunol. 33:313-319, 1977.
494. Kimchi, A., Shure, H. and Revel, M. Nature 282:849-851, 1979.
495. Sokawa, Y., Watanabe, Y., Watanabe, Y. and Kawade, Y. Nature 268: 236-238, 1977.
496. Sreevalson, T., Taylor-Papadimitriou, J. and Rozengurt, Biochem. Biophys.Res.Commun. 87:679-685, 1979.

497. Kimchi, A., Shure, H., Lapidot, Y., Rapoport, S., Panet, A. and Revel, M. FEBS Letters 134:212-216, 1981.
498. Sreevalson, T., Rozengurt, E., Taylor-Papadimitriou, J. and Burchell, J. J.Cell.Physiol. 104:1-10, 1980.
498a. Taylor-Papadimitriou, J., Shearer, M. and Rozengurt, E. J.Interferon Res. 1:401-409, 1981.
499. Inglot, A.D., Oleszak, E. and Kisielow, B. Arch.Virol. 63:291-296, 1980.
500. Oleszak, E. and Inglot, A.D. J.Interferon Res. 1:34-45, 1980.
501. Inglot, A.D. Texas Rep.Biol.Med. 41:402-410, 1982.
502. Waterfield, M.D., Scrace, G.T., Whittle, N., Stroobant, P., Johnsson, A., Wateson, A., Westermark, B., Heldin, C.H., Huang, J.S. and Deuel, T.F. Nature 304:35-39, 1983.
503. Marquardt, H., Hunkapillar, M.W., Hood, L.E., Twardzik, D.R., DeLarco, J.E., Stephenson, J.R. and Todaro, G.J. Proc.Natl.Acad.Sci. USA 80: 4684-4688, 1983.
504. Shimizu, N. and Sokawa, Y. J.Biol.Chem. 254:12034-12037, 1979.
505. Oikarinen, J. Biochem.Biophys.Res. Commun. 105:876-881, 1982.
506. Panet, A., Czarniecki, C.W., Falk, H. and Friedman, R.M. Virology 114:569-592, 1981.
507. DeClercq, E., Warrinier, J., Balzarini, J., Imai, J. and Torrence, P.P. Antiviral Res. ABSTR 1, No.2, p. 52, 1983.
508. Eppstein, D.A., Schryver, B.B., Marsh, Y.V., Larsen, M.A. and Kurahara, C.G. J.Interferon Res., 3:305-312, 1983.
509. Doetsch, P.W., Subadolnik, R.J., Sawada, Y., Mosca, J.D., Flick, M.B., Reichenbach, N.L., Dang, A.Q., Wu J.M., Charubala, R., Pfleiderer, W. and Henderson, E.E. Proc.Natl.Acad.Sci. USA 78:6699-6703, 1981.
510. Maroun, L.E. Biochim.Biophys.Acta 517:109-114, 1978.
511. Maroun, L.E., Franciskovich, P., Precup, J.W. and Degner, M. Antiviral Res., ABSTR 1, No.2, p.73, 1983.
512. Taylor-Papadimitriou, J. In: "Biology of the IFN System" (Eds., E. DeMaeyer and H. Schellekens), Elsevier, Amsterdam, in press, 1983.
513. Schneck, J., Rager-Zisman, B., Rosen, O.M. and Bloom, B.R. Proc. Natl.Acad.Sci. USA 79:1879-1883, 1982.
514. Schmidt, A. Ph.D. Thesis (M. Revel, Advisor), The Weizmann Institute of Science, Rehovot, 1980.
515. Gewert, D.R., Shah, S. and Clemens, M.J. Eur.J.Biochem. 116:487-492, 1981.
516. Lundblad, D. and Lundgren, E. Biochem.Biophys.Res.Commun. 105:1569-1576, 1982.
517. Lin, S.L., Garber, E.A., Wang, E., Caliguiri, L.A., Schellekens, H., Goldberg, A.R. and Tamm, I. 3rd International Congress for Interferon Research, Abstracts S42, Miami, Florida, 1982.
518. Oren, M., Reich, N.C. and Levine, A.J. Mol.Cell.Biol. 2:443-449, 1982.
519. Bendori, R., Resnitzki, D. and Kimchi, A. FEBS Letters, in press, 1983.
520. Sarkar, S., Glassy, M.C., Ferrone, S. and Jones, O.W. Proc.Natl. Acad.Sci. USA 77:7297-7301, 1980.
521. Besancon, F., Bourgeade, M.F. and Thang, M.N. 3rd International Congress for Interferon Research, Abstracts, Miami, Florida, 1982.
522. Lieberman, D., Volloch, Z., Aviv, H., Nudel, U. and Revel, M. Mol.Biol.Rep. 1:447-451, 1974.
523. Rossi, G.B., Matarese, G.P. and Grappelli, C. Nature 267:50-52, 1977.
524. Rossi, G.B., Dolei, A., Cioe, L. and Peschle, C. Texas Rep.Biol. Med. 41:381-387, 1982.
525. Dolei, A., Colletta, G., Capobianchi, M.R., Rossi, G.B. and Vecchio, G. J.Gen.Virol. 46:227-235, 1980.

526. Lotem, J. and Sachs, L. Int.J.Cancer 22:214-220, 1978.
527. Tomida, M., Yamamoto, Y. and Hozumi, M. Cancer Res. 40:2919-2924, 1980.
528. Tomida, M., Yamamoto, Y. and Hozumi, M. Biochem.Biophys.Res.Commun. 104:30-37, 1982.
529. Fisher, P.B., Miranda, A.F., Babiss, L.E., Pestka, S. and Weinstein, I.B. Proc.Natl.Acad.Sci. USA 80:2961-2966, 1983.
530. Matsuno, T. and Shirasawa, N. Biochim.Biophys.Acta 538:188-195, 1978.
531. Fisher, P.B., Mufson, R.A. and Weinstein, I.B. Biochem.Biophys. Res. Commun. 100:823-830, 1981.
532. Keay, S. and Grossberg, S.E. Proc.Natl.Acad.Sci. USA 77:4099-5003, 1980.
532a. Renton, K.W., Keyler, D.E. and Mannering, G.T. Biochem.Biophys.Res. Commun. 88:1017-1023, 1979.
533. Bastos, R.N., Volloch, Z. and Aviv. H. J.Mol.Biol. 110:191-203, 1977.
534. Fowler, A.K., Reed, C.D. and Giron, D.J. Nature 286:266-267, 1980.
535. Pouillart, P., Palangie, T., Jouve, M., Garcia-Giralt, E., Fridman, W.H., Magdelenat, H., Falcoff, E. and Billiau, A. Eur.J.Cancer Clin. Oncol. 18:929-935, 1982.
536. Chany, C. and Vignal, M. J.Gen.Virol. 7:203-210, 1970.
537. Brouty-Boye D. and Gresser, I. Int.J.Cancer 28:165-173, 1981.
538. Brouty-Boye, D., Cheng, Y.-S.E. and Chen, L.B. Cancer Res. 41:4174-4180, 1981.
539. Hunter, T. Cell 22:647-648, 1980.
540. Bourgeade, M.F., Rousset, S., Paulin, D. and Chany, C. J.Interferon Res. 1:323-330, 1981.
541. Duesberg, P.H. Nature 304:219-226, 1983.
542. Panniers, L.R.V. and Clemens, M.J. J.Cell Sci. 48:259-265, 1981.
543. Pfeffer, L.M., Wang, E. and Tamm, I. J.Exp.Med. 152:469-475, 1980.
544. Werenne, J., Edmonts-Act X. and Bartholuyns, J. Annal. N.Y.Acad.Sci. 350:623-624, 1980.
545. Brouty-Boye D. and Zetter, B. Science 208:516-518, 1980.
546. Bourgeade, M.F., Chany, C. and Merigan, T.C. J.Gen.Virol. 46:449-454, 1980.
547. Wang, E., Michl, J., Pfeffer, L.M., Silverstein, S.C. and Tamm, I. J.Cell.Biol. 91:238a, 1981.
548. Gresser, I. In: "Interferon 4" (Ed. I.Gresser), Academic Press, New York, pp.95-127, 1982.
549. Moss, J., Woodrow, D., Sloper, J.C., Riviere, Y. and Guillon, J.C. Brit. J.Exp. Pathol. 63:43-49, 1982.
550. Gresser, I., Aguet, M., Morel-Maroger, L., Woodrow, D. and Puvion-DuFilleul, F. American J. Pathol. 102:396-403, 1981.
551. Calvert, M.C. and Gresser, I. Nature 278:558-560, 1979.
552. Blalock, J.E. and Stanton, J.D. Nature 283:406-408, 1980.
553. Johnson, H.M. Proc.Soc.Exp.Biol.Med. 164:380-385, 1980.
554. Gisler, R.H., Lindahl, P. and Gresser, I. J.Immunol. 113:438-444, 1974.
555. Paraf, A., Phillips, N., Simonin, G., DeMaeyer-Guignard, J. and DeMaeyer, E. J. Interferon Res. 3:253-260, 1983.
556. Hooks, J.J., Moutsopoulos, H.M. and Notkins, A.L. Annals N.Y. Acad. Sci. 350:21-32, 1980.
557. Chousterman, S., Poitrine, A., Chousterman, M., Chaput, J.C. and Thang, M.N. Antiviral Res. ABSTR. 1, No.2, p. 50, 1983.
558. Merigan, T.C. In: "Interferon 3" (Ed. I.Gresser), Academic Press, New York, pp. 135-154, 1981.
559. Tyrrell, D.A.J. In: "Interferon 4" (Ed.I.Gresser), Academic Press, New York, pp. 155-175, 1982.
560. Levin, S. Eur. J. Pediatr. 140:2-4, 1983.

INDEX

GPSR Compliance

The European Union's (EU) General Product Safety Regulation (GPSR) is a set of rules that requires consumer products to be safe and our obligations to ensure this.

If you have any concerns about our products, you can contact us on ProductSafety@springernature.com

In case Publisher is established outside the EU, the EU authorized representative is:

Springer Nature Customer Service Center GmbH
Europaplatz 3
69115 Heidelberg, Germany

Batch number: 10755284

Printed by Printforce, the Netherlands